AF740000

Beef cattle nutrition and tropical pastures

M H Butterworth
International Livestock Centre for Africa, Addis Ababa, Ethiopia

Longman *London and New York*

For Juanita

Longman Group Limited
Longman House, Burnt Mill, Harlow
Essex CM20 2JE, England
Associated companies throughout the world

Published in the United States of America
by Longman Inc., New York

© Longman Group Limited 1985

All rights reserved; no part of this publication may be reproduced, stored in a retrieval system, or transmitted in any form or by any means, electronic, mechanical, photocopying, recording, or otherwise, without the prior written permission of the Publishers.

First published 1985

British Library Cataloguing in Publication Data
Butterworth, Martyn
Beef cattle nutrition and tropical pastures.
1. Beef cattle – Feeding and feeds
I. Title
636.2′13 SF203

ISBN 0-582-46340-8

Library of Congress Cataloging in Publication Data
Butterworth, M. H. (Martyn Holt), 1932–
Beef cattle nutrition and tropical pastures.

Bibliography: p.
Includes index.
1. Beef cattle – Feeding and feeds. 2. Beef cattle – Tropics – Feeding and feeds. 3. Pastures – Tropics.
I. Title.
SF203.B87 1984 636.2′13 83–794
ISBN 0-582-46340-8

Set in 10/11pt Linotron 202 Bembo
Printed in Singapore
by The Print House Pte Ltd

Coleg Prifysgol Cymru
The University College of Wales
Aberystwyth
Acc. No. 8501049
Class No. SF 203 B9

Contents

81 6509921 2
TELEPEN
WITHDRAWN FROM STOCK
UCW LLYFRGELL LIBRARY ABERYSTWYTH

Preface

Large areas of natural pastures exist in the tropical and sub-tropical regions of Africa, Asia, Latin America and Australia and these support many cattle which are used for beef production and other purposes. However, productivity of these herds is generally low when expressed on a per capita or per hectare basis.

The reason for this low productivity is predominantly nutritional; seasonal differences in precipitation cause wide fluctuations in both the quantity and quality of forage available and consequent deficiencies are often exacerbated by the poor nutrient status of many tropical soils.

Research to clarify the nature of the nutritional constraints to beef cattle production from tropical pastures, as well as to remedy them, has been carried out in different parts of the tropics and sub-tropics for several decades. The role of individual nutrients, particularly protein and non-protein nitrogen as well as major and trace minerals, has been investigated and the importance to increased production of improved pastures and their management has been demonstrated. A considerable body of information is thus available and this should be used as a basis for both development and further research.

Nevertheless, no attempt has been made as yet to prepare a comprehensive review of the various aspects of the nutrition of beef cattle grazing tropical pastures and it was with a view to meeting this need that this book was conceived. It is the intention to provide a literature source which is hoped will be useful to students, research workers and those engaged in development. In recognition of the fact that the solution of a technical problem does not automatically lead to increased productivity, a chapter on economic and sociological constraints has been included.

The author's approach has generally been historical and geographical. The earliest work carried out on a particular subject has been reviewed first, followed by work on similar lines in the same region. It is believed that this approach is more relevant than comparisons made across geographical boundaries. Each chapter concludes with an integrating discussion which provides a synthesis of findings in different geographical and managemental situations.

It is inevitable with a work such as this that chapters are finished at different times and some are thus more up-to-date with regard to the current literature than others. I apologize in advance therefore for any glaring omissions on this account.

It is a pleasure to acknowledge the assistance of many colleagues during the ten years or so that this book has been in preparation. I would particularly like to thank Roger Arnold for contributing the section on 'Enteque seco', Joe Conrad for reading the chapters on mineral deficiencies and supplementation, Bill Adams for reading that on the use of fertilisers and John Clatworthy those on legumes and management. Meg Norton's help in preparing an index of reprints etc. was very much appreciated.

M H Butterworth
International Livestock Centre for Africa
Addis Ababa, Ethiopia
8 April 1984

1 Tropical and subtropical pastures

Despite the fact that pasture and range land in the tropics and subtropics occupies some 14% of the earth's total land area (or about 58% of total permanent pasture) and supports some 60% of the world's population of bovines, relatively little experimental work has been done on aspects of beef cattle production related to nutrition and management and with few exceptions (see, for example, Norman 1974a, b; Smith 1974) results have not been reviewed.

The present work will comment briefly on the nutritive characteristics of tropical and subtropical pastures as well as the nutritional requirements of the beef cattle that obtain their sustenance from them.

The main body of the text, however, consists of a review of work on the production of beef from tropical and subtropical pastures. It includes a consideration of mineral deficiencies and their correction, the use of protein and energy as supplements to growing and breeding stock, establishment of legumes to improve tropical pastures, and the use of nitrogenous and other fertilizers to increase productivity. The role of various management systems such as deferment and rotational grazing are discussed and considerable emphasis is placed on the importance of correct stocking rate on pasture productivity. Finally, some of the factors which act as constraints on the development of a productive beef industry in tropical areas are considered. No attempt has been made to include work on genetic aspects of beef production nor on aspects of health and parasitology. Most of the literature reviewed refers to the grazing beef animal but additional references have been used where necessary to clarify or amplify particular points.

In most cases (exceptions have been made where the amount of information available does not warrant such treatment), an historical approach has been used and work is dealt with in chronological order. It has been classified on a national and regional basis.

To complete this introduction, it is necessary to define the area to be dealt with and to describe briefly some of its relevant characteristics.

P.C. ABERYSTWYT LLYFRGELL

Introduction

Strictly speaking, the tropics extend from the Tropic of Cancer at latitude 23° 28′ north of the Equator to the Tropic of Capricorn at the same latitude south. However, because of the many similarities between the true tropics and the transitional areas to the south and north of them, it is thought more appropriate to use the latitudes of 30° north and south of the Equator as boundaries. This is in agreement with the convention followed by Norman (1974a) for Australian grasslands. At the same time, certain examples have been drawn from both north and south of these latitudes (e.g. North Louisiana in the United States of America and Cape Province in the Republic of South Africa) because of particular relevance to the subject under discussion.

It is not the purpose of this chapter to set down exhaustive detail as to the composition and ecology of the area as this has been amply provided elsewhere (see Whyte 1974; UNESCO/UNEP/FAO 1979; and further examples for specific areas given below) but rather to define the area together with the major factors influencing its floristic composition and productivity.

The most important factor which controls both the composition and productivity of grassland is rainfall. Not only amount, but distribution is important and bioclimatic criteria have been used to define the main types of tropical and subtropical climate. The total precipitation defines the vegetation type to be found in any given area as well as the length of the dry season. Thus, at one extreme, a humid climate may have precipitation of more than 1500 mm and a dry season of less than four months. At the other extreme, a dry climate receives annual rainfall of between 100 and 500 mm with a dry season of seven months whilst a desertic climate receives less than 100 mm. True desertic climates contribute little to livestock production whilst a large proportion of the high rainfall areas of Africa, South America and Asia is occupied by dense forests of various types which are to all intents and purposes ungrazed.

Most of the rainfall in tropical and subtropical areas falls during the warmer months and thus temperature is not usually limiting to grass growth, although this may be the case where climate is further modified by altitude or in transitional areas (see, e.g. Butterworth, et al. 1973). Winter rainfall areas are more Mediterranean in type and occur in Africa north of the Sahara, north-east Mexico and the southern part of South Africa.

A further aspect of rainfall which is of great importance in the more arid parts of our area is the fluctuation from year to year. Drier areas have not only shorter grazing seasons but a greater chance of drought conditions occurring in years of less than average rainfall. In general, the lower the total precipitation, the higher its coefficient of variation.

Tropical rains often fall in brief storms of very high intensity. This not only limits their agricultural utility because of reduced time at maximum infiltration rate but increases run-off which causes erosion with all its consequences.

Tropical and subtropical areas are also characterized by minor fluctuations in day length which restrict floristic composition to species which do not require wide variations.

Soils play a major role in the vegetative composition and productivity of tropical grasslands. A majority are highly weathered and leached and deficient in nutrients (latosols) although some of the more recent alluvial and volcanic soils are less leached and higher in nutrients. Climax vegetation develops as the soil develops and is stable under moderate use. Misuse, however, leads to the predominance of less desirable plant species and eventually to soil degradation. Whilst a suitable method of management can be found for all vegetation climaxes on even the most unstable soil types, generalized misuse by overgrazing is leading to considerable soil degradation and loss in many parts of the tropics and subtropics and is certainly the major factor threatening the maintenance of productivity.

As would be expected from the large differences throughout the area insofar as climate and soils are concerned, there is a large variety of types of tropical and subtropical grazing lands. This is not only confined to the grass species to be found in such areas but also to the extent and type of woody species. Several systems have been proposed for the classification of these different types but for the purpose of this review the simplest will be adhered to. This differentiates between shrublands (or steppes), savannahs and grasslands.

Shrublands are characterized by rainfall of less than 500–600 mm and a dry season of eight to nine months. Gramineae are generally widely spaced and growth is entirely determined by the amount of rainfall during the short rainy season. Grasses are both annual and perennial, usually with narrow, rolled or folded leaves. Woody species are typically xerophytic in type and frequently thorny or succulent. Fire plays a negligible role in the maintenance of these ecosystems because of the limited amount of fuel available. They may, however, be the result of regression from a climax because of overgrazing.

Savannahs are usually in higher rainfall areas than shrublands, and have a more continuous grass layer generally characterized by more perennial species. A wide range of woody species is found and the type and density of these have been used as a basis for further subdivision (e.g. shrub savannah, savannah woodland). The habits of woody species range from dense thicket-forming shrubs (e.g. *Acacia gillettiae*) to relatively widely-spaced trees.

Clearly both the species and habit of the grasses to be found in the understory is influenced by the type of bushes and trees present. Some grasses are more shade tolerant than others, for example

Panicum maximum, and removal of the tree facie may cause them to diminish significantly in certain areas. Similarly, when shade increases, grass cover tends to become more sparse and the growth habit of certain species (e.g. *Hyparrhenia*) may change.

Because of their more open nature and more continuous grass cover, savannahs are more subject to fires, and these also affect the botanical composition by removing immature woody plants. In some areas, it is thought that open grasslands have been produced from woodland by the effect of fire (e.g. the highveld of South Africa, the llanos of Colombia and Venezuela).

Many savannah areas are also subject to change by man, cutting wood for fuel or construction materials, as well as cultivation.

The term grasslands as used in its ecological sense refers to associations of mesophytic or hydrophytic grasses or sedges and it is more usually found in temperate climates. However, it occurs at high altitudes as in the *Exotheca* grasslands in Malawi and Uganda and also in badly drained areas such as the 'vleis' of Southern Africa or 'dambos' of Malawi.

A further type of tropical grazing land which must be included in this discussion is improved pasture. Native species are often adapted to low levels of productivity and do not respond adequately to increased levels of management. However, introduced species may be more productive and give greater responses to periodic cultural treatments. At the same time, the introduced species is often higher in nutrients, particularly protein, phosphorus and calcium. The majority of grass species which have been successfully used in this way have had their origin in Africa. Examples which have met with wide acceptance in tropical and subtropical America and the Caribbean include Pangola grass (*Digitaria decumbens*), Star grass (*Cynodon nlemfuensis* and *C. aethiopicus*), Guinea or Coloniao (*Panicum maximum*), Jaragua (*Hyparrhenia rufa*), Rhodes (*Chloris gayana*), Buffel (*Cenchrus ciliaris*) as well as many others. Hybridization has also been used as a means of improvement. For example, Burton (1947) produced Coastal Bermuda as a result of crossing a local variety of *C. dactylon* with var. *elegans* from South Africa. Subsequently, Coastcross 1 was developed as a cross between cv. Coastal and *C. nlemfuensis* var. *robustus* and resulted in a forage of improved digestibility (Burton et al. 1967).

Tropical pasture legumes have been used to improve the productivity of grazing lands. Following the fortuitous introduction of *Stylosanthes humilis* to Australia, considerable attention was given to the search for other suitable legume species. A number of these have now been developed (after initial introductions were made, mainly from South America) and several are in commercial use. Examples include *S. guianensis*, *Centrosema pubescens*, *Macroptilium atropurpureum*, *Desmodium intortum* and *D. uncinatum*. Although most of the work on the development of tropical pasture legumes was carried out in Australia, these are now widely used elsewhere in the tropics and subtropics.

In some areas, grasses which were originally introduced have become so well adapted to local conditions that they have become resident and the predominant species. A good example of this is *Panicum maximum* in the Huasteca gulf coast area in Mexico and in the states of Yaracuy and Falcon in Venezuela.

Because of the rich variety of good natural pasture species in Africa, improved pasture has played a relatively minor role in livestock production (exceptions include the intensive use of Henderson No. 2 Star grass in Zimbabwe and *Eragrostis curvula* cv. Ermelo in South Africa). However, in large expanses of Australia, the Americas and Asia it is no exaggeration to state that livestock production has been based on the use of improved pasture.

The term artificial pasture has not been used in this classification because of the cases where introduced grasses have naturally become resident as mentioned above. However, a definition of the term is useful at this point because so much of the following review is of work carried out on such pastures. They have been defined by Ibrahim (1975) as 'grazing lands under relatively intensive management usually supporting introduced forage species and receiving periodic cultural treatment such as tillage, fertilization, moving and irrigation'. Cultivated and tame pasture are mentioned as synonyms. Intensive pasture may also be a useful term in this connection.

Grazing lands in tropical areas

Africa

The amount of grazing land in Africa is difficult to quantify because of the non-specific nature of the classification used. The FAO production year book distinguishes between permanent grasslands and pastures, arable land and other areas. However, this latter area includes large areas of semi-arid and arid land which may be used as range, even though of low productivity. Similarly, it may be supposed that in countries with a high population pressure (and most populations in Africa are increasing at a rate exceeding 2% per annum) grazing land is continually being put under cultivation. Even though some of this land will return to pasture when it is exhausted and abandoned, the secondary succession involved in the re-establishment of grazing land will be slow and productivity will be limited until a state near to climax is achieved. A further important factor which reduces the amount of grazing land is that of erosion. Whilst the major part of this is from overgrazing, additional causes may include trampling (e.g. routes to dip-tanks) and civil engineering work carried out without thought for the consequences on the ecology of surrounding areas. Finally, the presence of tsetse flies (*Glossina* spp.) in many areas of woodland savannah renders the

Table 1.1 Land use in some tropical African countries (ha × 10^3) (Data from FAO 1978a)

	Permanent grassland and pastures	*Arable and permanent crops*	*Other land*
Tropical humid Africa			
Burundi	435	1 272	858
Cameroon	7 380	8 300	31 264
Congo	14 300	667	19 183
Central African Republic	100	5 900	59 248
Benin (Dahomey)	442	2 950	7 670
Gabon	4 750	427	20 590
Gambia	330	265	405
Ghana	10 700	2 705	9 597
Guinea	3 000	4 170	17 416
Guinea Bissau	1 280	285	1 235
Ivory Coast	8 000	9 160	14 640
Liberia	240	371	9 021
Nigeria	20 830	23 990	46 257
Rwanda	530	955	1 010
Sierra Leone	2 204	4 098	860
Togo	200	2 285	2 875
Tropical dry West Africa			
Mali	30 000	9 800	82 200
Mauritania	39 250	199	63 591
Niger	3 000	15 000	108 670
Senegal	5 700	2 404	11 096
Chad	45 000	7 000	76 400
Tropical dry East Africa			
Ethiopia	64 600	13 730	31 770
Kenya	3 770	2 270	50 885
Somalia	28 850	1 066	32 818
Sudan	24 000	7 495	206 105
Tanzania	44 700	5 100	38 804
Uganda	5 000	5 538	9 433
Dry tropical Southern Africa			
Angola	29 000	1 830	21 180
Botswana	44 000	1 360	13 177
Malawi	1 840	2 278	5 290
Mozambique	44 000	3 080	29 473

grazing land unsuitable for unresistant cattle because of the presence of the genus *Trypanosoma*, a protozoa which causes a group of allied diseases (the trypanosomiases) such as Nagana (in humans, sleeping sickness).

On the other hand, the use of arable land in pasture rotations, made highly productive by fertilization and sometimes irrigation, may enable increased livestock production to be achieved far in excess of that which would be expected from natural pasture (see, e.g., Rodel and Boultwood (1967) for such work carried out at Henderson).

Despite these limitations on the extent to which grazing lands may be characterized, however, it is still useful to quote the land-use figures for some African countries (FAO 1978a, following the classification used by UNESCO/UNEP/FAO, see Table 1.1).

The botanical composition of African grasslands is widely varied as would be expected from the extreme differences to be found in both climate and edaphology. Many studies have been carried out on specific areas but the most comprehensive is that of Rattray (1960) which covers the whole of the Continent.

He identifies twenty-three distinguishing genera which are used to characterize various associations. The most important of these is *Hyparrhenia* which is to be found over large areas of central and southern middle rainfall areas from some 15 °N to as far south as Swaziland and Natal. The drier sahelian areas in the north are characterized by *Cenchrus* spp. changing to *Aristida* as rainfall decreases (*Aristida* spp. are also found in association with woodland in southern Zimbabwe adjoining Kalahari desert associations as well as in the shrublands and deserts of Cape Province, South Africa and South West Africa, Namibia). Drier areas in the Kalahari desert area of Botswana are mainly composed of *Eragrostis*-dominated shrublands whilst in the horn of Africa and northern Kenya, *Chrysopogon* spp. are found in association with desert of shrubland. Moister coastal areas in West Africa typically are characterized by *Pennisetum* associations whilst higher rainfall areas on the east costs of Tanzania and Kenya are dominated by *Panicum*. *Themeda* spp. characterize large parts of Tanzania and the coastal areas of Natal and the Transkei.

The above description is clearly a gross oversimplification; Rattray identifies more than 150 typical associations but these are also generalizations. Acocks (1975) has described 70 associations in South Africa alone.

North and Central America and the Caribbean

The subtropical south-east United States comprises the state of Florida, parts of Louisiana, the coastal plains of Georgia, and the gulf coast area of Texas. Rainfall is more than 1200 mm and mainly falls during the summer months. Large areas of the original vegetation were dominated by Longleaf and Slash Pine (*Pinus palustris* and *P. elliottii*) whilst Shortleaf and loblolly pine (*P. echinata* and *P. taeda*) were also common. Associated grasses included *Andropogon, Panicum, Paspalum, Sporobolus, Aristida* and *Muehlenbergia* spp. together with *Axonopus affinis* and common *Cynodon dactylon*. However, this vegetation was of low productivity and nutritive value and the establishment of improved species was started as early as the 1930s with the use of Carpet grass, Bahia grass (*Paspalum notatum*) etc. Early pasture work has been described by Blaser and his colleagues (Blaser et al. 1945). Later pastures were established using such introductions as Pangola grass and Limpograss (*Hemarthria altissima*) etc.

Northern Mexico presents a contrast, with the vegetation on the western side similar to that of the western United States, *Bouteloua* spp. being typical of the climax vegetation. On the eastern side, however, vegetation is subtropical and the two zones are divided by the deserts of Coahuila. Buller (1960) has described the tropical and subtropical areas of Mexico as tropical deciduous forest and disturbed tropical evergreen forest. The lower strata of the former include *Bouteloua filiformis, B. hirsuta* and *B. curtipendula, Opizia stolonifera*, and widespread herbaceous legumes including *Sebaria, Crotolaria* and *Calopogonium* spp. In lower rainfall areas, vegetation is more open and *Acacia* spp. are widespread. In the latter areas, grass species include the more tropical *Paspalum notatum, P. conjugatum, P. minus, Axonopus affinis* and *A. compressus; Leersia hexandra* and *P. fasciculatum* are found in river basins and flooded areas. However, large areas in the Gulf coast areas of the Huasteca have now been planted with improved species, notably *Panicum maximum, Digitaria decumbens*, Coastcross 1 and 'Ferrer'.

Similar patterns are to be found in Central America with the extensive cattle-raising area located on the drier Pacific coast region.

There is a wide variation in rainfall in the islands of the Caribbean but the more important beef-producing countries such as Jamaica and Cuba base the industry almost entirely on improved pastures. A large amount of the developmental work on Pangola grass (*D. decumbens*) was carried out in Jamaica and Trinidad (see, e.g. Nestel and Creek 1962).

South America

On the Caribbean coast of South America, the pattern of beef production from improved pastures, which in many areas have become naturalized, resembles that in Mexico, Central America and the islands of the West Indies. Thus, the states of Zulia, Falcon, Aragua and Valencia in Venezuela have a considerable beef and milk industry based on naturalized Guinea grass. Further inland, however, grazing lands are characterized by more marked wet and dry seasons and the 'llanos' or plains are flooded during most of the wet season.

Blydenstein (1965) has identified ten different savannah types in the Colombian llanos. These follow the humidity gradient and key species are *Trachypogon vestitus, Paspalum pectinatum, T. vestitus/Axonopus purpusei, T. ligularis, Leptocoryphium lanatum, Mesosetum* and *Andropogon* spp. from the driest to inundated gradients respectively. *Paspalum carinatum* dominates higher ground south of the Meta River whilst the areas above 600 m and up to 2000 m have been colonized by *Melinis minutiflora.* This grass was introduced to South America accidentally but has become perfectly naturalized, particularly in cooler, moister areas such as central Colombia where it plays an important role in livestock production.

The ecology of the llanos of Venezuela has recently been described by Medina and Sarmiento (1979). A classification which groups these savannahs into four floristico-ecological types has been proposed. *Trachypogon-Curatella americana* (Chaparro) savannahs occur on well-drained soils throughout the western, eastern and southern llanos. *Axonopus-Mesosetum* savannahs are pure grasslands which cover intermediate topographic positions and *Sorghastrum-Copernicia* associations occur, often with palms, on heavy soils with poor drainage. The fourth type is characterized by *Paspalum fasciculatum* and is typical of areas which are flooded during the rainy season. Many of the characteristics of the Venezuelan and Colombian llanos are to be found also in the grazing lands of Peru, Bolivia, Ecuador and the Guyanas.

The grazing lands of Brazil are to be found in the south in the states of Parana and Santa Catarina (the tribes to be found in Rio Grande do Sul are more typical of temperate areas). Precipitation ranges from 1300–1900 mm/year and typical grasses include *Andropogon, Panicum, Aristida, Paspalum, Tripogon, Elyonurus* and *Heteropogon* spp. Santa Catarina is characterized by scattered grasslands amongst *Araucaria* forests.

The central region comprises the states of Sao Paulo, Rio de Janeiro, Guanabara, Espirito Santo, Minas Gerais, Goias, Mato Grosso and Distrito Federal; rainfall varies from 1000–2000 mm/year and falls mainly in the summer months. Grasslands are 'pantanales' (marshes) which are low lying and flooded during the rainy season, 'campos cerrados' (closed pastures) or 'campos limpos' (clear pastures). Typical grasses of the first type include *Paspalum repens, P. fasciculatum* etc., *Leersia monandra, Oryza subulata, Echinochloa polystachya* and *Paratheria prostrata.* Campos cerrados are xeromorphic mixed savannahs of trees, shrubs and grasses. The latter include *Panicum, Paspalum, Tristachya, Setaria* and *Andropogon*, spp. which occur with a number of native legumes, notably *Desmodium* spp. Campos limpos are predominantly grasslands without trees or bushes. Large areas of the tropical parts of Brazil have been planted to improve pastures, notably *Hyparrhenia rufa* (Jaragua) and *Panicum maximum* (Coloniao). Improved pasture legumes have also been used

Table 1.2 Land use in some tropical American and Caribbean countries (ha × 10³) (Data from FAO 1978a)

	Permanent grassland and pastures	*Arable and permanent crops*	*Other land*
Central America			
Costa Rica	1 558	490	3 018
Guatemala	880	1 800	8 109
Honduras	2 000	915	8 294
Nicaragua	3 384	1 505	8 111
Panama	1 150	565	5 790
Caribbean			
Cuba	2 770	3 150	5 532
Dominican Rep.	1 480	1 230	2 163
Haiti	520	870	1 385
Jamaica	210	265	624
South America			
Bolivia	27 100	3 305	78 142
Brazil	166 000	40 720	638 931
Chile	11 800	582	62 498
Colombia	17 550	5 505	80 815
Ecuador	2 200	5 089	20 395
Paraguay	15 100	1 120	23 510
Peru	27 120	3 433	97 477
Venezuela	16 830	5 337	66 038

extensively. The grazing lands of Brazil have been described by Grossman et al. (1966).

Indications as to land use in some tropical American and Caribbean countries are given in Table 1.2. Figures are not given for the United States, Argentina or Mexico where pastures are predominantly temperate.

Australia

Some three-quarters of the continental land mass of Australia have been classified as arid or semi-arid (Wheeler and Hutchinson 1973) and are only lightly grazed by cattle or not grazed at all. Vegetation is characterized by lack of thorns in contrast with that found in other parts of the dry tropics and few *Cactaceae* or other succulents are represented. These areas may be classified as Saltbush steppe characterized by Saltbush (*Atriplex* spp.) in the presence of xerophytic species of such genera as *Astrebla*, *Panicum*, *Dichanthium*, *Danthonia* and *Stipa*. Mulga scrub is characterized by a top story of *Acacia* spp. with a grass stratum largely composed of ephemerals and short-lived perennials. Mitchell grass communities are mainly composed of isolated tussocks of *Astrebla* spp. (Mitchell grass) and *Iseilema* spp. (Flinders grass) which may be complemented by ephemeral species when rains are favourable. Compared with the other associations in arid or semi-arid areas, Mitchell grass associations have a relatively high carrying capacity, although this amounts to only one beast per 100–800 ha. Finally, the most widespread community to be found in arid Australia comprises *Spinifex, Triodia* and *Plectrachne* spp. However, in large areas these are completely unstocked.

Higher rainfall areas (750–1500 mm) occur in the northern parts of eastern Western Australia and the Northern Territory as well as the Cape York Peninsula and eastern Queensland. Rainfall above 1500 mm occurs in coastal areas of western Cape York Peninsula and isolated areas of Queensland and New South Wales.

In the Northern Territory, typical grass species include *Themeda australis* (probably a geographically isolated form of *T. triandra*), both annual and perennial *Sorghum* spp., *Chrysopogon fallax* and *Sehima nervosum*. These are accompanied by *Eucalyptus* spp. of different heights depending on the rainfall.

Tropical tallgrass communities in Queensland are dominated over large areas by Spear grass (*Heteropogon contortus*) which probably owes its dominance to repeated burning accompanied by grazing. Other grasses to be found in this area include Blue grass, *Dichanthium* and *Bothriochloa* spp. and *Themeda australis*. Zones of western Queensland resemble the more xerophytic areas referred to previously. Areas where Brigalow (*Acacia harpophylla*) predominate are virtually useless for grazing (grasses which occur after clearing include *Chloris acicularis, C. truncata, Eragrostis* spp.) but livestock production in these areas has been based on complete replacement of the natural vegetation by fertilized grass/legume pastures.

Large areas of Queensland have been invaded by the legume *Stylosanthes humilis* (originally known as Townsville lucerne, now Townsville stylo) which has become completely naturalized. Following this accidental introduction, more systematic searches for suitable plants have been made which has resulted in large-scale establishment of such grasses as *Panicum maximum, Cenchrus ciliaris, Digitaria decumbens* and many others. Pasture legumes which have been developed from introductions include cultivars of *Neonotonia, Leucaena, Stylosanthes, Centrosema*, etc. as well as Siratro (*Macroptilium atropurpureum*) which was line-bred from introduced material. A historical account of the introduction and evaluation of grass and legume species in Australia has been given by Burt and Williams (1975).

The southern limit of our area, northern New South Wales, is characterized by more evenly distributed rainfall intermediate between the summer rains of the northern tropical areas and the mediterranean climate of the south. Natural grasses are represented by *Bothriochloa macera* (Redgrass) and *Aristida ramosa* (Purple Wire-grass) but extensive use has been made of Kikuyu (*Pennisetum clandestinum*) in the establishment of improved pastures.

Asia

Great climatic and topographical difference occur in this area and these are also compounded with wide differences in land use. The Indian subcontinent is heavily stocked with bovines, both cattle and buffalo, and considerably degraded by overgrazing in many areas. On the other hand, vast areas of monsoon Asia, for example Indonesia, are undergrazed. Nevertheless, there is a similarity among the grazing systems of Asia, and a small number of genera play a preponderant role and are common to most ecosystems. Considerable detail on the floristic composition of the area has been given by Whyte (1968) and the greater part of the present brief description has been taken from UNESCO UNEP/FAO (1979).

Originally, four climax types could be recognized in India but these are now only represented in small areas, having largely been replaced by species of a lower ecological order as a result of degradation caused by overgrazing. *Dichanthium, Cenchrus* and *Lasiurus* spp. characterize semiarid shrubland areas in north-western India. However, these perennials rapidly give way to annuals (largely *Aristida* spp.) in overgrazing situations. *Sehima nervosum* and *Dichanthium annulatum* characterize a large area whose potential is all of peninsular India having a forest climax. *Themeda triandra* is favoured by erosion whilst further overgrazing leads to the establishment of *Heteropogon contortus* followed by usual annuals, e.g. *Aristida* and *Eragrostis* spp. Associated tree species in most arid areas include *Acacia* and *Ziziphus* spp. whilst pioneer species in humid zones include *Emblica officinalis* and *Pterocarpus* spp. Associations typical of badly drained soils near the Ganges and Brahmaputra are tallgrass savannahs consisting of *Phragmites, Saccharum* and *Imperata* spp; heavy grazing reduces the *Phragmites* whilst frequently reoccurring fire encourages *Imperata* spp. This type is found widely throughout south-east Asia. Finally, *Themeda* and *Arundinella* spp. occur with tropical species (*Bothriochloa, Chrysopogon, Cymbopogon,* etc.) on south-facing slopes of the Himalayas in rainfall areas of as much as 10 000 mm/year.

Sri Lanka consists of grassland associations caused by disturbance of forests by cultivation. *Heteropogon contortus* and *Sehima nervosum* are again common, with *Cymbopogon nardus* occurring in more humid areas, and *Imperata cylindrica* may be locally dominant. Improved pastures have been established with *Pennisetum clandestinum.*

Imperata cylindrica is the predominant grass species in most of south-east Asia, particularly where soils are deep red over basalt, whilst the *Andropogoneae* predominate where soils are shallower.

Table 1.3 Land use in some Asian countries and Australia (ha × 10^3) (Data from FAO 1978a)

	Permanent grassland and pastures	*Arable and permanent crops*	*Other land*
Southern Asia			
Bangladesh	600	9 125	3 666
India	13 000	169 400	146 359
Pakistan	5 000	20 300	52 572
South-east Asia			
Burma	362	9 999	55 527
Indonesia	12 046	17 200	151 889
Cambodia	580	3 046	14 258
Laos	800	963	21 317
Malaysia	27	6 480	26 348
Philippines	970	8 100	20 747
Sri Lanka	439	2 131	3 904
Thailand	308	17 650	33 219
Vietnam	4 870	5 500	22 166
Australia	450 700	44 900	266 193

Themeda and *Arundinella* spp. increase with altitude whilst degradation leads to predominance of the *Aristidaea*.

Some indications as to land use in different Asian countries are given in Table 1.3.

Summary

The area dealt with in this review comprises mainly those tropical and subtropical grasslands falling between 30° north and south of the Equator although isolated examples are drawn from outside this area. Both the amount and distribution of rainfall are the main factors to determine the productivity and floristic composition of tropical grasslands. However, temperature may occasionally limit grass growth. The coefficient of variability of rainfall and its intensity are also important considerations. Tropical soils differ widely and soil degradation occurs commonly in overgrazed areas.

Grazing lands may be classified into shrublands, savannahs and grasslands. The former are characterized by rainfall of less than 500–600 mm and a long dry season. Forage production results solely from rainfall during the short wet season and both grasses and shrubs are xerophytic in morphology. Fire plays a negligible part in their maintenance. Savannahs are normally found in higher rainfall areas, and more perennial species are present. Fire plays an important role in their maintenance. Grasslands occur in higher or badly drained areas.

A considerable amount of work has been carried out in tropical areas on the introduction and evaluation of suitable grass and legume species with which to establish improved pastures and examples of these are given.

Finally, some characteristics are given of the more important grazing lands occurring in tropical and subtropical Africa, America, Australia and Asia.

2 The nutritive value of tropical forages and nutrient requirements of beef cattle

The quantity and quality of the nutrients in tropical forages are influenced by a number of factors. These include stage and season of growth, climatic conditions, soils, etc. as well as the genotype of the forage itself.

Emphasis will be placed in this chapter on the content of protein as well as its digestibility and on the provision of energy by the forage (other nutrients are considered in appropriate chapters). This will be followed by a consideration of the different factors which influence the intake of pasture and brief comments on the rumen microflora of grazing animals under tropical conditions and the production of volatile fatty acids in the rumen (because of paucity of information on beef cattle, results obtained from other ruminants will be included in this discussion).

The chapter will conclude with a discussion of the nutrient requirements of beef cattle under tropical conditions and how these may be compared with those of cattle in temperate climates at similar physiological states.

Protein and energy

The crude protein (N × 6.25) contents of tropical forages have been tabulated by Butterworth (1967a) and Göhl (1975). The average crude protein (CP) content of the grass samples listed in the first work was 7.5 ± 3.5% (*n* = 477) with 235 samples (equivalent to 49.3%) containing less than 7% CP (this value is an estimate of the protein requirement at zero liveweight gain).

The average value for the CP content of 670 samples of tropical grasses listed by Göhl (1975) was 9.9 ± 5.8% of which 27% contained less than 7% CP. Although these values are higher than those given above, it is clear that a large proportion of samples analysed contained less protein than that required for maintenance.

A characteristic of these values is also their high standard deviation. Göhl has also listed values for 419 samples of leguminous browse for which the mean CP content was 14.3 ± 6.34 (n = 213) of which only 21 samples contained less than 7% CP and these were all pods, husks, haulms, stems or gum.

It is well known that the protein content of tropical grasses declines with increasing maturity as cell contents are 'diluted' by structural components and vegetative growth decreases. It was shown, for example, that the protein content of Pangola grass harvested at 10, 15, 21 and 42 days was 14.9, 13.8, 9.2 and 4.8% respectively (Butterworth 1961).

Similarly, as protein content declines so does the digestibility of the crude protein. Various authors have calculated equations to formalize this relationship and the most appropriate form has usually been shown to be logarithmic (see, e.g. Duckworth 1946; Butterworth 1963; Bredon et al. 1963; Glover et al. 1957). Butterworth and Diaz (1970) gave values for the constants in the equation:

$$CP_2 = A_o + A_1 \text{ Log CP}$$

where CP_2 is percentage apparent digestibility of crude protein and CP is crude protein content of the forage. Solutions were calculated according to the least square 'best' values and classifications were made according to species of grass (where more than ten sets of data were available), species of animal and as fresh material, hay or silage. Values of r^2 for individual grasses ranged from 0.49 (*Panicum maximum*) to 0.89 (*Cynodon plectostachyus*). The use of logarithmic form is justified biologically by the contribution of the minimum value of the endogenous faecal nitrogen fraction which also accounts for negative values for A_o with consequent negative apparent digestibility values at low levels of CP content. It should be noted, however, that Elliott and Fokkema (1960) do not consider this departure from linearity justified from a biological point of view.

It has been reported that increasing crude fibre level exerts a negative effect on the digestibility of crude protein (Glover and French 1957; Elliott and Fokkema 1960; Butterworth 1963) but this was not confirmed when a wider range of values was analysed (Butterworth and Diaz 1970), perhaps because of the empirical nature of this fraction and differences in technique among laboratories.

Butterworth (1967b) compared the digestible CP content of 473 samples of tropical grasses with the recommendations of the National Research Council and concluded that 58% contained less than the requirements for growth of beef cattle.

It must be concluded therefore that tropical grasses are frequently deficient in both crude protein *per se* and in digestible crude protein. Leguminous and other browse, however, tends to be adequate in this respect.

One aspect of protein provision which has been neglected is that of degradation in the rumen. Although the importance of the solubility of proteins with respect to rumen metabolism has been recognized since the 1950s, no attention has been given to this aspect of tropical forage utilization until quite recently. Aii and Stobbs (1980) demonstrated that wide differences in protein solubility occurred between different species and different parts of the plant but was unaffected by stage of growth. A further trial was carried out using the nylon bag technique to determine rate of disappearance from the rumen. Three forages were used, *Panicum maximum* cv. Gatton, *Desmodium intortum* and *Leucaena leucocephala*. There were marked differences in the rate of disappearance after 12 hours in the rumen in that 42.9% of the protein of *P. maximum* had disappeared compared with 16.1% of *L. leucocephala* and only 4.9% of *D. intortum*. It was suggested that the tannins in the legume depressed solubility. It is not clear what practical implications these findings have but it is clear that forages with a more rapid rate of disappearance in the rumen are more exposed to degradation before reaching the duodenum than those that are less soluble. Productivity may well be limited by the absence of 'by-pass' (or 'escape') protein in the duodenum. It has been suggested that this may be so in the case of dairy cows, and Flores et al. (1979) have shown that milk production of cows grazing young tropical grass could be increassed by the provision of a small amount of *L. leucocephala*. Whether this would occur with beef cattle is not known but the work of Falvey (1977, 1979), indicating a response to meat-and-bone meal, suggests that investigation along these lines might be worthwhile.

The energy value of tropical forages has been expressed in many ways including Digestible Dry Matter (DDM), Digestible Organic Matter (DOM; to compensate for differences in mineral components), Total Digestible Nutrients (TDN), Digestible Energy (DE) and Metabolizable Energy (ME). Many of these have been shown to be closely related e.g. DDM to DOM (Butterworth 1964; McCleod and Minson 1974), DDM to DE (Butterworth 1964) and DE to TDN (Butterworth 1964). The factor of 0.83 has been suggested by Blaxter (1962) to convert DE to ME but this has not been confirmed for tropical forages.

Despite these relationships, however, Butterworth and Diaz (1970) could demonstrate no equations of any value for the prediction of TDN from proximate analysis determinations. This is probably not surprising in view of the empirical nature of the crude fibre determination mentioned earlier and also because many cases have been mentioned where the digestibility of crude fibre is greater than that of nitrogen-free extractives in tropical forages (Butterworth 1967a) which leads to an overestimate of the TDN content.

However, more sophisticated methods of fibre analysis have been developed recently (van Soest and Robertson 1979) which may enable this apparent contradiction to be resolved. Perhaps the most commonly used system is that of separation of plant material into fractions corresponding to cell contents (soluble in neutral detergent) comprising lipids, sugars, pectin, starch, non-protein nitrogenous compounds and soluble proteins, and cell wall constituents (insoluble in neutral detergent) or neutral detergent fibre (NDF). The latter fraction may be further divided into that part which is soluble in acid detergent (fibre-bound protein and hemicelluloses) and acid-detergent fibre (ADF). ADF is composed of cellulose, lignin and lignified nitrogenous compounds. Details of the methodology involved are given by Goering and van Soest (1970).

Workers at the Instituto de Produccion Animal at Maracay in Venezuela used this method of analysis to clarify the relationships between the fibrous components of legumes and grasses and the content of 'indigestible' lignin in ADF and found a highly significant negative correlation between this and the digestibility of fibrous components (Parra et al. 1972). There were no significant correlations between cytoplasmatic components and lignin content which is to be expected as the former are not lignified. The possible effects of silica (Si) content were also investigated as these have been shown to be related to digestibility in temperate forages (van Soest and Jones 1968) but no significant relationships could be demonstrated.

The relation of chemical composition to digestibility of tropical grasses was investigated using similar methods by Kayongo-Male and his colleagues in Puerto Rico. In this work, ADF seemed more important than lignin in the determination of digestibility values, and the ratio of lignin to ADF was less useful for *in vitro* than for predicted estimates (from van Soest's summative equation). Silica was not significantly related to any measures of digestibility (Kayongo-Male et al. 1976).

There are indications therefore that the digestibility of the cell wall fraction and thus overall digestibility (as cell contents, once released, are highly digestible) is negatively correlated with the degree of lignification of the cell wall as measured by either lignin itself or by total ADF. There does not seem to be any causative or associative effect attributable to Si in tropical grasses or legumes. (But see van Soest and Jones, 1968).

More sophisticated methods of analysis were used to determine the effects of chemical composition of cell walls from both leaves and stems of *Eragrostis tef* by Morris (1980). Neutral sugars and uronic acids were determined as well as lignin and it was shown that marked differences were present between leaf and stem cell wall insofar as lignin and other components were concerned and that these were apparent in digestibility determinations using both fungal cellulase and the nylon bag technique. Both types of cell wall were equally degradable whilst the grass was in the vegetative state, that of the stem decreased markedly with the onset of maturity and the production of the inflorescence. This represents an advance on previous statistical methods in that biochemistry of cell wall degradation of the different structural plant components is clarified and it would seem that this will prove to be a useful technique for future work.

An elegant confirmation of the effects of lignification on digestibility was provided by Klock et al. (1975) working in Gainesville, Florida, who measured cell wall thicknesses by microscopic observation. It was established that among individual genotypes (of *Digitaria*), those with the highest cell wall thickness had the lowest *in vitro* dry matter digestibility and vice versa. This supports the theory that increased lignin is associated with decreased cellulose availability. Similar results were obtained with *Hemarthria* spp. (Schank et al. 1973).

Another line of investigation which is likely to yield interesting results is the determination of digestibility in plants grown at different temperatures. Several authors have indicated that plants grown at lower temperatures have a higher digestibility than the same plants grown at higher temperatures. Minson and McCleod (1970) drew the inference from temperate grasses whilst Wilson and Ford (1971) worked with seedlings. Working with regrowth of *Cenchrus ciliaris* and *Pennisetum clandestinum*, Ivory et al. (1974) showed that both leaf and stem digestibilities were significantly higher at low temperatures (15 °C) than at higher temperatures (25, 30 and 35 °C). Despite later work by Moir et al. (1977), who showed that effects directly attributable to temperature were less pronounced when results were corrected for stage of maturity, this is an important finding. In any event, if the effect of increased temperatures is to advance plant maturity, this is a factor in the consideration of the digestibility of tropical forages. Similarly, the effect of wilting on the increase of lignin in tropical forages is of interest.

Despite the inadequacies of TDN to describe the energy value of feeds, more values for tropical forages are available from this system than any other. It is of interest therefore to use this information to assess the extent to which energy deficiencies might be prevalent in grazing cattle.

The average value of TDN in the 312 samples for which determinations had been made in the author's tabulation was 54.0 ± 7.9% with 53% of the samples being below the value of 55% TDN which the National Research Council recommends as being adequate for growing cattle (Butterworth 1967a,b).

The corresponding value for grasses (271 samples) in the compilation by Göhl (1975) was 56.9 ± 10.3% with 43.5% of samples containing less than 55% TDN. Values for legumes were somewhat higher, being 62.3 ± 11.7% with 35 of a total of 122 samples containing less than 55% TDN. It may be concluded from the literature, therefore, that whilst energy deficiencies are not likely to be as prevalent as those of protein, they occur widely and may be expected to limit production under certain circumstances.

Intake

The intake of nutrients of the grazing animal depends on a complex interaction of a number of factors. These include herbage present, herbage allowance, legume content, canopy structure, protein and mineral content ('t Mannetje and Ebersohn 1980). Climatic factors, particularly temperature and humidity, may also be involved as may moisture content of the herbage. Breed, sex and physiological state of the animal is also important.

It is not the objective of this section to review comprehensively all the work on the intake of beef cattle under tropical conditions but rather to highlight points of particular interest, drawing examples from various species.

It should be strongly emphasized, however, that by far the most important factor influencing the intake of green forage in large areas of the tropics and subtropics is the restricted amount of forage available. This may be caused by limited rainfall, either seasonally or over a longer period (the higher coefficient of variation to be expected in low rainfall areas has already been mentioned), which may extend into a drought. When this occurs over a period of years, as in the recent case of the Sahelian droughts, it can lead to disastrous effects on the cattle population. The other principal reason for shortage of forage is that there are too many animals competing for the limited amount available, i.e. the liveweight per unit of available forage exceeds the grazing capacity of the pasture. This inevitably leads to the situation where the gain per animal decreases (although not necessarily the gain per unit of land area in the first instance) until finally only sufficient forage is available to provide maintenance requirements after which weight loss occurs. If this situation continues, pasture deterioration will ensue, thus effectively reducing the grazing capacity of the available land. If action is not taken at this point (e.g. by reducing the number of animals grazing the land), a vicious circle will start as the same number of animals will require more land and overstocking will be intensified. The reasons for this over-exploitation are many and include ignorance and rapacity but more frequently a system of land tenure where it is not in the interest of the individual to conserve the land. This will be dealt with in more detail in a later chapter.

Another important factor which limits intake in large parts of tropical and subtropical grazing lands is the presence of material which is unpalatable because of maturity or senescence. This is particularly true in areas where there is a long dry season. It is frequently removed by burning and, if this is done at an inappropriate time, range degradation results. This exemplifies the paradox of tropical pasture management; forage is either inadequate, or more than can be consumed at the moment when it is palatable and of high nutritive value.

Apart from the above rather general remarks, the main climatic factors which would be expected to influence intake are temperature and humidity (and possibly solar radiation). Situations in which these factors reach stressful proportions have been described by many authors (for a recent review, see Fuquay 1981) and it is not the intention to dwell on them here. However, many of the results which are frequently quoted have been obtained by the study of cattle in climatic chambers and doubt must arise as to how these can be applied to cattle in a grazing situation. Another unknown factor is that of acclimatization; can results obtained from European cattle confined for a relatively brief period of time be applied to 'native' cattle which have become adapted to an environment over a period of years? It should also be taken into consideration that tropical climates are rarely extreme for any length of time and that diurnal variation often ensures that even though heat stress may be great during the day, the animal has the opportunity to graze during the night. Wilson (1961) has shown that it is possible to alter the grazing habits of cattle by providing better grazing at night than during the day. The importance of providing good night grazing has also been emphasized by Payne et al. (1951) for conditions in Fiji, and Payne (1966) has reviewed the more important observations of grazing time made in the tropics. From the available evidence it was shown that grazing time amounted to a total of 8.7 ± 1.6 hrs of which an average of 25% was during the night. Incidentally, in many parts of Africa, cattle are not allowed night grazing as they are 'kraaled' for fear of predators and theft. The author has seen, in Mozambique, cattle taken from the kraal at 10 a.m. and returned at 2 p.m.!

Very few critical observations as to the combined direct effects of climate on cattle are available. However, the classical experiment of Payne and Hancock (Hancock and Payne 1955; and Payne and Hancock 1957) compared the performance of sets of identical twins, one of which was reared in a warm temperate climate in New Zealand and the other in the humid tropical climate of Fiji. The same hay and concentrates were fed to animals in both environments. There was a wide variation of intake among individuals but the hay intake in the tropical environment was significantly lower than in the temperate. However, intake of total starch equivalent was substantially the same for twins in both environments.

Comparison of intake between wet and dry seasons was made by Grant et al. (1974) working at Los Baños, Philippines, with cattle and buffalo steers under stall feeding conditions. Both species consumed more dry matter under dry season conditions than under wet, i.e. 2.1 v. 2.4% of body weight. Digestibilities of both feeds were comparable during the two seasons but there was a slightly greater dry matter content of the dry season forage (Napier grass in both cases).

Dry matter intake during wet and dry seasons was also compared by Butterworth et al. (1961) in Trinidad. The stock used were cross-bred Zebu/Holstein cows with a mean milk yield of about 6 litres/day. The determination of intake was carried out using the chromic oxide technique on the same pasture of well-established Pangola grass. Intake was higher in the dry season than in the wet (3.7 v. 3.0% body weight). The TDN content of the grass in the dry season trial was marginally higher than that during the wet season (62 v. 59%) whilst the content of digestible crude protein was markedly lower (3.2 v. 6.8%). The authors suggest that the difference in intake could have been attributable to an inhibitory effect of the higher moisture content of the wet season grass (76.6 v. 60.7%). In retrospect, however, this must have been an oversimplification in view of the confounded effects of climate, N content, digestibility and dry matter content.

It has been suggested that intake may be limited by reticulo/rumen 'fill'. This implies that forage consumed is equal to that which leaves the rumen. The only ways in which material can leave the rumen is either by absorption of digested material through the rumen wall or by passage through the reticulo-omasal orifice. Intake may thus be influenced by digestibility and by rate of passage through the reticulo-rumen or the entire digestive tract. Blaxter et al. (1961) and Conrad et al. (1964) have suggested that this is indeed the case at least up to levels of dry matter digestibility of 66%. However, this mechanism does not seem to operate with tropical forages. Butterworth and Butterworth (1965) could not demonstrate consistent relationships between intake and digestibility when Pangola and Coastal Bermuda grass hays were fed to sheep. Milford and Minson (1965) showed that the correlation coefficients between the digestibility of dry matter and intake in seven tropical grasses and two legumes were positive but did not account for sufficient of the total variation to enable them to be used in prediction. Similarly, González et al. (1972), using the results of digestibility trials carried out with twenty-three grasses and twelve legumes, demonstrated that positive correlations existed between dry matter digestibility and intake but that these did not reach significance. The use of multiple regressions, including cell wall constituents, crude protein and 'index of digestibility of fibrous fractions', did not improve the value of the equations for the purposes of prediction. Laredo and Minson (1973), using six tropical grasses fed to sheep, could not establish any relation between digestibility and intake for either leaf or stem fractions. Romero and Siebert (1980), although demonstrating that organic matter digestibility was the main factor limiting liveweight gain on pasture, did not obtain a consistent relationship between dry matter intake and digestibility using steers' grazing grass/legume pastures. Whilst these results indicate that the relationship between digestibility and intake is not exact enough to justify a prediction, the work carried out by Vohnout and his colleagues at Turrialba, Costa Rica, proves the exception. They carried out complete collection trials on cattle of various breeds using six tropical grasses, and related these to measurements of chromogens, lutein, indigestible protein, lignin, silica and *in vitro* dry matter indigestibility. Whilst chemical fractions and digestibility parameters could not be used to predict intake, it was shown that dry matter indigestibility at 48 and 96 and the potential indigestibility value accounted for up to 92% of the total variation (Pezo et al. 1977).

It has been suggested that low N content might limit the intake of tropical forages: Elliott and Topps (1963a), working with dry veld grass, obtained a significant reduction of intake when levels of crude protein in the forage fell below 5.5%, whilst similar reductions occurred in work carried out by Minson (1971) at levels below 6–8.5% crude protein. It may be that this effect is connected with the fact that cellulose-digesting micro-organisms (e.g. *Bacteroides succinogenes*) have a requirement for certain branched-chain fatty acids including iso-butyric and iso-valeric acids which result from the fermentation of certain amino acids (Hungate 1966). The effect of the N on intake would therefore be indirectly through its influence on digestibility. The observation that intake is limited by N content of the forage has not been invariable. Butterworth (1965) could demonstrate no reduction in intake when sheep were offered Elephant grass containing 4.6% crude protein; however, the crude protein content of the material actually eaten by the sheep was 7.6% and thus probably not low enough to influence intake. The deficiency of several minerals has also been shown to limit intake ('t Mannetje and Ebersohn 1980).

Venezuelan work could not demonstrate any consistent relation between rate of passage through the gut and intake (Butterworth 1965; Butterworth and Butterworth 1965). However, Laredo and Minson (1973) showed that the higher consumption of the leaf fraction in the experiment referred to above was associated with a shorter retention time in the reticulo-rumen than the stem of which less was consumed. Clearly this relationship is not straightforward, for whilst we may assume that more digestible forages have a shorter total retention time, more time spent in the reticulo-rumen enables micro-organisms to degrade the plant material further.

It has been shown quite conclusively that the structure of the forage has a profound effect upon its intake. Laredo and Minson (1973) chaffed samples of five tropical grasses (*Chloris gayana, Digitaria decumbens, Panicum maximum, Pennisetum clandestinum* and *Setaria splendida*) into lengths of approximately 2–4 cm and then separated into 'leaf' and 'stem' using a gravity separator. Although the stem was, surprisingly, slightly more digestible than the leaf fraction, mean voluntary intake of the latter was some 46% higher than the stem. As well as being associated with a shorter retention time in the reticulo-rumen, as mentioned above, this appeared to be caused by the large surface of the leaf available for bacterial degradation. There were no differences in rate of digestion *in vitro* as structural differences had been eliminated by grinding.

The importance of plant structure to the grazing animal has been confirmed by Stobbs (1973a). Having shown that the prehended bite size can limit intake (see below), it was further shown that treatment with growth regulators, which altered the growth habit of the grasses used, also influenced intake. Gibberellic acid and 2-chloroethyltrimethylammonium chloride were applied to pastures of *Digitaria decumbens* and *Chloris gayana*. The former resulted in the formation of a sward characterized by the presence of tall stemmy plants with long internodes whilst the latter produced a leafy dense sward with reduced plant height and internode length. Jersey cows fitted with plugged oesophageal-fistulae were used to determine the bite size. The bite size of the leafy sward was 0.52 g OM/bite whilst that of the stemmy sward was 0.40 g OM/bite. Different sward types were also promoted by nitrogen fertilization and the bite size on the N-fertilized *Setaria sphacelata* sward was 0.39 g OM/bite whilst that on an adjacent sparse unfertilized sward averaged 0.13 g OM/bite. The higher leaf to stem ratio was also characterized by a higher sward bulk density; the leafy and stemmy swards of Pangola grass referred to previously for example had bulk densities of 160 and 269 kg/ha cm respectively at six weeks. Stobbs also noted that the total number of bites recorded in one day rarely exceeded 36 000 and calculated the estimated number of bites required at various bite sizes to achieve intakes of 2.0, 2.5 and 3.0% body weight of a 400 kg animal. It was noted that at bite sizes of less than 0.3 g OM/bite, intake may be restricted. It is interesting that the high intakes achieved by Butterworth et al. (1961) of 3.0 and 3.7% of body weight referred to above were achieved on dense leafy swards with high leaf: stem ratio (personal observation).

In a subsequent paper, Stobbs (1973b) observed the influence of different states of growth of *Setaria sphacelata* and *Chloris gayana* pastures on bite size. Regrowth of 2, 4, 6 and 8 weeks was stratified into layers 15 cm thick and leaf and stem fractions were separated. The bite size at 4 weeks was greater than that at later dates and this was associated with a high sward leaf density. In other words, it corresponded to the phase when swards contained 'the highest proportion of accessible leaf'. In more mature swards, cows harvested leaves from layers 30–60 cm above ground level where they were relatively inaccessible due to a high stem content. It was concluded that in mature tropical pastures any possible advantages of selective grazing could well be offset by limitation of intake because of low bite size. 'Cows can be starving in the midst of plenty.'

In a later experiment, the effect of progressive defoliation of a *Setaria sphacelata* sward on the grazing behaviour of cattle was studied. In the early stages, cattle selected mainly leaf (i.e. more than 80%); as progressively less leaf became available, bites became smaller and total grazing time and bite rate increased. In later stages of defoliation, all these criteria again decreased. This experiment which was carried out during spring (early rains) and autumn (early dry season) established maxima for total grazing time (646 and 593 min. for autumn and spring respectively), bite rate (62.4 v. 59.4 bites/min.) and total bites (39 500 v. 34 000, this last value bearing a remarkable similarity to the values quoted in the last series of experiments). The finding that eating time increased as the quality of forage decreased is in agreement with the findings of Butterworth (1965) who fed green Elephant grass at various stages of growth to pen-fed sheep. Of the three main factors determining grazing intake, i.e. grazing time, bite rate and bite size, Chacón and Stobbs (1976) confirmed that bite size was the most important.

The above work has illustrated one important facet of animal behaviour which influences intake and has yielded interesting results. There is, however, very limited knowledge on other factors which influence preferences on mixed swards as well as improved pastures. Aspects of palatability, particularly of range plants, are imperfectly understood.

Microbiology of the rumen and production of volatile fatty acids

Relatively little information is available on the microbiology of the rumen of cattle, or indeed other ruminants, grazing tropical forages. Elliott et al. (1965) separated rumen contents of steers grazing tall grass Hyparrhenia veld into fractions representing protozoa and bacteria. This was done by low- and high-speed contrifugation respectively. The total dry matter from these two fractions accounted for approximately 15–35% of the solid material of the rumen contents, and was appreciably lower in winter than in summer. The writers fancifully suggest that 'one could imagine the microorganisms becoming emaciated along with the cattle during the winter'.

The total counts of rumen bacteria in hay-fed sheep were determined by Butterworth and Butterworth (1965). Pangola and Coastal Bermuda hays cut at different stages of growth were used and total bacterial counts (cf. viable counts) ranged from $10^9 \times 8.0$ to $10^{10} \times 1.9$/ml. Higher counts were associated with higher digestibility coefficients of dry matter. It is reasonable to suppose that more substrates would be supplied by the more digestible diets and this would lead to increased populations of rumen micro-organisms with consequent 'beneficial results to the internal economy of the animal'.

Types and relative quantities of bacteria present in Gram stains of rumen fluid smears were also given. The most frequently occurring organisms were Gram-negative cocci which were high in all diets with the exception of one sample of Coastal Bermuda hay when they were moderate. Small and large Gram-positive cocci were present in appreciable quantities in all diets. Large Gram-negative cocci, thin Gram-negative rods, short and thick Gram-negative rods, Gram-negative curved rods and *Selenomonas* spp. were present in all diets. Gram-negative and Gram-positive *Sarcina* spp., Gram-positive rods with differing characteristics, large *Cocci* in long chains (usually Gram-negative) and *Oscillospira* sp. were also recorded. It was noted that results obtained from tropical forages fell within the range of total bacterial population expected in temperate climates.

The occurrence of various cellulose digesters in the rumen of different species grazing tropical forages has been reported by Hungate (1966). *Bacteroides succinogenes* has been found in several African antelopes including eland, kongoni and Grant's gazelle. *Butyrivibrio fibrisolvens* was isolated on cellulose rumen fluid agar from starving Zebu cattle on very poor forage at Archer's Post, Kenya, as was a curved acetigenic rod.

A study of the ability of different micro-organisms to degrade the cell walls of *Eragrostis tef* was carried out by Morris and van Gylswyk (1980). Isolation was carried out on the basis of the ability to ferment xylan from a sheep fed a diet of *Eragrostis tef.* Organisms were classified as *Butyrivibrio*-like, *Bacteria ruminicola*-like, *Eubacterium*-like, Ethanol-producing rods, *Ruminococcus flavefaciens* and *R.7 albus*. Cultures were incubated with teff cell walls to determine the solubilization and utilization of pentose, hexose and uronic acid. There was a wide variation in the ability of hemicellulolytic and cellulolytic bacteria to degrade plant cell walls, cellulolytic bacteria generally being the most effective. As structural carbohydrates are linked to other cell wall constituents, especially lignin, through phenolic esters, other degradative enzymes are probably required. Organisms were classified into three types: those which produced freely diffusable enzymes, those which attached themselves to the cell walls and those that did both. It appeared that those organisms in the first category were more effective in attacking cell walls than those in the other two categories.

The rumen microbiology of wild ruminants grazing and browsing under tropical conditions is also probably relevant to this discussion in that general conditions in the rumen would be expected to be more or less the same as those to be found in cattle. Giesecke and van Gylswyk (1975) carried out a study of certain rumen functions in six species of South African wild ruminants including buffalo, blue wildebeest, gemsbock (all three predominantly grass eaters), impala, springbok and kudu (browsers). Microscopic counts of Quinn's Ovals, *Selenomonas* and *Oscillospira*, averaged 670, 602 and 61 $\times 10^6$/g. Total 'culturable' counts and cellulolytic counts averaged 210

$\times 10^7$ and 53×10^6 respectively. These values were similar to those obtained from parallel determinations carried out with sheep rumen fluid and the authors conclude that ruminal metabolism has definite limits and that it will be maintained within these limits with a more or less constant turn-over of micro-organisms and metabolites regardless of diet or of animal species.

Few data are available on the numbers of protozoa present in the rumen of ruminants fed tropical forages. Naga and el-Shazley (1969) investigated the activities of rumen micro-organisms in water buffalo and in Zebu cattle but unfortunately, out of a number of rations fed, the only one corresponding to a forage typical of our area was sweet Sudan grass on which only limited observations were made with buffalo. The total count of 0.45×10^5/ml was lower than that on the other diets (3.3×10^5/ml) of which an average of 81.1, 14.4, 2.1 and 2.5% were *Entodinium*, *Isotricha*, *Dasytricha* and *Eudiplodinium* respectively.

Intermediate values were observed by Giesecke and van Gylswyk (1975). Counts of protozoa ranged from 1.69 to 10.59×10^5. Differences were attributable to both species and diet; the browsers referred to previously appeared to have higher numbers than the grazers (2.92 v. 6.66×10^5/g) and it was suggested that this might have been associated with higher levels of rumen ammonia N.

The total numbers of protozoa observed in the experiments cited above are in general agreement with those experiments carried out elsewhere and reviewed by Hungate (1966). The predominance of *Entodinium* spp. is at variance with the observation by Hungate that the holotrich protozoa tend to predominate in the grazing animal but probably because of the readily available carbohydrate in the forage.

Several studies have been made of both the total quantity and the ratios of individual volatile fatty acids on ruminants either grazing or fed tropical forages and values are set down in Table 2.1 for convenient reference.

The mean value for the total volatile fatty acids VFAs in Table 2.1 was 9.39 ± 4.03 meq/100 ml. In general, the values were relatively similar. Those differences reflected changing nutritive values with more digestible forage leading to higher levels of VFAs (the average of 26 values for cattle, sheep and goats was 7.76 ± 2.55, an appreciable decrease in the SD). The relative proportions of individual fatty acids was also relatively constant. The mean value for acetic acid was 71.26 ± 5.85%, for propionic, 17.70 ± 3.85% and for butyric, 8.69 ± 2.75%.

In general, the acetic acid produced in the rumen reflects the level of crude fibre in the diet and this is generally true of the results in Table 2.1. It has been postulated that the decrease in efficiency of utilization of diets high in fibre has been associated with this increased percentage of acetic acid. However, recent work by Ørskov and McDonald (1980) using infusions of different proportions of

Table 2.1 The mean total concentration of volatile fatty acids (VFA) and relative molar proportions of individual VFAs in rumen liquor of ruminants eating different tropical forages

Forage and species	*Total VFA meq/100 ml*	*Acetic acid (%)*	*Propionic acid (%)*	*Butyric acid (%)*	*Reference*
Fresh veld grass	9.89	65.8	19.4	9.0	Topps and Elliott (1964)
Late summer veld					
Cattle	8.2	69.4	17.9	8.6	Topps *et al.* (1965)
Winter veld	7.5	70.2	19.8	7.1	Topps *et al.* (1965)
Early summer veld	9.8	68.5	18.1	10.4	Topps *et al.* (1965)
Elephant grass					
Sheep	7.42	60.6	25.8	8.7	Butterworth (1965)
	7.82	63.9	25.5	7.2	Butterworth (1965)
	7.48	58.6	26.4	11.5	Butterworth (1965)
Pangola grass hay					
Sheep	5.72	67.5	23.6	6.4	Butterworth and Butterworth (1965)
	6.24	65.3	22.7	8.2	Butterworth and Butterworth (1965)
	7.68	61.4	23.8	11.5	Butterworth and Butterworth (1965)
Coastal Bermuda hay					
Sheep	7.64	71.9	16.7	7.7	Butterworth and Butterworth (1965)
	7.86	71.4	15.9	9.0	Butterworth and Butterworth (1965)
	9.76	68.8	19.6	7.0	Butterworth and Butterworth (1965)
Mixed grazing					
Buffalos	15.2	74.2	16.0	9.9	Giesecke and van Gylswyk (1975)
Wildebeest	13.2	74.5	11.2	14.3	Giesecke and van Gylswyk (1975)
Gemsbok	13.7	74.8	13.0	10.2	Giesecke and van Gylswyk (1975)

Mixed browse					
Impala	15.2	76.3	16.8	5.6	Giesecke and van Gylswyk (1975)
Impala	13.0	77.3	16.3	5.8	Giesecke and van Gylswyk (1975)
Springbok	12.0	65.7	17.3	14.6	Giesecke and van Gylswyk (1975)
Kudu	14.2	73.8	16.4	8.2	Giesecke and van Gylswyk (1975)
Spear grass					
Penned steers	7.45	78.4	14.3	5.8	Playne and Kennedy (1976)
Townsville Stylo	7.77	72.7	17.4	7.6	Playne and Kennedy (1976)
Spear grass (wet)					
Grazing steers	13.71	69.7	16.7	10.6	Playne and Kennedy (1976)
Spear grass (dry)					
Grazing steers	8.98	75.8	15.5	6.9	Playne and Kennedy (1976)
C. nlemfuensis plus *C. pubescens*					
Steer	5.40	80.4	15	5	Mba *et al.* (1976)
Sheep	4.19	80.5	15	5	Mba *et al.* (1976)
Goat	3.45	80.9	15	5	Mba *et al.* (1976)
Bush grassland					
Haired sheep	12.4	77	12.4	7.6	Hoppe *et al.* (1977)
Massai goat	12.9	77	13.3	7.0	Hoppe *et al.* (1977)
Thompson's gazelle	14.3	76	14.0	6.0	Hoppe *et al.* (1977)
Grant's gazelle	14.1	76	16.0	5.4	Hoppe *et al.* (1977)
Impala	17.8	74	14.1	9.3	Hoppe *et al.* (1977)
Rhodes grass, frozen					
Sheep, high level	6.4	65.8	20.6	12.2	Tagari *et al.* (1977)
Sheep, low level	5.2	67.8	17.6	13.1	Tagari *et al.* (1977)
Rhodes grass hay					
Sheep, high level	6.4	65.2	20.4	12.8	Tagari *et al.* (1977)
Sheep, low level	4.5	68.2	17.6	12.5	Tagari *et al.* (1977)

acetic and propionic has shown that there were no differences in energy retention among mixtures as widely different as 35 molar % acetic acid plus 55 molar % propionic acid and 85% acetic acid plus 5% propionic acid.

Nutrient requirements of beef cattle in the tropics

Relatively little information is available on the requirements for nutrients of beef cattle in tropical environments and although a large amount of data is available from climatic chamber work most of this relates to the effects of heat stress on the animal and its extrapolation to the grazing beef animal is of doubtful validity. However, some reasonable assumptions as to the requirements of the grazing animal may be made from the limited evidence available. This enables qualitative comparisons to be made with accepted standards for other areas. Nutrient requirements may be estimated either empirically as a result of feeding trials or by the factorial method. For the purpose of this discussion, the latter approach will be followed.

Energy requirements for maintenance comprise expenditure on fasting metabolism which is related to body size and environmental temperature as well as that expended in activities involved in assimilation of food and water.

Body size may be expressed as weight or as metabolic body weight, i.e. $kg^{0.75}$. Although the latter expression has been shown to be valid over a wide range of species, there is some doubt as to whether it is applicable to cattle breeds at or near equilibrium weight. Frisch and Vercoe (1977) carried out measurements of fasting metabolism on European and Brahman and Afrikander cross cattle and found that the metabolism of the european cattle was consistently and significantly higher than that of both the tropical cross-breds, although this effect was not as great as that attributable to diet (cattle were fed high-quality lucerne hay or poor-quality tropical pasture hay) which amounted to some 12%. Nevertheless, the authors indicate that breed differences in maintenance requirements are 'important in northern Australia, where for several months each year grazing provides only a maintenance or sub-maintenance ration'. They also found that the value of the within-breed exponent derived from the regression of log-fasting metabolism on log-fasted liveweight was 1.087 ± 0.094 which was significantly different from 0.75 and suggested that the expression of values for fasted metabolism was more appropriate per kg than per $kg^{0.75}$. Breed differences in fasting metabolism appeared to be greater at lower growth rates; it was suggested that *Bos indicus* breeds that have evolved under conditions of poor nutrition have been automatically selected for low fasting metabolism and thus good survival ability but that this is linked with relatively low growth rates under better conditions. It is suggested that *Bos indicus* cattle are better 'buffered' against fluctuations in feed supply than european breeds. Later work (Frisch and Vercoe 1979) indicated that the fasting heat production was lower in cattle with a low inherent growth rate.

It also appears that energy requirements for maintenance may be reduced when animals are kept for some time at the same weight. This was first noted as long ago as 1915, by Trowbridge et al. and confirmed by Ledger and Sayers (1977) working at Muguga in Kenya. Groups of *Bos indicus* and cross-bred *Bos taurus* × *Bos indicus* steers were maintained at constant weights of 185, 275 or 450 kg for a period of 24 weeks by adjusting the daily ration. It was noted that as time progressed, less ration was required to maintain liveweight, indicating a progressive increase of efficiency of feed utilization as time at the same weight increased. When results where corrected for age of the various animals, the Boran steers were some 46% more efficient than would be expected from comparison with the normal standards whilst the corresponding value for cross-bred steers amounted to 33%.

This finding is extremely important in the consideration of the requirements of cattle grazing tropical grasslands which may spend considerable periods in states of maintenance or near-maintenance. The indication that *Bos indicus* cattle are more efficient in this respect is also interesting.

Cattle in cold climates must expend heat to maintain body temperature and whilst extremes of climate in tropical areas may occasionally involve stressful cold conditions, the extent of energy expenditure for this purpose is clearly less than in cold or temperate climates. On the other hand, energy is required to maintain homoeothermia at higher temperatures. The upper critical temperature, although subject to wide variation, generally lies between 24 and 27 °C. At temperatures above this, however, additional energy must be expended to maintain body temperatures at an acceptable level. It is also probable that acclimatized animals respond to high environmental temperatures by tolerating body temperatures higher than those which would be normal in unadapted cattle.

Another obvious factor which affects the maintenance requirement of cattle grazing tropical pastures is the distance that they must walk to drink and harvest adequate forage. This is extremely variable and its importance is merely noted here.

The protein requirements of cattle both for growth and for maintenance were studied by Elliott and Topps (1963a). Afrikander and Mashona steers were given isocaloric diets (energy was not limiting) with crude protein contents of 5, 7, 9 and 11%; this range was obtained by adjusting the proportions of cassava and cottonseed cake in the concentrate portion of the diet which accounted for 60% of the total ration. Measurements were made of intake, faecal and urinary excretion and retention. Computed regression lines for N retained v. apparently digestible N consumed enabled mean digestible N required for N equilibrium to be calculated by extrapolation. These values were closely related to metabolic body size (cf. Frisch and Vercoe 1977) by the equation N (mg) = 197 W $(kg)^{0.73}$. The requirement was approximately equal to three times the endogenous N output of the cattle and was considerably lower than other published standards. The low requirement was attributed to the endogenous N loss and the high efficiency of the N of the low protein diet. This was calculated to have a mean biological value of 82.5, which was higher than values generally assumed by other authors. It was taken that this could have been due to efficient use of N by increased recycling to the rumen via the saliva. However, the authors emphasized the necessity of carrying out longer-term trials to assess the validity of these results under conditions more typical of those to be found in the field.

The reports of these longer-term trials were reported in a subsequent paper (Elliott and Topps 1963b). In this experiment, Aberdeen Angus steers were included to represent european breeds in addition to the Mashona and Afrikander steers used previously. Animals were fed diets containing two levels of energy (150 and 100% of the levels recommended for maintenance by Brody 1945) and sufficient protein to provide only half of the level of apparently digestible N recommended by the same author. The diet was fed for a period of 15 weeks. The steers on the high-energy ration gained weight throughout the trial period and it was assumed that the quantity of N fed was adequate when energy was not limiting. The animals on the low-energy diet, however, lost weight during the trial and the difference was attributed to the inadequacy of readily fermentable carbohydrate in the latter ration.

These results are interesting in view of the frequently observed fact that cattle subsist and even prosper on diets appreciably lower in crude protein than the levels normally considered to be essential. This phenomenon occurs frequently in tropical areas subject to seasonal droughts and where grasses mature rapidly with extremely low levels of crude protein. The low-energy diet would be more comparable with pastoral conditions and the amount of weight loss may be lower in adapted animals than those that are not adapted (cf. Frisch 1973) although this is not consistent with the findings of Elliott and Topps (1963b) that Aberdeen Angus steers lost similar amounts to the *Bos indicus* cattle.

The above results have indicated that the levels of protein required by cattle in the tropics may be expected to be less than those generally found in temperate areas, at least when energy levels are adequate (although one of the main problems with tropical forages is that both energy, especially readily fermentable carbohydrate, and protein are limiting at the same time). However, there is some indication that exposure to heat stress might increase the excretion of urinary creatinine, especially in *Bos taurus* and cross-bred animals. Colditz and Kellaway (1972) suggested that protein catabolism might be higher during heat stress and that this might account for the lower N retention of non-*Bos indicus* animals. However, this work was carried out in a climatic chamber at the relatively high temperature of 38 °C and it is probably not generally applicable to field conditions.

There is some evidence that energy for growth is less efficiently used by Zebu than by european breeds of cattle. Rogerson et al. (1968) showed that gains, intake and efficiency of utilization of a low roughage–high concentrate diet were all better in *Bos taurus* than in *Bos indicus* steers and this was later confirmed using low and medium roughage diets (Ledger et al. 1970). Frisch and Vercoe (1969) showed that the intake of *Bos taurus* cattle was 21% higher than that of Brahmans on a similar lucerne hay diet, and in a later study (Frisch and Vercoe 1977) it was shown that *Bos taurus* × *Bos indicus* animals had intakes between the parental breed means. These differences in intake were also associated with differences in weight gain and efficiency of utilization as well as a high fasting metabolism. It was concluded that animals with a high fasting metabolism were also those which used feed more efficiently for growth. Relative feeding levels were calculated (voluntary feed intake over fasting metabolism) but these were found to be relatively constant both within and between breeds. The authors therefore concluded that differences in performance could not be ascribed to differences in relative feeding level unless differences also existed in the proportion of metabolizable energy (ME) diverted into different body tissues or the proportion of ME used for maintenance (although Rogerson et al. 1968, whilst confirming the lower fasting metabolism of the Zebu animal, state that no differences in the utilization of either digestible or metabolizable energy could be demonstrated between types of animal using calorimetry). It appears, therefore, that further work on breed differences in feed utilization are required before definite conclusions can be drawn.

Total nutrients required are closely related to rate of gain and it has often been noted that tropical cattle breeds generally have lower rates of gain than european breeds which would result in lower absolute requirements. A further factor to be taken into account, however, is the composition of the gain as the deposition of fat is less efficient than that of lean. In this context it should be noted that Zebu cattle generally contain less carcass fat than european breeds at an equivalent stage of maturity (see, e.g. Cole et al. 1964). The results of Rogerson, et al. (1968) and Ledger et al. (1970) provide the exception to this generalization, however, as in the former work, the fattest cattle were *Bos indicus* (Improved Boran) whilst in the latter trial, the leanest cattle were Herefords. However, carcass composition is also known to be markedly influenced by both age and rate of fattening as well as breed. In any situation, therefore, there are a number of variables which may influence the requirements of energy for weight gain including rate of gain, breed and composition of gain, all of which may be confounded.

A further factor which is undoubtedly of importance in assessing the efficiency of utilization of nutrients for gain is the phenomenon known as compensatory growth. This is the increased rate of growth which occurs when an animal is realimented after a period of under-nutrition. The extent to which this occurs depends on several factors including the nature, severity and duration of the period of under-nutrition as well as the pattern of realimentation. Compensatory growth is clearly of great importance in beef cattle grazing tropical pastures which are frequently subjected to periods of diminished forage supply. Recent work carried out by Ledger and Sayers (1977), previously referred to with respect to requirements of energy for maintenance, indicated that the increased efficiency accompanying reduced feeding (i.e. some 40%) persists, although it decreases in intensity throughout the period of realimentation, for as long as 15 weeks after the start of a more favourable feeding regime. This effect is clearly of great importance when it is considered that one of the main characteristics of tropical forages is a flush of growth soon after the onset of the rains. This permits the rapid realimentation of cattle which have been undernourished during the rainy season. Although, on theoretical grounds, overall efficiency may be improved by the maintenance of a constant rate of gain, in practice this is not normally economically viable. Such aspects of compensatory growth will be discussed in later sections. The field has been comprehensively reviewed by Wilson and Osbourn (1960).

It would not be expected that requirements for reproduction in tropical cattle would differ markedly from those in temperate areas insofar as protein and energy are concerned. Adequate nutrients must be available for spermatogenesis in the male and ovulation in the female and for subsequent development of the embryo. Additional requirements are necessary for the foetus itself, together with the uterus and its associated membranes, as well as a limited amount of protein for the development of the mammary gland prior to lactation. The basal metabolism of the dam is also higher as a result of pregnancy because of the heat increment of gestation and it is conceivable that this might impose an additional load under conditions of heat stress because of the need to maintain body temperature. Special cases of the effects of protein and energy deficiency on the reproductive performance of beef cows are discussed in a later chapter.

The nutrient requirements for lactation depend on a number of factors, the most important of which are the amount of milk produced and its composition. A considerable amount of data is available on the milk production of tropical breeds of cattle. However, most of these were obtained from cattle kept under semi-intensive conditions typical of dairy operations and few data are available from grazing animals. For a review of information on tropical dairy cattle, the interested reader is referred to the review by Mahadevan (1966).

Work on beef cows in Queensland was carried out by Holmes et al. (1968) at the National Cattle Breeding Station, Belmont and the Pasture Research Station at Samford near Brisbane. Brahman × Shorthorn and Afrikander × Shorthorn cows grazed on pastures predominantly comprising *Heteropogon contortus* at Belmont whilst Hereford cows at Samford grazed on a wide range of subtropical grasses and legumes of generally good quality. The milk yield of the Hereford cows ranged from 7.0 kg/day at the start of the lactation to 2.8 kg/day after 164 days. Corresponding values for milk fat were 5.1 and 4.4% and for solids-not-fat (SNF) 8.9 and 8.3%. The yield of Brahman cross cows ranged from 7.9 to 5.6 kg/day at the start of the lactation and at 120 days whilst the equivalent values for the Afrikander cross cows was 9.0 and 5.2 kg/day. Milk was not analysed in the first two periods but mean values for the rest of the lactation were 4.44 and 4.73% for fat and 8.7 and 8.7% for SNF for breed types in the order given above.

The milk production of Brahman cross-bred and Shorthorn cows was measured at Swan's Lagoon on both native pasture and a mixed sward of Townsville stylo and native pasture (presumably *Heteropogon contortus*) during both wet and dry seasons. A variety of supplements were also given in the dry season. There was no difference in milk production attributable to pastures (this was explained by bad establishment of the Townsville stylo and the imposition of a heavier stocking rate and confirmed by the fact that faecal protein levels on the two pastures were comparable indicating a similarity of protein intake). Pasture means were 3.3 kg/head/day and 4.4 kg/head/day respectively for dry and wet seasons. Breed differences were not significant during the dry season but there was a breed/season interaction during the wet season as milk production was 3.6 and 5.2 kg/day for Shorthorn and cross-bred cattle respectively. The effect of supplementation was significant in both dry and wet seasons, milk production being 2.6 and 3.4 kg/day and 4.0 and 5.2 kg/day for unsupplemented and supplemented cows during the dry and wet seasons respectively (Allan et al. 1972).

The milk yield of cows grazing veld and supplemented with various levels of concentrates was measured at the Mlezu Rural Technical Institute in Zimbabwe (Richardson et al. 1977d). Cows on the high plane of nutrition received 450 g/day of cottonseed cake from beginning of winter until early summer with additional amounts of 450 g/day maize meal added to the ration at the beginning of July,

August and September. The medium-plane cows received half this amount and the low-plane cows received no supplement. The 98-day milk yield was 737, 661 and 627 kg for the three groups in the order given above (i.e. 7.5, 6.7 and 6.4 kg/day respectively). The high-plane cows produced significantly more milk than the other animals but those on the medium plane were not significantly more productive than those on the low plane. Milk yield was positively correlated both with cow's weight at calving and with calf's birth weight. It was also directly related to changes in cow weight during early lactation. In a parallel trial, pen-fed heifers were given 133, 100 or 67% of their individual requirements for energy and milk production. Average daily milk yields were 6.5, 5.9 and 4.7 kg for high-, medium- and low-plane animals respectively. Fat content varied from about 3.2 to about 4.4% (results are expressed graphically), protein from 2.9 to 3.8% and SNF from 8.9 to 9.8%. Fat content increased with increased weight of grazing cows and those changes reflected *pre partum* nutrition. Content of SNF changed only slightly and inconsistently.

The milk yield of Brahman cows grazing *Panicum maximum* pastures in north-central Venezuela was measured by Neidhardt and his colleagues (Neidhardt et al. 1979). The estimated 24-hour milk yield was 6.2 kg corresponding to a lactation yield of 1340 kg in 216 days. Milk yield was significantly positively correlated with cow weight loss *post partum* as in the previous work. An interesting finding was that the relation between milk yield and average maximum temperature during the 14 days before the test weighing was positive. This shows that even though the average mean maximum temperature was 31.7 °C and the mean minimum 23.2 °C, the production of relatively high levels of milk was not interfered with.

Milk yield therefore appears to range between approximately 3 and 7 kg/day depending upon the plane of nutrition as well as the breed of the cow, stage of lactation, etc. Milk fat varies from 3.2 to values of more than 5 (although the review by Mahadevan previously cited suggests that Zebu cattle may generally produce milk with higher fat levels). In view of the finding by Neidhardt et al. (1979) that a relatively high environmental temperature has no deleterious effect on milk production, there seems no reason that nutrient requirements should not be based on the standards used in temperate climates taking yield and composition into account. It might have been supposed that the relatively high levels of acetic acid to be found in the rumens of cattle grazing tropical forages might have reduced the efficiency of milk production but, in view of the recent results of it Ørskov and McDonald (1980) previously referred to, this seems unlikely.

Summary

Many tropical grasses do not contain sufficient protein to supply maintenance requirements of beef cattle although higher levels are generally present in legumes and non-leguminous browses. The protein content of grasses declines with increasing maturity, and as the content of crude protein declines so does its digestibility. This is generally agreed to follow a logarithmic relationship and accurate predictions of digestibility may be obtained from crude protein content. The logarithmic relationship is a result of the effect of metabolic faecal N. The effects of crude fibre content of the grass on the digestibility of the protein do not appear to be clear-cut, possibly because of the imprecision associated with the empirical method of crude fibre determination. The importance of protein solubility and by-pass protein has recently been recognized in tropical forages and it appears that the lower solubility of legume proteins may be associated with the tannin content of the plant.

Many of the expressions of the energy value of tropical forages can be related but there has been no effective means of predicting the content of total digestible nutrients of a forage from its proximate analysis. However, newer methods of chemical and microscopical analysis may shed more light on existing relationships and enable satisfactory predictions of energy value to be made. One interesting aspect of the digestibility of tropical forages is the negative effect of higher temperatures during plant growth. Values of TDN were used to assess the probability of encountering deficiencies in the energy value of forages and it was concluded that although these would not be as common as those of protein, they might frequently limit growth.

There is still considerable uncertainty as to which of the many physiological and behavioural factors involved actually limits the intake of tropical forages by ruminants. However, the most critical factor in practice is the availability of forage. Tropical grazing lands are characterized by fluctuations in forage availability, either because of natural causes (e.g. drought) or because of uncontrolled over-stocking. Another restriction is the build-up of unpalatable mature material. Direct effects of climate are certainly involved in the limitation of intake but many studies have been carried out in climatic chambers and the relevance of the results of these to field conditions is dubious.

Direct climatic effects may also influence grazing behaviour so as to reduce intake, and a grazing management system which allows good night grazing is advantageous. On the other hand, grazing time is often artificially limited by the practice of night confinement (or 'kraaling') of cattle.

There is, however, certain evidence that intake may be higher during wet than dry seasons but this tends to be confounded with other factors.

The evidence that digestibility of the forage, or extent of reticulo-ruminal fill, limits intake has not been convincing and equivocal results are obtained from studies of rate of passage through the gut.

There seems to be no doubt, however, that the structure of the forage has an important role to play in its intake. This is associated with leaf-to-stem ratio but also with sward density which exerts a profound effect on the grazing behaviour of the animal with particular regard to bite size, number of bites, etc.

Little information is available on microbiology of the rumen in the tropics although some descriptive and quantitive studies have been carried out on a limited number of individuals of various species. It has been established that the number of bacteria in the rumen is positively correlated with the digestibility of the forage supplied. Little information is available on the protozoa of the tropical ruminant. Some values for the ruminal content of volatile fatty acids have been determined and these indicate that total values of VFAs are higher in ruminants fed diets of better quality. There is also some evidence that the molar proportion of acetic acid is higher with diets composed of tropical forages but the importance of this to the energetic economy of the animal may well be less important than was hitherto believed.

There is little available information on the nutrient requirements of the grazing beef animal in the tropics. but some reasonable assumptions may be made to enable us to compare these with the requirements of stock in temperate areas, at least from a qualitative point of view. There is some evidence that the utilization of energy for maintenance is more efficient in tropical than in temperate cattle as fasting metabolism seems to be lower in breeds which have evolved under tropical conditions. It has also been suggested that the use of the customary metabolic body size ($Wkg^{0.75}$) may not be appropriate to animals under tropical conditions although this is still a moot point.

There is probably little need for expenditure of energy in order to maintain body temperature under tropical conditions although there may occasionally be an expenditure of energy to reduce it. There is also the consideration that one possible response to high environmental temperatures might be toleration of a higher corporal temperature. The importance of the activity increment involved in watering and harvesting forage over long distances was also emphasized.

There is strong evidence that the requirement of N both for maintenance and body weight gain is lower than the corresponding value for cattle under temperate conditions. This seems to be associated with a lower endogenous N loss as well as more efficient utilization.

This may be caused by recycling of N to the rumen via urea in the saliva. Although it is probable from observation that these low N requirements are a true representation of the metabolism of ruminants in the tropics, they have only been conclusively demonstrated in high-energy rations and further work is required on this aspect of beef cattle nutrition.

It would appear that energy is not so efficiently used for growth as for maintenance in Zebu cattle and this is associated with lower intake of nutrients as well as lower fasting metabolism. The composition of gain must also be taken into account when discussing the requirements for growth of cattle in the tropics, and although Zebu cattle frequently finish with less fat than european cattle, this is by no means invariably the case. Another important factor which affects requirements is that of compensatory growth. This has been associated with a persistence of the increased efficiency that exists under conditions of subnormal nutrition.

There appears to be no evidence that indicates that requirements for reproduction are any different from animals kept under temperate conditions although it could be that the increased basal metabolic rate associated with gestation might impose an extra burden in maintaining homoeothermia at high ambiental temperatures.

The nutrient requirements for milk production are dependent on yield and composition. However, it appears that milk production in the beef cow is not influenced by temperature and therefore requirements are probably analagous to those obtaining in temperate climates.

3 *Phosphorus deficiencies and their correction*

The nutritional shortcomings to tropical grasses insofar as protein and energy are concerned have been detailed in an earlier chapter. However, possibly the most widespread and certainly the most reported mineral deficiency is that of phosphorus. Deficiencies of this element have been shown to cause reduced growth and appetite, abnormalities of bones and teeth, pica or depraved appetite, reduction of milk yield as well as reproductive disturbances (for full details see the review by Underwood 1966).

In this chapter, a brief historical outline will be given followed by a review of works indicating deficiencies, either by analyses of herbage or of blood or other tissues. Studies involving supplementation of various classes of stock and using different sources of the element will be discussed in the subsequent two chapters.

Historical background

Southern Africa

A great deal of the early work on phosphorus deficiencies was carried out in Southern Africa where it was intimately associated with the occurrence of the disease, gallamziekte or lamziekte (literally Lame sickness) or botulism. Sir Arnold Theiler has mentioned early references to the disease by some of the first visitors to the Cape.

Le Vaillant (1796), in his book entitled *Travels into the Interior Part of Africa*, noted lamsiekte (sic.) as a 'real palsy, which comes on all of a sudden'. He also mentioned the practice of bone eating or osteophagia: 'as these bones were of a calcareous nature, by gnawing them they blunted the force of the irritation and acidity which tormented them'. Lichtenstein (1812), in his book *Travels in Southern Africa in the Years 1803, 1804, 1805 and 1806*, mentioned that animals could be cured from lamziekte by moving them from an affected area to a valley where the disease had never occurred.

A certain amount of controversy on the subject seems to have reigned throughout the nineteenth century as Worcester (1889) in a letter to the *Agricultural Journal* of the Cape Colony commented that he had used 'a decoction of the red stem ferns' which had been referred to in earlier correspondence. He reported that 'the animal thereupon became inflamed and died' and concluded with the heartfelt cry 'I hope that this intimation will force some more discussion on the subject'.

The first mention of the use of bonemeal to correct phosphorus deficiencies appears to be in the editorial reply, also in the *Agricultural Journal* of the Cape Colony, to a query from LK dated 9 January 1894. In his reply, the editor advocated the use of bonemeal supplementation in cases of lamziekte and claimed that he advised its use twenty years previously in the district of Albany and that this 'was adopted with considerable success'.

Hutcheon (1892), veterinary surgeon of the Cape Colony, gave further examples of the use of bonemeal in the treatment of the disease and reviewed work reported in previous annual reports in an article entitled *Lamziekte, Adynamia or Paralysis in Cattle.* He attributed the simple form of the disease of adynamia to 'a deficiency of phosphates or earthy salts in the food'. He continued, referring to districts where lamziekte was prevalent 'no one can watch the cattle . . . and observe their craving for bone and all kinds of matter containing earthy phosphates without being convinced that there is a great deficiency of such substances in the general food available for the stock'. In the same article, Hutcheon recommended the use of bone ash as a phosphorus source rather than bonemeal because of its smaller bulk and greater ease of handling. He quoted a Mr J Cammack at about the same date as reporting that 'after the use, for only a short time . . . of a mixture of one pound of hydrochloric acid, a hundred pound of burnt bones and the same quantity of salt, lamziekte ceased on a farm near Kimberley although bone eating had been going on for some time'. He further quoted a recipe given by the Rev. Roger Price to the *Diamond Fields Advertiser* of 19 January 1886.

> Observing the intense fondness of cattle for chewing bones, my predecessor, Rev. Mr Mackenzie, conceived the idea of making cakes of bran and mealie (maize) meal with which he pounded bones, egg shells and salt. The meal or bran was cooked in a pot like ordinary porridge or 'pap' and the salt and powdered bones mixed with it. The pap has to be made very thick so as to give it adhesiveness. When it is cool, it is taken up with one hand and formed into a ball about the size of one's fist . . . once (cattle) have got the taste, they soon become tremendously eager for it.

Hutcheon concluded his summary by recommending several other specifics which he had used with success in the treatment of mild forms of adynamia. These included a drench consisting of 'mealie meal or oatmeal gruel: three bottle, strychnine: two grains and brandy or whisky: a tumblerfull'! Hutcheon was also apparently responsible for the first experimental work to be carried out on bonemeal supplementation.

These experiments were cited by Theiler (1920) as being carried out by Borthwick at Wittelclayrug and Spruell at Koopmansfontein and demonstrated that if bonemeal was supplied regularly at the rate of three or four tablespoonfuls per day the mortality from lamziekte was materially reduced.

On the other hand, Bowhill (1906) maintained that the disease had nothing whatsoever to do with a deficiency of phosphorus but was simply due to infection with pasteurella. Bevan (1906–07), writing in Rhodesia, quoted Robertson as reporting the complete digestion of crushed bonemeal (as opposed to pieces half an inch square which passed through the alimentary tract unaffected) by the digestive juice of an ox. The bonemeal was stained with methylene blue and could not be detected in the faeces. He went on to comment, 'as soon as the bone is digested, its constituent elements are as readily assimilated as if they came originally from a bag of wheat'. Bevan also 'confidently' recommended bonemeal as a preventive of what was recognized as real lamziekte and styfziekte, a similar condition characterized by stiffness in the joints.

It was left to Theiler (1920, 1927) finally to elucidate the links of the aetiological chain leading to the occurrence of lamsiekte. These he listed as follows:

1. The toxin that poisons the animal.
2. The toxicogenic saprophyte which elaborates the toxin.
3. The carcass material out of which the toxin is manufactured.
4. The pica (allotriophagia, osteophagia or depraved appetite) which impels the animal to eat decomposing matter which it would otherwise shun.
5. The vegetation which produces the pica, the soil upon which that vegetation grows and the climate of the district.
6. The susceptibility of the animal towards pica and towards the toxin. Thus vegetation containing low levels of phosphorus induces a deficiency in the animal which in its turn provokes a depraved appetite causing the animal to become infected with the toxicogenic organism subsequently identified as *Clostridium botulinum*.

Theiler (1920) has credited Hutcheon (1892) with the correct explanation of the disease styfziekte as a simple aphosphorosis uncomplicated by intoxication with the pathogen. Hutcheon (1903) makes two interesting comments on the susceptibility of animals to the two conditions: 'In our experience a beast which is growing and improving in condition appears to be much more likely to contract lamziekte than one which is simply maintaining the same uniform condition. The latter are more subject to styfziekte'; and 'acute lamziekte would be most abundant where the vegetation was abundant and chronic lamziekte and styfziekte most prevalent when the vegetation was dry and not so abundant'.

Theiler et al. (1924) also predicted the importance of phosphorus in metabolism. 'It is not unreasonable to assume that a 'growth impetus' responding with a weight increase . . . predicates more than a mere structural use of phosphorus; predicates a functional use in intermediary metabolism.' The authors also added in simpler terms for the farmer that 'cattle on a phosphorus deficient diet do not utilize their feed economically. Even if they eat a lot they waste it and do not give a profitable return in liveweight for it.' Whilst on the subject of depraved appetite and its role in the aetiology of lamziekte, it is interesting that Theiler et al. (1924) classified animals into 'marked cravers', 'slight cravers' and 'non cravers' and they outlined a technique for differentiating them by offering sterilized 'rotten bones'.

In a later paper, Theiler and Green (1932) philosophized on the development of osteophagie, a specific 'pica' in cattle 'is then a predilection for bones something instinctive, or is it a piece of acquired knowledge obtained by the sound scientific methods of trial and error – knowledge acquired most easily perhaps by the wiser ruminant chewing the cud of reflection after satisfying an obscure hunger on the skeletal debris of a departed brother'. It would appear that geophagia, or the eating of earth, cannot be associated with phosphorus deficiencies. French (1945), in review of the subject, pointed out that all the analyses publised on earths which were commonly eaten by cattle were too low in this element to make good any deficiency.

The Americas

Chicco and French (1959) cite Azara (1838) as reporting osteophagia in cows in Paraguay and claim that this was the first mention of indications of phosphorus deficiency to be noted in Latin America, whilst 'salt stiff' was attributed to a deficiency of phosphates in Florida in 1872 (Dodge 1874), although this was subsequently put down to iron deficiency (Becker et al. 1965). In the Gulf Coast area of Texas, Schmidt (1926) reported that a fatal disease of cattle, 'creeps', could be prevented by bonemeal and salt supplementation. Later, the disease 'stiffs' or 'sweeney' was shown to be caused by a deficiency of phosphorus in Florida (Becker et al. 1933).

Australia

The article by Theiler et al. (1924) was reproduced in its entirely in the *Queensland Agricultural Journal* of 1925. A subsequent letter to the editor by a Mr Tronson (Tronson 1926) was probably the first report of effective bonemeal supplementation to cure 'soft bone' in Queensland and Australia. Later, Turner et al. (1935) reported that this condition had been diagnosed as aphosphorosis in 1931 and that trials at the farm 'Helenslea' at Charters Towers had shown an increase equivalent to 47.5 and 37.0% of body weight in steers fed dicalcium phosphate and disodium phosphate respectively; control animals, without phosphorus supplementation, gained only 18.6% of body weight during the same nine-month period. Subsequent communications from Queensland bore witness to the efficacy of bonemeal supplementation under commercial ranching conditions (Bulcock 1938; Hirschfeld 1940).

Requirements for phosphorus

Estimates for the requirements for phosphorus by beef cattle have been published by the Agricultural Research Council (1965)* in the UK and by the National Research Council (1976) in the USA, and these differ widely.

When the requirements are expressed as concentration of the diet, the American estimates range from 0.18% for growing finishing steers and heifers and dry cows up to 0.23% for breeding bulls and lactating cows.

The British estimates, on the other hand, express value according to expected growth rate and energy concentration of the diet; these range from 0.21% for cattle growing at the rate of 0.7 kg/day on a diet with an energy concentration of 1.8 Mcal/kg (7.56 kJ/kg) to 0.60% for cattle growing at the same rate on a diet with an energy concentration of 3.4 Mcal/kg. With a diet containing 2.2 Mcal/kg, the corresponding value is 0.30%. Butterworth (1964) estimated the metabolizable energy of twenty-four tropical grasses to range between 1.85 and 2.66 Mcal/kg with a mean of 2.26 ± 0.195 MCal/kg. It is likely therefore that the discrepancies between the two sets of estimates would be greater at higher forage energy levels.

As there does not appear to be any published work indicating that expected phosphorus requirments of cattle in tropical and subtropical regions are in any way different from those in temperate areas (although the indication by the work of Reveron and Topps (1972) that blood phosphorus was reduced to 50% of the normal level and radiography demonstrated symptoms of osteodystrophy in young lambs affected with the gut parasite *Trichostrongylus colubriformis* is

* ARC (1980) not available when this was written.

interesting in this connection), we may assume that forage levels of less than 0.18% indicate a probable deficiency for most classes of stock.

Occurrence of phosphorus deficiencies

Deficiencies of phosphorus are widespread and have been identified both by analyses of vegetation as well as by analysis of various animal tissues, although both techniques have disadvantages.

Phosphorus levels in forage

Southern Africa

Early vegetation studies were carried out by Henrici (1928, 1930) in Vryburg in the Orange Free State (then in Bechuanaland) and also (1930) in higher rainfall areas in Ermelo in the Transvaal High Veld. It was shown in both places that the vegetation was deficient in phosphorus for the greater part of the year as only in spring (which coincided with the start of the rains in both locations) were values satisfactory.

Further studies in South Africa were made by du Toit et al. (1932, 1934) in eleven grass species planted in plots, as a result of which they concluded that 'phosphorus deficiency is a foregone conclusion'. The same group of workers (du Toit et al. 1935b, 1940a) based at Onderstepoort near Pretoria carried out an ambitious study of the mineral content and feeding value of natural pastures as a result of which a map showing areas where phosphorus deficiencies could be expected was published. Summer values ranged from 0.12 to 0.17% whilst the corresponding values for winter were 0.05 to 0.07%. Phosphorus analyses were carried out by the Animal Production Research Unit in Botswana (APRU 1976). Mean monthly herbage values from thirteen stations over a two-year period showed a range of 0.03 to 0.08% of dry matter; the authors pointed out that in the case of a herbage which contains 0.03% phosphorus an animal must eat 50 kg of dry matter daily to obtain sufficient!

Butterworth and his colleagues carried out determinations of herbage phosphorus of grass species typical of the three main ecological zones in Swaziland, high-, middle-, and lowveld, again over a two-year period. Maximum values were 0.37, 0.16 and 0.31% for the three zones respectively whilst corresponding minimum values were 0.03, 0.03, and 0.07%; once again, clear evidence of a phosphorus deficiency (FAO 1978b).

East Africa

The first work carried out in East Africa appears to have been about the same time as the South African surveys, as Orr and Holm (1931) reported low phosphorus levels in three out of four districts of Kenya. This was subsequently confirmed by Howard et al. (1962), who obtained a yearly mean of 0.145 with a range between monthly maxima of from 0.072 to 0.252%.

Similar results were obtained in Kenya by Dougall and Bogdan (1958b) who analysed fifty-eight samples of grass and obtained a mean of 0.174 ± 0.07% with a range of 0.05–0.37%. Work in Tanzania by Naik (1965) gave the high mean value of 0.319 ± 0.1679% for the phosphorus content of seventeen grasses of the Mbuli district but the high SD was partially attributable to several samples with phosphorus content of less than 0.05%.

In Uganda, Long et al. (1969) have given value for twenty-two samples of grasses taken from Ankole and the Queen Elizabeth National Park and obtained a mean value of 0.215 ± 0.064% with several lower values of less than 0.15%.

Latin America

De Alba and Davis (1957) reviewed work carried out in Latin America and identified a large area where low pasture phosphorus levels are to be expected which stretches from Argentina in the south, through Paraguay to the south of Brazil.

Low values for phosphorus in grasses were reported by Reichert and Trelles as early as 1922 in Argentina. De Alba and Davis (loc. cit.) quoted values ranging from 0.03 to 0.16% in samples of *Paspalum, Andropogon* and *Axonopus* spp. taken in Barrerito in Paraguay. These findings are supported by analyses carried out by Ramirez (1954) on *Paspalum rojasii*. Jardim et al. (1952) have published values for the phosphorus content of samples of *Hyparrhenia rufa* and *Chloris gayana* taken from the state of Sao Paulo which range between 0.07 and 0.20% of dry matter.

Da Silva Lopes and his colleagues (Da Silva Lopes et al. 1973a) analysed samples of forage taken in the Distrito Federal and found average values ranging from 0.03 to 0.06% phosphorus, clearly demonstrating the necessity for supplementation.

A further zone found to be deficient in phosphorus, according to analyses of vegetation, stretches from the llanos (plains) of Colombia to the east of the Rupununi of Guyana in the west and includes the greater part of the llanos of Venezuela as well as part of the Andes. Values have been determined by Grünwald et al. (1947), Ortiz (1952), cited by de Alba (1965) Azcarate (1954), Iljin (1955) and Butterworth (1968 and unpublished).

De Alba and Davis (loc. cit.) also commented on a further phosphorus-deficient zone in the Mexican states of Coahuila, Chihuahua, Sonora, Nuevo Leon and Durango, and pointed out that deficiencies of phosphorus may be provoked by high levels of iron in the soil which inhibit phosphorus uptake by the plant. Iron levels are frequently high in the tropical areas of Latin America.

The existence of deficiencies in the north of Mexico was confirmed by the work of Gonzalez (1966) who sampled fifteen grasses monthly over three years at various sites in Chihuahua. The three-yearly mean values for the various sites and grasses were 0.08 ± 0.02 and 0.04 ± 0.09% for summer and winter seasons with a maxima of only 0.11 and 0.06% respectively. These low values clearly indicate a year-long deficiency of the element.

Early indications of phosphorus deficiencies were also noted by Follett-Smith (1930) who showed that the phosphorus contents of four pasture grasses grown on the coastal soils of British Guiana (now Guyana) ranged from 0.02 to 0.12%.

Devendra (1977), in a study of the mineral content of Caribbean feeding-stuffs, presented values for the phosphorus content of 100 samples of grass which range between 0.18 and 0.33 (mean 0.34 ± 0.04%), i.e. considerably higher than most of the values which have been quoted above. However, it may be pointed out that the great majority of samples analysed by Devendra were introduced *Digitaria* and *Brachiaria* spp. and that few native species were represented.

Five hundred values for the phosphorus content of tropical grasses taken from a compilation of data from the Latin American and Caribbean region (or 32%) indicated that the mean was 0.21 ± 0.14% and that 160 samples contained less than 0.15% phosphorus indicating widespread deficiencies in the area. It is unfortunate that no details were given in this compilation as to the origin of samples (McDowell et al. 1974).

Australia

Robinson and Sageman (1967) cited earlier Australian work carried out by Beck and Underwood (1938), Snook and Durack (1946) in Western Australia and Hely and Hallsworth (1947) in Northern Australia who obtained low values for herbage phosphorus content. The values obtained by Robinson and Sageman ranged from 0.01 ± 0.001 to a maximum of 0.12 ± 0.003% indicating definite phosphorus deficiency. Determinations were made by Norman (1963) at Catherine in Northern Territory of Australia and similar low values which ranged from 0.03–0.13% were encountered (these are approximate values as they were read from the graph).

Schofield (1946), on the other hand, obtained much higher levels of phosphorus in introduced species in coastal northern Queensland, the only grass which manifested a deficiency being *Hyparrhenia aucta* with 0.09% phosphorus. The author also points out the wide variation among species with regard to phosphorus content.

India

Padmanabha et al. (1931) determined the mineral composition of some of the indigenous grasses grown in the Central Province of India. They reported that of a total of thirty-one samples relevant to this review, the average phosphorus content (converted from P_2O_5 as originally expressed) was 0.10 ± 0.06% with a range of 0.01–0.24% clearly indicating phosphorus deficiencies. Although the authors do not report any actual results of trials in which bonemeal is fed, they comment that 'preliminary experiments on feeding with bonemeal tend to show that the cattle are immensely benefited'.

From the preceeding review of the literature, it is clear that phosphorus deficiencies are to be expected in grazing animals in many parts of the tropics and subtropics. It is also likely that they exist in many other areas where no studies have as yet been carried out.

Blood phosphorus levels

Low levels of phosphorus in the blood were originally suggested as a symptom of phosphorus deficiency by Theiler et al. (1937); and although it is known that this is affected by many factors, it has been used by workers in various parts of the world as a criterion.

Underwood (1966) has stated that levels between 4.5 and 6.5 mg/100 ml blood may be considered normal although fluctuations outside that range are not uncommon.

Southern Africa

In Southern Africa, Malan et al. (1928) reported low blood levels of animals grazing phosphorus-deficient pastures. Considerable data was also presented by Bisschop (1964) who described work carried out at Armoedsvlakte in the Bechuanaland Protectorate over a period of several years. It was shown that blood phosphorus levels indicated deficiencies during the last month of pregnancy from September to December. This coincides with the end of the winter dry season and the beginning of the rains; levels of 2.4 mg inorganic phosphate % were reported.

The University of Edinburgh Veterinary Expedition to Botswana reported inadequate levels of serum phosphorus in various regions and suggested that only in bonemeal-supplemented cattle could levels be regarded as adequate (Reed et al. 1974b, c and d).

Hurrell and Dugdale (1958) confirmed aphosphorosis in lactating cows by blood analysis in the Shangani–Fort Rixon part of Rhodesia and treated this by the administration of sodium monophosphate.

Other parts of Africa

In Kenya, Gitter et al. (1975) reported that 23.9% of blood samples had less than 3.5 mg phosphorus % although there was no clinical evidence of aphosphorosis. They carried out a survey of cattle on six farms in an area limited by Lake Victoria in the east and Machakos in the west at altitudes between 1371 and 2134 metres.

In an extensive blood-sampling programme in which more than 6000 samples were taken in the north of Senegal, phosphorus deficiency was confirmed for cattle during the dry season (Friot and Calvet 1971). In the Yare Lao area, the mean value for fifty-nine samples taken during the dry season was 3.3 mg phosphorus %.

Latin America

Values for blood phosphorus are available for different parts of Latin America. Early literature from Brazil has been surveyed by Da Silva Lopes et al. (1973b). These workers also carried out a year-long analysis of blood samples in the Distrito Federal during 1972/3 and found that levels were critically low in May and June, these being 4.0 and 3.8 mg phosphorus % respectively.

In Venezuela, Chicco and French (1959) carried out an extensive survey of blood phosphorus levels in the cattle-raising areas in the central and eastern states. Deficiencies were encountered in all states, the highest incidence being in Bolivar where 78.6% cattle were affected. The second highest was in Monagas with 30.4%. In a subsequent study (French and Chicco 1961), blood samples were taken from animals in the Andean states of Venezuela where results were much more satisfactory, only 13% of animals in the state of Lara manifesting deficiencies.

Having reviewed some of the cases where deficiencies of phosphorus have been diagnosed by low blood levels, it should be pointed out that this method is rendered somewhat uncertain by the ability of the animal to withdraw both calcium and phosphorus from its skeletal reserves as a response to a dietary deficiency. Although low serum levels will be reached after a long period of severe deficiency, this may not be the case if deficiencies are less marked.

Such a case has been reported by Cohen (1973), who found that although pasture levels of phosphorus were within the range normally regarded as deficient, the concentration of plasma phosphorus was always within the range 4–8 mg% which must be regarded as normal. Furthermore, Cohen pointed out that the correlation coefficient between pasture phosphorus content and plasma inorganic phosphorus concentration was not significant, emphasizing 'the unreliability of single blood sample to reflect changes in pastures phosphorus and thus to determine the phosphorus status of cattle'. This supported the contention of Hemingway (1967, cited by Cohen 1973) that 'a normal blood phosphorus concentration must not be regarded as an index of adequate bone nutrition'. Little (1972) and Cohen (1973) further advocated the use of bone phosphorus content as the best estimate of the phosphorus status of grazing cattle because it significantly reflected variation in the phosphorus content of pasture. The latter author noted that blood (and hair) failed to do so, and indicates that levels below 13.7% in bone may indicate a phosphorus-deficient status in yearling steers.

Little and Minson (1977) showed that samples taken from the twelfth rib were less variable than those taken from the thirteenth, and Little and Shaw (1979) suggested that levels of 120 mg phosphorus/cm^3 indicated deficiency whilst levels of 150 mg or thereabouts indicated adequacy.

Finally, the work of Moir (1966) showed that the faeces of cattle grazed on low phosphate soils were lower in this element than those of cattle grazed on high phosphate soils and suggested that faecal phosphorus content could be used to indicate probable deficiencies.

Summary

The descriptions of aphosphorosis were made in South Africa by Sir Arnold Theiler and his co-workers who ascribed the disease 'lamsiekte' or botulism to the eating of bones infected with *Clostridium botulinum*. This is its turn was caused by the depraved appetite resultant from a phosphorus deficiency, Similarly, the less acute form of the condition, 'styfziekte', was attributed to a simple phosphorus deficiency uncomplicated by infection. It was also shown that the incidence of both diseases could be greatly reduced by the administration of phosphorus-containing compounds including bonemeal, bone-ash, etc. Workers of this school also predicted the importance of phosphorus compounds in intermediary metabolism as well as in its purely structural role. Descriptions of other similar deficiency conditions were also forthcoming from the Americas and later from Australia.

Analysis of forages from many parts of the world pointed to the widespread nature of deficiencies of this element and these were supported by serum analysis, although it was later pointed out that these could be equivocal because of the storage capacity in the bones and that analysis of bone itself could be a better indicator of a state of deficiency than analysis of forage or serum. Faecal phosphorus status has also been suggested as an indicator of the condition.

Phosphorus supplementation of growing stock

A considerable amount of work has been carried out on supplementation with phosphorus-containing substances in the tropics and subtropics. This will be dealt with with respect to differing classes of stock; i.e. calves prior to weaning, growing stock and breeding cows. Various methods of supplementation will then be discussed together with the different sources of phosphorus which have been used for this purpose.

Supplementation of calves prior to weaning

The first work on supplementation of young calves was carried out at Armoedsvlakte near Vryburg in the Orange Free State (then Bechuanaland). A programme of upgrading indigenous female stock with both exotic and indigenous sires, showed that $\frac{3}{4}$ and $\frac{7}{8}$ improved stock from bonemeal-fed dams did not do so well as first cross calves and it was thought that this could have been due to a necessity for bonemeal feeding (Bisschop et al. 1939). Calves were given bonemeal from 30 to 480 days of age and compared with control animals that received no bonemeal until they were weaned at the age of 271 days. There was no difference between either growth rate or body conformation (as expressed by such measurements as length of body, height at withers, etc.) between supplemented and control groups either before or after weaning. However, control animals showed marked osteophagia and low blood phosphorus levels until they were fed bonemeal. The authors concluded that bonemeal-fed cows supplied enough phosphorus in the milk to complement pasture levels and to support normal growth of calves and that the quantitative factor determining bodyweight gain and conformational development was milk intake. They also concluded that under range conditions in Bechuanaland, bonemeal-fed cows supplied their calves with sufficient milk to sustain normal growth until weaning age (although actual weaning weights were 136 kg at 270 days which can hardly be considered satisfactory).

Rudert and O'Donovan (1976) fed various mineral supplements to cows and their progeny grazing *Cynodon aethiopicus* pastures in a trial carried out at the Henderson Research Station in Rhodesia. Experimental treatments included salt and monosodium phosphate, a simple mix consisting of 66% coarse rock salt, 34% monosodium phosphate and a trace of potassium iodate. The control group of cows and calves received coarse rock salt only. Information on weaning weights was obtained on a total of 122 calves during a two-year period and average weaning weights were 199, 200 and 204 kg for the groups fed the complex mixture, the simple mixture and salt only respectively, a slight though non-significant advantage for the calves fed salt only. From this limited evidence we may conclude that under the conditions of the experiments carried out, the feeding of phosphate supplements to calves before weaning was not beneficial although, under certain circumstances, functional advantages could result.

Supplementation of growing stock with positive results

Southern Africa

Whilst little information is available on calves prior to weaning, considerable information is available on growing stock. The first experiments were also carried out at Armoedsvlakte and have been described by Theiler et al. (1924) in their classic paper entitled 'Phosphorus in the livestock industry'. When young stock of between nine and eighteen months of age weighing 133 kg on average were supplemented with bonemeal, they gained approximately 27.3 kg more than the unsupplemented group on quantitatively sufficient but phosphorus-deficient veld herbage during April and May (early winter or dry season). During the winter period, comprising June to October, the supplemented animals increased their advantage to 35.5 kg, and from October to April, grass growth provided relatively abundant and less phosphorus-deficient feed and the unsupplemented animals increased to 218 kg. However, the animals fed bonemeal increased to more than 273 kg; their final superiority with respect to liveweight gain was 61 kg.

It is interesting to note that phosphorus supplementation was effective in increasing weight gains not only during the winter period during which there was no doubt of the low phosphorus content of the herbage but also during the summer months. This has been observed on other occasions in South Africa and elsewhere and will be commented on in more detail below.

Theiler et al (1924) also carried out work using young oxen weighing about 295 kg. Weight was lost from June to September because of the poor quality of the winter pasture but the bonemeal-fed animals lost less weight than the unsupplemented controls and at the end of the period weighed 16 kg more. During the spring and summer growing season the bonemeal-fed animals increased their advantage to a clear superiority of 63 kg per head. Theiler comments that the fed animals would have realized £3 more per head than the unsupplemented group according to an auctioneer's evaluation against an expenditure of six shillings on 25 kg of bonemeal. It is an interesting comment on the extent of inflation that the equivalent increase in value at present would be about £35 per head using today's (1981) prices. The cattle were not in fact slaughtered but were carried through the succeeding winter period during which time they increased their advantage to 77 kg per head.

In addition to these two trials, Theiler and his colleagues obtained similar results with a mixed herd with an average weight of about 310 kg, said to be typical of an average lean herd in the Vryburg area at the end of winter.

Further studies at Armoedsvlakte were carried out by Bisschop and his colleagues between the years 1933 and 1955 (Bisschop 1964), and although much of this work refers to methods of feeding phosphorus as well as dosage rates and will be referred to later, one experiment with growing stock is of interest in the present context. One group of young Friesland oxen (tollies) were fed 85 g of bonemeal/day throughout the summer period from the beginning of October until the end of April and no bonemeal during the winter from the beginning of May until the end of September. A similar group was fed the same quantity of bonemeal throughout the experiment whilst a control group received no supplement. The author concluded that the two supplemented groups were superior (some 182 kg) to the unsupplemented group 'but that between the two bonemeal groups the liveweight differences were not of functional importance', suggesting that supplementation during the winter months was not necessary. Nevertheless, inspection of the growth curves indicates that the cattle which were not supplemented during the winter showed considerable losses of weight, particularly towards the end of the period when their reserves had become depleted. It was stated that they also showed pronounced osteophagia. Weight gains were made up during the summer by compensatory growth.

Becker (1932) obtained positive results by the supplementation of three-year-old oxen with 15 g per day of disodium phosphate, apparently the first record of the use of this compound in animal feeding in the tropics or subtropics. The gains of the supplemented animals was 154 kg as opposed to only 54 kg during a 14 month period from November to the end of December the following year, i.e. rather less than two full growing seasons. Becker also mentioned that both lamziekte and styfziekte were effectively controlled by these means.

Positive responses to bonemeal supplementation of growing stock were also demonstrated in Botswana (previously Bechuanaland) by Trail and his colleagues (APRU 1972). Twenty-nine Tswana (cattle of the Sanga type) steers were maintained under Kalahari sandveld conditions from 15 months to 34 months of age. One group was fed a bonemeal and salt mixture (exact proportions are not given) at the rate of 200 g/day. The supplemented animals gained 0.4 kg daily whilst the unsupplemented animals gained only 0.2 kg. Cold dressed weight, killing-out percentage and grade were all better for the supplemented animals. The consequent increased value of the latter animals amounted to R29.67 or £19.65, a value quite similar to that calculated for the gains of Theiler's animals when converted to modern-day terms!

In the province of Zambezia, Mozambique, positive results were obtained by the administration of approximately 30 g/day of a supplement containing 65% salt and 35% dicalcium phosphate. Although actual weight gains are not quoted, responses were noted in the development of weaners to 18 months and complete recovery of calves presenting clinical-deficiency symptoms (Paisana 1967).

More recently in Rhodesia, Grant (1976b) compared monocalcium phosphate with a commercial supplement (Rumevite 12) and obtained a marked response in bodyweight gains of both supplemented groups vis-a-vis the unsupplemented controls. Additional gains amounted to 33 and 31 kg/animal over the summer period of November to June.

Australia

Working in the Northern Territory of Australia, Norman (1960) obtained positive responses from feeding bonemeal and also investigated the effect of tick control on liveweight gain. Shorthorn steers running on native pasture were divided into four groups and allocated to the following treatments: supplemented (with bonemeal once fortnightly at the rate of 5 g/head/day) and sprayed (with 0.5% w/v pp′-isomer DDT at fortnightly intervals), sprayed only, supplemented only and no treatment (control). The experiment was carried out during two consecutive years. Daily gains were 0.83, 0.68, 0.32 and 0.16 kg respectively during the first year of the trial and 0.94, 0.72, 0.71 and 0.64 kg respectively during the second year. The response to supplementation as daily gain was approximately 68 g/day or 12 kg during the season. Unfortunately, a statistical analysis of the data was not presented and the author concluded that there was no appreciable interaction in either season between phosphate supplementation and spraying, although superficial examination of the means presented indicates a substantial difference. Data are also presented on the phosphorus content of the various grasses present in the native pasture (*Themeda australis*, *Sorghum plumosum*, *Chrysopogon fallax*, *C. latifolius* and *Sehima nervosum*). These ranged from a mean maximum of 0.21% in December, rapidly falling to 0.12% in January and as low as 0.03% at the end of the trial in May. The author concludes that the native pasture at Katherine, Northern Territory was deficient in phosphorus even in the wet season.

Moran (1973a) obtained positive results from feeding meat and bonemeal (4.8% phosphorus) to weaner Brahman cattle at Adelaide River in the Northern Territory of Australia. Responses were probably not entirely due to phosphorus, however, but also to the protein contained in the supplement; Moran attributed part of the increase to phosphorus because it was accompanied by increased serum levels of this element, although it has been noted elsewhere that this frequently occurs without a corresponding weight increase.

McDowell *et al.* (1982) obtained positive results by feeding bonemeal to growing cattle in Bolivia.

Supplementation of growing stock with ambiguous results

Southern Africa

On other occasions, however, such unequivocal results as those quoted above have not been obtained. Murray and his co-workers (Murray *et al.* 1936) fed dicalcium phosphate at the rate of some 20 g/day to Lincoln Red and Shorthorn steers grazing sandveld pastures at Matopos in Southern Rhodesia. No difference in growth rate between supplemented and non-supplemented animals could be demonstrated, although chemical analysis indicated probable deficiencies and blood phosphate levels were higher in the supplemented animals. In a further experiment, dicalcium phosphate was compared to bonemeal. Weaned grade Hereford steers were fed the same quantity of dicalcium phosphate as before, 85 g of bonemeal or 21 g blood meal/day, and compared with a similar group of calves grazing unsupplemented veld; no difference in growth rate could be demonstrated (Murray and Romyn 1937b).

This was confirmed in a later study at Matopos (Rhodesia Ministry of Agriculture 1967) in which both Hereford and native weaners were supplemented with bonemeal and no advantage with respect to growth rate could be demonstrated. However, it had previously been observed that the rate of dental attrition in exotic cattle grazing sandveld was higher than that of native cattle. In the trial referred to it was noticed that bonemeal feeding resulted in a marked improvement in the quality of teeth.

In studies carried out in the Mashonaland region of Rhodesia at the station at Marandellas, Rhodes (1956) carried out a series of experiments on the effect of feeding mineral, protein and carbohydrate supplements to growing cattle. In the first trial which was carried out during the winter, a comparison was made between various levels of supplementation with common salt with and without iron oxide (this will be referred to later), as well as additional maize meal, groundnut meal or 85 g of bonemeal (no combination of both phosphate and protein supplementation was made). It was shown that although the administration of bonemeal increased the phosphorus content of the blood (this was, however, not significant), weight losses during the experimental period were not affected. This

Table 3.1 Liveweight changes (kg) in supplemented and unsupplemented growing beef cattle (Marandellas) (adapted from Rhodes 1956)

Supplement	*First winter loss*	*First summer gain*	*Second winter loss*	*Second summer gain*
Control (veld grazing alone)	−4.5	77.7	−56.4	140.0
Protein plus veld grazing	−1.4	91.8	−36.4	138.6
Minerals plus veld grazing	−5.6	118.6	−62.7	183.2
Protein/minerals plus veld grazing	−5.9	135.0	−40.5	184.5

was also the case with the animals supplemented with groundnut meal. Rhodes therefore carried out a second trial to test the effects of supplementation with both minerals and protein. Four groups of Shorthorn weaners were allowed unsupplemented veld grazing, this plus protein, plus minerals and plus both protein and minerals. The protein supplements consisted of 680 g of groundnut meal/day and the mineral supplement of 57 g bonemeal with 28 g of common salt and 7 g of iron oxide. Results are shown in Table 3.1.

Over the approximately 27-month period from June 1949 to September 1951, the combination of minerals turned the first winter losses of the other groups to a slight gain, whereas in the second winter it appeared that only protein was effective in reducing weight loss. However, during the summer periods, the groups fed minerals were superior with respect to growth rate. Overall, gains were very much in favour of the two groups supplemented with minerals. Clearly mineral levels during the summer, although higher than those to be found during the winter, were not sufficient to sustain maximum growth rates.

In an experiment carried out in Muguga in the Kenya highlands, Lampkin (1961) could show no difference in weight gain of paired calves fed 19 g/day of dicalcium phosphate compared with the control calf. Average weight gains of the two groups were 0.14 kg/day, which would indicate that there were no major nutrient deficiencies to be expected and also blood inorganic phosphorus levels were relatively high although supplementation caused a significant increase (5.9 v. 4.9 mg/100 ml respectively).

Australia

Cohen (1972) in New South Wales found that there was no difference in weight gains of animals drenched with 30 or 35 g/week of phosphorus given as a drench of sodium dihydrogen orthophosphate (monosodium phosphate) compared with those of control animals given no phosphorus supplement when grazing non-fertilized *Axon-*

opus affinis dominant pasture; the pasture was apparently deficient in phosphorus, containing between 0.04 and 0.11% during the period December 1969 to December 1970.

Winter et al. (1977b) could demonstrate no difference in growth rate when steers grazing sown pastures in the Cape York Peninsula were given phosphorus supplements. This was shown in an experiment comparing various rates of fertilizer application and different stocking rates on pastures of either *Brachiaria decumbens* var. Basilisk or *Panicum maximum*, each sown in with *Stylosanthes guianensis* cv. Endeavour and *Macroptilium atropurpureum* cv. Siratro. The supplement given was a 2:1 mixture of monocalcium phosphate and dicalcium phosphate containing 21% phosphorus and 15% calcium fed so as to give the equivalent of 10 g phosphorus/day, with 80 g of common salt/day. The supplemented cattle had a significantly higher blood phosphorus level than the unsupplemented ones, these being 4.3 and 4.7 mg/100 ml respectively, but there were no differences in weight gain, although in an earlier paper (Winter et al. 1977a) results were presented showing that phosphorus levels in the forage were frequently below 0.15%.

An example of phosphorus supplementation actually increasing liveweight losses during the winter period was provided by the work of Winks and Laing (1972) carried out at Swan's Lagoon near Townsville in Queensland. Brahman Shorthorn cross-bred steers were fed 0.23 kg molasses/head/day with or without 5 g of phosphorus supplied as commercial black phosphoric acid (63% H_3PO_4). Weight losses for those animals fed both molasses and phosphoric acid were 0.33 kg/day whilst those fed molasses alone lost only 0.25 kg/day. However, the authors attribute the depressions in liveweight performance to a disturbance in rumen acid-base balance.

A similar although slighter depression in liveweight was obtained by the same group using weaners. During the dry period control animals gained 7 g/day whilst the animals supplemented with phosphoric acid (to supply 5 g/head/day) and molasses lost 64 g/day (animals fed molasses alone in addition to pasture gained 14 g/day). However, during the post-supplementation phase, some slight supplementary growth took place, the gains for the three groups of animals in the same order as above being 63, 61 and 66 g/day respectively. When the overall rate of gain was calculated, there was no difference, the rates of gain per day for the three groups being 44, 43 and 44 g/day (Winks et al. 1976).

Summer v. winter supplementation

It is appropriate at this point to discuss the merits of supplementation during winter and/or summer periods. It has been shown in the section on the phosphorus content of forages that there is a pronounced seasonal variation and that winter values are usually lower than those encountered during the summer season when the grass is in its active growing stage. This has given rise to the practice of feeding phosphorus supplements during the dry season 'when the farmers are acutely aware of the necessity of some form of supplementary feeding' (van Niekerk 1974).

Van Niekerk (1974) has reviewed early work carried out in Southern Africa and has cited the studies of various workers (Theiler et al. 1924; Bisschop 1964; Ward 1968; Schur 1968; van Schalkwyk and Lombard 1969) to justify the contention that the benefits derived from phosphorus supplementation are almost exclusively associated with the summer season when phosphorus levels are at their highest. Elsewhere, the same author (van Niekerk 1973) states that 'phosphorus supplements alone are almost valueless in preventing weight loss or in promoting weight gains in cattle and sheep during the dry season'. This is in agreement with the results of Rhodes (1956) already referred to. The comparable results obtained by Schur (1968, cited by Van Niekerk 1974) and van Schalkwyk and Lombard (1969) are given in Tables 3.2 and 3.3.

Schur (1968, cited by Van Niekerk 1974) fed cattle year-round from weaning to three and a half years of age with 3.6 g P/day, dosed as monosodium phosphate twice weekly and found that winter losses were similar to those of the controls which were not fed phosphorus.

Van Schalkwyk and Lombard (1969), working at Vaalharts in the Kalahari thornveld shrub bushveld, used Afrikaner-type steers ranging in age from 18–24 months given four treatments: no supplement, supplement during the winter (1 April to 30 September), supplement during the summer (1 January to 31 March and 1 October to 30 September) and year-long. The supplement used was composed of 55% dicalcium phosphate, 33% common salt and 12%

Table 3.2 Seasonal liveweight (kg) response to year-round phosphorus supplementation in grazing cattle (adapted from van Niekerk 1974, after Schur 1968)

Season	*Negative control*	*Phosphorus 3.6 g/day*
1st winter	− 16.4	− 18.6
1st summer	+ 97.7	+119.1
2nd winter	− 37.7	− 50.0
2nd summer	+108.7	+120.9
3rd winter	− 61.8	− 69.1
3rd summer	+174.5	+187.3
Net weight change	+265.0	+289.6

Table 3.3 Weight changes (kg) of cattle fed phosphorus seasonally v. year-long (adapted from van Schalkwyk and Lombard 1969)

Weight changes	*Control*	*Winter only*	*Summer only*	*Yearlong*
1 Jan.–31 Mar. (summer)	+35.2	+41.6	+ 61.6	+ 57.1
1 Apr.–30 Jan. (winter)	+ 5.3	+18.1	+ 14.0	+ 6.8
1 Jul.–30 Sep. (winter)	– 29.8	–30.3	– 36.3	– 30.6
1 Oct.–31 Dec. (summer)	+45.7	+61.6	+ 77.5	+ 88.1
Net weight change	+56.4	+91.0	+116.8	+121.4

binder; stocking rate was 4.28 ha/head; the animals fed during the summer period only gained almost as much weight as those fed year-round. Van Niekerk (1974) concludes that 'it is therefore evidence that phosphorus is probably the major limiting nutrient in the diet on natural pasture during the summer months in many areas of South Africa'.

Similar results were obtained in the Chobela region of Mozambique by Morgado (1961) who demonstrated that no additional effects were obtained by the feeding of bonemeal during the winter period to either Africander or Landim (one of the Sanga types of cattle) heifers but that the Africanders gained 20.4 kg more during the summer period than non-supplemented controls. This effect, however, could not be demonstrated in the Landim animals. Total gain by the latter was, however, much lower both in supplemented and non-supplemented animals and this might have accounted for the difference in the effect of phosphorus.

This was confirmed in Northern Queensland by Winks et al. (1977b). The effect of a phosphorus supplement (as black commercial 63% phosphoric acid) on native pasture as well as on Townsville Stylo (invaded by *Digitaria ciliaris* and *Brachiaria meliniformis*). Cross-bred Brahman shorthorn steers were used and the stocking rate was 0.49 head/ha. Overall increases were produced by phosphorus supplements during the wet season but not during the dry season. Phosphorus supplements were also ineffective when given with Townsville fertilized with superphosphate.

Whilst van Niekerk's hypothesis is clearly sound as far as it goes, it does not explain all the information available. In more recent studies, carried out in Zimbabwe (Richardson et al. 1975b), weight gains of steers fed only common salt were compared with those fed a commercial phosphorus supplement (Rumevite 12P) and mono-calcium phosphate with salt. The results are shown in Table 3.4.

Table 3.4 Weight changes (kg) of steers offered different phosphorus supplements while grazing on veld (adapted from Richardson et al. 1975b)

Treatment	*Winter gains*	*Summer gains*	*Net weight change*
Control	4.2	132.0	136.2
Rumevite 12P	7.7	125.8	133.5
Monocalcium phosphate	19.3	126.1	145.4

Steers fed during the winter gained significantly more weight than either controls or those given the commercial supplement; phosphorus supplementation had little effect on body weight gains during summer. The authors conclude that animals whose winter growth was restricted by a presumed lack of available phosphorus were unable to compensate for this during the subsequent summer growing season.

Several situations therefore can be encountered; that suggested by van Niekerk (1973, 1974) where, although phosphorus levels during the dry winter period are low, supplementation with phosphorus alone is not sufficient to promote increase in weight because of the simultaneous deficiency of another nutrient, presumably protein.

Also Cohen (1973), postulated normal bone phosphorus contents by extrapolating the multiple regression between pasture calcium and bone calcium and phosphorus. He found that these normal values were not reached, despite supplementation of steers in an earlier experiment (Cohen 1972) and suggested that this contributed to the lack of liveweight response.

Lack of response, particularly in bushveld areas, could be due to additional intake of phosphorus from sources other than grass (du Toit et al. 1940a). Dougall and Bogdan (1958a) in Kenya and Piot (1970) in Cameroun have demonstrated relatively high phosphorus values in browse plants (although reported analyses of some Australian trees and shrubs showed that 14 of 21 samples had values of 0.15 or below, McLeod 1973). Similarly, winter response to phosphorus supplementation may be due to protein adequacy to support growth rates, either because of unusually 'sweet' veld grasses or because of availability of high protein browse.

Phosphorus defieiencies may even be confused with those of protein in certain cases; thus, Siebert et al. (1975) showed that osteomalacia-like symptoms could be produced in cattle fed an adequate supply of phosphorus but inadequate protein.

Summary

Work carried out with calves creep-fed with a phosphorus supplement before weaning has not generally shown an improved growth rate (although functional improvements have occurred in some cases) presumably because sufficient phosphorus is supplied in the milk of the dam.

As might be expected, given the wide range of conditions where experiments have been carried out, responses to supplementation of phosphorus by growing stock after weaning have been variable. In cases where no beneficial responses have been obtained to phosphorus supplementation during the winter or dry season, it must be assumed that this element is not the nutrient most limiting to growth. Experiments involving protein have indicated that this might be more critical in many cases (see relevant discussion elsewhere). Alternatively, in many bushveld areas it is likely that although the major grass species represented may be deficient, additional phosphorus sufficient to maintain body weight or even sustain limited growth rates may be obtained by the consumption of browse species which are frequently richer in this element. It is also logical that in some cases where phosphorus is deficient year-round, supplementation with this element during the summer or rainy season, when the availability of other nutrients is adequate to support high growth rates, may lead to dramatic responses.

Phosphorus supplementation of breeding stock

Phosphorus has been implicated in various productive and reproductive functions in the breeding cow. These include milk production (and thus weaning weight), birth weight, reconception, duration of postpartum anoestrus, calving interval, calving percentage and weaning percentage.

Effects of phosphorus supplementation on milk production

Africa

Sir Arnold Theiler and his colleagues carried out the first studies on the effects of phosphorus supplementation on cows. After pointing out the particular strain placed on the phosphorus reserves of an animal by lactation and confirming that this frequently gave rise to osteophagia, he described a trial carried out at Armoedsvlakte (Theiler et al. 1924). Ordinary veld cows were selected and milked from alternate halves of the udder on alternate days, leaving the other half for the calf. After taking preliminary readings for a few weeks, the cows were divided into two groups so that total milk yields in the two groups were approximately equal. After administration of 225 g of bonemeal the yield of the supplemented cows increased to an average of 40% more milk than the control animals. The milk yields were about 5 litres/day which can be considered reasonable for non-improved, range Africander stock. It was concluded that low phosphorus was a limiting factor for milk production. Similarly, it was stated (albeit on the basis of meagre data) that calves from bonemeal-fed mothers averaged from 10 to 20% heavier at birth than those born to controls; they were also stronger, got more milk from their mothers and so got a better start in life. However, the capacity for maternal sacrifice was emphasized and 'many a scarecrow cow throws a fine calf'.

In a later study, du Toit and Bisschop (1929) stated that calves from supplemented mothers grew 16.5% better than controls because of increased milk production (although improved birth weights were not demonstrated). This was also accompanied by better skeletal growth and improved disease resistance.

Bisschop (1964), in a summary of a long series of investigations carried out at Armoedsvlakte, described an experiment where cows fed differing levels of bonemeal (85, 113 and 142 g/diem) were compared with controls which were not fed supplements. No significant differences in liveweight gain, inorganic blood phosphorus, milk yield, milk quality or reproductive parameters could be demonstrated between cows fed different levels of bonemeal; however, in all characteristics save that of milk quality, the supplemented cows were superior to those which were unsupplemented.

Australia

Increased milk production as a result of phosphorus supplementation has been reported more recently by Allan and his co-workers from the Swan's Lagoon Cattle Research Station at Millaroo in Queensland (Allan et al. 1972). The milk production of cross-bred (Shorthorn × Brahman) heifers was measured by calf weighing once every twenty-eight days. Heifers fed 28 g/head/day phosphoric acid (as commerical black phosphoric acid, 63%), twice weekly with molasses in the wet season and molasses and urea in the dry season and water in roller drum lick feeders were compared with unsupplemented animals.

Cattle grazed either native pasture alone or native pasture with Townsville stylo and supplements were fed during both wet and dry seasons. Effects of phosphorus supplementation were significant during the wet season, daily milk yields being 4.6 kg compared with 4.1 kg of the unsupplemented controls. Wet season milk production was also increased by dry season supplementation. This is in agreement with the contention already mentioned that the greatest response to phosphorus supplementation occurs in the wet season, but the author also comments that 'there are practical limitations to the effective wet season supplementation of cattle in Northern Australia', and adds that 'There would be more widespread use of phosphorus supplements if they could be effectively fed in conjunction with other dry season supplements', thus exploiting the carry-over effect.

Effects of phosphorus supplementation on reproductive performance

The effects of phosphorus supplementation on the various measurements of fertility in beef cattle have not been so conclusive as those on milk production. Although earlier work in Southern Africa showed positive responses, experiments carried out elsewhere have not always been quite so clear-cut.

Africa

An early South African study was carried out by du Toit and Bisschop (1929). They showed that 66% of cows fed bonemeal produced three calves in three years whereas none of the non-supplemented cows did so. Supplemented cows produced 87.3% of total possible calves during the three-year period whereas control, non-supplemented cows only produced 56.5% of total possible calves. The same authors also studied the effect of phosphorus supplementation on sexual maturity and showed that 84.4% of supplemented heifers calved before two and a half years old, whereas in non-supplemented controls the comparable value was 73.7%. The positive effects on fertility mentioned in this study are interesting in the light of information by Theiler et al. (1937) that cattle fed low levels of phosphorus over long periods continue to present normal oestrus and ovulation.

Lack of positive effects of phosphorus supplementation were recorded in Rhodesia. Murray (1938, 1940) working at Matopos could demonstrate no beneficial effects of supplementary phosphorus in either pregnancy or lactation of beef cows.

Later, in the Shongani region of Rhodesia (Zimbabwe), Bembridge (1966) carried out a trial on Hereford, South Devon and Tuli cattle grazing pastures predominantly composed by *Hyparrhenia* spp. and *Heteropogon contortus*. *Ad libitum* monosodium phosphate and common salt were fed with or without 450 g cottonseed cake/day. Percentage of cows in calf for the various treatments were as follows: salt only, 64.7; cottonseed only, 82.4; phosphate only, 47.1; and cottonseed with phosphate, 94.1. It was concluded that phosphorus supplementation alone was not effective in improving reproductive performance and that both phosphorus and protein supplementation were required to increase calving rate under the conditions of the experiment.

Ward (1968) fed various supplements to Mashona heifers grazing *Hyparrhenia* veld in the Rhodesian midlands. Treatments were either vitamin A fed at the rate of 40 000 IU vitamin A/45 kg body weight/month in winter, or 900 g groundnut cake/head/day in winter; these were fed either with or without a mixture of two parts of bonemeal to one part of salt *ad libitum*. Control animals were given salt and bonemeal alone or no supplement. The mean consumption of this mixture was 80 g/day when groundnut cake was fed but somewhat higher (105 g) in the other groups. The reproductive and weaning performance of each animal over the next six years was recorded; it was considered that this represented lifetimes performance.

More calves were weaned as a consequence of bonemeal feeding (120 v. 104) although this was less than from heifers fed with groundnut cake. The former effect was due to increased number of calves born and decreased mortality and was barely significant. More cows in the bonemeal-fed group had three or more calves than in the control group. Bonemeal appeared to have no effect on post partum anoestrus nor annual weight changes of cows and calf birth weight, but bonemeal-fed cows increased calf-suckling gains over the six-year period presumably because of increased milk production (although this was not measured). Both groundnut meal and bonemeal, fed singly or together, increased the total weaner production/cow. The effect due to bonemeal amounted to some 21% but this was not of the same order as those encountered by Bisschop (1964). It has been suggested that this may be explained by normal blood phosphorus levels in growing stock during the summer (Ward unpublished).

To complete this review of work from Rhodesia, Rudert and O'Donovan (1966), in a study carried out at Henderson Research Station, could demonstrate no difference in calving rate, growth rates and body mass changes of cows fed either a simple or a complex mineral supplement when they were compared with unsupplemented controls.

Equivocal results were obtained by Lampkin and his colleagues in Kenya who investigated phosphorus supplementation in areas where the more striking pathological symptoms of phosphorus deficiency were absent (Lampkin 1961). However, twenty-six of 52 cattle used were culled because they did not come on heat before 40 weeks and the average calving interval of the animals that did conceive was 516 days and the level of blood inorganic phosphorus was 3 mg/100 ml. In two experiments carried out at Muguga in the Kenya highlands, 30 g of dicalcium phosphate was fed daily to Northern Frontier Zebu cows either from calving until they calved again or were culled for infertility, or for the 16 weeks before they were due to calve. No significant differences could be demonstrated in weight changes between the supplemented and control animals. However, slight but non-significant superiority in various criteria of fertility were evidenced. Forty-nine of 54 supplemented cows were diagnosed pregnant compared with 43 of 52 unsupplemented animals. Number of days to conception for the two groups were 128 and 136 respectively, whilst average number of services per conception and average number of services per animal with active ovaries were 1.98 and 2.28, and 2.00 and 2.81 respectively (this latter difference was significant). It should perhaps be pointed out, however, that forage phosphorus could not be considered low, being on average 0.35% with a range of 0.19 to 0.57%. The authors conclude that at every point where there appeared to be a beneficial effect from feeding a phosphate there was also evidence of the lack of other nutrients. Nevertheless, no conclusive evidence of such deficiencies was presented as mean crude protein content was 10.8% which cannot be considered low in comparison with other situations in tropical and subtropical areas, and although a 'high' Ca : P ratio of 3.2 :1 during the drought was mentioned, no further values to support this were presented. (This latter point was not endorsed by Little (1970) who, reviewing factors of importance in the phosphorus nutrition of beef cattle in Northern Australia, concluded that whilst Ca : P ratio may exert a slight impairing effect on the growth rate of young animals, no effect on fertility would be expected.)

Australia

Playne et al. (1973) working at Lansdown near Townsville in Queensland could demonstrate no improvement in fertility when cows were given 11 g of phosphorus/day as phosphoric acid, the mean conception rate of cows who were rearing calves when mated were 81 and 75% for unsupplemented and supplemented cows respectively. However, conception rates were increased to 90 and 86% when supplements or urea and sodium sulphate were given. The authors comment that the significance of the differences could not be estimated because only small numbers of cows were used in the first two years of the experiment. Similar results were later obtained by Siebert et al. (1976) when two- and three-year-old heifers were given supplements of either 11 g phosphorus/head/day (as phosphoric acid) or the same amount of phosphorus fed with urea, and sodium sulphate. However, in three years out of the five during which the study was carried out, sufficient rain fell during the dry season for regrowth of pasture to take place and for animals to gain in weight. In one other year when no rain fell, supplements had no significant effect on weight change or pregnancy rate.

Only a slight effect on *post partum* anoestrus was demonstrated by Little (1975) in an experiment to determine the effects of dry season supplementation of phosphorus at the rate of 10 g/day (as NaH_2PO_4) to multiparous Brahman/Shorthorn and Hereford cross cows grazing grossly phosphorus-deficient *Digitaria didactyla* and *Axonopus affinis* pastures at Samford in south-east Queensland. The proportion of cows in oestrus by eight weeks after calving was 70% compared with 38% for unsupplemented cows. However, the percentage for the supplemented cows was not improved at twelve weeks after calving and the superiority over the controls was significant only at the 20% level; calf gain to four months was not improved by phosphorus supplementation.

Holroyd et al. (1977) at Swan's Lagoon investigated the effect of supplementary phosphorus on reproductive performance. Lactating breeding cows, grazing either native pasture (mainly *Heteropogon contortus*) or native pasture oversown with Townsville stylo and fertilized with superphosphate, were observed over a four-year period. Phosphoric acid was given twice a week in drum rollers to the supplemented animals. There was a diminishing response in the conception rate of the cows grazing natural pasture and supplemented with phosphorus during the wet season. The response on improved pasture was non-significant.

Underwood (1966) has stated that subnormal fertility is associated with phosphorus deficiency manifested through a 'disturbance in the oestrus cycles, reflected in inhibition, depression or irregularity of oestrus. Conception is thus prevented or delayed and fertility is reduced. *This is a specific effect of phosphorus deficiency*' (my emphasis).

Nevertheless, more recent work does not bear this out. Teleni et al. (1977) fed penned cows with hay cut from native pasture at intervals throughout the wet season (predominantly *Heteropogon contortus*; they were subsequently fed hay in yards and finally allowed to graze. Although half the cows were supplemented with the relatively high level of 25.8 g phosphorus/day (100 g NaH_2PO_4), none of the cows exhibited oestrus and the level of progesterone in the plasma declined as did the protein content of the hay. The cows were refractory to treatment with luteolytic prostaglandin. After supplementation of both groups with protein, however, cows had high plasma progesterone levels and all came on heat after treatment with prostaglandin. The authors, after conceding that the study did not employ sufficient animals for definitive reproductive studies to be carried out, concluded that under situations of poor protein nutrition, phosphorus supplements have no effect on ovarian activity and that 'the reported association between phosphorus and fertility is an indirect rather than a specific one'.

Summary

Despite many opinions to the contrary, there appears to be little concrete evidence of consistent beneficial effects on reproductive behaviour following supplementation with phosphorus. It has been shown that normal oestrus cycles can occur even in prolonged phosphorus deficiency and there is little to indicate that supplementation can exert any shortening effect on the duration of *post-partum* anoestrus. It seems likely that in most cases, when phosphorus is deficient, this is also the case with protein and if the deficiency of the former is remedied without correction of the latter, no improvement in performance will result. These conclusions from field trials have also been supported by studies of plasma hormone levels and treatment with prostaglandins. Presumably, the skeletal storage capacity of the mature cow plays an important role in enabling it to survive through periodic phosphorus shortages. On the other hand, it would appear that phosphorus supplementation has a positive effect on milk production and also, therefore, on the growth of calves prior to weaning. It also seems that this effect may also be more pronounced under certain circumstances during periods of abundant pasture availability. A similar effect on growth has been commented upon in the previous chapter. There is also some evidence that phosphorus-deficient animals have a reduced resistance to disease.

Different sources of phosphorus and methods of administration

Use of phosphorus from different sources

Various sources of phosphorus have been used for supplementation in different areas largely depending on availability and price. Results on the efficacy of the supplements will be briefly reviewed in this section.

Theiler et al. (1924) compared the effects of feeding various forms of phosphorus on the reduction of osteophagia in cattle at Armoedsvlakte. Wheat bran, bonemeal, sodium phosphate and phosphoric acid were used. The authors pointed out that phosphoric acid was not a good source because of its tendency to disturb the acid-base equilibrium of the body; they further commented that the bonemeal was the most effective form because of its palatability and ease of administration. Ground rock phosphate proved too difficult to digest and superphosphate 'productive of digestive disturbance'. Ground mineral phosphates were said to be cheaper but much less effective than bonemeal whilst precipitated calcium phosphate, although having similar effects to bonemeal, was more expensive, less palatable and more difficult to administer. Theiler et al. (1924) also commented on the effects of manuring pasture with phosphate. A manured camp contained three times the level of phosphorus than an untreated camp and was effective in controlling osteophagia. It was pointed out, however, that at the prevailing land prices (5–15 shillings an acre!), this could hardly be expected to be an economically viable proposition whereas, 'at least 80% of the bonemeal fed to an animal ultimately passes out again in the excreta . . . and remains on the farm, manuring the veld, slowly increasing its carrying capacity for vegetation and stock and slowly increasing the phosphorus content of the pasture'.

Becker (1932), also working in South Africa, compared disodium phosphate dissolved in water and fed at a rate of 40 g of the supplement (20% P_2O_5) with bonemeal and disodium phosphate given in the solid form to supply approximately similar quantities of phosphorus. In a trial carried out during a fourteen-month period (from October 1929 to December 1930) gains in weight were 183, 151, 135 and 53 kg for disodium phosphate in water, bonemeal, dosed disodium phosphate and control animals respectively. Becker attributed the superiority of the liquid-fed animals to the fact that they were handled less.

Also working at Armoedsvlakte, Malan and du Toit (1932) compared two types of bonemeal, precipitated calcium phosphate, degelatinized bone flour, superphosphate, ground egyptian rock phosphate and disodium phosphate, both fed solid and in solution with water, as phosphorus supplements for male and female cattle. On weight change and efficacy in controlling 'pica', they were put in the following order of 'phosphorus availability': sodium diphosphate, precipitated calcium phosphate, bonemeal and degelatinized bonemeal. Sodium diphosphate fed in the water was also good but rock phosphate and superphosphate provoked toxicity (this was probably due to fluorine).

Murray and Romyn (1937b) compared bonemeal with dicalcium phosphate as sources of phosphorus for Hereford weaner steers grazing the Rhodesian sand veld at Matopos. No difference in efficacy could be demonstrated between the two sources but, on the other hand, it was apparent from blood phosphorus values that unsupplemented animals were receiving sufficient from pasture. It is an interesting point that these authors mention what might still be a record in bonemeal consumption, 'one cow was found to consume nearly 7 lbs in one day and was, and remained, in perfect health and condition'.

Disodium phosphate was compared with bonemeal in a series of trials carried out on the Encino division of the King Ranch near Kingsville in the Gulf Coast part of Texas from 1941 to 1946 (Reynolds et al. 1953). It was shown that cows of the unsupplemented control group weighed 364 kg and weaned calves weighed 222 kg, whilst cows given bonemeal or disodium phosphate weighed 459 and 468 kg and weaned calves weighed 243 and 246 kg respectively. This would indicate that the two sources of phosphorus were equally effective. Similarly, calving percentage for the three groups over the four calving seasons of the study were 76.4 93.4 and 98.7% respectively (in the same order as given previously). The authors concluded that the cows which received disodium phosphate in water had a small but consistent advantage over those fed bonemeal. In contrast to the results of the previously cited authors, however, the cows given disodium phosphate ingested slightly more phosphorus than their bonemeal-fed counterparts (6.7 v. 6.2 g/day).

Disodium phosphate was also shown to be slightly superior to bonemeal by Martinez (1974) in Coahuila, Mexico, when growth of heifers was used as an indicator; animals fed the former supplement gained 408 g/day v. 344 g for the bonemeal-fed group ($P. < 0.05$). In this work, there was a marked superiority of the disodium phosphate-fed animals insofar as phosphorus intake was concerned. The disodium phosphate was not administered in solution but in solid form together with cottonseed meal and grain sorghum. Disodium phosphate was shown to be superior to phosphoric acid in a trial carried out in Tamaulipas, Mexico, González (1976) fed the supplements to 14–16-month-old heifers grazing *Cenchrus ciliaris* pastures at a stocking rate of 2 ha/beast; those fed disodium phosphate gained at a rate of 481 g/day whilst those fed phosphoric acid gained 455 g/day. The blood phosphorus level of the former animals was also significantly higher.

It is possible that the superiority of disodium phosphate when compared to other phosphorus sources which also contain calcium may be due to an improved ratio of calcium to phosphorus. Certainly, the calcium to phosphorus ratio is high in Tamaulipas where high levels of disodium phosphate were used effectively (Zavala et al. 1972).

Winks and his colleagues (Winks et al. 1976) compared phosphoric acid with superphosphate or mono-ammonium phosphate as a supplement to weaners grazing native Speargrass pasture in Northern Queensland. In a trial carried out in 1971–2, phosphoric acid was shown to be inferior to superphosphate. Animals fed the former lost 22 g/day whilst those fed the latter gained 23 /day. Although there was no evidence of compensatory growth, this difference diminished in the post-supplement period and the overall gains were 34 v. 37 g/day. The following year a similar trial was carried out using nono-ammonium phosphate as the phosphorus source for comparison with phosphoric acid. In this case, the inferiority of the phosphoric acid was less marked, both groups losing more weight, those supplemented with phosphoric acid losing 78 g/day and those fed mono-ammonium phosphate, 170 g/day. In all cases, phosphorus supplements were fed in conjunction with molasses. The authors conclude that although there was a general lack of response to phosphorus supplements in this series of experiments, either superphosphate or mono-ammonium phosphate may be preferable to phosphoric acid as a source of phosphorus. However, they also draw attention to the fact that, in economic terms, allowance must be made for the labour costs involved in preparing the supernatant liquid from the alternative sources.

Summary

Several phosphate sources have been used in the tropics and subtropics and there seems to be some difference in their effectiveness.

Bonemeal has been widely used as a supplement and is clearly palatable and easily utilized.

Disodium and mono-sodium phosphates have been widely investigated, both in solid and solution form, and have been generally effective especially in cases where herbage calcium levels are high.

The use of both phosphoric acid and diammonium phosphate appear to be associated with some problems, possibly, in the case of the former, because of a disturbance of acid-base balance.

Similarly, the use of rock phosphates and superphosphates must be treated with caution because of the provocation of digestive disturbances as well as the presence of fluorine which may be toxic.

Methods of administration and dosage rates

Although phosphorus sources havr traditionally been given free-choice, usually mixed with salt or some other carrier to increase palatability, various other methods of administration have been employed. Theiler et al. (1924) first discussed different methods of bonemeal feeding. They advocated the use of a large number of small cheap wooden troughs so that each animal was able to obtain its share and so that stronger animals could not rob the weaker ones. They also advocated the construction of kraals (pens) at watering places to contain the small troughs; thus each animal readily learned the purpose and location of its trough and passed through rapidly without waste of time. However, the allocation of the precise amount required by the animal into the trough was suggested rather than the practice of *ad libitum* access. The authors also commented that instead of using troughs, some farmers find it sufficient to place the ration in little heaps on the ground.

Theiler and his colleagues also commented on free-choice feeding but pointed out that this may lead to overconsumption and mentioned a 'gourmand' in the Armoedsvlakte herd which regularly consumed 1.8 kg at a 'sitting' (although this 'record' was later bettered by a cow cited by Murray and Romyn (1937b) as mentioned above!). Feeding with salt was not recommended because 'the craving for the two does not run parallel' although this comment may be found controversial (see the review by Pamp et al. 1976, for example). If the bonemeal was to be mixed with salt, then the ratio of two parts of bonemeal to one of salt (and never less than one to one) was recommended.

The authors continued to comment on the 'crush' method where animals are dosed individually with quantities of up to 140 g by means of calibrated metal spoons or tobacco tins. This ensures the right dosage for each type of animal and therefore maximum economy. Part of the procedure is worthy of quoting verbatim: 'The native assistant holds a horn and pulls the tongue aside while the white man wipes off the dose on the back of the tongue. The animal blinks contentedly and passes on as if given a sweetmeat!' It was claimed that up to 200 animals/hour could be dosed. Bisschop (1932) subsequently claimed that as many as 300 head/hour could be dosed by this method.

In a later paper, Bisschop (1964) compared the relative merits of the 'lick' and 'dose' methods of application with interesting results. Although the 'lick' method gave as good results as did the 'dose' method with regard to liveweight development, beef production and calving percentages, there were several disadvantages of the former. It was wasteful and uneconomic in that steers consumed 92% and breeding cows 87% in excess of their requirements and, furthermore, it appeared that both steers and cows developed a progressive liking for bonemeal quite out of proportion to their requirements (this in itself contradicting the 'craving' hypothesis of Theiler et al. (1924) mentioned earlier). Notwithstanding the higher *average* intake of bonemeal (and inorganic phosphorus level in the blood) of the 'lick' cattle, there were symptoms of depraved appetite in 13.5% of the steers and 4.9% of the cows as compared with only 4.6 and 2.9% respectively of the dosed animals. It is therefore to be supposed that the former animals would be more likely to contract lamsiekte.

Before leaving the subject of the 'dose' method of application, it is appropriate to consider the quantity to be given as well as the frequency of dosing. First indications on the quantity to be given were by Theiler et al. (1924) who demonstrated that it was necessary to feed 85 g of bonemeal six times a week (every day except Sunday) to growing and dry cattle in order to prevent osteophagia and thus increased exposure to lamsiekte. The corresponding quantity for cows and heifers in calf and cows in milk was 140 g. However, because of the introduction of an efficient vaccine against this disease, Bisschop (1964) was led to investigate lower dosage levels in order to find out whether economies could be made. Using Sussex weaner 'tollies' (young steers) grazing Armoedsvlakte veld, he dosed them with 30, 60 or 85 g of bonemeal six times a week. No differences among treatments could be demonstrated for growth rate, conformational development, quantity or quality of beef produced, or wet weight or wet surface area of the hides. It was concluded that 170 g of bonemeal a week (equivalent to approx 18 g phosphorus) is sufficient to sustain normal growth in cattle up to a liveweight of 340 kg approx.

In a subsequent experiment, Bisschop (1964) confirmed that in normal or better than normal seasons, 30 g of bonemeal equivalent to 3 g available phosphorus daily six days a week was sufficient, but that during subnormal seasons 60 g/day from weaning onwards was necessary.

Bisschop (1964) went on to determine whether it was possible for breeding cows and heifers to maintain their productivity when given less than 140 g of bonemeal daily, six times a week (equivalent to about 15 g phosphorus) using the whole of the Armoedsvlakte herd, some 400 cows and heifers in all. These comprised Afrikander, Sussex, Redpoll and Fries (Freisian). There were no differences of functional significance insofar as liveweight development, inorganic blood phosphorus, average calving percentage, or average weight of calves at birth or weaning. Milk yields of the two milk breeds, Fries and Redpoll, were also measured in addition to the other criteria mentioned above. There were no differences of consequence in the milk yields of the three experimental groups, independent of breed or within the Fries breed. However, the groups of old non-pregnant cows given 140 and 113 g/day in the Redpoll herd gave significantly more milk than did the corresponding group given only 85 g per day. Insofar as milk quality was concerned, the only significant differences which could be demonstrated was that, in the Fries breed cows, the milk of those given 140 g was richer in fat than the milk of those cows given 113 or 85 g.

It was concluded that for conditions at Armoedsvlakte for breeding beef cattle, 85 g of bonemeal/day (except on Sundays) was sufficient but that for milking cows, 140 g were necessary.

With regard to the frequency of dosing required, Bisschop (1964) investigated the difference between feeding bonemeal to growing steers at the rate of 170 g three times a week instead of 85 g six times a week. The results indicated that the differences between the two experimental groups were of no functional or economic significance. A more extreme method of dosing was that of Norman (1960) who found that giving the equivalent of 5 g phosphorus/head/day as bone-meal once every fortnight was sufficient to maintain growth rates superior to those of the controls.

Another frequently used method for the administration of phosphorus is by dissolving soluble phosphates in the water supply. This was first described by Becker (1932) who dissolved disodium phosphate in water so as to supply more or less 40 g daily and compared this with a similar quantity given as a 'dose' as a supplement to nine-month-old calves over a period of about thirteen months. During this period (which included two wet seasons), the animals given phosphate in the water gained 183 kg whilst those which were dosed gained 135 kg. Becker attributed the difference to the fact that the former animals were handled less. Despite the proven efficacy of the method, Bisschop (1964) stated that 'due to increasing costs and insufficient supplies, not only of disodium phosphate but also of other soluble phosphates such as monocalcium or mono-ammonium phosphate, his method was not at the time adopted by farmers'.

Malan and du Toit (1932), comparing various sources of phosphate, also found that disodium phosphate was efficiently utilized when administered in the water. However, in an attempt to utilize a cheaper source, they found that both rock phosphate and superphosphate were unsatisfactory because of their content of toxic fluorine. However, in a later paper, du Toit et al. (1939) showed that a non-toxic solution could be prepared by dissolving superphosphate in water: virtually all of the phosphate and only a small proportion of the fluoride was dissolved; this could be improved by subsequent dilution and the addition of slaked lime. They calculated a concentration of 0.4 g P_2O_5/litre could be attained. The authors also gave a practical recipe for the method and described a variety of devices by which the solution could be metered into the drinking water.

More practical applications of the method were given in a later paper (du Toit et al. 1940b). The method was further refined by Louw and his colleagues (Truter and Louw 1952; Reinach and Louw 1952) when the application of heat was introduced and this latter method is recommended to farmers (see, for sample, *Handbook for Farmers in South Africa*. Dept of Agriculture, Pretoria), which is quoted below:

1. Add 50 kg superphosphate (18.5% P_2O_5) to 180 litres water in a 200 litre oil drum.
2. Stir thoroughly for 5 minutes.
3. Allow to settle for 15 minutes.

4. Stir again for 5 minutes; the soluble phosphates and fluorides are now in solution.
5. Leave overnight.
6. Siphon off the clear supernatant (if a tap is fixed about 20 cm above the bottom of the drum, the clear solution can be drawn off without disturbing the sediment).
7. The supernatant is then brought to the boil over a fire.
8. It is then left to cool overnight when the sediment containing fluorine will settle out. The supernatant is siphoned off and this is the concentrated phosphorus supplement.
9. This is mixed with the drinking water at the rate of 1 : 70 in winter and 1 : 100 in summer (the Reinach-Louw Formula).

Bisschop (1964) tested the efficacy of the above method using weaner tollies and concluded that because supplementation of young growing non-producing cattle was not necessary in the winter months (because other factors are limiting as discussed earlier) provided summer supplementation is adequate, then the factor of 1 : 100 during the summer also is adequate for this class of stock. Bisschop (1964) points out, however, that no experimental evidence was available for other classes of stock.

Work on supplying disodium phosphate through the drinking water was also carried out by Black and his colleagues working at the King Ranch in Texas from 1941 to 1946 (Black et al. 1943; Reynolds et al. 1953). Disodium phosphate sufficient to supply approximately 1 g phosphorus/5 litres was added to the drinking water of heifers and the performance of these animals was compared with cattle with access to bonemeal in self-feeders. The authors concluded that both methods were practical and that the bonemeal-fed cows produced 133 kg weaned calf/ha whilst those administered disodium phosphate in the drinking water produced 164 kg/ha (this compared with 107 kg for cows which received no supplement and cows on pasture fertilized with triple superphosphate). The cows which yielded the highest return/acre after the costs of supplementation had been deducted were those fed disodium phosphate, followed by those on fertilized pasture, those fed bonemeal and the unsupplemented group, the returns being \$20.30,19.97, 16.91 and 13.91 respectively. However, it should be pointed out that whilst the stocking rate on the non-supplemented treatment and the disodium phosphate supplemented treatments was 0.16 cow/ha, that on the bonemeal supplemented treatment was 0.14 cow/ha and that on the fertilized treatment, 0.19 cow/ha.

Finally, before leaving methods of administering disodium phosphate, the work of Zavala et al. (1972) may be mentioned. The compound was fed solid in combination with salt and grain sorghum (the latter feed was found necessary in order to combat the acidity of the disodium phosphate). In a trial in which two levels were fed

30 and 60% of the total mixture respectively, the mixture with the higher level gave significantly higher daily gains than the lower level and the control, values being 549, 445 and 389 g/day respectively.

Summary

To sum up this section, it has been shown that 'dosing' with bonemeal is more effective than free-choice feeding. The precise amount required by an animal may be administered and overconsumption is avoided. The amounts required are approximately 3 g supplementary phosphorus/day for growing, non-reproducing stock (although as much as 6 g supplementary phosphorus may be required in a bad year), whilst lactating and reproducing cows need 9 g supplementary phosphorus.

Nevertheless, notwithstanding the clearly demonstrated advantages of individual 'dosing', it is probable that by far the majority of farmers stick to the *ad libitum* system of providing a phosphorus supplement free-choice, usually mixed with salt, water or some other carrier, because of the simplicity and ease of operation of the system.

Many forms of phosphate are suitable for supplementation but several authors have reported a superiority of disodium phosphate, particularly when given dissolved in water. Care must be taken in the use of rock phosphate and fertilizer superphosphate because of the frequent presence of fluorides.

4 *Deficiencies of other major minerals and their correction*

Deficiencies of phosphorus occur more frequently than those of any other major mineral in tropical and subtropical areas. Deficiencies of calcium, magnesium, sodium, chlorine and potassium have, however, been reported from various areas and these will be discussed below.

Calcium

Underwood (1966) has pointed out that calcium deficiencies would not normally be expected to occur in the grazing animal except under conditions of acid, sandy or peaty soils 'or humid areas where the grasses are quick growing and are free from leguminous species'.

A few comments will first be made on the occurrence of deficiencies and secondly on manifestations of calcium excesses and the disturbance of its metabolism.

Levels of calcium in forage which would appear to be conducive to deficiencies have been quoted in the literature far less frequently than those of phosphorus but, nevertheless, they do exist. Out of 504 selected samples of Latin American tropical grasses quoted by McDowell et al. (1974), only 88, or 12.7%, had a calcium content of less than 0.17%. However, of these, only 24 or 4.8% were unaccompanied by deficiencies of phosphorus.

Also, although calcium is traditionally associated with phosphorus because of their frequent occurrence together in chemical and biological compounds and in bone metabolism, no statistical relationship appears to characterize their relative proportions in plant materials. When the value of r^2 was calculated for the calcium and phosphorus contents of 100 selected samples of tropical grasses (McDowell, et al. 1974), the value was as low as 0.02, i.e. only 2% of the total variation could be accounted for by the relationship between the two elements. It is also worth commenting that the mean calcium content of the 504 samples mentioned above was as high as 0.42% associated with the very high standard error of ±0.86. The high values of calcium indicate that the ratio of Ca : P may be disturbed in grass species.

Deficiencies of calcium in forage have also been reported in East Africa (Dougal and Bogdan 1958b), in Southern Africa (FAO 1978b), in Venezuela (Grünwald et al. 1947) (this was the only case mentioned by de Alba and Davis (1957) in a review of mineral deficiencies in Latin America) and in Australia (Schofield 1946).

In various areas, deficiencies have been suspected as a result of blood analyses. These include Kenya (Gitter et al. 1975), Botswana (Reed et al. 1974b), Senegal (Friot and Calvet 1971), Venezuela (Chicco and French 1959; French and Chicco 1961). Authors comment that although calcium levels may be low, no clinical evidence of hypocalcaemia could be demonstrated (see, e.g. Gitter et al. 1975).

Theiler et al. (1937) have also pointed out that feeding 'calcium-low diets with a sufficiency of phosphorus leave the animals apparently undisturbed and clinically healthy for much longer periods'. They contrast the relatively simple physiological functions of calcium with the varied range of roles played by phosphorus (in intermediate metabolism, for example). Reed et al. (1974) observe, on the other hand, that if values as low as that reported are common in Botswana, 'then during times of stress the occurrence of calcium deficiency collapse symptoms would seem to be a possibility especially amongst young stock'. However, Underwood (1966) has pointed out that because of the physiological mechanisms controlling serum calcium, this measure is not either a prompt nor a particularly accurate indicator of calcium deficiency. Reed et al. (1974) noted that the lowest serum calcium levels occurred in areas where bonemeal was fed and suggest that increased management pressures may have decreased these levels.

Although forage calcium deficiencies and low serum calcium levels have been reported on several occasions, clinical symptoms directly attributable to calcium deficiencies have apparently not been encountered. Undoubtedly, one reason for this is that calcium deficiencies more often than not occur in conjunction with those of phosphorus and the correction of the phosphorus deficiency with a supplement which also contains calcium 'kills two birds with one stone'. I have only encountered one example where a supplementation with calcium may have resulted in increased growth rate and this was in a paper by Kaiser (1975), who suggested that the growth rate of calves grazing nitrogen-fertilized Kikuyu grass was increased by the use of a mixed supplement containing salt and phosphate as well as calcium. The author based this assumption on the 'lower' calcium levels (0.35%) found in the grass samples but also admitted that from his data it was impossible to partition the mineral response.

Enteque seco

Enteque seco is a disease described by Carrillo and Worker (1967) in Argentina. It is also known as Manchester wasting disease (Arnold and Fincham 1950) in Jamaica, Naalehu disease (Arnold, *pers. comm.*) in Hawaii, enzootic calcinosis in Germany (Dirksen et al. 1970) and Papua-New Guinea (Arnold, *pers. comm.*) and Espichamento in Brazil (Döberreiner et al.) 1971).

The condition is basically a disturbance of mineral metabolism, being a form of naturally occurring metastatic calcification in ruminants but also known and producable artificially in other herbivores and even poultry. Enteque seco was originally described in Argentina by Lignieres (1912) who distinguished it from 'Enteque diarretico' (due to parasitic infection) and believed it to be caused by a Pasteurella infection. Although the disease is widespread in the grazing areas of South and Central Argentina, Brazil and Uruguay, it was apparently little studied over the succeeding years and although the aetiological role of *Pasteurella* was not confirmed, no satisfactory explanation of the cause of pathogenesis was given until 1956. Arnold and Bras (1956) published a comparative study of the disease in various countries of the Americas and identified the pathological process as that of metastatic calcification. The precipitating cause was not identified until 1967 when Worker and Carrillo (1967) showed that a common pasture weed *Solanum malacoxylum* contained a glycosidic derivative of 1, 25 dihydroxycholecalciferol, a form of vitamin D. Experimental work had shown that lesions could be produced by dosage with other precursors of vitamin D, particularly when associated with imbalanced dietary intake of calcium and phosphorus. Although other plants have been identified as sources of similar calcium-mobilizing agents, their sources in some of the countries where the disease occurs have still not been identified and it is possible that in the tropics, sufficient vitamin D may be available directly or indirectly from solar irradiation for this to produce the pathogenesis in the presence of a diet unbalanced in calcium and phosphate.

The disease is characterized by a progressive loss of weight and stiffness of the joints, and with symptoms associated with impaired circulation. Affected animals stand with arched back and slightly bowed forelegs, the hindlegs being somewhat straighter than usual; they are reluctant to move and may spend some minutes on their knees while rising. The symptoms usually lead to death through debility. At autopsy there is a characteristic deposition of calcium in the lining membrane of blood vessels. This process appears to originate in the proximal aorta or left ventricle and to extent peripherally through the arterial system in severe cases, small capillaries in all parts of the body being converted to rigid calcified tubes. In such cases the lungs take on the appearance of a 'loofah' sponge and deposits of calcium salts occur in the spleen and sometimes kidneys. Ulceration of articular surfaces is usual, especially in the weight-bearing joints.

No successful treatment of the disease has been described but in areas where it is well recognized it is customary to remove animals from 'affected' grazings when some degree of recovery occurs and they can be sent for slaughter. Control is by the elimination of *Solanum malacoxylum* or other responsible plant where this has been identified.

Vitamin D

Before leaving the subject of phosphorus and calcium nutrition, it is perhaps appropriate to mention the role of vitamin D. It is well known that this vitamin is involved in the efficient utilization of both calcium and phosphorus and also facilitates the metabolism of these elements when the Ca : P ratio is less than optimum. However, it has also been established that it can be formed by the irradiation with sunlight of 7-dehydrocholesterol in the animal or of ergosterol in plant materials (particularly hays, etc.). It seems most unlikely therefore that deficiencies of this vitamin would occur in tropical or subtropical areas where solar radiation is normally intense and no evidence has been presented to implicate vitamin D in problems of calcium or phosphorus utilization in these areas.

Magnesium

A deficiency syndrome associated with forage levels of magnesium of less than 0.2% is known as hypomagnesic tetany. This has been reported from most countries where dairying is highly developed (Underwood 1966) as well as from some of the developing countries (McDowell 1976) including, in the tropics or subtropics, Haiti, Honduras, Jamaica, Peru and Surinam, whilst Boyazoglu (1976) has suggested that magnesium deficiencies may be encountered in the Eastern Cape, Orange Free State and Natal in subtropical South Africa as well as in South West Africa.

De Alba (1958), however, considers that magnesium deficiencies are most unlikely in the grazing animal in tropical and subtropical areas and puts the content of magnesium below which symptoms of deficiencies are likely to occur as low as 0.05% and very few forages contain levels of less than 0.1% (see, e.g. McDowell et al. 1974). It has also been pointed out (ARC 1965) that the availability of magnesium in the forage decreases with the maturity of the animal but increases with the maturity of the forage. When recalculated in terms of percentage of dry matter intake, the values recommended by the ARC (1965) range from 0.07 to 0.08% in animals between 200 and 400 kg growing at rates of 0.33 to 1.00 kg/day. This assumes an availability of only 20%. The rapidity with which tropical grasses mature and the relatively low levels of production normally attained support the contention that deficiencies are likely to be rare. It is also well known that deficiencies are likely to be provoked by high levels of fertilization with potassium or nitrogen (Kemp et al. 1961) but the extent to which this is practised under tropical or subtropical conditions is most limited.

Sodium and chlorine

Both sodium and chlorine are required in the body for the regulation of fluid volume as well as a wide variety of other functions. Thus, chlorine is important in the maintenance of stomach acidity and sodium is associated with nerve impulse transmission. Because of the often large losses of these elements induced through sweating in high temperatures, it would be expected that particular attention would have been paid to them under tropical and subtropical conditions.

Deficiency symptoms may include loss of appetite, staring coat, listlessness and decreased productivity both of meat and milk. These symptoms would be expected to be more acute in relatively highly producing animals. For a fuller discussion of aspects such as metabolism, absorbtion and transport etc., a review such as that by Mills (1969) should be consulted.

Requirements

The quantitative requirements for sodium and chlorine in the grazing animals have not been accurately established for tropical and subtropical regions. After an extensive investigation of pasture samples taken from all parts of South Africa, du Toit et al. (1940a) concluded that requirements for growth in cattle could be met by 0.02% sodium and 0.07% chlorine in forage, and that for the lactating cow the requirements would be 0.15 and 0.19% respectively. Underwood (1966) commented that the values for growth appear to be too low when compared with those for sheep and for the growing calf. Later estimates suggested by the ARC (1965) indicate that for growing cattle, sodium requirements range from 1.6 to 7.5 g/day. This would be equivalent to approximately 0.11 to 0.07% in forage consumed at a rate of 2.5% of body weight. The corresponding values for chlorine are slightly higher (0.14 and 0.11% respectively).

Occurrence of deficiencies

Forage levels which are likely to cause deficiencies of sodium have been reported by several authors in both tropical and subtropical areas. The extensive investigations undertaken by du Toit and his colleagues have already been mentioned in connection with the determination of requirements (du Toit et al. 1935–40). After analysing more than 5000 samples, collected by veterinary assistants from farms in all parts of South Africa, they quoted average maximum values of 0.08% and average minimum values of 0.01% and concluded that pastures composed mainly of grasses might be deficient in sodium during five to nine months of the year. They also pointed out that sodium values in bush species were exceptionally high (from 1.0 to 1.86%) and concluded that deficiencies are not likely to be encountered in 'bushveld' areas. This is supported by personal observation of the author (unpublished) that cattle grazing the Swaziland bushveld evince little if any interest in a supplement of salt and bonemeal.

Confirmatory evidence of the probable existence of sodium deficiencies in Southern Africa has been given by Jones. In studies carried out in the Mt Makulu area of Northern Rhodesia (now Zambia) near Lusaka on natural veld grasses, values of less than 0.02% were encountered in both wet and dry seasons. The grasses analysed included *Hypparhenia dissoluta, H. filipendula, Heteropogen contortus, Urochloa pullulans, Sporobolus pyrimidalis* as well as *Cynodon dactylon* (Jones 1963). In a further study on some cultivated grasses in the same area, values of less than 0.05% were encountered in all species with the exceptions of *Chloris gayana* (0.31%) and *Panicum coloratum* (Katambora strain) (0.15%) (Jones 1964).

Early analysis of pastures in East Africa carried out by Orr and Holm (1931) in Kenya and by French (1933) in Tanzania indicated that lower levels could be expected in the dry than in the wet season herbage.

Subsequent work carried out by Howard and his colleagues indicated the existence of mineral deficiencies at Muguga in the Kenya Highlands (Howard et al. 1962) and it was calculated that more than 90% of all samples studied contained less than 0.015% sodium whilst 60% contained less than 0.007% sodium.

In Uganda, Long et al. (1972) determined the mineral status of the dominant pasture species, which largely comprised *Choris gayana, Hyparrhenia rufa, Panicum maximum, Pennisetum purpureum, Centroscema pubescens* and *Stylosanthes guianensis.* All samples were judged to be deficient in sodium, the mean value being 0.056 + 0.016%.

In South America, Dirven (1963) reported that grasses grown on clay soils in Surinam contained as little as 0.01% sodium.

Playne (1970) investigated the levels of sodium in grasses and legumes in the Townsville area of Queensland and characterized *Panicum maximum, Chloris gayana* and *Cenchrus setigerus* as 'high sodium' species with levels of more than 0.40% sodium. *Urochloa mozambicensis* had a mean intermediate value of 0.27% but *Cenchrus ciliaris, Heteropogon contortus* and the legumes *Stylosanthes humilis* and

Phaseolus atropurpureus contained mean values of 0.09, 0.06 and 0.02% respectively. The author concludes that deficiencies could occur with these levels particularly in the case of grazing cows and adds that levels are likely to be even lower during the dry season (June to December) than in the growing season for which results are reported.

Also in Australia, Murphy and Plasto (1972) have reported low values in New South Wales of 0.008, 0.011, 011 and 0.019% in *Themeda australis, Bothriochloa decipiens, Dichanthium affine* and *Heteropogon contortus* respectively. Murphy and Connell (1970) have suggested that the sodium level in saliva can be used as an indicator of deficiency and devised a simple technique for its collection. This technique was subsquently used by Hennessy and Sundstrom (1975), also in New South Wales, who suggested that the ratio of sodium to potassium in the saliva was lower in animals grazing low sodium kikuyu and clover pastures (containing 0.006–0.014% and 0.040–0.064% sodium respectively) than in those which were given salt licks.

It is appropriate at this point to mention geophagia in the context of salt deficiency. This has been exhaustively reviewed by French (1945) who concluded that there was no evidence that earth eating satisfied salt deficiencies (although Hudson (1944), confirming this, does mention two samples of licks from the Chelbi area of the Northern Frontier District in Kenya which contained 15 and 78% sodium chloride).

A typical early investigation was carried out by Anderson (1933) who investigated the phenomenon of 'pica' in animals in Nigeria and implicated sodium because of trials carried out with a native salt 'kanwa' which contained sodium· chloride and sodium carbonate. However, animal growth rates did not confirm this and it was also most unlikely because of the high sodium levels in the herbage (0.26%).

Supplementation with sodium chloride

Although supplementation with salt has been traditional for many centuries, particularly in Europe (Underwood 1966), it is only relatively recently that experimental work has been carried out on this subject under tropical or subtropical conditions.

Africa

Working in Marandellas in Rhodesia, Husband (1932) reported a condition described as 'nutritional anaemia' which he compared with 'salt sickness' (cf. Becker et al. 1965); this was characterized by a low blood haemoglobin level and responded both to administration of salt alone and salt with iron. However, as only two cows were used in the trial, this can hardly be taken as conclusive!

Subsequently (Husband 1934), more details were given on the condition, which included progressive emaciation, loss of appetite, staring coat, and it was observed that the condition was more common in breeding cows where an additional symptom was decreased milk production. The syndrome was attributed to a deficiency of chlorine as the main difference between the composition of the herbage at Marandellas and that at Matopos, where the condition did not occur, was that less of this element was present in the former. The administration of common salt caused an increase in calving percentage (57% vis-a-vis 37% in unsupplemented animals) but no response to common salt supplementation could be demonstrated in growing calves.

Husband concluded his series of notes on nutritional anaemia with the description of clinical cases leading to death. Autopsy confirmed the existence of severe anaemia. A further extreme case was successfully treated with a mixture of sodium chloride, iron and copper therapy (Husband 1935; Husband and McChlery 1936).

Work carried out in Rhodesia by Murray and Romyn (1932) at Matopos suggested that indeed these pastures might be deficient in both sodium and chlorine throughout the year.

They used Lincoln Red Shorthorn steers aged between 16 and 18 months to compare the use of 30 g daily of sodium chloride ('Liverpool salt') as a supplement to natural grazing overlying a granite formation ('sandveld'). The supplements were individually hand-fed every day except Sunday. The study was carried out during a period of almost three years. The authors conclude that any differences between the salt-fed and the unsupplemented animals were not significant and therefore that 'the grazing, including browse, supplied a sufficiency of both sodium and chlorine for cattle maturing at the rate these did'. (They were sold at a weight of 473 kg at an average age of 4 years and 4 months, i.e. weight per day of age of approximately 300 g).

Walker studied the effects of supplementation with salt in Mazabuka, Zambia. A weight increase was demonstrated when salt was fed to grade Sussex cattle, both with or without protein, but not with 'native' cattle. This was attributed to the lesser heat stress on the latter animals with consequent lower requirement for sodium chloride (Walker 1948). Several notes on the same subject followed (Walker and Rhodes 1951–4), and in 1952 a more comprehensive study was initiated which continued until mid-1956. This involved the feeding of common salt at two levels with or without groundnut meal as an additional source of protein. The stocking rate was approximately one animal to 6 ha and the cattle used were all 'native' breeds, i.e. Angoni, Barotse and Tonga. The levels of salt fed were 30 and 60 g/day and these quantities were individually administered to the cattle with a spoon. Both levels of salt gave significantly improved performance over the unsupplemented animals, total liveweight gain being 296, 419 and 334 kg for the unsupplemented animals, those fed 30 g/day and those fed 60 g/day respectively. There was no benefit of the groundnut meal but there was an unexpected effect in that the animals fed 30 g of salt grew significantly better than those fed 60 g (Walker 1957). Continuing the work started by Husband at Marandellas, Rhodes (1956) carried out trials supplementing exotic and native steers with salt during both winter and summer periods. It was found that during the summer (rainfall) periods, the feeding of salt had a markedly beneficial effect on liveweight gains of supplemented and non-supplemented high-grade Sussex weaner steers of 119 v. 89 kg, 138 v. 121 kg and 179 v. 164 kg respectively. Carcasses of the former animals also graded better than those of the latter. Salt supplementation, however, had no effect on winter (dry) season loss. In a subsequent experiment, where Afrikander–native type animals also grazing natural veld were employed, salt was fed both with and without groundnut meal, at two levels, 30 and 60 g/day. No benefits could be demonstrated from supplementation with salt at either level except for a slight advantage conferred by both levels during the first summer season. No difference could be demonstrated which could be attributed to dosage rate. The author attempted to explain this lack of effect to an adaptability of the indigenous animal to light level (through the control of the adrenal cortex by the interior pituitary) but also concluded that wide seasonal differences in the natural nutritional level and possibly an inadequate protein supplementation in the latter trial could have played a role.

Australia

Some work has also been carried out in Australia on supplementation with salt. Murphy and Plasto (1973) carried out investigations on a property on the eastern edge of the Darling Downs west of Brisbane in Queensland where a severe sodium deficiency had already been demonstrated as a result of pasture analyses. (Murphy and Plasto 1972).

The cattle used were Hereford cows of between three- to six-year-old suckling young calves. Groups of forty were run in 160 ha paddocks and one group was given common salt *ad libitum*. Positive results of salt supplementation were noted in both the adult cattle and the calves. The supplemented cows showed a significantly better rate of liveweight increase (0.35 v. 0.14 kg/day) and their calves grew significantly faster than those of the unsupplemented group (0.77 v. 0.62 kg/head/day). Salivary sodium values of the unsupplemented cows were indicative of severe deficiency. Blood and milk concentrations of sodium and potassium, however, were not influenced by sodium intake.

As has been mentioned earlier, Murphy and Plasto (1973) were working in an area where natural pastures were dominant. However, relatively close by in Legume, New South Wales (NSW), Hennessy and Sandstrom (1975) carried out experiments on the supplementation with common salt of Hereford weaner steers grazing pastures dominated by *Pennisetum landestinum* (Kikuyu) and also containing *Paspalum dilatatum* and *Trifolium repens* (White clover). The results obtained were more equivocal than those obtained by Murphy and Plasto (1973). The steers fed a salt lick were apparently heavier than the unsupplemented animals at the end of the trial (324 v. 297 kg), but the authors point out that paddocks were not replicated and rotation was carried out only at infrequent intervals.

The work carried out by Kaiser (1975), also in NSW, may be briefly mentioned. This author supplemented calves grazing fertilized Kikuyu grass with a mixture containing sodium calcium and phosphorus and suggested that the liveweight response obtained was attributable to a combined response to sodium and calcium. However, the author admitted that it was not possible to partition the mineral response and this must only be taken as a possible effect of sodium.

Dramatic results were obtained in Papua New Guinea when lactating cows were given uniodized salt (Leche, 1977a). Cattle were Brahman cross-breds which had not been given any mineral supplements throughout the previous three-year period. Supplemented cows were 30% heavier than their unsupplemented counterparts after a twelve-week period and their calves gained 40% more liveweight than those from unsupplemented dams. Gains of supplemented cows were 61 kg during the period whilst unsupplemented cows lost 23 kg on average; calves of the former cows gained 49 kg compared with only 35 kg for the control group. Analyses of saliva and urine confirmed severe deficiency in the non-supplemented animals. The author concluded that a condition known as 'nutritional lactation stress', which is characterized by loss of weight and condition and sometimes death, during the first months of lactation, may be attributed to sodium deficiency. This disease only occurs in some but not all lactating cows and not in other classes of stock. In a subsequent experiment, Leche (1977b) showed that supplementation with salt was ineffective in increasing liveweight gain in steers grazing native pastures in the same area suggesting that response was limited to animals subjected to the additional stress imposed by the secretion of significant quantities of sodium in the milk.

Summary

There appears to be little doubt that there are many years in both the tropics and the subtropics where, based on forage sodium levels, it would be expected that a deficiency of sodium would exist. A chlorine deficiency is less likely. However, many trials that have been carried out on the supplementation of cattle with common salt have not given positive results. One reason for this may be that browse provided a higher level of sodium than the pasture or that the sodium content of the pastures were underestimated. The sodium-conserving mechanisms in the body are also doubtless effective under these circumstances.

In any event, it would seem that in view of the inexpensiveness of common salt and also its usefulness as a carrier for other supplements (e.g. iodine, bonemeal, etc.) its provision is worth while where animals show an appetite for it.

Potassium

Although McDowell (1976) has cited cases of relatively low potassium levels in Brazil, Haiti, Panama and Uganda, it would appear that a deficiency of potassium caused by an inadequate intake is unknown except on specially-prepared synthetic diets (Mills 1969). Further, it has been stated that the 'likelihood of deficiency given conventional diets is remote' (ARC 1965). The ARC (1965) do not commit themselves to recommending forage levels adequate to prevent deficiencies for this reason but Ward (1968) has suggested 0.5% of dry matter. Sutmöller et al. (1966) implicate potassium in the dramatic 'mal de cai' which they observed in the Amazon Valley in Brazil. This was reported to cause the deaths of many cattle but was also associated with deficiencies of sodium and imbalances of serum calcium, phosphorus and magnesium, so the role of potassium is not clear.

Sulphur

This element has recently been implicated in the intake of tropical forages. It had not been reviewed here but see Gartner et al. 1980).

Summary

Although deficiencies of calcium are indicated by forage and, in some cases, blood levels, there appear to have been no cases where clinical symptoms directly attributable to deficiencies of this element have been reported. However, the aetiology of metastatic calcification in the presence of vitamin D derivatives and associated with imbalanced diet intake of Ca and P has been described. It is unlikely that deficiencies of vitamin D would occur in tropical and subtropical areas because of the high rate of formation of this vitamin from its precursors, both of animal and plant origin, in the presence of sunlight.

Cases of magnesium deficiencies have been reported in the literature but it is unlikely that these would be a severe constraint to production of the grazing animal as few forages contain levels of magnesium low enough to induce symptoms of deficiencies even in more highly productive stock.

Sodium and chlorine are normally considered together and, because of high losses associated with sweating at high temperatures, deficiencies are to be expected. These have been confirmed by analyses of forages in various parts of Africa and Australia. However, trials carried out with salt supplementation have given equivocal results possibly because dietary intake was underestimated and because of the conserving mechanisms of the body.

Deficiencies of potassium are unlikely to occur under normal grazing conditions.

5. Deficiencies of trace elements and their correction

A number of minerals are required by the body in smaller quantities than those previously reviewed, but deficiency conditions have been identified in many parts of the tropics and subtropics. Because these trace elements are frequently involved in enzyme systems, rather than subserving a structural or regulatory function, their deficiency symptoms may be of considerable complexity. In addition, deficiencies of several trace minerals frequently occur in the same animal simultaneously; the matter is further complicated by the fact that antagonisms between minerals may give rise to 'conditioned' deficiencies. Other trace minerals become toxic when their concentration exceeds a certain level.

Copper

Copper is present in the respiratory enzyme, cytochrome oxidase and also in uricase. It plays a role in the formation of the porphyrin nucleus of haemoglobin and in the maintenance of function of bone osteoblasts. It is also implicated in melanin pigment generation in hair and in keratin formation in wool. The exact nature of these roles are as yet not completely understood (Mills 1969). Deficiencies of the element in cattle give rise to bone changes and fragility, anaemia (hypochromic, macrocytic) and defective pigmentation (achromotrichia) as may be inferred from the above-mentioned functions. It also causes fibrosis of the myocardium (falling disease or sudden death) and diarrhoea. The symptoms of these conditions have been reviewed by Underwood (1966). The element has also been mentioned in connection with infertility (van Rensburg 1961).

Occurrence of deficiencies

There is no doubt that deficiencies of this element are widespread throughout the world and McDowell (1976) has mentioned their occurrence in Argentina, Brazil, Colombia, Costa Rica, Haiti, India, Kenya, Malaysia, Mexico, Panama, Peru, Surinam, Uganda, Republic of South Africa and Zaire. Absolute deficiencies of the element are found, as are 'conditioned' deficiencies; these are caused by interference with the absorption of copper by high levels of molybdenum and/or sulphate (Dick and Bull 1945; Dick et al. 1975).

Australia

Copper deficiencies have been widely documented in Australia, considerable evidence having been reported from Queensland. In addition to the trials where copper supplementation was carried out as reported below, early references include Lee and Moule (1947), Harvey (1952) and Sutherland (1952), all of whom report hypocupraemia. Later work by Harvey and his colleagues (Harvey et al. 1961; Ryley et al. 1961) was carried out on sheep and cattle grazing pastures predominantly consisting of *Paspalum dilatatum* at Rocklea. Although no clinical symptoms of copper deficiency were apparent. levels of the element in liver and blood were low; this was not caused by molybdenum in the presence of inorganic sulphate and some unknown factor was implicated.

Africa

In Africa, several workers have reported that copper deficiencies are common. In the Republic of South Africa, Boyazoglu (1976) has indicated that deficiencies may be expected to occur in the Northern Transvaal (north of Warmbaths), in the Highveld area (Central and Southern Transvaal, Orange Free State as far as Bloemfontein) in Natal and in South West Africa. These conclusions were based on the results of liver assays.

Earlier, Boyazoglu (1970) suggested that copper might also be deficient in Swaziland as a result of the analysis of several liver samples taken in various parts of the kingdom. In West Africa, Friot and Calvet (1971) reported that copper deficiencies occurred during the dry season in the Ferlo and Casamance districts of Northern Senegal. The diagnosis of the deficiency was made on the basis of blood analyses.

In East Africa, Howard and his colleagues (Howard et al. 1962) found occasional levels of below 5 ppm, said to be critical, in pastures near Muguga in the Kenya Highlands. In later papers (Howard 1963; and Howard 1969, cited by Tilman 1970), the author indicated that there was an area along the Rift Valley from Mount Longenot to Nakuru, Njoro, Rongai, Solai regions, in Mwiga, Naro, Moru, Nanyuki and Timau districts and in Thomson's Falls, Ol Kalou and Ol Joro Narok areas. Of all the samples analysed, approximately 35% contained less than 5 ppm copper. Howard (1969) also reported a 'conditioned deficiency' in the Marigat region where forage copper levels were above average but molybdenum and sulphate were also high. Similarly, Naik (1965), working in the Mbulu district of Tanzania, found various copper deficiencies when forages were analysed but these were largely confined to mature grasses. See also Roeder (1980) for descriptions of deficiency symptoms in Ethiopia.

The Americas

Various workers have reported copper deficiencies in the Americas. de Alba and Davis (1957) cite Tabusso (1942) as having reported enzootic ataxia (or 'Renguera peruana') in lambs in Peru but point out that no cases involving cattle were described. In Surinam, Dirven (1963) observed that of twenty-four samples of forage analysed, thirteen were deficient in copper using 4 ppm as his criterion of adequacy. (Incidentally, ARC (1965) indicate that 10 ppm may be a more appropriate value.)

Bingley and Carrillo (1966) reported hypocuprosis of cattle in the Argentine; this was found in low areas and was 'conditioned', being associated with high levels of molybdenum.

Sutmöller and his colleagues (Sutmöller et al. 1966) carried out a large number of analyses of various tissues of cattle in various parts of the Amazon Valley in Brazil. These workers concluded that liver copper status and, in some cases, hair status provided a good indication of whether or not a copper deficiency was present. They classified deficient cattle on the basis of the soil types on which they were kept and found that most deficiencies occurred in cattle kept on class II fertility soils which were periodically deeply submerged under rain or brackish water. These are found in the lowlands of Marajo Island and the coast of the Amapa territory.

Deficiencies were also found in the lowlands of the Lower Amazon region. It was noteworthy that in some areas, deficiency or a tendency thereto occurred in the rainy season, a trend that does not agree with results obtained in other parts (see, e.g. Friot and Calvet 1971). The authors also commented on the fact that copper deficiency was absent on soils classified as group III although these soils had the lowest copper content. Similar results have been reported more recently by Camargo et al. (1979).

Ammerman et al. (1974) reported indications of copper deficiencies in cattle from the El Real area of Eastern Panama as a result of liver analyses of fourteen mature non-pregant Brahman cows.

Becker et al. (1965) carried out autopsies on animals which had died from falling disease in Florida and concluded that this was attributable to copper deficiency. Other conditions were found, particularly on muck soils, called variously 'paces', 'limehided' and 'marsh sickness', all of which could be put down to copper deficiency either in the presence or absence of high levels of molybdenum. Suitable mineral mixtures to overcome this deficiency were described elsewhere (Becker et al. 1957; Chapman and Kidder 1964, Cunha et al. 1964).

Finally, Hill et al. (1962) reported on the copper status of cattle and buffalo in Malaya and found that relatively low levels of copper were to be found in cattle from all states of Malaya and although no specific clinical symptoms could be directly attributed to this deficiency, bone fragility and emaciation were indications that it could exist.

Supplementation with copper

Australia

Early studies with supplementation of copper were carried out at Rocklea in south-east Queensland by Sutherland (1956) who could demonstrate no effect although deficiencies had already been shown (as mentioned above, Sutherland 1952). However, later work at the same station (Harvey et al. 1963) showed some evidence of a response to copper coinciding with low blood copper levels during a prolonged drought period (June 1957–October 1958). Hereford heifers were dosed with 0.125% solution of copper sulphate sufficient to provide from 30–120 mg copper every four days. It was concluded that copper sulphate was not effectively utilized.

Copper glycinate was used as a copper source in later work carried out in Cromarty near Townsville in northern coastal Queensland (Donaldson et al. 1964) when copper and cobalt supplements were compared. Shorthorn heifers, ten to eleven months old, were dosed with 400 mg of this compound every three months. Gains during a period of approximately six months (April to September 1960) amounted to 20 kg and 23.1 kg for the groups supplemented with copper and copper plus cobalt respectively, whilst the control animals and those supplemented with cobalt alone gained 0.6 kg and 2.7 kg respectively. There were also significantly more pregnancies in the copper-supplemented animals.

Similar comparisons of copper and cobalt were carried out in the Broadside Peninsula in central coastal Queensland by the same group (Alexander et al. 1967). In a series of four experiments with male and female weaners, cows and steers, positive effects of copper supplementation (again using copper glycinate as the source) were demonstrated both in terms of increased weight gain during the growing season and diminished weight loss during the winter. No response to cobalt supplementation could be demonstrated.

Results contrary to those mentioned above were obtained by Gartner et al. (1968b) when steers grazing *Digitaria decumbens* and *Lotononis bainesii* were injected with 400 mg copper glycinate every three months at the Coolum Research Station north of Brisbane in south-east Queensland. In this experiment, no positive response to either copper or cobalt supplementation could be demonstrated although liver reserves of the former element were low (cv. the results of Sutherland 1952, 1956 mentioned above). An interesting finding in this trial was that liver copper concentration was significantly depressed by cobalt therapy. The authors offered no explanation for these findings.

Similarly, no response to copper supplementation could be demonstrated by Winter et al. (1977c) when animals in the Cape York Peninsula were treated with copper glycinate. Pastures were composed of *Panicum maximum* cv. Petrie and *Stylosanthes guianensis*. Although herbage copper levels were below those recommended and the molybdenum levels resulted in ratios likely to be predisposing to copper deficiency, no deficiency symptoms of impairment of growth were apparent. Liver copper levels indicated that treatment with cobalt or copper increased the storage of copper. This seems to be in direct contrast to the results of Gartner et al. (1968b) quoted above, who found that treatment with cobalt significantly decreased liver copper levels.

To conclude this review of the work carried out in Queensland, it may be mentioned that although low levels of liver copper were not apparent, Winter et al. (1977b) demonstrated a positive relationship between liveweight gain and blood copper concentration in heifers and steers grazing either *Brachiaria decumbens* or *Panicum maximum* pastures each sown with *Stylosanthes guianensis* cv. Endeavour and *Macroptilium atropurpureum* cv. Siratro.

Africa

Responses to copper therapy have been demonstrated in East and Southern Africa. In Kenya, Howard (1970) demonstrated a positive response to 120 mg of copper as copper glycinate in Zebu and crossbred Zebu steers grazing *Cynodon dactylon* pastures at Nakuru. Copper-treated steers gained 43 kg from August 1964 to March 1965 whereas the non-treated controls gained 8 kg. However, these results could not be replicated in a second trial.

Native and Hereford cattle were fed copper and cobalt supplements for 165 days at Matopos Research Station in Rhodesia (Ministry of Agriculture: Rhodesia 1967). However, no significant effects of this supplement could be demonstrated insofar as effectiveness in preventing weight loss was concerned.

Conflicting results were obtained by Butterworth and his colleagues in Swaziland (FAO 1978b). In a trial carried out in the lowveld at Lavumisa, copper (as copper glycinate) and cobalt and resorcylic acid lactone were administered to male Nguni animals weighing between 200 and 465 kg. Average daily gains over a period of 146 days (calculated by regression analysis using monthly weights) showed that animals supplemented with both copper and resorcylic acid lactone gained significantly more than any other group. All groups administered with resorcylic acid lactone were significantly better than those to which this compound was not administered. Amongst the latter animals, copper and copper plus cobalt groups ranked higher than control animals and those given cobalt alone, but differences did not reach significance. In another trial carried out in the sandy lowveld at Impala, comparing copper and cobalt without the adminstration of resorcylic acid lactone, no significant difference in ADG could be demonstrated.

In order to clarify these results, a further trial was carried out to compare copper and resorcylic acid lactone administration (Ffoulkes 1977). No significant effects due to either of the two treatments could be demonstrated. However, in the last experiment, a drought occurred which reduced the overall rate of gain considerably and it is likely that copper was not the limiting nutrient in the herbage. Roeder (1980) has demonstrated positive effects of copper supplementation to growing animals in Ethiopia.

Summary

Copper deficiencies, as diagnosed by forage, blood, liver and hair contents of this element, have been reported from many parts of the world and it is likely that these are very widespread.

Nevertheless, responses to supplementation have been equivocal and positive effects have not always been associated with apparent deficiencies as diagnosed above.

Some of the contradictions are clearly associated with the effectiveness of different supplements (e.g. copper sulphate v. copper glycinate) but this does not fully explain the anomalies.

Similarly, the interaction between cobalt and copper supplementation on liver storage of the latter element have not been fully clarified.

Cobalt

The role of cobalt in the ruminant is in the formation of vitamin B12 and its derivatives by the micro-organisms of the rumen. Mills (1969) has summarized the roles of this vitamin in transmethylation, erythropoesis and purine biosynthesis as well as in tissue propionate metabolism which is of particular importance for the ruminant animal.

The symptoms and effects of deficiencies have been listed by Underwood (1966); these include loss of appetite and body weight, rapid muscular wasting (marasmus), depraved appetite and severe anaemia culminating in death. However, the same author points out that subclinical deficiencies of cobalt may also be of considerable economic importance.

Deficiencies of cobalt may be diagnosed by the levels of this element or of vitamin B_{12} levels in the liver, and Underwood (1966) has reviewed the levels which may be associated with severe, moderate or mild deficiency. However, this author has also (loc. cit) pointed out that the diagnostic value of the latter measurement may be limited by other conditions that reduce appetite as these tend to increase liver vitamin B_{12} level.

The dietary requirement for cobalt is lower than for the other trace elements, the ARC (1965) suggesting 0.1 mg/kg feed dry matter.

Occurrence of deficiencies

McDowell (1976) has stated that with the exception of phosphorus and possibly of copper, cobalt is the mineral most likely to be deficient to the grazing animal in developing countries and lists the following countries in this context: Argentina, Brazil, Colombia, Costa Rica, Egypt, El Salvador, Guyana, Haiti, India, Katanga, Kenya, Mexico, Nicaragua, North Africa, Peru, Surinam, Uganda, Union of South Africa (sic.) and Zaire.

Australia

The original diagnosis of cobalt deficiencies were made in Western Australia by Underwood and Filmer (1935). Since that time, elsewhere in Australia, Bryan et al. (1960) suggested that in view of the differences in cobalt status among plant species it might be possible for animals grazing certain pasture types in the Queensland Wallum district to become deficient. In the northern Cape York Peninsula, also in Queensland, Winter et al. (1977c) reported that animals became emaciated and listless after about fifteen months of grazing improved pastures and they showed dramatic improvement after being dosed with an anthelmentic and treated with supplementary cobalt.

Malaysia

Devendra et al. (1973) reported that Friesian cross-bred and Kedah-Kelantan cattle failed to thrive. The animals were listless, had rough coats and lost weight, leading to severe emaciation, often resulting in death.

After it was shown that infectious disease was not involved, evidence was presented by 't Mannetje and his colleagues to show that the disorder was due to cobalt deficiency ('t Mannetje et al. 1976) (see below).

The Americas

Cobalt deficiency was implicated in the condition 'salt sick' in Florida as early as 1937 (Neal and Ahmann, cited by Becker et al. 1965). This was confirmed by Becker and his colleagues who showed that cobalt supplementation increased appetite and blood haemoglobin of a twelve-month-old grade Freisian heifer and that some pasture samples contained undetectable quantities of this element (Becker et al. 1965).

De Alba and Davis (1957) reviewed other cases of cobalt deficiency reported from American continents. In Brazil, Machado Magalhaes (1949, cited by de Alba and Davis 1957) described a disease known amongst other things as 'Chorona' ('Weeper') or 'Pela rabo' ('bald tail') which was more frequent during the rainy season and which was cured by moving the animals to another pasture for as short a time as fifteen days. De Alba and Davis (1957) commented that this was undoubtedly a cobalt deficiency and that it was probably complicated by the toxicity of some plant which might account for the lachrymation and loss of hair from the tail as these are not normally associated with the former.

Davis observed symptoms definitely associated with cobalt deficiency in animals in the Rio Paraiba and Aracatuba basins in the state of Sao Paulo and stated that these caused significant losses in animal production. The symptoms were confirmed by Correa (1956, cited by de Alba and Davis 1957). The existence of cobalt deficiencies in Brazil was confirmed by the studies of Sutmöller et al. (1966) who concluded that, in the Amazon Valley, they could be expected in animals grazing pastures on ground water laterite soils, yellow latosols and humic gley soils, provided that the cattle have only limited access to pastures which are liable to flooding by silt-containing river water. Deficiencies in this area have also been recently reported by Camargo et al. (1979).

Elsewhere in South America, Dirven (1963) reported cobalt deficiencies as a result of analysis of indigenous grasses in Surinam.

Africa

In Africa, the first recognition of cobalt deficiency appears to be by Hudson (1944) who referred to 'Nakuruitis' ('a clumsy name') which manifested itself in the Nakuru area after the rains when grazing was plentiful causing cows to become 'a bag of bones'. Hudson (1944) mentioned that this condition had been recognized by workers from the Rowett Institute as early as 1926 when it had been confused with a deficiency of iron. I have not been able to trace this report but it is probable that affected animals responded to small quantities of cobalt impurity in the iron supplement given as has happened elsewhere.

Howard (1969, cited by Tilman 1970) confirmed the importance of cobalt deficiencies in Kenya and pointed out that large areas in the country produce forages which are deficient in the element. The main incidence of the disease according to this author appeared to be along the Rift Valley from Mount Longenot to Nakuru.

Naik (1965) reported that cobalt deficiencies occured in the Mbulu district of Tanzania.

Also, in East Africa, Long et al. (1970), working in the Buganda and Busoga districts of Uganda bordering Lake Victoria, showed that cobalt deficiencies were widespread.

Finally, as a result of his extensive investigations of the mineral content of liver, Boyazoglu (1976) identified cobalt deficiencies in the Republic of South Africa but these were limited to that area of the Cape Coast between Cape Town and East London seaward of the mountain ranges, which strictly speaking is out of the area covered in the present work, being at approximately 32 °S.

Supplementation with cobalt

Although various cobalt supplements have been used with varying degrees of success, it would appear that the most popular is the cobalt bullet, a mixture of cobalt oxide and ferruginous clay. These are extremely dense and after oral administration lodge in rumen or reticulum where they may remain for many years and constantly release minute quantities of cobalt oxide which is sufficient to provide the requirements of the rumen micro-organisms for the synthesis of vitamin B_{12}. They also satisfy the important need for a continuous supply of the element. The efficiency of this type of administration may be increased by administering the bullet together with a steel screw or 'grinder' which removes the impermeable coating of calcium phosphate which tends to form round the bullet thus reducing its effectiveness. Full details of these techniques have been given by Underwood (1966).

Various workers have investigated cobalt supplementation in countries which fall within the scope of this review. Many of these studies have been unable to demonstrate a positive response in weight gains of cattle; this may have been associated with the difficulties of determining the element and the imprecise nature of the diagnosis which finally depends on a positive response to a supplement in order to establish a deficiency state (see, e.g. Sutmöller et al. 1966).

Negative trials

In Australia, Donaldson et al. (1964) were unable to demonstrate any positive effect when Shorthorn heifers were dosed with 5 g 90% cobalt oxide every six months near Townsville in northern coastal Queensland. This may have been because this type of dosage did not give the constant provision of the element already mentioned as necessary.

In central coastal Queensland on the Broadside Peninsula, Alexander et al. (1967) carried out four experiments to investigate the effects of copper and/or cobalt on various classes of Hereford cattle grazing predominantly Para grass (*Brachiaria mutica*) pastures. Cobaltic oxide was administered in the form of pellets but only in one experiment was there a significant response by castrated male weaners during the dry season.

Similar results were obtained in Southern Queensland, at the Coolum Research Station, by Gartner et al. (1968b). Hereford steers grazing Pangola grass (*Digitaria decumbens*) and *Lotononis bainesii* pastures were dosed every three months with pellets containing 30 g of 60% cobalt oxide. No significant effects of the supplementary cobalt could be demonstrated but it was found that this treatment significantly depressed liver copper content.

The absence of positive response to supplementation with cobalt has also been noted in various governmental annual reports. In Rhodesia, no effects could be demonstrated by the addition of both copper and cobalt (neither element was fed alone) to either Hereford or native cattle during the winter period at Matopos Research Station (Ministry of Agriculture: Rhodesia 1967). In Swaziland, copper and cobalt were compared both severally and together in combination with resorcylic acid lactone for growing steers grazing sweet lowveld under both wet and dry season conditions but no positive effect could be attributed to the use of a cobalt bullet (FAO 1978b). At the St Ignatius Livestock Station in British Guyana, there was a slight indication of beneficial effects but these 'were not really significant and certainly hardly economic'. There was some 10 kg difference between treated and untreated animals over a twelve-month period (British Guiana 1962).

Positive trials

It was Howard (1970) who was the first to demonstrate really positive effects of the administration of cobalt pellets to cattle in the old 'Nakuruitis' area of Kenya. Two trials were carried out on a farm eight miles south-east of Nakuru. The altitude was above 1800 m and the pastures grazed were predominantly Star grass (*Cynodon dactylon*). The first trial was carried out using typically unthrifty Zebu and Zebu-cross cattle during a seven-month period from August 1964 to March 1965 during which adequate precipitation occurred and growth of pasture was normal. The response to cobalt therapy was dramatic as treated animals gained approximately 90 kg more than the controls; only a moderate improvement resulted from injection with copper glycinate whilst treatment with both elements resulted in a growth rate slightly inferior to that obtained with cobalt alone. Blood haemoglobin levels were higher in cobalt-treated animals.

The second trial was carried out using heifers and was similar to the first insofar as treatments were concerned. Responses to the treatment with cobalt were not so dramatic in this trial and only reached significance during the rainy period from May to December 1966. During the previous period, June 1965 to April 1966, semi-drought conditions prevailed and it was concluded that requirements for other nutrients were not being met. Cobalt significantly increased blood haemoglobin levels during the second period but not during the first.

Dramatic results also were obtained by 't Mannetje and his colleagues working in Johore, Malaysia ('t Mannetje et al. 1976). Kedah-Kelantan bulls grazing *Panicum maximum* (77%), *Pueraria phaseoloides* (16%) and *Stylosanthes guianensis* (5%) showed a highly significant negative relation between liveweight changes and time spent on pasture. This, combined with previous observations that animals grazed on these pastures for a prolonged time failed to thrive, were listless and had rough coats and severe emaciation, suggested the possibility of cobalt deficiency. Administration of cobalt bullets, with grinders, permitted satisfactory liveweight gains (more than 100 kg from April 1975 to January 1976) whilst control animals lost more than 20 kg during the same period (weight changes taken from graph). Copper and terramycin treatments were ineffective in preventing weight loss but injection of vitamin B_{12} (1000 μg intramuscularly) was effective until October 1975 after which weight losses occurred. In a subsequent experiment carried out with animals in good condition, response to treatments were not so consistent as in the first group although animals which had received cobalt appeared to gain at a slightly higher rate than the others. Chemical analysis of the herbage on offer confirmed that levels were around 0.01 ppm.

Finally, dramatic results have been reported (Winter et al. 1977) from the northern Cape York Peninsula in Queensland although it will be remembered that other experiments reported from other parts of that state could not demonstrate positive effects. Cobalt deficiency was suspected after several animals became emaciated and listless after about fifteen months of grazing improved pastures (*Panicum maximum–Stylosanthes guianensis*). During the period from January 1974 to February 1975, animals supplemented with cobalt alone gained 142 kg whilst those given both copper and cobalt gained 121 kg; control animals and those supplemented with copper alone, on the other hand, lost 21 and 2 kg respectively during the same period. Cobalt treatment also increased serum vitamin B_{12} levels three to four-fold and liver levels of the same vitamin ten to twenty-fold. McDowell et al. (1982) also obtained positive results when growing cattle were supplemented with cobalt in Bolivia.

Summary

Cobalt levels in herbage which would appear to be sufficiently low to provoke deficiency symptoms have been reported from many tropical countries both on natural and improved pastures. However, it would appear that biologically and statistically significant responses to supplementation have only been reported from areas where clinical symptoms of cobalt deficiency have been present. Although sub-clinical deficiencies of this element may be of considerable economic importance, as has been suggested by Underwood (1966), this has yet to be demonstrated.

Zinc

Zinc forms a part of various enzyme systems, notable carbonic anhydrase but also many of the dehydrogenases, alkaline phosphatase and carboxypeptidase (Vallee 1962). It is thus essential for both digestion and cellular metabolism. The biochemical functions of zinc in animals have recently been reviewed by Chester (1978).

Although there appear to be considerable areas where soil and pasture levels of zinc would appear to be insuficient to support normal growth of cattle, only one case of clinical symptoms of a deficiency of this element has been described in the literature. Legg and Sears (1960) described a condition of parakeratosis of cattle grazing the intermediate savannas (containing 18 to 42 ppm) of the Ebini Livestock Station in Guyana which could be cured by the administration of zinc. Otherwise, McDowell (1976) has reported that deficiencies could be expected in Brazil, India, Panama, Puerto Rico and Uganda whilst Boyazoglu (1976) has suggested that zinc deficiencies may be expected throughout the Republic of South Africa, South West Africa and Swaziland (Boyazoglu 1970).

Deficiencies of this element have been reported from the Guanacasta and San Carlos areas of Costa Rica (McDowell et al. 1978) and probable deficiencies were found in the northern Mato Grosso of Brazil (Conrad et al. 1978). In both cases, levels of the element in soil, plant and liver were taken as criteria. Liver levels of zinc were also found to indicate a probable deficiency in both Kenya and Tanzania (Mkwakatundu, personal communication). It would appear therefore that zinc deficiencies are more widespread than was originally supposed.

The ARC (1965) has suggested that diets for cattle should contain 50 ppm to avoid deficiencies and it would seem logical therefore to add appropriate quantities of zinc salts to the mineral supplements in areas where it is suspected that sub-clinical deficiencies of this element may exist.

Iron

Iron has been implicated in 'salt sick' or nutritional anaemia in cattle since 1901, when Stockbridge (cited by Becker et al. 1965) treated animals with iron sulphate and noted that they recovered. Becker and his colleagues reviewed experiments initiated in 1929 which apparently support this contention. Unfortunately, the number of animals involved was small and conditions not always as controlled as might be desirable. For example, one Jersey heifer 'improved with the iron supplement, but she broke into a sweet corn field and died of bloat'. On the way to one farm, 'a pedestrian told of using Dawson's non-iron tonic with a salt-sick animal . . . his animal died of salt sick'.

A deficiency of iron was also implicated in 'Nakuruitis' by a team from the Rowett Research Institute in 1926 (Hudson 1944) as the disease has some symptoms in common with nutritional anaemia, and in 1932, Husband observed a condition similar to 'salt sickness' in Marandellas, Rhodesia. This latter author (Husband 1932) observed a marked response to iron fed with salt and with salt alone (27 and 18 kg respectively) but, as in the Florida work, insufficient animals were used, there being only one cow to each treatment. Later work was carried out in Rhodesia by Murray (1938) who could obtain no response to iron supplementation in young stock at Matopos and later at Marandellas by Rhodes (1956) who obtained similar results.

Despite reports by Sen and Ray (1964) of the occurrence of iron deficiencies in India and by Chicco (1972) in Panama, it would appear that the statement made by Underwood (1962) generally retains its validity: 'Uncomplicated iron deficiency in grazing stock has never been clearly demonstrated under naturally occurring conditions. Earlier claims for such occurrences were invalidated by the finding of cobalt deficiencies in the affected areas.' The latter comment refers to the frequent presence of cobalt in the iron compounds successfully used in the treatment of the conditions observed.

It would appear, however, from conversation with the Florida school, that Becker's work did in fact demonstrate the efficacy of iron therapy rather than both iron and cobalt as the salts used were free of the latter element, although this is not stated in the original paper (Conrad, personal communication).

Iodine

The nutritional role of iodine is as a component of the hormone thyroxine which is responsible for the control of energy exchange and metabolic rate. For this reason, this element has a particular interest in areas with a high ambiental temperature where the animal is faced with problems of temperature control.

It is likely that iodine is deficient in all areas where human goitre is encountered and these occur in all the continents of the world and in almost every country (Allman and Hamilton 1949). The requirements of the element are related to energy expenditure if it is assumed that thyroxine is broken down proportionally to metabolism, as has been suggested by Levine et al. (1933); this might indicate that the requirements of the element might be lower in hot climates than the 20–40 mg/100 cal energy intake suggested by these authors.

Very little, if any, work seems to have been carried out on iodine deficiencies or supplementation of cattle *per se* in tropical or subtropical areas, and little remains to add except that the use of iodized salt which is generally to be recommended (see previous remarks on sodium and chlorine) would be expected to reduce the incidence of deficiencies except where common salt is exposed to the light for too long a period when a significant loss of iodine is to be expected.

For further details on the occurrence and utilization, etc. of iodine, the reader should refer to the reviews of Underwood (1962) or Mills (1969).

Manganese

Manganese is essential for growth and reproduction and acts as a cofactor for enzymes important in the biosynthesis of glycoproteins. The metabolism and functions of the element have recently been reviewed by Leach and Lilburn (1978).

Manganese deficiency has been produced artificially and is characterized by skeletal abnormalities and reproductive problems. However, there is little evidence that deficiencies occur in beef cattle in the tropics and subtropics.

Bisschop and Groenewald (1963) have reported low manganese levels in natural pastures and clinical deficiency symptoms in young cattle in South Africa. McDowell et al. (1974) reported that 5.2% of 293 samples where manganese levels were given contained less than 20 ppm, and Chicco (1972) reported low manganese concentrations for seven out of twenty-eight locations in Panama. The role of manganese in ruminant nutrition has recently been reviewed by McDowell et al. (1979). Requirements for the element are between 1 and 20 ppm (NRC 1976) and it may be supplied as manganous carbonate, chloride, oxide or sulphate.

Selenium

Selenium has long been known for its toxicity and Rosenfeld and Beath (1964) suggest that Marco Polo may have encountered this

U.C.W. ABERYSTWYTH LIBRARY

when travelling in China in 1295; if this is the case, it is one of the earliest recorded observations on trace minerals. In any event, selenosis has been widely reported and has been reviewed elsewhere (see, e.g. Ammerman and Miller 1975).

The recognition of selenium as an essential trace element is more recent and dates from 1957 when it was shown that it prevented exudative diathesis in chicks (Patterson, Milstrey and Stokstad 1957). Deficiency of selenium and/or vitamin E gives rise to muscular dystrophy (white muscle disease) and selenium is present in the enzyme tissue glutathione peroxidase. Although Underwood (1971) has stated that the total areas of the world affected by selenium deficiency are far greater than those affected by its toxicity, the reports of deficiency in the literature are few. In a review of selenium in ruminant nutrition, Ammerman et al. (1979) have cited evidence of its deficiency in Brazil, Honduras, Ecuador, Mexico, Paraguay, Peru and the Bahamas. Requirements for the element are between 0.05 and 0.10 ppm (NRC 1976) and deficiencies may be corrected by supplementation with sodium selenite which contains 45% selenium. Caution must be employed in its use because of possible toxicity. McDowell et al. (1982) obtained positive results when growing cattle were supplemented with selenium in Bolivia.

6. Energy and protein deficiences: I. The use of protein and energy supplements with growing stock

The protein deficiences found in tropical grasses, particularly when they are mature during the dry or winter season, have already been detailed. The requirements for protein of growing cattle appear to be about 6% crude protein on a dry matter basis (see e.g. Langlands and Bowles 1976) and it is logical therefore to supply protein to the grazing animal in an attempt to correct this deficiency.

All sources of protein (as opposed to non-protein nitrogen which will be dealt with in the following chapter) provide the body with energy, particularly so in situations where glucose availability limits metabolism. It is difficult, therefore, to disassociate entirely protein from energy and, indeed, it was thought by some early workers that it was energy rather than protein that was the most important constraint to beef cattle performance during the dry season. In any event, as has already been discussed, protein deficiency may also have a deleterious effect on both energy consumption and on the digestibility of nutrients. In the present discussion, deficiencies of protein *per se* will generally be referred to. However, several studies involving energy supplementation of intensively managed, cultivated or summer pastures have also been included.

Southern Africa

Zimbabwe

Although work on supplementation with protein lagged behind the investigations of mineral deficiencies and disease factors started by Sir Arnold Theiler and his school, the first trials in the field were also carried out in Southern Africa. Murray et al. (1936) in Rhodesia followed up the suggestion of du Toit and his colleagues (du Toit et al. 1932) that a possible protein deficiency might exist. Cattle kept at Rhodes Matopos Estate were given 454 to 680 g of groundnut cake (containing 45% crude protein) each per day. The sandveld on which the Lincoln Red – Shorthorn steers grazed was said ‘to be of extremely poor quality during the winter months' and this comment was borne out by the values for the crude protein content of the herbage which were 3.7, 3.1, 3.3, 2.9, 3.1 and 1.9% for the months of May to October respectively. The trial was carried out for two successive years and winter weight changes were −47 kg and approx. 2 kg, and −99 kg and −43 kg for unfed and fed animals respectively. Some compensatory growth was shown during the first summer, gains being 69 and 59 kg unfed and fed animals respectively, but this was not the case during the second summer when gains were 100 and 104 kg for the same two groups of cattle. In any event, the authors describe the difference between the two groups as 'simply astounding' and support this by the statement that the unfed steers reached finishing weight, 456 kg, 13.5 months after the supplemented animals and 'they by far did not have the quality, good beef conformation and finish' of the other animals. Murray et al. (1936) also suggested that 'the main deficiency in the earlier part of the winter is protein but later in the winter there is a deficiency of energy as well as protein'. This work at Matopos was followed by several other studies of a similar nature.

Murray and Romyn (1937a) carried out an experiment to determine the effect of supplementation during the first winter on subsequent growth, compared with additional supplementation during second and third winters. The final weights of the control animals and those supplemented during the first only were 433 and 455 kg respectively, which were not significantly different. Animals fed throughout, however, weighed 503 kg at the end of the experiment and this was significantly greater (level not stated) than the gains of the other groups, as were heights at the withers.

Murray and Romyn (1937b) also carried out a study to compare different winter supplements for young growing stock and to compare these from the point of view of subsequent development and costs of production. The same level of groundnut cake as used previously was compared with a daily supplement of 2–3 kg of cowpea hay, and a further group were given groundnut cake with the addition of from 0.5 to 1.5 kg of maize meal plus 1 to 2.5 kg of silage. Final weights for the three groups in the order given were 513, 482 and 511 kg and it was concluded that cowpea hay was essentially as effective as groundnut meal and that there was no advantage to the extra supplementation.

Further work was carried out at the Grassland Research Station at Marandellas in Mashonaland, Rhodesia. Studies on supplementation of winter sandveld grazing were started by Rhodes in 1948 (Rhodes 1956). Sussex weaner steers were given either 680 g daily of groundnut meal or maize meal and compared with unsupplemented animals over a three-year period. The groundnut meal reduced winter loss in all three years whereas the supplementation with maize meal was not effective. Losses were approximately 38, 26 and 39 kg for the three years for control, groundnut meal and maize meal fed animals respectively. The author commented that protein appeared to have been the only nutrient to which a response was noted during winter but also added that this difference did not reach significance.

Continuing work (Rhodes 1956) confirmed that winter loss was reduced by the same quantity of groundnut meal as in the previous trial, values being 3 and 46 kg for unsupplemented and supplemented animals respectively for the two years of the trial. It should be mentioned, however, that even despite supplementation, weight losses were quite substantial during several years of the two trials and that this could presumably be attributed to deficiencies of other nutrients including absolute lack of forage. It is also interesting to note that the supplemented animals gained more during the summer than the unsupplemented animals in this latter trial, a finding contrasting with that of the first trial where compensatory gains occurred.

Rhodes (1956) carried out a third trial to investigate the combined effects of protein and salt; protein did not check the winter weight loss in this case and the author concluded that the level used was possibly not sufficient (although this was the same as that used in the previous two trials).

It was also noted that the killing-out percentage of supplemented animals was slightly (although not significantly) lower than that of non-supplemented cattle (51 v. 52%). If this is a consistent effect, it would tend to reduce any advantage to supplementation measured by weight change only (this will be referred to later).

The author concluded that differences in responses among the three trials and from year to year within trials ‘could include a consideration of wide seasonal differences in the natural nutritional level’.

More recent work carried out at Marandellas (Grant 1975) sheds more light on the effects of compensatory growth after winter supplementation with protein and various sources of NPN. Crossbred Hereford steers fed 500 g cottonseed meal daily from the beginning of June until the end of October gained a little weight during this period (3 kg) vis-à-vis a loss of 24 kg by unsupplemented control animals. However, the subsequent summer gain of the supplemented animals was only 80 kg as against 97 kg of the control animals. Weight differences between supplemented and unsupplemented animals had approximately halved by the end of the summer period. However, when subjected to a further feed-lot period, supplemented animals grew somewhat faster than controls, taking a shorter time to reach a common slaughter weight.

A subsequent experiment (Grant 1976a) substantially confirmed the above findings. However, when the results were interpreted in terms of carcass weight, it was shown (cf. Rhodes 1956) that losses by the supplemented animals were more marked than when only body weight was taken into account. Thus, whilst the body weight change of supplement v. unsupplemented animals was only –4 kg compared with 28 kg, carcass losses were –10 and 27 kg respectively.

'The response in terms of beef production by grazing cattle to protein-rich concentrates is likely to vary with the quality and quantity of the grass available to the animal.' This hypothesis was formulated by Richardson et al. (1976). Since it may be expected that differences in stocking rate might influence both the amount of grass on offer and its quality (by the exercise of selective grazing), these authors started an experiment with different stocking rates (2, 3 and 4.5 ha/weaner steer) and different levels of cottonseed cake supplementation (0, 250, 500 and 750 g/day).

During the first season, body weight changes during winter were positively related to the amount of cottonseed cake on offer, this being −1, 5, 4, 12 and 15 kg between July and October for the four treatments in the order given. It was concluded that increments became smaller at levels above 500 g (although this statement would have been more convincing had a level of 1 kg/day been included). Anomalous results were obtained with regard to stocking rate but these were attributed to carry-over grass growth being available in all paddocks. However, during the second year, more lightly stocked animals gained more than their more heavily stocked partners both during the summer (144 v. 133 kg) and the winter (−17 v. −22 kg) as would be expected. Contrary to the results obtained by Smith (1966) which will be discussed below, no interaction between amount of protein and stocking rate could be demonstrated at any stage (Richardson et al. 1977c). At the same time, the authors conceded that stocking rate *per se* may provide a poor indication of the amount of protein and energy available to an animal and they therefore carried out determinations of non-esterified fatty acids (NEFAs), plasma albumin, plasma urea nitrogen and nitrogen in faecal organic matter. Although only two years' results have been reviewed in the present work, it was tentatively concluded that elevated plasma NEFAs were observed in both high stocking rate animals and those given no protein supplement, indicating that these animals were in energy deficit in comparison with animals on other treatments. Plasma albumin, plasma urea nitrogen and faecal nitrogen were all increased by cottonseed supplementation and albumin was higher in animals at the lower stocking rate.

In a further experiment designed to investigate the effect of the protein content of the diet and the consumption of energy on various metabolites, it was confirmed that both plasma urea and plasma albumin were influenced by the protein content of the diet. It was stated that the amount of energy administered did not influence any of the metabolites studied; however, it was not made clear whether

NEFAs was one of these (Richardson et al. 1977b).

One further winter supplementation trial carried out at Marandellas should be noted here although it was not carried out on veld but in intensively cultivated Sabi panic. Barnes (1966) showed that there was an interaction between level of supplementation and stocking rate, at least during the latter part of the winter. At a stocking rage of 1.6 beasts/ha, weight change was −18 kg when no supplement was fed and +12 kg when a supplement of green oats and maize meal was given. At a stocking rate of 3.2 beasts/ha, however, the corresponding values were −10 and +7 kg, the interaction being significant at the 5% level.

Republic of South Africa

The conditions at Mara in the Northern Transvaal are not entirely dissimilar to those of Rhodesia but early work in that area was limited to creep feeding. Preliminary trials carried out by Harwin and Venter (1970) indicated that creep feeding significantly increased the growth rate of calves from 860 g to 1 kg/day and the authors concluded that this was economically justifiable at the prices ruling at the time. The creep supplement consisted of a mixture of 75% maize meal, 15% sunflower cake, 7% lucerne meal and 3% fishmeal.

A later summary of Mara work by Lüdemann (1971) suggested that creep feeding could be profitable in areas of rapidly deteriorating veld-grazing towards the end of the season and during periods when the grass was low in protein and energy, and the additional carcass value of animals slaughtered at weaning might justify the feed costs. Even if this were not the case, then creep feeding could be justified if the animals finished better during a subsequent supplementation or stall-feeding period.

Results from a trial to determine whether in fact this was so, however, indicated that no advantage would be obtained. Daily weight gain during a feed-lot period was 0.92 kg for animals that had previously been creep-fed v. 0.97 kg for animals which had not. The final weights for the two groups when slaughtered on attaining super-grade condition was 477 v. 468 kg for creep-fed and non-creep-fed cattle respectively, but the increased carcass weight did not justify the cost of the creep feed (Lüdemann 1973).

The question of year-round supplementation was also considered with a view to the maintenance of near maximum growth throughout so as to reach marketable weight before their second winter. This would also produce a 'sparing' effect on the veld enabling more animals to be kept per unit area.

The final conclusions, however, were that supplementation of stock both before and after weaning could not be recommended. This was because inexpensive good-quality veld grazing was substituted by relatively high-priced concentrate without any compensating response. Although gains were observed, both creep feeding and subsequent supplementation had a detrimental effect during subsequent finishing period.

Experiments on winter feeding in the high veld area of South Africa started as early as 1937. Several early experiments by Joubert have been reported by Verbeek and von la Chevallerie (1958).

During the period 1940 to 1942, the following treatments were given to oxen between 12 and 14 months of age:

1. Grazing alone.
2. Grazing plus 680 g 35% crude protein concentrate/day.
3. Grazing plus 2.3 kg maize meal plus 450 g groundnut meal/day.
4. Grazing plus 2.3 kg maize meal plus 1.3 kg bean haulms/day.

The change in weight during the first winter (May to October) was −26, 5, 63 and 54 kg for the four groups respectively.

These results indicate that there was little effect of supplementary protein when compared with a supplement which provided relatively large quantities of energy as well as protein. This could well be because, at Potchefstroom, where these experiments were carried out, the typical grasses are 'sour' and of low palatability and thus intake of dry matter would be limited. Acocks (1975) has described the typical species of this type of veld as *Themeda* and *Cymbopogon*.

Interesting work on both winter and summer supplementation at Potchefstroom was carried out by Pieterse and Preller (1965). Cattle were fed on high, medium and low planes during the winter so that their bodyweight changes would be 31, −6 and −14% respectively. During the following summer, half of each group were given 11 kg each of maize meal weekly for a period of 107 days. Although considerable compensatory growth occurred, as would be expected, this was not sufficient to equalize final slaughter weights. A summary of the results obtained is given in Table 6.1. It was concluded that summer supplementation was justified regardless of previous treatment 'because of improved carcase grade'. It will be noted that effects on killing-out percentage are not consistent with results quoted earlier.

In an essentially similar experiment conducted over two successive years, Leach et al. (1968) confirmed the view that supplementation was economically justifiable because of the better condition at slaughter of the supplemented animals.

Several of the studies quoted above have made it clear therefore that judgement of the effectiveness of supplementation must not be based solely on weight gain over an isolated period of an animal's life but rather on life-long performance combined with an evaluation of the economic conditions pertaining at slaughter.

The equivocal nature of the results to be expected from supplementation during the summer is emphasized by the work of Reyneke (1976) also from Potchefstroom.

Table 6.1 Effects of summer supplementation of cattle subjected to different levels of nutrition during the previous winter (Potchefstroom) (adapted from Pieterse and Preller 1965)

	*High**		*Medium**		*Low**	
	With supp.	*Without supp.*	*With supp.*	*Without supp.*	*With supp.*	*Without supp.*
Gain/day (kg)	0.73	0.47	1.04	1.00	1.10	1.30
Final weight (kg)	424	416	373	363	357	332
KO%	46	48	49	49	50	51
Grade	Prime 1	1	1	2+	1	3+

* Explained in the text

Reyneke designed an experiment to study the growth response of steers supplemented 'strategically' over the summer grazing period only. Strategic supplementation was carried out with each of the supplements whenever daily weight gain dropped below 700 g/day. Cattle were Friesian steers of comparable age (about eighteen months) and conformation. Treatments were as follows:

1. Grazing only; stocking rate of 0.6 beasts/ha (no details are given of the location or type of veld grazed but it has been described by Acocks (1975) as being a highland sourveld to *Cymbopogon–Themeda* veld transition and thus similar to the type predominating at Potchefstroom itself).
2. Grazing as above plus 1.36 kg/steer/day of maize meal providing TDN 1115 g and CP 122.4 g; depending on the response to this supplement, additional maize meal was given to a maximum of 2.3 kg/head but this is not entirely clear.
3. Grazing plus 1.36 kg/steer/day of a concentrate mixture providing TDN: 1123 g and CP 367.2 g. The concentrate contained equal parts of yellow maize meal and groundnut cake meal.
4. Grazing plus 1.36 kg/day of groundnut cake meal alone.

Average daily gains over the three summers of the trial were 538, 598, 645 and 647 g for the four treatments in the order given above. Statistically-improved gains of the supplemented animals over the controls during two of the three years was noted, although final weights were not significantly different. It was suggested that the absence of significant differences in the second year of the trial was attributable to a higher crude protein content of the sward during this year although this is not borne out by the results given.

No significant differences could be demonstrated among groups with respect to carcass characteristics (e.g. cold carcass weight, fat cover over *M. longissimus thoracis*, etc.).

It was clear from this experiment that no conclusive advantage could be expected during summer supplementation from either energy, protein or a combination of the two.

East and Central Africa

In comparison with the large body of work that has been carried out in the Republic of South Africa and Zimbabwe, comparatively little information is available for the rest of Africa.

Zambia

Earlier work was carried out in Mazabuka in Zambia (then Northern Rhodesia) on winter supplementation of steers grazing *Hyparrhenia* veld. Smith (1961) compared different systems of management, and the effect of these on weight gains with and without supplementation with 900 g/day of groundnut meal. Management treatments involved cutting as hay and feeding during the winter of the whole or half the area or leaving all the forage as standing hay. Although there were no statistical differences among management treatments, supplementation consistently reversed weight losses, overall means being −700 and 280 g/day for unsupplemented and supplemented animals respectively. The feeding of protein resulted in a 140% increase in hay intake. This was later confirmed in stall-feeding trials (Smith 1962a) in which protein-supplemented animals consumed 8 kg of hay/day v. 6.3 kg in the case of the non-supplemented animals, corresponding to approximately 130%.

In later work, Smith (1966) compared different stocking rates with and without supplementation. The author confirmed that a protein supplement prevented weight loss but commented that further work was necessary 'on the interaction between supplementary feeding and stocking rate as this differed in different seasons for no apparent reason'.

The results of several years' trials were combined by Smith and Hodnett (1962) to calculate a regression of weight gain during the rainy season on gains during the preceding dry season for various age groups of cattle. The average value of eliminating both years and age groups amounted to some −280 g; this figure indicates the extent to which the gain during the rains could be expected to be increased in compensation for each 450 g lost as a result of a lower plane of nutrition during the dry season. The authors suggest that this relationship 'provides a basis for an economic assessment of alternative systems of management'. Thus the costs of dry season feeding can be evaluated against the final mature live weights of the cattle.

A further trial was carried out at Mazabuka on winter supplementation with both energy and protein (Anon 1968). Supplements consisted of ground snapped corn, ground snapped corn plus cottonseed (795 and 114 kg respectively), and maize bran plus cottonseed in the same proportions. They were fed during the period July–November and all supplements were available *ad libitum*. Weight changes during this period were −5, 30, 61 and 46 kg for control cattle and supplemented cattle in the same order as given above. The author commented that poor grazing conditions prevailed but this is hardly justified in view of the fact that the control animals only lost some 5 kg during the period under consideration and the fact that the daily *per capita* consumption of the three supplements was 4.4, 4.6 and 4.2 kg respectively in the order given above. Thus the trial was rather one of feeding than supplementation *per se*.

Tanzania

Owen (1968) noted that weaner steers grazing natural or improved pastures on the Kongwa plain in central Tanzania lost weight prior to the onset of the rains. A trial was carried out to investigate whether it was practicable to arrest this loss by feeding either hay alone as a supplement to newly-established *Cenchrus ciliaris* pastures or hay with 450 g daily of a high-protein concentrate (a distinction is drawn between old and newly-established pastures but there was no control group of animals grazing newly-established pastures with neither hay nor supplement; however, hay alone fed to animals on the old established pastures was ineffective in increasing weight gain). Stocking rate was approximately 0.4 beast/ha.

Steers fed new pasture plus hay gained 23 kg during the first period from November 1960 to February 1961 whilst those supplemented with hay plus protein gained 32 kg. However, during the second trial period from February to June, when fresh grass was available, considerable compensatory growth occurred and the advantage gained during the first period was completely lost, overall gains being 93 and 92 kg for hay-fed and supplemented groups respectively.

Walker (1969a), working at Ukiriguru in the Mwanza province of Tanzania, noted that the weight changes of cattle followed a cyclical pattern, consisting of 30–35 weeks during which the animals gained weight (most favourable compared with many parts of the tropics and subtropics; see, e.g. Norman and Arndt 1959), some 12–16 weeks of weight maintenance and 4–6 weeks during which the cattle lose weight. This weight loss occurs during the first part of the dry season. The trial was concerned with an attempt to combat this loss by feeding relatively poor-quality hay (mean CP content 5.7%) during this period. It was concluded that this practice was not effective in increasing body weights, because although the extent of the weight loss was decreased in the animals fed hay, these were counteracted by compensatory gains once sufficient natural grazing was available. The author attributes the loss of weight to the fact that the dry matter content of the grazing was too low to allow acceptable intake to be attained.

Malawi

In Malawi, Addy and Thomas (1977) fattened steers on fertilized *Chloris gayana* pastures during the summer on the Lilongwe plain in the Central Region.

Different stocking rates were compared as was the use of a supplement composed of equal parts of madeya (maize bran; the residue of village production of maize flour) and maize at rates of 0.6 and 1.5% liveweight per day (there was not treatment without supplement); no effect of an increased level of concentrate on growth rate could be demonstrated.

Thomas and Addy (1977) used leucaena meal (*Leucaena leucocephala*) to replace cottonseed cake in supplements for Malawi Zebu yearling steers grazing Rhodes grass standing hay. Cottonseed cake was fed at the rate of 0.34 kg/day and leucaena was fed so as to provide the same amount of nitrogen. The liveweight changes during the winter period are shown in Table 6.2.

Table 6.2 Liveweight changes (kg) of steers grazing Rhodes grass with and without supplements (Chitedze) (adapted from Thomas and Addy 1977)

Rhodes grass No supplement	*Rhodes grass Plus leucaena*	*Rhodes grass Plus cottonseed cake*
4.5	19.1	20.9

Sudan

In the Western Sudan, Mukhtar (1967) supplemented groups of nine yearling Sudanese cattle with 2 kg of grain sorghum and 1 kg cottonseed meal/day. Whilst it was clearly demonstrated that the feeding of concentrates had a beneficial effect on the growth of the cattle, no compensatory growth was noted and there was a positive carry-over effect of supplementation which lasted until the following dry season.

West Africa

Upper Volta

Cabaret (1975) carried out trials fattening Zebu steers and bulls in Saria in Upper Volta. A comparison was made between animals grazing natural pasture (no details of botanical composition or stocking rate are given) during both dry and wet seasons, with animals supplemented during both wet and dry seasons. The supplement given was cottonseed *ad libitum* and 250 g/day of groundnut meal. Results are summarized in Table 6.3.

Although very considerable compensatory gains were made by the non-supplemented animals during the wet season, these were not sufficient to overcome the advantages of the animals fed supplement during either the dry season alone or during both wet and dry seasons. These advantages were also accompanied by slightly improved killing-out percentages. As no details are given of stocking rate, it would appear that the considerable quantities of supplement given must have been eaten at the expense of pasture consumption.

Table 6.3 Daily gains and consumption of concentrate by steers during wet and dry seasons (Saria) (adapted from Cabaret 1975)

	No supplement	*Supplement dry season only*	*Supplement wet and dry season*
Starting weight (kg)	209	206	206
Daily gain; dry season (g/day)	69	456	456
Daily gain; wet season (g/day)	859	472	673
Daily gain; overall	385	462	543
Consumption of cottonseed (kg/day): dry	—	2.0	2.0
Consumption of cottonseed (kg/day): wet	—	—	1.5
Killing-out percentage	56.6	57.6	60.3

Senegal

Calvet et al. (1976) noted that the end of the dry season and the first three weeks of recovery of grazing after the first rains were the critical periods of weight loss. They supplemented animals at Labgar in the 'ancient desert' of Ferlo in Senegal with various protein and mineral supplements during this period. Apparently the most effective supplements in the reduction of weight loss were of groundnut cake (300 g/AU (Animal unit) day). However, the authors themselves question the validity of the results because of the 'Le Caractère Indiscipline' of the stock raisers, the limited number of replicates of treatment herds and differences due to herding practices. Any assessment by the reader is complicated by the fact that only results of statistical analyses are given in many cases.

Australia

Supplementation of tropical pastures with energy or protein was taken up later in Australia than in Southern Africa, but trials have been carried out over a wide area comprising Queensland, Northern Territory, part of Western Australia and New South Wales.

Queensland

Sutherland (1959) supplemented steers in south and central Queensland with grain sorghum, 1 kg lucerne hay/day or a 20% crude protein lick based on molasses and meat meal. The sorghum had no effect but the lucerne hay increased gains to 33 kg over the control (unsupplemented animals) and the lick 22 kg. Cattle grazed unimproved pastures.

Alexander et al. (1970) carried out eight trials in south Queensland involving both weaner and yearling cattle. Supplements were either of grain sorghum or cottonseed meal fed during the dry season. Treatments are summarized in Table 6.4.

Table 6.4 Supplements used in experiments in south Queensland (adapted from Alexander et al. 1970)

Experiment	*Supplement (kg/day)*	*Class of stock*	*Days fed*
1	0.45 cracked grain sorghum	Calves	95
2	0.45 cracked grain sorghum	Calves	228
3	0.45 cottonseed meal	Yearlings	161
4	1.25 cottonseed meal	Weaners	201
5	1.13 cottonseed meal	Weaners	172
6	0.45 cottonseed meal plus 0.45 cracked sorghum	Weaners	112
7	1.36 cracked sorghum	Weaners	113
8	0.45 cottonseed meal	Weaners	188

The first five experiments were carried out in the Eidsvold district on pastures principally composed of *Heteropogon contortus*, *Dichanthium* spp. and *Themeda australis*. Experiment 6 was carried out in the Springsure district with pastures composed principally of *Dichanthium* spp., *Panicum* spp., *Aristida* spp. and *Heteropogon contortus*, whilst in the pastures used in experiments 7 and 8, the predominant grasses were *Astrebla squarrosa*, *Chloris* spp., *Bothriochloa decipiens* and *Dichanthium* spp. Stocking rates were 4.9, 7.3 and 3.2 ha/beast for the three pasture types respectively.

The supplement in experiment 1 had a significant effect on growth rate of calves (i.e. when creep-fed) whereas that in Experiment 2 did not. However, in the first trial, the control group lost weight whilst in the second they gained 0.5 kg/day. There was no difference between groups when supplementation was carried on after weaning.

In experiment 7, the only other trial where sorghum was used, results were also positive although the animals used were weaners.

The cottonseed meal-fed animals in experiment 3 gained significantly more than the controls. In all the other experiments, the cottonseed meal-fed animals gained significantly more than the unfed control animals. A summary of results is given in Table 6.5.

In all cases where cottonseed meal was fed to weaners or yearlings, supplementation caused significantly higher gains in weight than in control animals. In several cases, weighings were carried on after supplementation had finished; only in the case of experiment 6 was

Table 6.5 Weight changes of cattle during feeding period (adapted from Alexander et al. 1970)

Experiment	*Treatment*	*Gain to weaning* (kg/head/day)	*Gain after weaning* (kg/head/day)	*Significance of difference*
1	Control	−0.05	0.51	H
	Supplement	0.09	0.47	
2	Control	0.12	0.50	N
	Supplement	0.12	0.54	
3	Control	—	−0.16	H
	Supplement	—	0.08	
4	Control	—	0.12	H
	Supplement	—	0.48	
5	Control	—	0.08	H
	Supplement	—	0.36	
6	Control	—	0.04	H
	Supplement	—	0.34	
7	Control	—	−0.05	H
	Supplement	—	0.08	
8	Control	—	0.15	H
	Supplement	—	0.27	

N = not significant H = highly significant ($P < 0.01$)

there any significant compensatory gain by the unsupplemented cattle and these were the most rapid gains during the whole series of trials (1.22 and 1.07 kg/head/day for control and supplemented animals respectively).

The authors conclude that the creep-feeding of grain to cattle may show beneficial results when pastures are of low quality. When yearlings and weaners were fed, beneficial results were obtained when the daily weight changes of the control cattle ranged from −0.15 to 0.15 kg but little benefit was obtained when control cattle gained in excess of 0.45 kg/day. It was also concluded that liveweight response was more closely related to protein intake than to that of energy. These effects were enhanced by the severity of the dry season but this also reduced the magnitude of subsequent compensatory gains. The authors cite the work of Auty (1964) to support this view but also comment that the effect may have been due to the relatively high levels of supplementary feeding employed (this would be in better agreement with the regressions obtained by Smith and Hodnett (1962) already mentioned).

Morris and Gulbranson (1970) also investigated the use of both energy and protein supplements for cattle grazing oats near Brisbane in south Queensland. Steers were allocated to the following treatments:

1. Control – no supplements.
2. Low energy – 1 kg rolled low-protein sorghum grain/head/day.

3. High energy – 2 kg rolled low-protein sorghum grain/head/day.
4. Protein – 1 kg soybean oil-extracted meal/head/day.

The sorghum contained 6% crude protein and the soybean oil meal 51% crude protein. The stocking rate was 8 animals/3 ha in each case.

The daily rates of gain were 0.87, 1.00, 0.95 and 0.95 kg for the four groups respectively and differences did not reach significance. The authors concluded that substitution (presumably of oats by supplement) appeared to have been almost complete although no evidence in the form of grazing behaviour, etc. was presented to support this contention. In any event, the lack of difference among treatments would appear to be explained by the facts that at all times the steers had a large amount of feed available to them and that the analysis of the oat forage did not indicate that nitrogen was likely to be the limiting nutrient to growth. Crude protein was 14% at the beginning of the trial and 8% at the end.

Copeman et al. (1977) investigated the effect of a supplement containing molasses (86.5%), meat-and-bone meal (8%), urea (1.2%), water (3.5%) and mono-ammonium phosphate (0.8%) when given to lean Brahman cross steers ranging in age from 14–18 months. Cattle were grazing a grass–legume pasture of *Panicum maximum*, *Stylosanthes guianensis* and *Centrosema pubescens* at a stocking rate of 2.5 beast/ha.

Although no details were given of botanical composition, it was stated this was visually similar among the various paddocks used. Treatments were also combined with copper, cobalt and anthelmitic therapy. The trial was carried out at Tully, north Queensland.

The daily weight gains of supplemented and non-supplemented cattle are given in Table 6.6.

The only period during which supplemented animals demonstrated a higher rate of gain than their unsupplemented counterparts coincided with that of minimum rainfall. There was no significant difference between weight changes in the two groups after the conclusion of supplementation. In this case, therefore, there was no compensatory growth (in fact, there was a slight indication that the supplemented animals continued to be superior).

The results in this study also conflict with those of Alexander et al. (1970) who could demonstrate little benefit from supplementation when weight gains were between 0.15 and 0.50 kg/day. Copeman et al. (1977) suggest that their supplement also provided specific factors which would otherwise have been limiting to growth at that time.

Copeman et al. (1977) also reported highly significant interactions between cobalt treatment and supplementation and interpret this as evidence of 'enhanced growth and activity of micro-organisms propagating in response to intake of supplement and consequently speeding up digestion and increase [sic.] in intake'. In other words,

Table 6.6 Mean daily weight gains of animals with and without supplementation (adapted from Copeman et al. 1977)

Period	*Gain per day (kg)*		
	With supplement	*Without supplement*	*Significance of difference*
10–31 July	0.06	0.14	N
31 July–4 Sep.	0.64	0.39	H
4–28 Sep.	0.12	0.21	N
28 Sep.–20 Nov.	0.62	0.53	N
Supplementation ceased 20 Nov.	0.62	0.53	N

N = not significant H = highly significant ($P < 0.01$)

the supplement precipitated the next limiting deficiency.

Llewelyn et al. (1978) investigated the effect of a supplement of meat meal and molasses on three-year-old pregnant Hereford heifers, having previously noted that molasses and urea was ineffective in preventing weight loss. Weight losses in supplemented cattle were reduced to 0.08 kg/day v. the loss of 0.25 kg/day in the control group. The authors attributed this effect to the presence of by-pass protein; it has generally been shown elsewhere, however, that molasses and urea is effective in preventing weight loss whilst supplementation with by-pass protein stimulates weight gain (see following section).

Northern Territory

Work was initiated in the Northern Territory by Norman (1960), after noting seasonal changes in the liveweight of cattle grazing native

Table 6.7 The effect of protein supplementation of dry season performance of cattle on native pasture (Katherine) (adapted from Norman 1960)

Supplement	*Season*	*Weight change (g/head liveweight)*
Nil (control)	1958	–554
	1959	–445
900 g/day guar	1958	195
	1959	177
1800 g/day guar	1958	450
	1959	468

pasture (Norman and Arndt 1959). Groups of Shorthorn steers of mixed ages were grazed on native pasture (*Themeda australis, Sorghum plumosum, Sehima nervosum* and *Chrysopogon latifolius* were mentioned among the species present) at a stocking rate of 0.17 beast/ha with the following treatments:

1. No protein supplement.
2. 900 g/head/day of crushed guar (*Cyamopsis tetragonoloba*).*
3. 1800 g/head/day of crushed guar.*

Trials were carried out during two successive dry seasons for sixteen and fifteen weeks. The results obtained are summarized in Table 6.7.

Although no statistical analysis is offered, the results appear to be fairly conclusive. The utilization of low-quality native pasture was improved by feeding relatively small quantities of protein. The authors point out that this could be either as a direct remedy for protein deficiency or as a means of increasing the dry matter intake and digestibility of the pasture.

Subsequent work (Norman 1963) demonstrated a linear relationship between the quantity of supplementary crude protein fed and the increase in body weight. The supplement was peanut (groundnut) meal. Levels fed were 0, 0.45, 0.90 and 1.35 kg/head/day and cattle were stocked at a rate of 0.2 beast/ha native pasture. Feeding was continued for 126 days.

Summarized results are given in Table 6.8.

Moran (1973a) also carried out trials in the Northern Territory but on improved pasture. These were Townsville-style dominant (50–60%) containing *Pennisetum pedicellatum* (20–30%) at the

Table 6.8 Responses to supplements by cattle grazing native pasture (Katherine) (adapted from Norman 1963)

Level of supplement (kg peanut meal/head/day)	*Efficiency of utilization (kg peanut meal/kg weight change compared with controls)*	*Weight change (kg/head/day)*
0	—	–0.26
0.45	1.03	0.12
0.90	0.75	0.45
1.35	0.66	0.59

* The author comments that 'Guar is a bushy legume, introduced from India that is well adapted to the Katherine environment. It produces an average of approximately 500 lb/acre of seed which may be harvested with a normal grain crop header. The seed has a crude protein content of 25–30%.'

Adelaide River Experiment Station, 76 km south of Darwin. The cattle used were Brahman Shorthorn heifers seven months old with a starting weight of 156 kg. The trial started during the dry season (July) and continued until 'it was considered that there was sufficient green material in surrrounding native pasture paddocks to promote liveweight gain'. Stocking rate was 2.6 beast/ha throughout.

Groups of animals were allocated to each of the following treatments:

Group A	Pasture alone.
Group B	0.21 kg oven-dry (i.e. 0.5 lb air dry) meat-and-bone meal (MBM)/head/day.
Group C	0.43 kg MBM/head/day.
Group D	0.85 kg MBM/head/day.
Group E	1.71 kg MBM/head/day.

The above quantities of MBM were on offer; actual consumption is indicated in Table 6.9.

There were no differences attributable to level of feeding during the first period. During the following three periods, however, significant differences were demonstrated. The gains in all cases were relatively low even though large quantities of meat and bone meal were fed and appreciable rain fell during the months of August, September and October (25, 28 and 74 mm respectively). The author

Table 6.9 Intake of supplement and weight changes (LWC) of steers offered different quantities of meat and bone meal during the dry season (Katherine) (adapted from Moran 1973)

Period	*Group**					*Significance of difference*
	A	*B*	*C*	*D*	*E*	
7 July–5 Aug.						
intake (kg)	0	0.21	0.43	0.64	0.61	
LWC†	0.22	0.28	0.28	0.34	0.34	N
5 Aug.–2 Sep.						
intake (kg)	0	0.21	0.37	0.76	1.03	
LWC	−0.02	0.16	0.31	0.30	0.33	S
2–30 Sep.						
intake (kg)	0	0.21	0.43	0.85	0.36	
LWC	−0.14	−0.06	−0.08	−0.02	0.10	S
30 Sep.–28 Oct.						
intake	0	0.21	0.43	0.85	1.20	
LWC	−0.13	−0.16	−0.21	0.02	0.02	S

* For explanation, see text † Liveweight change N = Not significant
S = Significant (P > 0.05)

comments that as all supplemented animals received sufficient protein from the supplement (which was assumed to contain 40% CP), some other nutrient was probably limiting growth and it was suggested that this might have been energy.

In fact, it would appear that the limiting factor was the high stocking rate used.

It was also noted that a Mitscherlich response to supplementation would be expected (i.e. where the animal reacted up to a certain level above which no further response would occur). The author contrasts this with the linear response obtained by Norman (1963) mentioned earlier but this was also probably attributable to a difference in stocking rate.

Cattle were condition-scored throughout the trial but no significant differences could be noted except for a marked inferiority of the unsupplemented cattle during the third period.

Trials on meat-and-bone meal supplementation of improved pastures were also carried out in the Northern Territory by Falvey (1977). Cross-bred Brahman Shorthorn bullocks were grazed on pastures which were predominantly composed of Townsville stylo (*Stylosanthes humilis*), Sabi grass (*Urochloa mozambicensis*) and some annual grasses. Stocking rate was approximately 2 beast/ha and supplemented animals were given 0.45 kg of meat-and-bone meal/beast daily during four successive dry seasons. At the end of the four-year period, supplemented steers weighed 80 kg more than the unsupplemented controls. Gains were made during the first three years but not in the fourth as the increased size of the supplemented animals led to a deficiency of DM (Dry Matter) intake. No compensatory growth occurred during wet seasons. Analysis of blood, bone and pasture and the fact that the grazing area had regularly been dressed with superphosphate supported the belief that the response could reasonably be attributed to protein rather than to phosphorus or energy.

Later (Falvey 1979), these data were used to derive the relationship between liveweight change and crude protein intake which has already been referred to.

Western Australia

After Armstrong et al. (1968, cited by O'Kelly and Robinson 1968) and Robinson and Steward (1968) had demonstrated that whole cottonseed was effective in arresting dry season weight loss in cattle in the Kimberley region of north-western Australia, O'Kelly and Robinson (1968) carried out a trial to determine the effects of feeding this supplement for different periods.

The supplement was combined with the administration of vitamin A to half of the cattle. Animals were on feed for 100, 70 or 40 days and all were grazed on natural pasture. The mean voluntary daily intake of the whole cottonseed which was offered *ad libitum* was higher in those animals fed for 100 days (2 kg) than in the other two groups (1.7 and 1.5 kg for animals fed for 70 and 40 days respectively).

Animals receiving cottonseed for 40 days were significantly lighter than those supplemented for 70 ($p < 0.05$) or 100 ($p < 0.01$) days respectively, but there was no significant difference between the 70 and 100-day feeding groups. After supplementation was discontinued after the onset of the rains, rapid liveweight gains were made by all animals but the differences caused by the supplementation during the dry season persisted throughout the wet season, i.e. there was no compensatory growth. Blood serum levels of total, free and esterified cholesterol, triglycerides and phospholipids were elevated as a result of supplementation with cottonseed whereas the level of free fatty acids or the ratio of free to esterified cholesterol were unaffected.

No significant effects of vitamin A administration could be demonstrated for any of the parameters measured.

New South Wales

Climatic conditious in the Northern Tablelands of New South Wales are more temperate than tropical but because Glen Innes (latitude

Table 6.10 Final liveweight of yearlings fed different supplements during the winter (Glen Innes) (adapted from Hennessy and Robinson 1978)

Supplement	*Stocking rate (weaners/ha)*	*Final liveweight (kg)*	*Economic returns (\$/ha)*†
None	2.0	211 a*	–1
None	2.7	200 a	–3
Oat grain (2% body weight)	2.0	252 c	–50
Oat grain (2% body weight)	2.7	257 c	–69
Forage oats‡	2.0	250 c	–1
Forage oats	2.7	235 b	–3
Forage oats plus oat grain (2% body weight)	2.7	280 d	–20
Oat grain plus clover hay (ratio 25:75)	2.7	249 c	–20
Forage oats plus oat grain (% body weight)	2.7	301 c	–10
Oat grain plus silage (ratio 25:75)	2.7	235 b	–48
Oat grain plus silage (ratio 50:50)	2.7	234 b	–54

* Different letters indicate significant differences ($P<0.01$)
† Calculated at a return of 26c. (Australian)/kg
‡ *Avena sativa* (cv. Klein) sown into 45% grazing area

29° 42′S) is situated within the latitudes dealt with here, the work of Hennessy and Robinson (1978) will be briefly mentioned.

Yearling Hereford calves grazing improved pastures mainly composed of *Phalaris tuberosa* and white clover (*Trifolium repens*) at a stocking rate of 2.75 weaners/ha were fed supplements of oat grain, *Sorghum vulgare* (var. Sudax) silage and red clover (*Trifolium pratense*) hay. Further details of treatments are given in Table 6.10. Experiments were carried out over a period of three years (1973–5) during the months of March to October.

The final liveweight of animals offered oat grain was significantly higher than that of other animals. Grain supplements were less effective when offered with silage but hay plus grain (ratio 25 : 75) was as effective as the full grain ration.

The authors comment that in spite of grain feeding, only 24% of weaners were suitably finished for slaughter (the 'target' weight was 270 kg) and the results were therefore uneconomic. It was also concluded that grain was utilized efficiently and did not reduce herbage intake unduly.

United States, Latin America and the Caribbean

Florida

Supplementation of growing beef cattle on improved pasture in south Florida (Belle Glade) has been practised for some time. Early work (Kidder and Beardsley 1952) carried out on organic soils indicated that steers grazing Roselawn St Augustine grass (*Stenotaphrum secundatum*) fed ground snapped corn gained weight as satisfactorily as those fed cottonseed meal. These results were confirmed by Chapman et al. (1961) who fed 2.7 kg daily of citrus pulp, cane molasses, ground snapped corn or a mixed feed comprising ground snapped corn, citrus pulp and cottonseed meal. Results are summarized in Table 6.11.

Table 6.11 Weight gains of steers fed different supplements at the rate of 2.7 kg/head/day (Belle Glade) (adapted from Chapman et al. 1961)

	No feed	*Citrus pulp*	*Cane molasses*	*Ground snapped corn*	*Mixed feed*
Daily gain (g)	476	776	649	722	749
Dressing percentage	54.9	57.8	56.5	56.3	55.8
Cooler shrink (%)	1.2	0.8	0.7	0.8	0.6

Thus not only was the daily gain increased by the feeding of supplements but killing-out percentage was markedly increased and cooler shrink was lessened, thus providing a more valuable carcass. However, the large quantities of concentrate fed must have exerted a significant sparing effect on the pasture and the economics of the practice must be queried. Although in a previous paper Chapman and Haines (1960) state that 'the primary goal . . . should be to utilize available pasture forage to the maximum and to have the concentrate feed serve as a supplement and not as a substitute'.

Subsequent work carried out at Belle Glade (Haines et al. 1965) investigated the effects of supplementing during the first, first plus second and first plus fourth quarters of the year on the subsequent feedlot performance of yearling steers. The treatments were as follows:

1. No supplemental feed.
2. An average of 2.3 kg of supplemental feed/day during the first quarter (November, December, January).
3. An average of 2.3 kg of concentrate/head daily during the first quarter and 1.4 kg/head daily during the second quarter (February, March, April).
4. An average of 2.3 kg of concentrate daily/head during the first quarter and 1.4 kg/head during the fourth quarter (August, September, October).

The stocking rate was 5 yearlings (average starting weight: 200 kg) per ha. A summary of the results during the period of pasture is given in Table 6.12.

The composition of the supplement was ground snapped corn (2 parts), citrus pulp (2 parts), cottonseed meal (41% CP) (1 part) and a mineral mix was given to each group free-choice. In each of the

Table 6.12 Weight gains of yearling steers supplemented at the rate of 2.3 kg/head/day during different quarters (kg) (Belle Glade) (adapted from Haines et al. 1965)

	Supplementation			
	None	*1st quarter*	*1st & 2nd quarter*	*1st & 4th quarter*
1st quarter gain	6	33	33	31
2nd quarter gain	19	21	39	20
3rd quarter gain	57	57	49	52
4th quarter gain	16	16	18	27
Total gain pre-feedlot	98	126	140	122
Total gain in feedlot	95	99	103	94

three years during which the trial was carried out, the supplemented animals outgained the unsupplemented ones during the first and second quarters; all animals made satisfactory gains during the third quarter when forage was plentiful and there was some evidence that some compensatory gain took place. When steers were fattened in a feedlot, those fed a supplement during the first quarter only, yielded the highest net returns, and it was concluded that this treatment was to be recommended on financial grounds.

In a later series of trials at the same station Pate et al. 1972), work was carried out to determine the effect of concentrate supplementation at various levels of steers from weaning to feedlot weights. Effects were measured in terms of response whilst on pasture and during a subsequent feedlot period.

In each of the three trials carried out, 3.4 ha pastures of Roselawn St Augustine grass were used; stocking rates in trials 1 and 3 were 20 steers to each pasture but in the second trial, stocking rates were adjusted so as to equalize pasture-grazing pressure by steers fed different quantities of supplement. The stocking rates in this trial were 20, 24, 28 and 32 steers/pasture for the four treatments respectively. The treatments were as follows:

1. Pasture only.
2. Pasture plus 2.3 kg concentrate daily.
3. Pasture plus 4.6 kg concentrate daily.
4. Pasture plus concentrate free-choice.
5. Pasture plus a molasses/soya, cottonseed and corn oil stock mixture free-choice.

After the period on pasture, treatments 1, 2 and 3 were full fed in a feedlot to finishing whilst treatments 4 and 5 continued on pasture.

The supplement fed at pasture was composed of ground snapped corn (40%), dried citrus pulp (40%), cottonseed meal containing 41% crude protein (20%). This would give an approximate crude

Table 6.13 Weight gains and feed consumption of steers at pasture fed different levels of concentrate (Belle Glade) (adapted from Pate et al. 1972)

	Level of supplementation (daily)			
	0	*2.3 kg*	*4.6 kg*	*Ad libitum*
Average daily gain (g)	336	486	613	726
Daily intake of supplement (kg)	—	2.2	4.2	6.5
Average daily gain in feedlot (g)	858	835	817	649
Killing-out %	56.6	59.3	59.3	59.9

protein content of 13%. The average initial weight was 230 kg. Results are summarized in Table 6.13.

Although responses to supplementation were to all intents and purposes linear, an economic analysis of both stocker and feedlot periods showed that supplementation was uneconomic despite the slightly improved performance in both growth rate and carcass characteristics. The authors conclude that 'pasture is the cheapest source of nutrients for stocker cattle'.

The same group of workers carried out additional trials to determine the effect of different levels of concentrate expressed as percentage of body weight (Pate et al. 1975). Yearling Brahman cross-bred steers with an average initial weight of 220 kg grazed Roselawn St Augustine grass pastures each of 3.2 ha. Treatments were as follows:

1. Pasture with no additional concentrate.
2. Pasture plus concentrate at a rate of 0.5% of body weight/day.
3. Pasture plus concentrate at a rate of 1.0% body weight/day.
4. Pasture plus concentrate at a rate of 1.5% body weight/day.

The concentrate fed, both during the supplementary phase and during the subsequent feedlot phases, was composed of 36.5% ground snapped corn, 36.5% dried citrus pulp, 18.0% cottonseed meal (41% crude protein), 7.5% blackstrap molasses with added minerals and vitamin A. The protein content of the concentrate would appear to be approximately 12%. Summarized results are given in Table 6.14.

Although the responses to supplementation were not so nearly linear as in the previous trials reported they were still marked. However, response during the feedlot period indicated that these

Table 6.14 Weight gains, days spent on pasture and subsequent feeder performance of steers feed different levels of concentrates on pasture (Belle Glade) adapted from: Pate et. al. (1975)

	Supplementation level (% bodyweight per day)			
	0	*0.5*	*1.0*	*1.5*
Days on pasture	443	327	282	247
Average daily again (g)	376	527	608	676
Concentrate/steer (kg/day)	—	1.5	2.9	4.3
Average daily gain in feedlot (g)	1 182	1 027	1 068	1 082
Killing-out %	59.2	59.9	60.8	60.4

differences were reduced by compensatory gain. Carcass measurements showed that non-supplemented animals were only slightly inferior insofar as killing-out percentage was concerned whilst other measurements (rib-eye muscle area, fat over rib-eye, etc.) did not differ markedly among treatments.

A full economic analysis of the data was carried out and it was concluded that the profitability of continuous supplementation of steers on pasture was questionable during times of steady cattle prices but that this could be justified at times of high prices.

However, the principal advantage of supplementation was the reduction in time on pasture or 'reduced inventory time'. It was stated that supplementation was of little value if the pastures were not restocked at the end of each period.

Work on supplementation was also carried out at the Range Cattle Experiment Station, Ona about 50 miles south-east of Tampa, Florida. However, in this series of trials, animals were fed so as to gain a predetermined amount daily and effects of treatments on subsequent feedlot fattening were also determined. (Peacock et al. 1964).

Calves used were predominantly grade Brahman, approximately 7.5 months old and with an average starting weight of 170 kg, grazing Bahia grass (*Paspalum notatum*) pastures at a stocking rate of 6 beasts/ha. The winter feeding phase of the trial lasted from October to February and the trial was repeated during three successive years.

Supplements to attain the decided daily gain are given in Table 6.15 as are rates of gain during both winter and feedlot period.

Responses to supplementation were again linear to all intents and purposes although growth rates were overall lower than in the other Florida experiments mentioned previously. There was also marked compensatory growth although this was not sufficient to overcome

Table 6.15 Average daily rations and weight changes of cattle fed at different levels (adapted from Peacock et al. 1964)

	Lot			
	1	*2*	*3*	*4*
Cottonseed meal (41% CP) (g)	480	580	713	785
Citrus pulp (g)	404	790	1 800	2 930
Daily gain (g) (winter period)	41	177	350	549
Daily gain (g) (feedlot period)	1 021	962	922	826
Killing-out %	56.3	57.7	58.2	59.4

Table 6.16 Steer gains on frosted pasture with limited and full-fed concentrate (Quincy) (adapted from Baker 1966)

	Limited feed	*Full feed*
First half winter		
Daily gain (g)	445	899
Feed consumption (kg)	3.4	7.2
Second half winter		
Daily gain (g)	953	1 090
Feed consumption (kg)	1.2	7.4

the treatment differences imposed during the wintering period. The carcass yield of the unsupplemented animals expressed as killing-out percentage was also slightly lower than the supplemented animals which would serve to emphasize differences.

The authors commented that an average daily gain of 350 g was necessary to maintain the animals' slaughter grade and that although low supplementation did not affect skeletal growth, it resulted in excessive loss of condition. There was a positive relation between rate of gain during the winter and efficiency of feed utilization during this period but a negative relation between winter gain and utilization of feed during the feedlot phase which resulted in no overall significant differences being apparent.

Baker (1966) working at Quincy in North Florida made some interesting comparisons which although the location is theoretically just out of our area, are relevant to the present discussion and will be mentioned briefly. Yearlings on frosted grass pasture with reseeded red clover gained more on full feed than on limited feed during the first part of the winter before regrowth of the clover occurred but this difference was reduced during the second half of the winter. Results are summarized in Table 6.16.

The efficiency of feed utilization by the animals on full feed clearly declined markedly as pasture quality improved.

Baker (1966) also compared the growth of steers on oats and rye pastures with and without concentrates and found that over the entire winter, daily gains were 704 and 831 g, and 535 and 767 g for unsupplemented and supplemented oats and rye respectively, this showing a clear superiority of the animals which received between 3 and 4.5 kg of corn/day. However, in a subsequent feedlot period, 'small end-of-winter differences largely disappeared'. Final feed costs per 45 kg of gain of the supplemented animals (after taking the feedlot period into account) were US$20.05 whilst the corresponding costs for the unsupplemented animals were US$20.12, a dramatic demonstration of the questionable advantages of supplementary feeding even under the relatively stable price structures of the 1960s.

Puerto Rico

Work has been carried out in Puerto Rico with the objective of achieving more intensive land use and more efficient use of pasture (Carlo et al. 1972). Four treatments were compared at the Corozal Substation of the University of Mayaguez. These comprised:

1. Grazing plus 1.4 kg maize/head/day.
2. Grazing plus 1.4 kg molasses/head/day.
3. Grazing plus a bulky feed at the rate of 1.4 kg/head/day during periods when grazing was poor only.
4. Grazing alone.

Animals used were grade Brahman and Charolais heifers aged 7 to 11 months with an average weight of 170 kg. Grazing comprised improved fertilized pastures (no species given) and the stocking rate was 5 beasts/ha. Results are summarized in Table 6.17.

Table 6.17 Weight changes and killing-out percentage of heifers fed different supplements during the winter (Carozal) (adapted from Carlo et al. 1972)

	Treatment			
	1	2	3	4[1]
Gain/day (g)	554	480	440	422
Killing-out %	60.7	60.4	59.1	59.0

[1]For details, see text

Although the animals fed the corn supplement gained significantly more weight than those fed no supplement, this was not economically beneficial when feed and labour costs were taken into account. There were no advantages of the other two treatments over grazing alone.

Carlo et al. (1972) also carried out a similar trial at the Isabela Substation. Treatments were:

1. 'Bulky' feeding *ad libitum.*
2. Grazing with a bulky ration finishing period.
3. Grazing alone.
4. Pangola grass soilage *ad libitum.*

The bulky formula used in both trials comprised (%):

ground yellow corn	35.00
sugarcane bagasse	25.00
cane molasses	32.50
fish meal	4.00

urea	1.75
salt	1.00
dicalcium phosphate	0.75

This would give a crude protein content of approximately 9%. Daily weight gains for the four groups in the order given were 535, 527, 458 and 408 g/day respectively. There were no advantages of any of the treatments over grazing alone except for a slight but significant advantage in killing-out percentage which was, however, not sufficient to make supplementation economically attractive.

Mexico

Carrera et al. (1963) carried out a trial using supplementation with molasses during both wet and dry seasons at Rodriguez Clara, Veracruz. Steers grazed Guinea grass (or 'Privilegio', *Panicum maximum*) pastures at a stocking rate of 1 animal/ha.

Trials were continued for 120 days during both wet and dry seasons. The results obtained are summarized in Table 6.18. Supplementation made a significant difference to growth rates during the dry season but the slight improvement produced during the wet season was not significant. Consumption of molasses was 730 and 1007 g/day for wet and dry season respectively.

Table 6.18 Response to supplementation with molasses of steers grazing Guinea grass during wet and dry seasons (Veracruz) (adapted from Carrera et al. 1963)

	Supplemented (g/day)	*Unsupplemented* (g/day)	*Significance difference*
Wet season	900	746	N
Dry season	724	397	S

N = Not significant S = Significant ($P < 0.05$)

Brazil

A considerable amount of work on supplementation has been carried out in the state of Sao Paulo, Brazil. Tundisi et al. (1966) carried out an experiment to determine the effects of different management systems on the rate of gain of entire male and female Nelore (sic.) cattle about 9.5 months of age. The experiment was carried out at Sertãozinho in the north of the state on non-fertilized *Hyparrhenia rufa* (capim Jaragua) pastures which were classified to be of medium to good quality. One of the treatments consisted of pasture only, and another of pasture with 300 g daily of a supplement consisting of 50% maize meal and 50% cottonseed meal. The period during which supplementation was carried out was the dry season from June to October but weighings were continued into the following summer wet season. Stocking rates were not given but it was implied that adequate pasture was available during the wet season.

Weight changes for the supplemented animals during the period of supplementary feeding were 34 and 26 kg for males and females respectively, whilst those for the unsupplemented animals were −8 and −10 kg. However, during the summer feeding period the weight gains were 127 and 98 kg, and 144 and 117 kg in the same order as above. Overall weight gains were 500 and 388 kg, and 498 and 384 kg, indicating that differences due to supplementation were to all intents and purposes negated by compensatory growth.

A further study was carried out in the district of Guararapes but on this occasion the grass used was Guinea (*Panicum maximum* Jacq.). The treatments were as follows:

1. Pasture only (no stocking rate given).
2. Pasture plus 0.5 kg cottonseed meal daily (28% crude protein).
3. Pasture plus 0.5 kg cottonseed meal plus 2 kg molasses daily.

Animals were groups of twelve Nelore yearling steers. The weight gains during the dry season when supplementation was carried out were 828, 820 and 750 g/day for the three treatments respectively. The lack of any effect due to the supplementation was ascribed to high forage availability due to favourable rains. The indication of a lower rate of gain on one of the supplemented groups could be due to reduced grazing time because of supplementation (Aguiar de Mattos et al. 1967).

Bisschoff and his colleagues carried out a series of experiments to determine the effects of various types of supplement of steers grazing improved pastures in Araçatuba in north-western Sao Paulo (21° 12′S) and Matao in the central part of the state (Bisschoff et al. 1967).

The first trial compared the effects of various levels of ground-ear-corn fed with or without a high protein supplement to steers grazing Colonial Guinea grass (*Panicum maximum*). The high protein supplement contained cottonseed, peanut and corn-gluten meals, molasses, urea and minerals; the content of crude protein was 44% and that of TDN, 57%. Supplements were given during both winter and summer.

Five animals were stocked in each 11 ha pasture as tester animals and other animals were added or removed from the pasture so as to maintain the forage in as near optimum grazing condition as possible. Treatments and weight changes are given in Table 6.19.

Both yearling and two-year-old steers were used in the experiment but as no significant interactions between age and treatment were

Table 6.19 Daily gains of steers fed different levels of supplement on Colonial Guinea grass pasture (Araçatuba) (adapted from Bisschof et al. 1967)

	Daily gain (winter) (g) (*112-day period*)	*Daily gain (summer)* (g) (*112-day period*)	*Overall* (g)
Pasture only	150	940	547
1 kg ground-ear-corn (gec)	130	990	558
2 kg ground-ear-corn	220	1 020	620
4 kg ground-ear-corn	450	950	701
1 kg gec with H meal (9:1)	160	990	572
2 kg gec with H meal (9:1)	250	1 000	621
4 kg gec with H meal (9:1)	410	940	678

apparent, results have been combined for the purposes of the present discussion. During the winter period, there was little response to the lower levels of ground-ear-corn supplementation indicating that the animals were substituting this for pasture on an isocaloric basis. The level of 2 kg/day increased weight gains but this was still at the expense of grass consumption as the weight increases were not concomitant with the amount of concentrate fed. At the 4 kg/day level, however, gains were more in agreement with the amount of grain fed and the efficiency of utilization was better.

There was no evidence that an increase in protein intake above that provided by the ground-ear-corn improved performance in any way. There was little if any advantage in feeding the rations during the summer period when forage was readily available and no statistical difference between treatments if gains for the summer and winter periods were taken together (although advantages for the cattle fed 4 kg ground-ear-corn appeared to be real even if not significant).

Replicates at both Araçatuba and Matao were used to determine the effects of high (H meal) and low protein supplements (L meal) on cattle grazing Colonial Guinea grass and *Hyparrhenia rufa*/*Paspalum notatum* (Jaragua/Batatais) pastures respectively. The high protein concentrate had the same composition as that used in the first trial whilst the low protein concentrate contained cottonseed meal, mandioca, molasses, urea and minerals and contained 18% crude protein for the first part of the experiment and 24% crude protein for the second (the change in composition was to improve palatability according to the authors). The levels fed were 0.5 and

Table 6.20 Influence of high and low protein supplements on Nellore steers grazing Guinea Grass (Araçatuba) (adapted from Bisschof et al. 1967)

	Daily gain (winter) (g) (*105-day period*)	*Daily gain (summer)* (g) (*195-day period*)	*Overall* (g)
Pasture only	190	710	526
0.5 kg L[1] meal	131	759	539
1 kg L meal	137	725	519
0.5 kg H meal	173	698	515
1 kg H meal	250	695	539
2 kg H meal	304	756	599

[1]For details, see text

1.0 kg/steer daily for the low protein supplement and 0.5, 1.0 and 2.0 kg for the high protein supplement. The stocking rates in the former trial were not stated (as the 'put and take' system was used as before) and in the latter 20 three-year-old Zebu steers were allocated to pastures of about 30 ha each.

The results are presented in Table 6.20.
The only statistical difference in growth rate was with the 2 kg supplement of high protein meal and this difference was maintained when winter and summer gains were considered together even though there were indications of compensatory growth on the other treatments.

Table 6.21 shows a summary of the results obtained in the trial

Table 6.21 Influences of high and low protein supplements on Nellore steers grazing Jaragua/Batatais grass (Matao) (adapted from Bisschof et al. 1967)

	Daily gain (winter) (g) (*140-day period*)	*Daily gain (summer)* (g) (*168-day period*)	*Overall* (g)
Pasture only	– 95	738	359
0.5 kg L[1] meal	– 25	795	422
1 kg L meal	– 44	834	435
0.5 kg H meal	54	734	425
1 kg H meal	207	704	478
2 kg H meal	334	751	562

[1]For details, see text

with the same treatments as above but carried out on Jaragua/Batatais pastures at Matao.

In this trial, pastures were of a lower quality and animals which were not supplemented or only supplemented at a low level lost weight; it was also strongly indicated that the high protein supplement was more effective than the low protein supplement which was not the case in the previous trial. The daily gain supported by the 2 kg level of high protein supplement was strikingly similar in the two trials. Again, when both winter and summer gains were analysed together, the only significantly superior group was that given 2 kg of the high protein meal, and compensatory growth occurred in the other groups.

A further experiment was carried out using different levels of salt with a view to limiting concentrate intake. The high protein supplement used previously was mixed with salt in the ratios of 20 : 1, 10 : 1, 5 : 1 and 2.5 : 1. None of these mixtures was effective in limiting intake during the winter period although there appeared to be a slight limiting effect during the summer season. There were no significant differences among groups in so far as weight gains were concerned. The coefficient of variation during this period was 108%, presumably reflecting high between animal variation insofar as salt tolerance was concerned.

Finally, an experiment was carried out using the different levels of high protein meal used previously, but supplemented with vitamin A, and vitamin A and terramycin. This confirmed the effects of the different levels of concentrate but did not demonstrate any effect which could be attributed to either of the additives.

In their final integrating discussion, the authors concluded that the extra protein supplied in the supplement was not effective in increasing weight gains over that which could be explained by an increased caloric intake. Consumption of concentrate took place at the expense of forage consumption and it was inferred that the carrying capacity of the pasture would be increased by supplementation and that a farmer feeding concentrates would not obtain the full benefit from them if this increased stocking capacity was not exploited.

This hypothesis was confirmed in a subsequent experiment (Mott et al. 1967) when it was shown that the administration of 2 kg of molasses daily/head decreased pasture consumption so that the carrying capacity was increased by 20% (other aspects of this trial will be discussed in the following chapter).

Finally, the effect of feeding whole-ground ear-corn to Gir and Sindhi heifers during the dry season at Ribeirão Prêto in the state of Sao Paulo was studied by Benintendi et al. (1974).

The treatments to each of which ten heifers aged between 9 and 22 months were allocated were pasture alone (Coloniao, *Panicum maximum*) and pasture plus 1.5 kg/day of ground whole-ear corn. The stocking rate was approximately 2 animals/ha and the duration of the experiment was from July to October. The protein content of the pasture (average of collection made each month of the experiment) was 7%. The supplemented animals gained 268 g/day whilst those unsupplemented gained 127 g/day ($p < 0.01$). The authors commented that weighings of the animals continued for 140 days after the termination of supplementation and no compensatory growth was observed.

Venezuela

A trial was carried out at Calabozo on an irrigation scheme in the Ilanos of Venezuela to determine the effectiveness of supplement administered with or without salt to limit intake. Brahman crossbred steers averaging 22 months of age were uniformly assigned to three treatments:

1. Pasture only (*Digitaria decumbens*), stocking rate 2 animals/ha.
2. Pasture as above plus basal 30% salt supplement fed *ad libitum*.
3. Pasture as above plus basal supplement fed to equal the intake less salt of the proceeding group.

The basal supplement was composed of 57% rice polish, 40% sesame meal and 2.8% bonemeal (sic.). For treatment 2, 30% of the ingredients were replaced by salt. The duration of the trial was 120 days. Intake of the 30% salt supplement and the basal supplement were 850 and 865 g/day respectively. Both groups of supplemented animals gained significantly ($p < 0.05$) more weight than the non-supplemented animals. Daily gains were 325, 474 and 510 g for control, 30% salt-supplemented and basal-supplemented animals respectively (Chicco, Shults, Rios, Plasse and Burguera, 1971).

Chicco and his colleagues (Chicco et al. 1973, cited by Chicco and Shultz 1975) carried out a trial to determine the effectiveness of supplementing at three- or six-day intervals compared with daily supplementation. The supplement was equivalent to 500 g of a

Table 6.22 Effects of supplementation at different intervals of steers during the dry season in the Venezuelan 'llanos' (Calabozo) (adapted from Chicco and Shultz 1975)

	Control: no supplement	*Supplement fed*		
		Daily	*Every 3 days*	*Every 6 days*
Dry season	176	277	224	201
Wet season	902	870	848	892
Overall	539	573	536	547

concentrate containing 42.5% crude protein daily. The trial was carried out during the dry season on native grasses in the Venezuelan 'llanos'. The results are summarized in Table 6.22.

The authors pointed out that the beneficial effect of the dry season supplement was almost completely overcome by the compensatory growth shown by the control animals when forage became abundant. However, the point was also raised that a more permanent effect of dry season supplementation might occur in cases when the dry season was longer than 100 to 120 days.

Trinidad

A particularly interesting study was carried out in Trinidad in the West Indies, as it is one of the few cases where studies of weight gain of supplemented animals was combined with a determination of intake from pasture. Cross-bred Holstein male and female calves about six months old were grazed on good-quality Pangola grass (*Digitaria decumbens*) pastures at the University field station. The stocking rate used was 0.08 ha/head. Half the calves were given 1.36 kg of a supplement containing maize meal, citrus meal, wheat offals, rice mill feed, rice bran, dried brewers grains, copra meal, soya bean meal, minerals and vitamins with a crude protein content of 16.4% and the other half was given 2.04 kg of the same supplement. The first group gained 340 g/day during the 180 days of the experiment whilst the second group gained 484 g/day. The difference was highly significant. The relatively low rates of gain are attributed to low intakes of pasture by the younger calves (Archibald and Cowlishaw 1972).

Two castrated male calves and two female calves were used to determine the pasture intake using chromic oxide administration and faecal sampling and *in vitro* digestibility. This procedure was carried out twice. The intake of herbage for the high and low levels of supplementation was 7.84 and 7.40 kg for the first determination and 4.83 and 4.62 kg for the second determination. Although the differences between the two trials was appreciable, in neither case was there any significant reduction of herbage intake which could be ascribed to the higher level of concentrate (Cowlishaw and Archibald 1972).

Costa Rica

Vohnout and his colleagues have carried out a large body of work in the humid tropical conditions of Turrialba to investigate the relations between supplementation and stocking rate. Supplements used were based on molasses (with 15% cottonseed meal and 1.5% urea added) or waste green bananas and pastures were composed of Guinea grass (*Panicum maximum*) in all cases.

Ninety cross-bred Charolais bulls were used to determine the effect of restriction of grazing time on the consumption of the molasses mixture. It was shown that 3 hours' grazing was sufficient for cattle to gain weight at a rate of 0.73 kg/day similar to that obtained by grazing alone; molasses consumption with this treatment was 7.5 kg/day. The biggest weight increments were obtained by a combination of 7 to 23 hours of grazing and molasses consumption of 7.3 and 0.5 kg/day/head. It was concluded that an additive relations existed between the ration and the nutritional value of the pasture and also a substitutive effect of the grass by the ration (Vohnout et al. 1973).

A further trial evaluated the effect of varying grazing intensity with both fixed and *ad libitum* molasses consumption (Vohnout et al. 1973a). Romo-Sinuano heifers grazed Guinea grass pastures at stocking rates of 2, 4, 6, 8 and 10 animals/ha. Molasses consumption by both animals fed molasses *ad libitum* and restricted was 1.15 kg/day at the stocking rate of 2 animals/ha and the growth rate was similar at 630 and 600 g/day respectively. In the highest stocking rate group, growth was markedly attenuated in the animals fed restricted molasses compared with *ad libitum* growth rates being 220 and 470 g/head respectively whilst molasses consumption increased to 1.74 kg. The maximum productivity per ha, obtained with the stocking rate of 10 head/ha, was thus 2.3 kg (sic.)/ha/day with restricted molasses consumption and 4.4 kg (sic.)/ha/day with the *ad libitum* system. The authors concluded that when sufficient protein was available the molasses exerted an additive and complementary effect to the pasture, increasing production without any apparent effect on stocking rate. Unfortunately, the abstract does not give details of gains of individual animals nor of intermediate stocking rates.

A similar trial was carried out using waste green bananas. Steers and heifers grazed Guinea grass pastures at stocking rates of 2, 4, 6, 8 and 10/ha for 220 days with or without *ad libitum* bananas. The consumption of bananas ranged from 4.4 to 5.5 kg/head/day. Supplementation made only a slight difference to animals on the low stocking rate, rates of gain being 610 and 520 g/head/day respectively. Maximum production per ha without supplementation in this experiment was attained with the stocking rate of four animals/ha and was 1.6 kg/ha/day. However, with supplementation at the stocking rate of 10 head/ha, this was increased to 4.3 kg/ha/day and the authors commented that abundant pasture still remained. It was concluded that administration of bananas exerted a marked sparing effect on the pasture and that for this reason should be carefully regulated so as to avoid waste of forage (Alpizar and Vohnout 1973).

A further experiment was carried out to investigate the effects of feeding different levels of bananas (Ruiz et al. 1973). Levels used were 0, 7, 5, 15, 22.5 and 30 kg/animal/day and to achieve these levels it was necessary to limit grazing time. An increase in banana consumption from 0 to 5 kg/day increased weight gains from 460 to 590 /head/day and further improvements were not noted above this level. It was confirmed that increased quantities exerted a sparing effect on the forage and increased wastage. These results were discussed in an article on the intensification of grazing systems (Cubillos et al. 1975), in which it was emphasized that stocking rate must be increased if supplementation is practised so as to make the most efficient use of available forage. The substitutive effect of the concentrate otherwise leads to inefficient use of pasture, the least expensive resource. This relationship becomes important in tropical conditions of wide fluctuations in forage availability because supplementation may be increased as forage availability decreases so as to maintain a high production per unit area.

Discussion

The results of the various trials carried out on supplementation of growing cattle during both wet and dry seasons on both improved and native (extensive or unimproved) pasture are summarized in Table 6.23.

Protein supplementation of extensive pastures in winter

In general, positive effects of this form of supplementation have been reported. This is hardly surprising in view of the low protein levels in the grass during the day or winter season.

This is not only attributable to the correction of a nutrient deficiency *per se* but also because of several associated effects.

Protein digestibility in ruminants is related to the absolute amount of protein present in the diet (see, e.g. Glover et al. 1957; Butterworth and Diaz 1970) and thus the addition of a small amount of protein will improve the overall digestibility (in many mature grasses, the digestibility of crude protein is likely to be negative; i.e. the animal excretes more protein than it consumes and is thus in negative nitrogen balance; see, e.g. Butterworth 1967a). The increase in digestibility might in itself bring about an increase in intake of dry matter because of more rapid transit time through the rumen (Blaxter and Wilson 1962) although there is some doubt that this is the case with tropical forages (Butterworth 1964).

Many protein supplements add substantial quantities of 'by-pass' protein to the diet of the animals. This is particularly true of meals which have suffered heat-treatment during processing. By-pass proteins are those which are not attacked in the rumen but escape digestion until they reach the duodenum and small intestine. They

Table 6.23 Responses to supplements in various trials

Reference	*Intensive*	*Extensive*	*Winter*		*Summer*		*Wt gain*	*Carcass*	*Comp. Growth*	*Pasture Sparing*	*Remarks*
			Protein	*Energy*	*Protein*	*Energy*					
1 Murray et al. (1936)		+	+				+	+	+		
2 Murray & Romyn (1937a)		+	+				+				
3 Murray & Romyn (1937b)		+	+	+			+				Response to protein only
4 Rhodes (1956)		+	+	+			+				Response to protein only but inconclusive
5 Grant (1975)		+	+				+				
6 Grant (1976)		+	+				+	–			
7 Richardson et al. (1976)		+	+				+				
8 Barnes (1976)	+			+			+			+	
9 Harwin & Venter (1970)		+	+	+			+				Creep feeding
10 Lüdemann (1971)		+	+	+			+			+	Not economic
11 Verbeek & Vonla Chevallerie (1958)		+	+				+				Response to energy (sour veld)
12 Pieterse & Preller (1965)		+			+		+	+	+		Economic
13 Lesch et al. (1966)					+		+	+	+		Economic

Reference	*Intensive*	*Extensive*	*Winter*		*Summer*		*Wt gain*	*Carcass*	*Comp. Growth*	*Pasture Sparing*	*Remarks*
			Protein	*Energy*	*Protein*	*Energy*					
14 Reyneke (1976)		+			+		+	+	—		
15 Smith (1961)		+	+				+				
16 Smith (1966)		+	+				+			ND	
17 Smith & Hodnett (1962)		+	+				+		+		
18 Anon (1968)		+	+	+			+				Responses were to feeding rather than supplementation
19 Owen (1968)	+	+	+	+			+		+		
20 Walker (1969)		+		+			+		+		Poor hay supplementation not effective because of comp. gain
21 Thomas & Addy (1977)	+		+				+				
22 Addy & Thomas (1977)	+			+			—				
23 Cabaret et al. (1976)		+	+		+		+				
24 Calvet et al. (1976)		+	+				+				
25 Sutherland (1959)		+	+	+			+				Response to protein

26 Alexander et al. (1970)		+		+		+				When creep fed weaners
		+		+		+				
		+	+			+		+		
27 Morriss & Gulbranson (1970)	+		+	+		—				Pasture oats
28 Copeman et al. (1977)	+		+	+		+		—		
29 Norman (1960)		+	+			+				
30 Norman (1963)	+	+	+			+				Linear response to increasing levels
31 Moran (1973)	+		+			+				Response to high levels limited by high stocking rate
32 Falvey (1977)	+		+			+		—		
33 O'Kelly & Robinson (1968)		+	+			+		—		
34 Hennessy & Robinson (1978)	+			+		+			—	Not economic
35 Kidder & Beardsley (1952)	+		+	+		+			+	Response to both energy and protein
36 Chapman et al. (1961)	+			+		+	+		+	Economic
37 Haines et al. (1965)	+			+	+	+	+	+		

Reference	*Intensive*	*Extensive*	*Winter*		*Summer*		*Wt gain*	*Carcass*	*Comp. Growth*	*Pasture Sparing*	*Remarks*
			Protein	*Energy*	*Protein*	*Energy*					
38 Pate et al. (1972)	+			+		+	+	+		+	Not economic
39 Pate et al. (1975)	+			+		+	+	+		+	
40 Peacock et al. (1964)	+		+	+				+	+		
41 Carlo et al. (1972)	+			+			+	—			Not economic
42 Baker (1966)	+			+			+		+		Not economic
43 Tundisi et al. (1965/6)	+		+	+			+		+		
44 de Mattos et al. (1967)	+		+	+			—				Large amounts of forage available
45 Bisschof et al. (1967)	+		+	+			+		+	+	No response to protein
46 Mott et al (1967)	+			+			—			+	
47 Benintendi et al. (1974)	+			+			+		—		
48 Chicco et al. (1971)	+		+				+				
49 Chicco et al. (1973)		+	+				+		+		
50 Archibald & Cowlishaw (1972)	+			+						—	

thus provide a source of essential amino acids which may not be synthesized in adequate amounts by the micro-organisms of the rumen and which may also serve as a source of glucogenic precursors under circumstances when glucose may be limiting to growth. It has been shown that addition of small quantities of by-pass protein to the diet can affect the overall dry matter intake of the animals (Hennessy, cited by Leng et al. 1977) by increasing the rumen 'through-put' of the animals. The field has been reviewed by Leng et al. (1977).

It is not surprising that the response to supplementation varies considerably from one experiment to another in view of the variation in conditions under which the trials were carried out. The ability of an animal to selectively graze forage of a higher nutritive value than that which would be expected from sampling and subsequent chemical analysis has been noted by workers in Zimbabwe and elsewhere (Gammon and Roberts 1978a, b) and similarly it has been remarked previously that browse may be of considerably higher nutritive value than forage. In this respect it is interesting to note that the faecal nitrogen excretion of a ruminant is closely correlated with the intake of this element (Butterworth 1963) and that this has been used to indicate the point at which a diet becomes inadequate and supplementation must be started (Winks and Laing 1972).

Improvements in carcass quality as a result of supplementation

In several trials an improvement of carcass quality or killing-out percentage as a result of supplementation has been noted. It is important when evaluating the extent of this improvement in economic terms to take into consideration whether the animals are to be slaughtered at the conclusion of the supplementary feeding period and, if so, whether the conditions prevailing in the market place allow for payment of premium prices for improved carcass quality.

Compensatory growth

When the growth of an animal has been retarded because of nutritional deficiencies, it is able to recover when ample feed becomes available by growing at an accelerated rate. This accelerated rate of growth was described as 'compensatory growth' by Bohmann (1955) and it was probably observed for the first time in beef cattle by Waters (1908). The subject has been fully reviewed by Wilson and Osbourn (1960) who stated that some of the factors governing the ability of the animals to recover were as follows:

1. The nutrient causing undernutrition (e.g. protein, energy, etc.).
2. The severity of undernutrition.
3. The duration of the period of undernutrition.
4. The stage of development at which undernutrition occurs.
5. The pattern of realimentation.

The cases in which supplementary feeding was followed by compensatory growth are summarized in Table 6.23. Differences in the degree to which this occurs can readily be explained by differences in the factors mentioned above.

In many cases, any advantage of supplementation was nullified, or almost so, by compensatory growth. Ledger and Sayers (1977) carried out stall-feeding trials of *Bos indicus* (Boran) and cross-bred Hereford Boran steers at Muguga and suggested that 'abnormally high efficiencies could be expected to persist in progressively decreasing degrees up to 15 weeks of realimentation'. It is clear then that when the effects of supplementary feeding are evaluated, performance and efficiency during the post-supplementary period must be considered (if animals are not to be slaughtered immediately) and that the cost of the additional supplement per weight of additional *net* gain must be taken into account.

Pasture sparing

McClymont (1956) has suggested that supplementation would not be expected to be effective as the supplement is eaten instead of pasture. Several of the studies quoted indicate that this is indeed sometimes the case. But few studies have been directed at the measurement of this effect. Pate et al. (1975) have emphasized that full advantage cannot be made of supplementation unless cattle are taken off and replaced by store cattle as soon as they are ready for finishing. Similarly, Bisschof et al. (1967) suggested that the pasture sparing effect of feeding supplements could best be exploited by increasing the stocking rate and this was confirmed by Mott et al. (1967) who demonstrated an additional carrying capacity as a result of supplementation of 20%.

Unfortunately, little work has been done on the grass consumption of supplemented animals but Cowlishaw and Archibald (1972) could not demonstrate any appreciable difference in pasture intake by supplemented or unsupplemented calves grazing good-quality Pangola grass. These aspects of supplementation deserve further investigation.

Feeding of grazing animals

The provision of extra feed, or energy if other nutrients are not limiting, will result in increased growth rates provided that the animal fed is not already achieving its genetic growth potential. Feeding should not be confused with supplementation which is the correction of specific nutrient deficiencies. Several trials mentioned in this review have been concerned with feeding, particularly those carried out in the southern United States and Brazil. Whether or not feeding is economically attractive depends on the price structure with regard to feedstuffs and slaughter cattle and also whether full advantage is taken of the increased stocking rate (as mentioned earlier) as well as the reduced time of the animals on inventory which will result in lower interest charges on capital. These considerations imply that management is already at a relatively high level of efficiency and also that land availability is an important factor. Where there is no limitation to the amount of grazing available, then considerations such as increased stocking rate potential become less important.

Creep-feeding

Creep-feeding, or the feeding of calves prior to weaning, is a special case of the above. Whilst usually effective, it obscures the ability of the breeding cow to rear a heavy calf at weaning and makes selection of female stock on performance impossible. Creep-feeding is often a practice of pedigree breeders anxious to achieve impressive weaning weights for commercial purposes and cannot be recommended to the commercial breeder under normal conditions.

Summary

Experiments involving the supplementation of growing stock using both protein and energy under both extensive and intensive grazing situations have been reviewed.

In the great majority of cases, such supplementation results in weight gains which could usually be attributed to correction of a protein deficiency in the pasture and less frequently to the provision of energy *per se*.

In a number of trials there was also evidence that carcass quality was improved as a result of supplementation, but this factor would not be of importance unless cattle were to be slaughtered at the end of the supplementary feeding period and only then when payment for the improved carcass would justify the costs of supplementation.

In several trials where cattle were weighed during the rainy season following supplementation, it was demonstrated that compensatory growth occurred which resulted in non-supplemented animals weighing the same by the end of the growing season as those which had previously been supplemented.

There is some evidence that supplementation of grazing cattle exerts a 'pasture sparing' effect, i.e. the supplement given serves as a substitute for forage. It is doubtful whether this is economic under normal conditions when supplement tends to be more expensive than pasture. In any event, management must be sufficiently flexible to enable stocking rate to be increased so as to utilize the additional forage which has been made available.

Creep-feeding of calves is a special case which is only justifiable in cases where maximum weight gain is required throughout the growing period or when particularly high weaning weights are justified by prices paid for the calves.

7 Energy and protein deficiencies: II. The use of non-protein nitrogen

Hageman showed in 1891 that rumen micro-organisms could utilize non-protein nitrogen (NPN) in a simple form such as urea or ammonium bicarbonate. However, it was not until the 1930s that the work of Hart and his colleagues at the University of Wisconsin showed that its use in practical rations for dairy cattle was feasible (Hart et al. 1939). It was discussions with this group that led to the first use of urea for beef cattle in the tropics by Murray and Romyn (1939) at Rhodes Matopos Estates near Bulawayo in Rhodesia. Apart from an unsuccessful trial by Groenewald and van der Merwe (1941), nothing else was done in Southern Africa until the 1950s when a considerable amount of work was carried out by workers at Potchefstroom and elsewhere. The history of the use of urea in stock feeding has been summarized by Strangel (1967).

The first utilization of molasses and urea was in areas of low rainfall to supplement poor-quality roughages low in nitrogen. It was used to counter weight loss as a supplement to maintain the activity of rumen micro-organisms.

It was clearly an attractive possibility as a substitute for more expensive protein of plant or animal origin and many trials were carried out under extensive conditions. Later, however, its use to supplement more intensively-managed pastures was investigated, particularly in the Americas and Australia.

The use of urea to supplement grazing has been reviewed by Pieterse (1967), Topps (1972), Alexander (1972) and more recently by Leng et al. (1977). The role of urea in ruminant protein synthesis was reviewed by Buttery (1976) and the proceedings of a symposium on the use of urea as a protein supplement was edited by Briggs (1967).

Southern Africa

Republic of South Africa and Zimbabwe

Murray and Romyn (1939) calculated that the cost of the nitrogen in urea was less than half that in peanut cake. They therefore compared the two nitrogen sources as isonitrogenous supplements to rations composed of veld hay containing 1.5% digestible protein, maize and sunflower silage (1.4% digestible protein) and molasses (which was fed so as to make the rations isocaloric). The trial lasted for three months. Three groups of pure-bred and grade heifers were fed as follows:

450 g peanut cake
225 g peanut cake plus 37 g urea
74 g urea

Weight gains for the three groups were 20, 17 and 18 kg respectively and, although no statistical analysis was carried out, the authors state that 'at the end of the trial, competent judges could not see any difference between the groups in regard to sleekness of coat or general health' and concluded that urea can be utilized at least in part as a substitute for peanut cake.

A somewhat less successful experiment was carried out by Groenewald and van der Merwe (1941) who attempted to overcome a nitrogen deficiency by the use of a solution in water of 0.066% urea and 2% molasses. This fermented to acetic acid and the authors commented that the animals were 'forced to drink the daily acetic acid equivalent of one and a half bottles of strong vinegar'. Clearly the role of the volatile fatty acids in ruminant nutrition was not appreciated at that time.

Clark (1952) initiated a long series of trials at the Experimental Station of Nooitgedacht at Ermelo in the Eastern Transvaal. This is the region classified by Acocks (1975) as north-eastern sandy highveld where sourer species (i.e. those not readily eaten during the winter dry season) predominate. Animals were fed veld hay only or veld hay plus 560 g molasses and 57 g urea/day. The latter animals gained 20 kg during a twelve week winter period whereas the unsupplemented cattle lost 27 kg during the same period. Clark (1952) also noted that the supplemented animals ate 36 kg hay/week whilst those which were not given supplement ate only 29 kg which was said to indicate the effect on the utilization of fibre.

This finding that animals fed molasses and urea ate more roughage was confirmed in a later trial (Clark and Barrie 1954). Animals fed veld hay alone ate only 8 kg/day whilst those fed veld hay plus a supplement of urea and molasses with mealie-cob (maize-cob) meal ate 9 kg/day. The supplement consisted of 4.5 kg urea dissolved in

11 litres water and added to 22 litres molasses; this was then sprayed over 45 kg of ground mealie-cob meal and fed at the rate of 1.8 kg/day. Gain for the two groups during a fourteen-week period were 450 g and 50 kg respectively. Clark and Barrie (1954) also mentioned a product called mola-meal which consists of molasses absorbed onto cane fibre (presumably bagasse) and to which urea can be added and which can be fed at the rate of 900–1350 g/head/day. The authors described a method of mixing urea with molasses without a carrier and emphasized the necessity for good mixing to avoid toxicity. They concluded that weight maintenance during the winter is economically justified but that fattening during this period is not.

In a subsequent trial, Barrie and Clark (1959) investigated the effect of added phosphorus and cobalt on the utilization of urea nitrogen and included a group of cattle which was not given urea. Degelatinized bone flour and disodium phosphate were compared as sources of phosphorus and both were fed with and without cobalt as a supplement to groups of animals fed on poor-quality grass hay with additional silage. Molasses including 7% urea was given to all groups except the control at the rate of approximately 0.5 litres/day. There was a significant advantage to the addition of cobalt, and without cobalt, soluble phosphate gave significantly better growth than bone flour. However, the mean weight gain for all the groups given urea was 37 kg during the fifteen-week period whilst the animals that were not given urea lost 13 kg, i.e. an overall difference due to urea of 50 kg.

Later work at Nooitgedacht was carried out by Kreft. In an early trial (Kreft 1963), cattle were fed licks with and without urea from the start of the summer in November until the end of the experiment in February more than a year later. The licks contained two parts of degelatinized bonemeal, two parts urea and one part common salt; and the same composition without urea. Average daily consumption of urea during the whole period of 456 days was 42 g and, as would be expected, the consumption was higher during the dry than the wet season. At the end of the experimental period, the difference between the two groups was 45 kg in favour of the animals fed urea.

The following observation was also made: 'A severe fall of snow effected a marked decrease in the body weights of the control groups of oxen while the urea group was not affected at all.' Clear evidence of the advantages of urea feeding under tropical and subtropical conditions!

In a further experiment, Kreft (1966) compared the results of feeding a lick during the winter only with year-long; control animals were given no supplement. The lick was composed of 35% dicalcium phosphate, 40% salt, 10% molasses and 15% urea. The results are summarized in Table 7.1.

Table 7.1 Feeding a urea mix year-long as opposed to winter only: weight changes and carcass characteristics (Nooitgedacht) (adapted from Kreft 1966)

	Urea lick throughout	*Urea lick winter only*	*No supplement*
Summer increase (kg)	135	132	136
Winter loss (kg)	27	23	43
Carcass wt. (kg)	257	253	235
Carcass grade (% head)			
Prime	30	20	0
1	60	60	60
2	10	20	40

Feeding the lick during the summer was without effect on gain. However, winter feeding significantly reduced winter loss, increased final carcass weight and improved carcass grade, an important factor where carcasses are paid for according to quality. It is interesting that more animals of those fed year-long graded prime than those fed during the winter only.

At about the time that work started at Nooitgedacht, Altona et al. (1960) carried out some empirical trials on the feeding of urea at Frankenwald (near Johannesburg). In 1952, Sussex steers were fed a ration of 9 kg veld hay daily which was supplemented with 198 g urea mixed with 0.6 litre of molasses dissolved in water. This group was compared with similar animals fed the same quantity of veld hay plus 9 kg maize silage/day 'plus a green bite of oats'. The animals fed urea gained 25 kg during the winter period of 145 days whilst the other group lost 31 kg. Despite the uncontrolled nature of this experiment, it showed that feeding of urea was beneficial and that a level approximately twice that normally recommended was not toxic. The same treatments were given during the following winter except that half the quantity of urea was given.

The results during the second winter were less dramatic but still significantly different as the urea-fed stock gained 9 kg whilst the controls lost 1.5 kg during the 150 days period.

The use of urea as a supplement to different hays commonly available in the high veld of South Africa was tested in 1956 (Altona et al. 1960). The veld hay consisted of *Trachypogon spicatus*, *Heteropogon contortus*, *Tristachya hispida* and *Hyparrhenia hirta*; the other hays used were good-quality Eragrostis hay (fertilized with N and P) (*Eragrostis curvula*), and good-quality Teff (*Eragrostis tef*) hay. Supplemented animals were given 0.4 litre of molasses containing 70 g urea. All animals were allowed hay *ad libitum*. The results obtained are summarized in Table 7.2.

The best results of urea supplementation were obtained with the

Table 7.2 Weight changes of steers fed different hays with and without urea (Frankenwald) (adapted from Altona et al. 1960)

	Weight change (g/day)	*Hay consumption (kg/day)*
Veld hay (6.8% CP)		
With urea	240	6.5
Without urea	−381	7.1
Eragrostis (9.5% CP)		
With urea	418	6.2
Without urea	295	7.7
Teff (6.1% CP)		
With urea	263	8.0
Without urea	−9	8.2

hays of low crude protein content, as would be expected. However, the authors make the surprising observation that hay consumption was less with urea supplementation than without. There was, however, no statistical evidence to support this claim and it was only marked in the case of the Eragrostis hay which also had the highest content of crude protein. Presumably this was sufficient to ensure adequate microbial activity and that this was not a restraint to further intake.

The effectiveness of urea as a supplement for young calves from six to nine months of age was investigated using silage and Eragrostis hay; 56 g of urea dissolved in 0.5 litre of molasses were sprayed onto the feed and the experiment lasted ninety days. Although there was slight indication that urea improved the growth rate, this was not significant and the trial was repeated with varying amounts of urea. Neither 14, 28 or 57 g of urea were effective, however, and the authors concluded that the utilization of urea by young calves is poor and that the practice is not to be recommended. Unfortunately, no details of the calves or their management is given as it would be expected that their rumens would be fully functioning by this age and that urea would be as effective as in a mature animals. The difference between this and the other experiments might not have been in the age of the animals but in that increase and not merely maintenance of weight was involved.

A further trial was carried out to determine whether either elemental sulphur or magnesium sulphate improved the utilization of urea when used as a supplement to Eragrostis hay. Sulphur in either form did not improve the utilization of urea and it was concluded that the rations fed were not deficient in this element. The weight gains for the whole experiment were low, however (180 g/day), and it is probable that some nutrient other than sulphur was lacking in the diet.

Finally, the incorporation of urea in a salt block for summer grazing was investigated. Salt blocks were manufactured, one of which contained 30% urea (the ingredients other than salt are not stated). Over the forty days of the trial, the animals given urea gained 43 kg. Although the difference was statistically significant, the results are inconclusive as treatments were confounded with possible pasture effects. However, they did provide an indication that summer feeding of urea might be beneficial.

In a later trial (Mackenzie and Altona 1964a), urea was compared with biuret as a nitrogen supplement. Treatments consisted of 130 g of crude biuret or 130 g urea fed with 0.45 kg of molasses/head/day. These were compared with cattle fed a mixture where 70% of the nitrogen originated from biuret and 30% from urea and a control group fed molasses only. The basic ration was either Teff (7.9% crude protein (CP)) or veld grass hay (4.2%CP) *ad libitum*. During the first phase during which the hay was of reasonable quality, improvements attributable to the nitrogen supplements were not significant. However, when a poor-quality hay was fed, significant benefits were obtained. Table 7.3 summarizes the results.

Biuret was subsequently included in a bonemeal lick (three parts biuret, two parts bonemeal, one part salt) for cattle and compared with a similar lick containing salt and bonemeal only (two parts bonemeal, one part salt). Those animals fed the biuret-containing lick gained 23 g/day whilst the animals which were given no supplementary nitrogen lost 136 g/day. The intake of the two licks was 71 and 46 g/day respectively and it was estimated that the nitrogen intake from the biuret was 10 g, a lower value than that normally obtained from a conventional lick (Mackenzie and Altona 1964b).

Although strictly out of our area at 32° 32′S, Dohne (which is north-west of East London in the Cape Province) will be included here for completeness. This is also justified because the veld type at Dohne is a modification of the Highland sour veld which occurs as far north as about 28° S (Acocks 1975), and important grass species

Table 7.3 The effects of urea or biuret supplementation on two hays of differing quality (Frankenwald) (adapted from Mackenzie and Altona 1964a)

Supplement	*Weight change (kg)*	
	Teff hay (7.9% CP) (70 days)	*Veld hay (4.2% CP) (84 days)*
Biuret	39	6
Urea	39	5
Biuret/urea	38	9
Control	30	–20

represented include *Themeda triandra, Heteropogon contortus* and *Tristachya hispida* which are common in the tropics and subtropics. Bishop (1957, 1959) described an experiment in which a mixture of 2.2 kg urea, 22 kg molasses and 13.5 litres water was sprayed on winter veld grazing at a rate sufficient to provide 85 g urea and 850 g molasses/head/day. Cattle on this treatment were compared with those fed an equivalent quantity of molasses without urea and unsupplemented cattle. Losses during the 71-day experiment were 16, 48 and 72 kg for the three groups in the order given. It was also noted that the veld sprayed with the molasses/urea mixture was better utilized than that sprayed with molasses alone.

Work in the highveld was continued by the Potchefstroom Agricultural College Workers. Verbeek and von la Chevallerie (1958) used two rations of urea to molasses, 1 : 20 and 1 : 10, with molasses only (all given at the rate of 0.9 kg/day) and compared the growth rate of oxen with unsupplemented animals. During a 141-day period, weight losses for the four groups in the order given were 22, 0.2, 42 and 37 kg. The value of the increased urea content was clearly demonstrated and it was also evident that molasses alone was without effect. In a subsequent, extension publication (von la Chevallerie 1959), the use of 0.9 kg of the 1 : 10 mixture was recommended to maintain steer weights on dry winter sour veld. The dangers of inappropriate feeding were emphasized. The use of a mixture of urea : water : molasses (1 : 8 : 8) to spray hay or veld was also recommended. von la Chevallerie (1965) returned to the comparison of 1 : 20 and 1 : 10 urea : molasses mixtures for both winter and summer supplementation on veld at Potchefstroom. Supplemented eighteen-month-old steers were compared with control animals without supplement and animals fed molasses alone. During the winter period, both levels of urea checked weight losses significantly whilst molasses alone had no effect. When winter and summer gains were added, the 1 : 20 mixture still had the advantage despite compensatory growth by animals on the other treatments. Results are summarized in Table 7.4.

The author concluded that whilst urea stimulates the appetite,

Table 7.4 Summer and winter gains of animals with and without urea supplements (Potchefstroom) (adapted from von la Chevallerie 1965)

	Liveweight changes (kg)		
	Winter	*Summer*	*Combined*
Veld only	–37	111	73
Veld + 1:20 urea/molasses	–22	103	81
Veld + 1:10 urea/molasses	–0.2	90	90
Veld + molasses	–42	114	72

molasses depresses it. This has the interesting corollary that there must be some level of urea where appetite is neither increased nor decreased and could help to explain some of the equivocal results obtained in trials where mixtures of urea and molasses are fed.

In 1961, Pieterse carried out an experiment in which urea was fed without mixing with either maize meal or molasses, but with these carbohydrate sources fed separately. The loss of weight of the animals fed bone meal and salt only was 39 kg during a 98-day period but animals fed urea with molasses or whole maize only lost 6 and 3 kg respectively. A second trial was carried out to determine whether the extra supplementary feeding with maize or molasses was necessary. Two groups of six yearling cattle were fed either two parts bone meal and one part of salt, or two parts of bone meal, one part of salt and one part of urea, *ad libitum*. The group consuming urea maintained their weight whilst the control animals lost an average of 20 kg (the duration of the trial is not given). Apparently this was the first trial to be carried out in Southern Africa on the use of urea without an additional readily-fermented carbohydrate energy source to aid in nitrogen fixation after ureolysis.

Investigation into licks without carbohydrate components was continued by Pieterse and de Kock (1962). A lick of deglutinated bone meal or dicalcium phosphate and salt (1 : 1) was compared to a similar lick to which one part of urea was added, and the same lick with one part of yellow mealie meal (cobalt, iron, copper and manganese were also included in the lick without mealie but they had no effect on weight gain). The weight losses for the control animals were 25 kg during the 113-day experiment whilst those given urea without mealie meal lost approximately 14 kg; the inclusion of mealie meal reduced the weight loss to 4 kg. The role of the mealie meal was discussed and it seems likely that its inclusion was effective by increasing the level of consumption. In a later extension paper, Pieterse (1964) advocated the use of the mealie meal-containing lick.

These findings prompted the start of more fundamental studies to clarify the role of urea and of additional energy sources. Use of penned steers to determine the rate of passage through the digestive tract of poor veld hay showed that this was markedly reduced by the addition of 50 g/day of urea. Mean retention times for the two treatments were 81.9 and 64.6 hrs and the difference was highly significant. Further supplements of maize starch or white maize meal did not cause further reductions. The urea supplement increased voluntary consumption of both hay and water by approximately 50% (Pieterse et al. 1963).

Simultaneous nitrogen balance trials showed that urea supplements improved nitrogen balance (retention) from −12 g in the animals fed veld hay alone to −3 g in animals fed urea. The further addition of either starch or white maize meal did nothing to improve the nitrogen balance. The addition of urea improved the intake of hay (as already noted) but although the contribution by the hay increased nitrogen intake, this was still not sufficient to achieve a positive nitrogen balance. It is clear, therefore, that a mere increase in the quantity of poor-quality hay consumed is not sufficient in itself to correct a nitrogen deficiency. The authors also suggested that nitrogen-conserving mechanism operated under conditions of nutritional stress. This appeared to be associated with a diminishing urinary nitrogen excretion (Pieterse and Lesch 1963, cf. Elliott and Topps 1963a, already referred to).

Simultaneous work related the supplementation of urea to an increase in the effectiveness of utilization of energy (Lesch et al. 1963). When urea alone was added to a ration of poor veld hay, it was shown that the fraction 'Digestible energy minus urine energy' (an approximation to metabolizable energy when the measurement of methane is not possible) was increased by 74%. Neither the addition of starch nor white mealie meal improved this significantly (although the latter increased the value of 'DE minus urine energy' by 79%).

The same balance trial was used to determine the effects of urea supplementation on the production of volatile fatty acids (VFAs) in the rumen of the steers. As would be expected from the previously cited experiments, the total production of VFAs increased from 51 to 58 moles/ml although this increase was not significant. There were no significant difference among the various proportions of individual fatty acids although there was an indication that slightly more acetic acid was produced in the rumens of animals supplemented with urea. This would be in agreement with the finding of the previous paper that the digestibility of crude fibre was slightly increased (du Toit et al. 1963).

A further series of digestibility trials (Lesch and van Schalkwyk 1967) indicated that the efficiency of urea as a supplement to poor-quality hay was solely as a result of increased intake and not of improved digestibility of energy *per se*. There is a slight inconsistency in the results of this series of experiments when they are compared with the hypothesis of Blaxter and Wilson (1962) that improved digestibility is associated with more rapid rate of passage (Pieterse et al. 1963), but not when compared with results obtained with tropical grasses by Butterworth (1965) who could not relate more rapid rate of passage to improved digestibility.

Experiments on a different highveld type were carried out at Glen Agricultural Station north of Bloemfontein in the Orange Free State. This is described by Acocks (1975) as dry *Cymbopogon-Themeda* veld. Heydenrych (1962) fed 56 g urea/head/day using either molasses alone or molasses and ground maize cobs as a carrier. The experiment was repeated in two consecutive years during the period from

June to November. Average weight gains for the supplemented animals were 26 and 23 kg for the molasses/maize cobs groups respectively whereas unsupplemented animals gained 17 kg. No statistical analysis of the data is presented but the authors suggest that this type of supplementation may be useful in the winter prior to marketing.

In a subsequent extension publication, Lombard and Greeff (1964) recommend the use of a lick consisting of 10% urea, 10% salt, 10% dicalcium phosphate, 10% lucerne meal, 55% mealie meal and 5% binding agent. The energy supplement is given in sour veld areas because of the rapid decline in digestible energy content. For sweet veld areas, however, an energy-free lick consisting of 25% urea, 30% salt, 35% dicalcium phosphate and 5% binding agent is recommended.

Most of the work on NPN supplementation in South Africa was carried out in the sour veld areas characterized by rapid decline in both nutritive value and palatability of herbage after the end of the rains. Little emphasis was given to investigating the effects of the urea supplementation during winter and summer in the lower 'sweet veld ' areas. However, Skinner (1964) working at Mara in the northern Transvaal gave supplement to Bonsmara heifers and cows grazing 'Arid sweet bushveld' (Acocks 1975). This veld type is characterized by the presence of *Eragrostis* and *Urochloa* spp., and *Panicum maximum*, in association with *Acacia tortilis*. Although the nutritive value of these grasses drops off sharply after the rains (results quoted indicated 7.7% CP in the summer and 4.2% in the winter), they retain their palatability and are readily grazed year-round. The supplement given consisted of one part of urea to ten parts molasses (as recommended by the Potchefstroom school) and was given at the rate of 600 g/head/day for young stock and 900 g/head/day for mature stock. Feeding was started in August and finished in November after the first rains. The heifers were weighed again in April to determine the residual benefits of previous supplementary feeding whilst the cows were slaughtered at the end of the feeding period. No significant differences in weight change occurred throughout the experiment.

It was noted that supplemented cattle started to graze indiscriminately after the end of the feeding period whilst the unsupplemented animals took from 30 to 45 minutes before they started to graze and were more selective. It appeared that supplementation increased both hunger and thirst. It was also noted that midway through the experiment, the supplemented animals started shedding their hair faster and that their dung was more fluid in consistency; they were also more active in disposition than the unsupplemented animals. However, as soon as cattle could eat enough green grass, they would not eat the supplement. It was concluded that winter supplementation was of limited value in sweetveld areas.

Louwrens (1964) carried out a collaborative trial with Mr Jamieson of Bulawayo in Zimbabwe. The subsequent extension publication made a point well worth quoting verbatim: 'While supplementary feeding is essential to raise our present low levels of production and reproduction, it should be primarily designed to correct the protein deficiency of natural supplies of roughage during winter at a level to keep animals in a strong and thrifty condition and NOT to cover up bad management of stock and grazing during the summer months in particular.'

The lick used was formulated as follows:

Molasses	65 kg
Urea	7 kg
Phosphoric acid	3 kg
Water	17 kg
Salt	8 kg
Total	100 kg

Copper sulphate and cobalt sulphate were added to the mix at the rate of 220 and 11 g respectively. The mixture was fed as a lick using a rotating tyre and it was emphasized that providing the animals free access to the lick proved uneconomic due to excessive intake. No control animals were included in the trial but weaners were fed enough of the mix to provide 90 g protein equivalent/head/day whilst long weaners received 180 g protein equivalent/head/day. Weight gain during the winter for the two groups was 9 and 2 kg respectively. The cost of supplementation was compared with that of cottonseed cake and found to be highly satisfactory.

Richardson and his colleagues at Matopos initiated a trial on the use of cottonseed cake cubes or cubes containing 5% urea (both supplements contained 37% CP) given either twice or six times/week during the winter. Both supplements were given to cross-bred Hereford weaner steers at the rate of 2.7 kg/week; weight gains were similar for the animals fed six times/week at 12 and 10 kg during the 113 days of the experiment for the cottonseed cake and urea supplements respectively. However, for the twice/week only treatment, those animals supplemented with cottonseed cake gained significantly more than those fed urea, i.e. 14 v. 2 kg (Richardson et al. 1977a) and when animals were finished at the end of the following wet season, there was no significant difference in carcass weights although those of the natural protein-fed group were slightly heavier (Richardson et al. 1977a). Results of the trial are summarized in Table 7.5.

It was concluded that, unless supplementation with urea was cheap enough to cover the increased costs of supplementation six

Table 7.5 Weight changes and carcass weights of steers given urea and natural protein based supplements during the winter (Matopos) (adapted from Richardson et al. 1977)

Weight change	*Cottonseed cake cubes*		*Urea cubes*	
	Twice weekly	*Six times weekly*	*Twice weekly*	*Six times weekly*
First dry season	14.0	12.1	2.3	10.2
Second dry season	–14.8	–14.5	–25.9	–19.3
Final body weight (kg)	498.7	499.7	488.8	488.8
Carcass weight (kg)	248.5	249.8	241.3	241.7

times/week, then it was not to be recommended (Richardson et al. 1977a) although even this was not justified by the similarities of carcass weight.

No explanation is given for the reduced efficiency of urea utilization in the twice-weekly feeding treatment but this is probably because ureolysis is always more rapid than uptake, and under the twice-weekly feeding regime, the available substrate would be rapidly used up leading to ammonia loss after absorbtion into the blood and subsequent excretion in the urine.

Grant (1979) working at the Grasslands Research Station at Marandellas also compared urea with natural proteins. Isonitrogenous quantities of cottonseed cake were replaced by levels of urea of 2, 4, 6 and 8% in a 40% CP equivalent supplement offered at the rate of 500 g daily. Cattle used were Hereford × Afrikander weaner steers weighing approximately 200 kg and the experiment was carried on for three successive dry seasons. The grazing was *Brachystegia* woodland with predominantly *Hyparrhenia* spp. Although all supplements reduced the body weight losses of unsupplemented animals, the response decreased as the level of urea increased. This was less marked with carcass weight than with body weight but the decline was still highly significant. A summary of results is given in Table 7.6.

This decline could be represented by the following equation:

Body weight change: $Y_1 = 28.60 - 1.86X$ ($r^2 = 0.94$)
Carcass weight change: $Y_2 = 16.38 - 0.76X$ ($r^2 = 0.92$)

A quadratic equation was also derived. The authors calculated the efficiency of utilization of urea using both the linear and quadratic models and obtained values for 2% urea of 35.6 and 70.6% respectively and for 8% urea of 35.0 and 32.6% respectively. Thus, the results concur with the commonly-held view that urea should not replace more than 30% of the total nitrogen in the diet (i.e. the 4%

Table 7.6 Body and carcass weight changes of steers given different levels of urea to replace cottonseed meal (3-year means) (Marandellas) (adapted from Grant 1979)

	Body weight change (kg)	*Carcass weight loss (kg)*
Control (no supplement)	−23.5	22.0
Cottonseed meal	+ 4.8	6.2
2% urea	+ 1.8	6.3
4% urea	− 3.6	8.9
6% urea	− 3.7	9.5
8% urea	−10.6	12.5

urea level in the present study). However, it is not entirely consistent with other experiments in which urea has apparently been as effective as natural protein supplements in weight maintenance if not in weight gain (gains were only minimal in the present trials). This may be because of the increased precision of Grant's work due to the large number of animals used (540 head) and also because control animals were slaughtered at the start of the experiment to enable accurate prediction equations of initial carcass weight to be calculated.

Eastern and Central Africa

In comparison with the large amount of research carried out in Southern Africa on NPN for grazing cattle, little has been done in Eastern and Central Africa. Smith (1961) at Mazabuka, Zambia, compared the effects of spraying veld hay (2.5% CP) with a mixture of one part urea to six parts molasses, with low-quality (4.0% CP) Rhodes grass (*Chloris gayana*) hay and the same hay sprayed with the urea/molasses mixture. Unfortunately, veld hay (mainly *Hyparrhenia* spp.) alone was not used as a control treatment as previous studies had shown an 'excessive' weight loss of the order of 450 g/day when the hay was fed without supplement. The cattle fed the veld hay sprayed with the urea/molasses mixture gained 400 g/day. There was no difference in weight gain between supplement or non-supplemented Rhodes grass hay as both groups of animals gained 320 g/head/day. This is a level at which no advantage would be expected from supplementation (see previous discussion). Although it would appear that the protein content of the Rhodes grass hay would be too low to support this level of weight gain, the author states that there was some selective eating.

In a later trial, it was shown that the addition of urea at the rate of 110 g/head/day increased the intake of veld hay (Hyparhenia spp., 42%; *Brachiaria brizantha*, 20%; *Setaria sphacelata*, 20%; *Heteropogon contortus*, 13%) by approximately 41% (Smith, 1962a).

Owen (1968) carried out several experiments aimed at increasing the productivity of cattle grazing the Kongwa plains in central Tanzania. In one of these, three-year-old Boran steers of 400 kg mean liveweight were allocated to treatments of three × three factorial to investigate the effects of hay or urea supplements to poor *Cenchrus ciliaris* pastures. Treatments were as follows:

No urea
Urea block I (40% urea, 3.8% phosphorus, molasses and other minerals)
Urea block II (35% urea, 45% common salt, 10% bonemeal, 10% molasses)
No hay
Medium-quality hay (6% CP, supply limited)
Good-quality hay (10% CP, supply limited)

The results obtained for the various treatments are shown in Table 7.7.

Table 7.7 Effects of supplementary feeding on weight loss of Boran steers (Kongwa) (adapted from Owen 1968)

	No urea (kg)	*Urea supplement*	
		Urea block I (kg)	*Urea block II (kg)*
No hay	–37	–44	–49
Medium hay	–35	–25	–32
Good hay	–20	–22	–26

The feeding period lasted 140 days. There was no response to the feeding of urea nor was there any interaction between urea and hay feeding. This result is not in agreement with others which have indicated that urea supplementation will counteract dry season weight losses. However, these results were probably because of increased intake and it is likely that there was insufficient pasture or hay available in the present study to allow this.

North and Central America and the Caribbean

United States of America

Surprisingly, little work has been found from the United States on supplementation with NPN and molasses, although Wagnon and Goss (1961) started experiments on spraying dry rank forage with urea and molasses in 1946–7 with positive results insofar as forage utilization was concerned. The supplement self-fed, however, was ineffective insofar as weight maintenance was concerned. As this work was carried out at the San Joaquin Experimental Range, which is out of our areas, it will not be mentioned in more detail here.

Workers at the Iberia Station in Louisiana used biuret to replace some cottonseed meal in rations for wintering beef heifers in the Gulf Coast area of the state (de Rouen et al. without date). Although this compound has not been employed as widely and as long as urea, it has the advantage of being less toxic and mixing is therefore less critical. Biuret was fed at the rate of 90 g/day and the authors concluded that it was used efficiently as a source of NPN because animals fed the ration containing biuret maintained similar rates of growth to those given only cottonseed meal, the average daily gains for the two groups being 163 and 168 g/day respectively. Unfortunately, no details are given of the type of pasture used or of the stocking rate.

Mexico

Carrera et al. (1969) supplemented cross-bred Zebu steers grazing a *Pancium maximum* pasture with a urea and molasses mixture containing the equivalent of 30% crude protein during both the wet and dry seasons in Paso del Toro, Veracruz, Mexico. The performance of the cattle is shown in Table 7.8.

The supplement was fed for 120 days in both seasons and the stocking rate was 1 beast/ha. Differences were significant for the dry season period only ($P < 0.05$).

Butterworth et al. (1973) supplemented Zebu/Criollo bulls grazing Pangola grass pastures with molasses alone or molasses containing 3 or 6% molasses. Cotnrol animals were given no supplement. The experiment was carried out in the district of Aldama, Tamaulipas, and lasted 112 days from February to May. The stocking rate was approximately 1.5 animals/ha. The results are summarized in Table 7.9.

Table 7.8 Gain of steers supplemented with urea/molasses during wet and dry seasons (Paso del Toro) (adapted from Carrera et al. 1969)

	Gain per day (g)		*Intake* (g)
	With supplement	*Without supplement*	
Wet season	808	746	1 000
Dry season	571	397	1 375

Table 7.9 Daily weight gains (g) and supplement consumption (g) of bulls supplemented with molasses with or without urea (Aldama) (adapted from Butterworth et al. 1973)

	Control	*Molasses*	*Molasses 3% urea*	*Molasses 6% urea*
1st period	388	528	603	740
2nd period	564	691	864	639
3rd period	751	925	592	968
4th period	781	1 066	888	960
Overall	621	802	736	826
Consumption of supplement	—	889	901	799

Gains of supplemented animals were significantly superior to unsupplemented animals. The influence of the supplements decreased as more herbage became available and the correlation between ambiental temperature and weight gain indicated that temperature was the most important limiting factor in the growth of the grass (weight gain/day increased by 30.3 g for every 1 °C increase in ambiental temperature). There was an apparent response to urea when forage availability was at its lowest but this could not be demonstrated during later periods. Response to supplements declined as the growth rate of the unsupplemented animals increased.

In a paper which combined the results of the two previous experiments, Butterworth and Carrera (1969) derived the relation between the growth rate of control animals and the expected benefit to be obtained by supplementation with 1 kg of molasses. Expected increments are summarized in Table 7.10.

Table 7.10 indicates that the efficiency of utilization of supplement is inversely related to the growth rate of unsupplemented animals.

Garza et al. (1975b) investigated the effects of fertilization and supplementation on the productivity of Zebu heifers grazing Pangola grass (*Digitaria decumbens*) at Paso del Toro, Veracruz. The heifers

Table 7.10 Benefits to be expected by supplementation with 1 kg daily of molasses related to growth rate of unsupplemented animals (adapted from Butterworth and Carrera 1969)

Growth rate of unsupplemented animals (g/day)	*Incremental growth expected from supplementation (g/day)*
300	290
500	236
700	183
900	129

Table 7.11 Weight gains of cattle with and without supplementation on fertilized and unfertilized Pangola grass during the wet season (Paso del Toro) (adapted from Garza et al. 1975b)

	Stocking rate (head/ha)	*Daily gain* (g)	*Liveweight gain per ha* (kg)
Control	3	529	223
Fertilized	5	408	286
Supplemented	5	449	315
Fertilized plus supplemented	7	427	419

were fifteen months old and weighed 226 kg. Treatments were 0 and 50 kg N/ha and 0 and 2 kg/head of a supplement containing 90% molasses and 3% urea (CP 8.5% and TDN 48.6%). The stocking rates were 3, 5, 5 and 7 head/ha for the treatments respectively. The experiment was carried out during the wet season from the beginning of August to the end of November (140 days). The results are summarized in Table 7.11.

Unfortunately, the fact that stocking rate and treatment were confounded in this experiment makes it impossible to distinguish what effects were due to treatment and which were due to stocking rate. It is clear that although there was an increase in beef production per ha on both supplemented and fertilized treatments, reduced forage availability due to increased stocking rate more than nullified any individual increases in weight gain.

The trial was carried into a second, dry season, phase which lasted 140 days during the months of February to June (Garza et al. 1975a). Similar stocking rates were used to those of the previous trial. Treatment effects were clearer in that fertilization supported both an increased stocking rate and increased gain per animal whilst supplementation supported the same gain as the unsupplemented animals

Table 7.12 Weight gains of cattle with and without supplementation on fertilized and unfertilized Pangola grass during the dry season (Paso del Toro) (adapted from Garza et al. 1975)

	Stocking rate (head/ha)	*Daily gain* (g)	*Liveweight gain* (kg/ha)
Control	3	309	130
Fertilized	5	406	284
Supplemented	5	309	216
Fertilized plus supplemented	7	464	355

but at a heavier stocking rate. Supplementation of animals grazing fertilized pastures gave highest gains per ha. Results are summarized in Table 7.12.

The evident pasture-sparing effect of supplementation is interesting in view of the discussion at the end of the last chapter where similar cases were mentioned (see Table 6.23).

Panama

Although carried out with stall-fed animals, the work of Roux and Parada (1969) and Roux and Rodriguez (1971) may be included in this section. Animals were given hay (*Hyparrhenia rufa, Digitaria decumbens, Echinochloa polystachya* and *Panicum maximum*) *ad libitum*. Experimental treatments were *ad libitum* molasses (nine parts) plus urea (one part), a similar mixture to the above but administered with a wooden 'duckboard' (this allowed the animals to lick the mixture but not to drink it and thus restricted intake). Control groups were given hay alone and hay plus molasses without urea, both supplied *ad libitum*. Weight gains for the four groups respectively were 500, 419, 126 and −117 g/day. Steers on the diets supplemented with urea and molasses consumed more dry matter than those animals on the other two diets but this was entirely accounted for by the consumption of the mixture itself. The intake of hay by the animals given molasses without added urea was significantly reduced compared with that of the other groups. It was concluded that the supplementation of molasses and urea during the dry season would probably be economically viable under Panamanian conditions but the authors also admitted that the possible effects of compensatory growth should be considered.

Cuba

The first work carried out in the Caribbean area appears to be that of the Cuban workers. Perez Infante (1969) reported gains of up to 0.9 kg daily of bulls on fertilized Pangola grass pastures supplemented with molasses/urea and concentrates, and Morciego et al. (1970) obtained gains of 0.84 kg daily on unfertilized natural pastures during the dry season at Camaguey.

Porres (1971) gave Holstein/Brahman bulls grazing highly fertilized Pangola grass, supplements of either molasses (84 °Brix) *ad libitum* or a mixture of molasses containing 2% urea and 6% water and compared with the growth rates of unsupplemented animals. The stocking rates for the three treatments were the same, i.e. 4.8 bulls/ha during the dry season and 7.3 bulls/ha during the wet season. The growth rates during the two periods are given in Table 7.13 as are intake of the molasses, and transit shrink and killing-out percentage.

There was no response to supplementation during the dry season

Table 7.13 Weight gain and carcass characteristics of cattle fed molasses and molasses/urea as a supplement to heavily fertilized Pangola grass (adapted from Porres 1971)

	Pasture alone	*Pasture plus molasses*	*Pasture plus molasses plus urea*
Weight gain dry season (g/day)	520	530	570
Weight gain wet season (g/day)	910	980	1010
Weight gain overall (g/day)	640	690	730
Molasses intake (kg/day)			
dry season	—	2.37	2.76
wet season	—	4.08	4.76
Transit shrink percentage	12.0	10.5	7.4
Killing-out percentage	47.4	49.0	50.6

but significant responses were obtained when both periods were considered together. The author concluded that the greater availability of forage during the wet season was the main reason for this. However, this explanation is unsatisfactory because of the different stocking rates used during the two periods. Better protein quality was also given as a contributory factor. The addition of urea increased the intake of the molasses mixture and this could also have been a reason for improved response although the authors suggest that increased consumption was at the expense of pasture. It was not likely that increased intake of nitrogen was the cause as the content of the grass was between 8 and 15% which would be more than adequate to maintain the gains reported. The lack of response during the dry season was attributed to the lack of true protein (or 'by-pass protein'; see discussion in previous chapter). The differences in shrink and killing out percentage were not significant but the trends were to be expected because of the substitution of roughage by concentrate.

A subsequent experiment confirmed that addition of urea was without effect during the wet season. Molasses (85%) containing 0, 1, 2, 3, 4 and 5% urea was given to Brahman and Brahman × Friesian bulls weighing approximately 200 kg. It was shown that different levels of urea did not affect gain but reduced the molasses intake and reduced the killing-out percentage (Veitia et al. 1972).

Ugarte and Preston (1975) considered that one of the disadvantages of molasses/urea supplementation was the necessity to add true protein in order to achieve high rates of gain. However, an experiment was carried out to investigate the effects of molasses and urea when lower rates of gain were acceptable (i.e. in young female cattle where high rates of gain were unnecessary). The treatments were fertilized Pangola pasture without supplement (control), Pangola pasture plus *ad libitum* molasses/urea (molasses 91.5, urea 2%, water 6.0%, common salt 0.5%) and Pangola pasture restricted to 2 hrs grazing/day plus the molasses/urea mixture as before *ad libitum*. These treatments were applied during the wet season. During the dry season, half of the animals of each treatment were given free and the other half restricted grazing. The crude protein content (% dry matter) of the Pangola grass was 8.6 and 5.8% for wet season and dry seasons respectively. Daily liveweight gain, consumption of the molasses/urea mixture and conversion are given in Table 7.14.

Table 7.14 Liveweight gain, molasses consumption and utilization efficiency by female Holstein calves grazing fertilized Pangola grass (Havana) (adapted from Ugarte and Preston 1975)

	Free grazing	*Free grazing plus ad libitum molasses/urea*	*Restricted grazing plus ad libitum molasses/urea*
Wet season			
Liveweight gain (g/day)	350	310	290
Molasses/urea consumption (kg/day)	—	2.87	3.96
Conversion	—	9.10	13.50
Dry season			
Liveweight gain (g/day)	—	420	180
Molasses/urea consumption (kg/day)	—	3.60	4.88
Conversion		8.60	27.10

Conversion equals kg molasses/kg liveweight gain

There was no response to supplementation during the wet season. There was a carry-over effect, however, because calves which consumed molasses during the wet season gained more weight during the dry season than calves which did not. Consumption was higher and conversion was poorer during both periods when grazing was restricted. The authors suggested that part of the poor growth of the restricted grazing animals during the dry season was because of a higher stocking rate (8.0 v. 4.8 head/ha). Animals were thus forced to consume lower-quality herbage. They further concluded that there was no economic advantage in the use of molasses/urea when pasture was available and that this was due to the substitution of pasture by molasses.

Much of the work carried out by the Cuban school has been directed towards the maximum use of molasses/urea under feedlot conditions. However, one experiment on the use of restricted grazing as a roughage source is of interest in the context of this review. Holstein steers five months old and weighing 140 kg gained 860 g/day given 1 kg/day of Pangola hay and the same when allowed access to a Pangola pasture for two hours each day.

The consumption of molasses of the two groups was 6.7 and 6.2 kg/day respectively, indicating that the latter animals had a higher consumption of roughage than the former. The higher efficiency of the grazing animals could have been due to their opportunity for selective grazing (Ugarte and Preston 1974).

Martín and Alonso (1978) investigated beef production/ha from Pangola grass fertilized with 100 kg/ha N supplemented with restricted levels of molasses/urea. The cattle used were cross-bred ($\frac{3}{4}$ Holstein × $\frac{1}{4}$ Zebu) yearling calves and either 1.5 or 2.5 kg/day/head were given of a supplement consisting of molasses containing 7% urea; the stocking rate used was 5 beasts/ha. Live-weight gains/ha/year for control (unsupplemented) and low and high levels of molasses/urea were 368, 541 and 595 kg respectively. Although higher gains were achieved by feeding more of the supplement, it was concluded that this would not be economically advisable and that the efficiency of utilization of the supplement declined with increasing levels. As will be discussed later, these positive effects of urea supplementation are unusual in the context of Latin American conditions with improved pastures and this may be associated with the relatively high levels of urea used.

The Dominican Republic

Campion (1972, 1973) carried out similar work on the use of pasture to provide the roughage required for maximum molasses consumption. High-grade Holstein bulls were restricted to two hours' grazing (Pangola?) pastures and were given three different sources of nitrogen: urea plus fishmeal, urea plus peanut meal plus soybean meal and peanut meal without urea. The nitrogen intake of each group was more or less similar. Growth rate was 500 g/day and there were no significant differences among groups. It was concluded that urea could effectively provide 60% of animal requirements when fed with a vegetable or animal protein concentrate. Growth rates were slightly lower than those obtained by the Cuban workers and this was attributed to the higher nitrogen content of the Cuban molasses.

South America

Brazil

Mott and his colleagues carried out a comprehensive series of experiments on grazing steers in Matao and Araçatuba in the state of Sao Paulo (Mott et al. 1967). This has already been referred to in connection with molasses supplementation in a previous chapter. In one experiment, molasses alone and molasses with urea were compared with ground maize as a supplement for cross-bred Zebu steers grazing Colonial grass. Trials were carried out both during summer and winter. The results are given in Table 7.15.

Table 7.15 Responses of steers grazing Colonial grass to supplementation with molasses and molasses/urea during winter and summer (Matao) (adapted from Mott et al. 1967)

	Gain (kg)		
	Winter (112 days)	*Summer (196 days)*	*Winter and summer (308 days)*
Control	– 7.5	132.3	124.8
Molasses alone (1.3 kg)	19.2	141.9	161.1
Molasses (1.3 kg) plus 81 g urea	23.0	138.2	161.2
Ground maize (1.0 kg) plus 69 g urea	27.0	126.1	153.1

Supplemented steers grew significantly better than unsupplemented steers but there was no advantage to the addition of urea. It appeared that there was a carry-over effect as animals supplemented during the winter grew faster than unsupplemented animals during the summer period.

Villaca and his colleagues carried out an experiment at the Experimental Station of 'Aqua Limpa' in the state of Minas Gerais. Cross-bred Holstein Zebu steers with an approximate starting weight of 265 kg were allocated to the following treatments:

Pasture plus molasses *ad libitum*
Pasture plus molasses containing 5% urea *ad libitum*
Pasture alone

The pasture was mainly Capin Gordura (*Melinis minutiflora*) and the stocking rate was 1 animal/ha. The molasses was given in roller feeders. The experiment continued for 140 days during the second half of the wet season. The daily gains were 636, 738 and 562 g/day for the three treatments in the same order as above. Although the authors claim that the mixture containing urea was more efficient than molasses alone, this conclusion was not borne out by the statistical analysis which did not demonstrate any statistical difference between the two supplemented groups although these were superior to the unsupplemented group (Villaca et al. 1976).

Venezuela

A more detailed study on the effects of molasses and molasses/urea supplementation was carried out in Maracay by Carnevali et al. (1970). Cattle grazed successively Pangola (*Digitaria decumbens*), Star (*Cynodon plectostachyus*) and Guinea (*Panicum maximum*) grass pastures at a stocking rate of 2.5 animals/ha. Average protein content was 9.5%. Groups were given pasture alone, pasture plus 4 kg/day of molasses, or pasture plus 4 kg/day of molasses plus 150 g urea. Weight gains for the three treatments were 598, 589 and 600 g/day respectively. There were no significant differences among treatments.

Cellulose utilization was determined by using the nylon bag technique and this was significantly lower in the animals given molasses. On the other hand, rumen fermentation rates were higher in animals given urea. The authors conclude that the addition of molasses reduces the fermentative processes responsible for digestion of roughage in the rumen but that the addition of urea re-established the balance between energy and protein necessary for bacterial activity. These results would indicate that not only does molasses feeding reduce the intake of pasture but that it also makes roughage utilization less efficient (see also Carnevali et al. 1968, which apparently was the first time this work was reported).

Chicco and his colleagues later carried out a trial to compare the effects of biuret and urea fed as supplements to Brahman cross-bred steers with an initial weight of 257 kg grazing natural pasture during both wet and dry seasons, at Calabozo (Chicco et al. 1975). The pastures used were approximately composed of:

Paspalum plicatulum 50%
Leersia hexandra 30%
Other (principally *Sporobolus indicus* and *Panicum laxum*) 20%

The treatments used were as follows:

Control (pasture plus salt mixture).
Pasture supplemented with salt mixture containing 45% biuret.
Pasture supplemented with 2 kg molasses/day plus urea containing an equivalent amount of nitrogen to that consumed by the animals in Group 2.

Pasture supplemented with 2 kg molasses/day plus biuret containing an equivalent amount of nitrogen to that consumed by the animals in Group 2.

The results are summarized in Table 7.16. There was a significant superiority of the animals fed molasses/urea during the dry season but no significant differences were noted during the wet season or overall. The authors point out that possible differences might have been obscured by the low stocking rate used which was 0.33

Table 7.16 Weight gains of supplemented steers grazing natural pasture in both wet and dry seasons in the Venezuelan llanos (Calabozo) (adapted from Chicco et al. 1975)

	Control	*Biuret*	*Molasses/ Urea*	*Molasses/ biuret*
Dry season gain (g/day)	493	442	593	465
Wet season gain (g/day)	597	630	597	605
Overall gain (g/day)	525	536	595	535

animals/ha, but compensatory growth undoubtedly occurred. The protein content of the pasture was 5.4% during the dry season and 7.0% during the wet.

Columbia

Workers in Carimagua investigated the effect of supplementing *Melinis minutiflora* pastures with molasses plus urea (400 g plus 80 g) during the dry season (CIAT 1974). Two stocking rates were used: 0.44 and 0.88 head/ha. A summary of results is given in Table 7.17.

Supplementation effectively reduced weight losses during the dry season. However, the unsupplemented animals showed considerable compensatory gain during the subsequent wet season at the lower stocking rate when ample forage was available, but this was not the case at the higher stocking rate. This interaction between compensatory growth and stocking rate explains some of the cases when compensatory growth has not occurred after supplementation as mentioned in the last chapter.

The supplementation with molasses and urea had a greater effect during a severe dry season (CIAT 1977). Highly efficient conversion ratios were obtained during this period. One kilogram of additional

Table 7.17 Weight changes of animals during wet and dry season, with and without molasses/urea at two stocking intensities (Carimagua) (adapted from CIAT 1974)

	Weight change per day (g/head)		*Gain/ha (kg/335 days)*
	Dry season	*Wet season*	
Control (no supplement)			
0.44 head/ha	–529	731	51
0.88 head/ha	–539	451	11
With supplement			
0.44 head/ha	112	532	59
0.88 head/ha	93	395	57

weight gain was produced by 0.28 kg urea and 1.40 kg of molasses in the low stocking rate treatment whilst the corresponding values for the high stocking rate treatment were 0.71 and 3.50 kg.

Animals grazing native savannah at Carimagua were also supplemented with urea plus cassava meal. Urea alone was ineffective in counteracting dry season losses and pronounced compensatory growth nullified the dry season effects of urea plus cassava (CIAT 1977).

Australia

Trials on urea supplementation have been carried out both in north and south Queensland and in north New South Wales since the late 1950s. Different grass associations, notably *Heteropogon contortus/Axonopus affinis* and some improved pastures, have been characterized by differing responses to the supplement.

The first work to be carried out in Australia was by Morris (1958) at Yeerongpilly. Yarded heifers were fed bush hay (*Paspalidium globoideum*, 20%; *Eriochloa* spp., 20%; *Dichanthium sericeum*, 15%; *Aristida leptopoda*, 10%; etc.) *ad libitum*. The hay was harvested in the Central Queensland Highlands and contained 4.6% CP. Animals were given the following treatments:

Bush hay *ad libitum*.
Bush hay *ad libitum* + lick *ad libitum* (bone flour + salt + Cu + Co).
Bush hay *ad lib* + lick *ad lib* + 450 g crushed grain sorghum/head/day.
Bush hay *ad lib* + lick *ad lib* + 450 g grain sorghum + 57 g urea + 5 g Na_2SO_4/head/day.

Weight changes and hay consumption over the twenty-one weeks of the trial are summarized in Table 7.18.

The lick was without effect in counteracting the weight loss but

Table 7.18 Weight changes and hay consumption of yarded heifers fed bush hay and various supplements (Yeerongpilly) (adapted from Morris 1958)

	Weight change (kg)	*Hay intake (kg/day)*
Hay only	−9	5.2
Hay + lick	−11	5.3
Hay + lick + sorghum	0	5.7
Hay + lick + sorghum + urea	+36	6.6

sorghum reduced it significantly ($P < 0.05$); the weight gain by the animals fed urea reached significance at the 1% level, however. The urea also significantly increased hay consumption.

Beames (1959) at Yeerongpilly fed supplements for molasses alone and molasses plus urea to maiden Hereford heifers stall-fed pasture hay. The hay was similar to that used in the last experiment and contained 3.5% CP. Three groups of four animals each were given one of the following rations:

Hay *ad libitum*.
Hay *ad libitum* + 680 g molasses/head/day.
Hay *ad libitum* + 680 molasses + 100 urea/head/day.

After the pen-feeding trial which lasted for sixteen weeks, cattle were turned out to graze on a predominantly *Paspalum* pasture for a further fifteen weeks.

The daily weight change on the three treatments was −545, −385 and −40 g/head in the same order as above. The addition of urea significantly reduced weight loss. After fifteen weeks on pasture, the weights were 432, 385 and 414 kg respectively, thus maintaining the trend established during the stall-feeding period although the latter differences were not significant. Hay intake was measured and it was observed that addition of molasses decreased hay consumption by approximately 200 g/day/head (although this was not significant), whereas with the incorporation of urea the hay intake was increased by some 39% which was highly significant. Both molasses + urea tended to increase the digestibility of dry matter but no statistical analysis was carried out in these results.

Subsequently, Beames (1960a) carried out trials to determine the intake of molasses containing different levels of urea fed daily or twice weekly. Levels of urea were increased by small increments at weekly intervals until urea was 50% of the weight of the molasses. After the third week, the consumption of the mixture decreased so that the intake of urea was maintained more or less constant. The treatment was stopped after twelve weeks and molasses was fed alone. After sixteen weeks, the 50% urea mixture was recommenced and one animal died. The obvious conclusion was drawn that precautions should be taken whilst introducing animals to so strong a mixture. A further trial confirmed that as the level of urea increased so consumption decreased. Unfortunately, neither of the two previous trials were controlled and data were not analysed.

The last trial in this series was to evaluate the effects of feeding 680 g of molasses plus 100 g urea either daily or twice per week. Hereford steers were fed hay *ad libitum* with or without the supplements. Both supplemented groups weighed significantly more than the unsupplemented animals after sixteen weeks but there was no significant difference between daily and twice weekly supplemented animals. There was, however, a strong indication that daily supplementation was more effective as judged by both weight change and voluntary hay intake. Results are summarized in Table 7.19.

Table 7.19 Weight changes and hay consumption of steers supplemented daily and twice weekly (adapted from Beames 1960)

	Weight change (kg)	*Hay intake (kg/day)*
Hay only	−30	4.5
Hay + molasses/urea daily	+25	7.3
Twice weekly	+11	6.2

In a further communication, Beames (1960b) noted that urea and molasses could be fed in aqueous solution by ensuring that the solution was alkaline. This was done by saturating the water with slaked lime and providing a surface protection with cetyl alcohol which prevented the absorption of CO_2 and ensured that the pH would remain between 10 and 11 for at least fourteen days. Although this treatment prevented the formation of acetic acid, as mentioned elsewhere, it does not seem to have achieved popularity.

Later, Beames (1963) compared cattle fed a urea block with animals fed molasses-treated hay and hay alone. The hay fed was composed of *Bothriochloa intermedia* (78%) and *Paspalum dilatatum* (89%) and contained 3.5% CP.

Hay was sprayed with molasses at the rate of 12 kg to every 100 kg air-dry hay. The composition of the block was: urea, 40%; molasses, 10%; trisodium phosphate, 2.5%; salt, 47.5%.

After eighty-five days of the trial, several animals were withdrawn from the first two treatment groups because of emaciation. Results are summarized in Table 7.20.

Molasses alone was ineffective in preventing weight loss; however, weight loss of animals fed urea was significantly less than those on the other two treatments. There was an indication that molasses reduced hay intake but that urea increased intake by some 30 to 45%.

Meanwhile, O'Bryan (1960), also working in south-eastern Queensland (60 miles north of Brisbane) obtained negative results spraying Carpet grass (*Axonopus affinis*) pastures with molasses and urea. The mixture used contained: urea, 12.5%; molasses, 78%; monosodium phosphate, 4.4%; cobalt chloride, 0.005% (the balance was water) in 1958; in the following year, the mixture was the same except that it contained 75% molasses and 18% urea. The treatment was ineffective in preventing the autumn/winter (March through September) weight loss as that of the two groups appeared (results were shown in graphical form) to maintain the same relation to each

Table 7.20 Weight changes and hay intake of steers fed treated hay and urea block for 161 days (adapted from Beames 1963): results from animals withdrawn because of emaciation have not been included

	Weight change (kg)	*Hay intake (kg/day)*
Hay only	–41	4.8
Hay + molasses	–51	4.5
Hay + molasses + urea	–14	6.2

other throughout the experimental period. It was suggested that the low palatability of the Carpet grass, combined with the high degree of selective grazing practised by the animals, were the main factors responsible for the lack of treatment effect. This was enhanced by heavy dews and the occurrence of intermittent rains. The extent to which selective grazing was practised is indicated by the fact that the protein content of the pasture as sampled averaged 5.4% whereas when faecal samples were analysed to calculate the protein content of the diet actually selected, the average values were 11%.

Coombe and Tribe (1962) carried out a series of experiments using penned and grazing sheep and cattle; however, only one of these was within our area and this is not particularly edifying. Pastures at Canovie Station (north Queensland) comprising predominantly Mitchell grass (*Astrebla* spp.) were sprayed with sufficient molasses/urea mixture to provide 120 g urea and 600 g molasses/head/day. The spraying was carried out from 10 October to 24 December. The crude protein content of the unsprayed grass during this period was 2.8%. It was not possible to weigh the cattle but subjective assessment by local cattlemen indicated that cattle grazing sprayed grass were 'more alert, more thrifty and had rounder flanks' than cattle grazing unsprayed grass.

Negative results were obtained by Jones (1966) working at Samford, south-eastern Queensland. A urea-containing block (urea, 35%, salt, 48%; molasses, 10%; pollards, 5%; trisodium phosphate, 2%) was fed as a supplement to planted pastures of *Setaria sphacelata* ('Nandi') or *Chloris gayana*; weaners weighing on average 170 kg were stocked at the rate of 5 beasts/ha. Mean consumption of the block was 65 g/day by the cattle grazing Rhodes grass and 57 g for the cattle grazing Setaria. The supplemented animals on Rhodes grass lost 3 kg/head during the ninety-eight days of the trial whilst unsupplemented animals lost 7 kg/head. The corresponding figures for the animals on the Setaria pastures were 9 and 8 kg/head. The differences were not significant. Figures for the nitrogen content of the two pastures indicate that the lowest value was 0.8% N (of the Rhodes grass during August) and for the most of the experimental period, values exceeded 1.0% N; it was suggested that no response to urea supplementation may be expected when animals are able to select a diet containing 1.0% N or above. This is consistent with the results obtained by O'Bryan (1960).

However, despite the preceding negative results, there is some evidence that urea and molasses was already widely used by farmers in the area by the mid-1960s. Mutch (1966) states 'the nitrogenous substance urea is being used extensively in both production and survival rations for cattle and is providing a substitute for protein in supplementary feeding throughout Queensland'. The author goes on to describe the highly successful lick feeder and recommends a mixture composed of urea, 48 kg; molasses, 120 kg; water, 320 kg for use therein. It was reported that heifers at 'Arcturus Downs' consumed approximately 450 g/head/day of the mixture.

In a subsequent extension publication, Weller (1969) describes the effectiveness of urea/molasses fed from a roller drum lick in increasing the appetites of cattle so that they browsed the normally unpalatable sandalwood saplings and cypress pine.

Pollard (1970) described an easy way of mixing molasses licks. This employed a divider (a third of 200 litre drum with 100 holes punched in the bottom) to separate the molasses into thin streams as it is run into the mixing vat; this is then stirred with a shovel and fed in a roller drum lick in the usual way. An interesting point in favour of urea/molasses supplements is made: 'The lick can be given without causing a stir among the cattle. When hay is fed, the cattle will run to the feeding point for fear of missing their ration.'

Perhaps the most effective use of molasses and urea mixtures has been made in the Spear grass zones of northern Queensland where the rainfall occurs during the summer months and the resulting forage rapidly becomes mature. The workers at the Swan's Lagoon cattle field research station have carried out a number of trials demonstrating the effectiveness of such supplements. The area is described as open canopy forest with Black Spear grass (*Heteropogon contortus*), Giant Spear grass (*H. triticeus*) and Kangaroo grass (*Themeda australis*) (Condon et al. 1970). The average carrying capacity is one livestock unit to 8 to 10 ha.

The first trial compared results of feeding molasses alone and molasses with urea as a supplement to native pasture for weaner Brahman × Shorthorn calves at a stocking rate of one beast to 2 ha. The average consumption was 250 g of molasses in both groups whilst urea consumption was 50 g/day. During the dry season, the urea-supplemented animals gained more weight than unsupplemented animals or those supplemented with molasses alone and the advantage persisted through the wet season. Results are summarized in Table 7.21. A complete analysis of costs was made and it was concluded that daily cost per head was about 1.5 cents/day and that

Table 7.21 Effects of feeding molasses alone and molasses with urea on weight gains of steers grazing native pastures (adapted from Condon et al. 1970)

	Weight gain (kg) (June–January) (223 days)	*Weight gain (kg) (June–May) (328 days)*
Pasture alone	4	82
Pasture + molasses	4	84
Pasture + molasses + urea	32	106

the increase in value of the cattle was at least $(Australian)7.50/head.

Winks and Laing (1972) carried out an experiment to compare molasses, phosphoric acid and urea as supplements to native grass during both wet and dry seasons at Swan's Lagoon. The treatments which concern us here (others have already been mentioned in the chapter on phosphorus) were as follows:

No supplement
Molasses (actual consumption: 0.21 kg/head/day)
Molasses + urea (consumption of urea: 51 g/head/day)
Molasses + urea + phosphorus

The supplements were fed from January until October. The changes in liveweight (Table 7.22) show a period of gain from January until June followed by a period of losses from June to October. Groups fed molasses during the wet season gained significantly more than other groups. However, during the dry season from May to June, the urea-fed group gained significantly more than all other groups, and from June to October both groups fed urea lost less weight than the other groups. Weight changes after supplemen-

Table 7.22 Weight changes of steers fed different supplements during wet and dry seasons (Swan's Lagoon) (adapted from Winks and Laing 1972)

	Mean weight changes during and after supplementation (kg/day)			*Final weight (kg) (May 1971)*
	Jan–May (1970) (wet)	*May–June (1970) (dry)*	*Jun–Oct (1970) (dry)*	
Nil	0.65	0.13	–0.23	365.4
Molasses	0.70	0.17	–0.25	366.8
Molasses + urea	0.64	0.35	–0.01	391.5
Molasses + urea+ P	0.68	0.18	–0.08	383.5

tation had finished did not differ significantly but the urea-fed animals retained their advantage until slaughter, producing a heavier carcass.

Faecal analysis showed that the response to urea feeding occurred when faecal protein levels were less than 8%. Results also agreed with those of Alexander et al. (1970) that responses to supplementation cannot be expected when cattle are already gaining between 0.15 and 0.50 kg/head/day.

It was concluded from faecal output data that urea caused increased intake of roughage.

Having proved the efficacy of feeding urea to cattle during the dry season, a subsequent trial investigated the best level to feed. Brahman × Shorthorn and Shorthorn cattle aged nineteen months were allocated to the following treatments:

No supplement
Molasses (0.23 kg/head/day) + urea (28 g/head/day)
Molasses (0.23 kg/head/day) + urea (56 g/head/day)
Molasses (0.23 kg/head/day) + urea (84 g/head/day)

The supplements were fed using roller lick feeders and the cattle were stocked at a rate of 1 beast/2.7/ha.

Weight changes during and after the feeding period are summarized in Table 7.23.

The experimental period lasted eighty-three days from July to October and the final liveweight was taken in May. Responses to urea declined as the level of urea fed increased and there was a marked difference in response attributable to breed. Compensatory growth did not occur during the rains and the supplemented animals maintained their advantage to the final weighing (Winks et al. 1972).

Winks et al. (1976) carried out further trials at Swan's Lagoon over a three-year period in which urea and biuret were compared as sources of nitrogen for cross-bred (Brahman × Shorthorn and Sahiwal × Shorthorn) weaners about six months of age and about 150 kg in weight. Molasses was fed at the rate of 0.23 kg/head/day

Table 7.23 Weight changes of cattle fed different levels of urea (Swan's Lagoon) (adapted from Winks et al. 1972)

	Weight change (g/day)		*Final weight (kg)*	
	Brahman cross	*Shorthorn*	*Brahman cross*	*Shorthorn*
Pasture only	−45	−68	404	386
28 g urea/day	250	60	419	398
56 g urea/day	286	98	424	397
84 g urea/day	299	169	426	406

Table 7.24 Effects of supplements on liveweight gains of weaners over three years (Swan's Lagoon) (adapted from Winks et al. 1976)

	Weight gain (g/day)			
	1970	*1971*	*1972*	*Average*
No supplement	7	7	–177	–54
Molasses	14	19	–151	–39
Molasses + urea	133	134	–39	76
Molasses + biuret	69	209	–23	85

through roller lick feeders. Urea and biuret were fed at the rate of 56 and 70 g/head/day respectively. Animals grazed native (Spear grass) pastures at a stocking rate of 0.57 beasts/ha.

Results for the relevant treatments are summarized in Table 7.24. Animals supplemented with urea or biuret gained significantly more than those given no supplement or supplemented with molasses alone.

However, these advantages were nullified in two of the three years by compensatory growth. Faecal protein levels were measured and low levels in non-supplemented animals confirmed that supplementation was necessary.

The authors concluded that biuret and urea were equally effective as nitrogen supplements and there were no indications that adaptation to biuret was slower than that to urea. The disadvantage of biuret insofar as cost was concerned was emphasized.

Having established that both urea and biuret improved liveweight gain during the period from July until the start of the rainy season, Winks and his colleagues carried out two further trials to determine at what time during the year responses to NPN supplements would occur and to what degree, and the effect of different levels of urea on dry season performance (Winks et al. 1979).

In the first experiment, Brahman × British cross-bred steers with an approximate weight of 174 kg were fed supplements from the date they were introduced to the experiment (January, December and February in the three years of the experiment respectively) until the first effective rains at the beginning of the rainy season. Molasses was fed at the rate of 0.23 kg/day/head through roller lick feeders. Urea and biuret were fed at the rates of 56 and 70 g/head/day respectively. From May to July in all years, both urea and biuret significantly improved performance and this trend continued until the end of the feeding period but was only significant on one of the three years of the trial. Results are summarized in Table 7.25. Overall differences were small because of a slight (but generally insignificant) tendency in the groups not fed urea or biuret to compensatory growth.

Table 7.25 Effect of urea and biuret supplements on weight gains of steers at different times of year (Swan's Lagoon) (adapted from Winks et al. 1979)

	Liveweight changes (kg/head/day)		
	May–July	*July–Oct*	*Overall (Jan–May)*[1]
No supplement	–0.15	–0.15	0.34
Molasses only	–0.10	–0.13	0.33
Molasses + urea	0.05	–0.01	0.38
Molasses + biuret	–0.03	0.02	0.36

[1] i.e. from the start to the end of the experiment (e.g. from Jan 1970–May 1973); means of three years' results

In the second experiment, Brahman × Shorthorn or Sahiwal × Shorthorn steers between eighteen months and two years of age weighing about 280 kg were drafted each year to one of four treatments: no supplement (control), 0.23 kg molasses plus 28, 56 or 84 g urea/day. Supplementation was given from April or May to October or November each year. A summary of the results obtained is presented in Table 7.26. Supplemented animals gained more than the controls in every year of the experiment. Responses to urea levels were not consistent and it appeared that there was no advantage to the feeding of higher levels. Compensatory growth occurred in non-supplemented cattle during the post-supplementation period. Responses to increased levels of urea appeared to be inversely related to the faecal nitrogen level, i.e. responses were higher when animals were more nitrogen deficient.

Table 7.26 Weight gains of steers fed different levels of urea (Swan's Lagoon) (adapted from Winks et al. 1979)

	Mean weight change (kg/day)		
	Supplementation period mentation	*Post-supplementation*	*Overall*
No supplement	0.08	0.66	0.33
Molasses + 28 g urea	0.17	0.61	0.36
Molasses + 56 g urea	0.17	0.60	0.36
Molasses + 84 g urea	0.18	0.62	0.37

Ernst (1973), also working at Swan's Lagoon, carried out a behaviour study of cattle given licks in roller feeders which were designed to provide 57 g urea and 0.23 kg molasses/head/day and 84 g urea and 0.23 kg molasses/head/day. It was observed that cattle visited the feeders for 'breakfast' between 0630 and 1000 hours and later, for a shorter time, for 'lunch' between 1200 and 1400. Although there was little difference in the aggregate time spent in the feeding area, the animals fed the lower level of urea spent longer at the feeder and in general the group fed the higher level of urea did not achieve the desired level of intake. The intake was effectively controlled in the former group by adding water to reduce the palatability; more water was required for this purpose as the dry season continued and the quality of the dry standing feed deteriorated. The advantage of regular ingestion on urea utilization is stressed and supports the use of the roller feeders. Playne et al. (1973) reviewed intake and digestibility studies carried out at the Lansdown Pasture Research Station. When cattle were fed *Heteropogon contortus* hay, supplements of urea increased intake of hay (and therefore digestible energy) on three occasions out of five by an average of approximately 16%, on one occasion intake was unchanged and on one occasion it was reduced by 9%. On two occasions there was a positive response to DM (dry matter) or OM (organic matter) digestibility amounting to approximately 13%.

Ernst et al. (1975) also carried out stall-feeding trials to investigate the effect of molasses and urea supplementation on the utilization of poor-quality hay. The composition of the hay (which was cut at Millaroo, near Townsville) is given below:

Heteropogon contortus	53%
Bothriocholoa intermedia (forest blue grass)	20.8%
Dichanthium spp. (blue grasses)	18.3%
Themeda australis, *Chloris* and *Pancium* spp.: the balance	

The treatments employed were:

1. Hay *ad libitum*
2. Hay *ad libitum* + 227 g molasses/head/day
3. Hay *ad libtium* + 57 g urea/head/day
4. Hay *ad libitum* + 57 g urea + 227 g molasses/head/day
5. Hay restricted to intake of Group 2 2 + 57 g urea + 227 g molasses/head/day

It was shown that the treatment with molasses and urea significantly increased consumption of roughage from 1495 g/head/day to 2473 g/head/day. Molasses alone also increased intake although less dramatically and this was put down to the relatively high protein content (6.1%). None of the supplements increased digestibility of dry matter or organic matter. Urea caused a highly significant improvement in crude protein digestibility (from −40 to 51%) but this would be expected because of the increased protein level (Butterworth and Diaz 1970).

Romero et al. (1976) carried out a trial at Townsville to investigate the effect of different methods of applying urea on the utilization of poor-quality *Heteropogon contortus* hay by eighteen-month-old steers. The treatments were as follows:

Heteropogon contortus hay without supplement
hay + 100 g urea every second day
hay + 50 g urea every day
hay + 25 g urea twice per day
hay + 50 g urea (as 25% solution) + Na_2SO_4 on a N : S ratio of 10 : 1 (sprayed on basal diet).

Urea was given as a drench and Na_2SO_4 was added to give the same N : S ratio.

All animals given supplements consumed more organic matter than the control animals, but these differences were only significant in the treatments where urea was given every second day or sprayed on the ration. There were no differences in apparent digestibility of organic matter between the treatments but urea supplementation significantly increased nitrogen balance. Results are summarized in Table 7.27. There was an indication that the utilization of urea was more efficient with increased frequency of ingestion; positive results accrued on nitrogen balance whatever the method of administration, and on intake when the urea was given every second day or sprayed on the basal ration.

Although Armidale (New South Wales) is slightly out of our area, some work carried out there will be mentioned for the sake of completeness and because of some similarities to work carried out in northern Queensland. Langlands and Bowles (1976) carried out interesting work comparing urea and molasses supplements with

Table 7.27 Effect of frequency of urea supplements on intake and nitrogen balance of steers (adapted from Romero et al. 1976)

	Organic matter intake (g/head/day)	*Nitrogen balance (g/head/day)*
hay only	2 451	−5.0
hay + urea every 2 days	3 092	5.7
hay + urea daily	2 897	5.7
hay + urea twice daily	2 949	7.0
hay + sprayed urea	3 151	8.1

linseed meal and linseed meal treated with formaldehyde. The treatments were as follows:

grazing alone (no supplement)
grazing + molasses and urea (457 g OM and 11.6 g N)
grazing + molasses and urea (518 g OM and 39.7 g N)
grazing + linseed meal (559 g OM and 37.1 g N)
grazing + linseed meal treated with HCHO (formaldehyde) (582 g OM and 38.6 g N)
(values in brackets represent mean consumption/head/day).

The grazing consisted of *Themeda australis, Bothriochloa macera* and *Danthonia* and *Poa* spp. and forage was available *ad libitum* (no stocking rate is given). The animals used were Hereford heifers aged twenty-three months with a mean liveweight of 384 kg and the experiment ran from 5 July to 13 September. All heifers lost weight but the weight loss of unsupplemented animals was significantly greater than that of the supplemented animals. Supplements reduced weight loss by 258, 323, 180 and 386 g/day respectively, and differences among treatments were not significant. It was suggested that the animals responded to energy rather than any form of N supplement. Supplementation increased total volatile fatty acids but had no consistent effect on the individual proportions of the fatty acids. It was suggested that the effectiveness of urea as a supplement to native pastures could be limited by the availability of higher fatty acids (although none of the supplements increased these).

Two later studies were carried out on Carpet grass (*Axonopus affinis*) in Grafton, New South Wales. These confirmed the earlier work of O'Bryan (1960) that molasses and urea were not always effective particularly when used as a supplement to Carpet grass.

Cohen (1974) investigated the effect of adding molasses (500 g), molasses plus 60 g urea, or molasses plus 120 g urea to a basal ration composed of hammer-milled Carpet grass given to yearling Angus steers weighing on average, 162 kg. The weight loss of unsupplemented steers was 0.54 kg/day and was significantly reduced (to approximately 0.12 kg/day) by the supplements, but no significant differences could be attributed to either level of urea. Similarly, the intake of dry matter from hay was increased by all the supplements (from 2.79 to 3.67 kg) but this was not changed by the addition of urea. There were slight upward trends in the values for apparent digestibility of dry matter and energy with successive increments of urea but these were not significant. The increase attributable to molasses feeding is surprising when compared with other studies, but may perhaps be associated with the higher digestibility of the crude protein fraction (24.6%) compared with other studies (cf. Ernst et al. 1975) where the digestibility of crude protein was −40.2%. The author comments that the faecal crude protein (5.4%) was lower than that level suggested by Winks and Laing (1972) as an indicator below which response to urea was likely.

Hennessy and his colleagues (Hennessy et al. 1978) studied the effects of infused supplements of urea and molasses on Hereford steers fed poor-quality (3.6% CP) hay. The hay was composed of *Axonopus affinis* (75%), native grasses (e.g. *Heteropogon contortus*) (16%) *Paspalum* spp. and herbs. The following infusions were given via rumen fistulae during a 24-hour period:

1.5 water
56 g urea in 1.5 litres water
56 g urea + 500 g molasses in 2 litres water
112 g urea + 500 g molasses in 2 litres water
112 g urea + 1000 g molasses in 2 litres water

All animals lost weight during the experiments; whilst apparently, supplemented animals lost less, this difference was not significant. None of the treatments increased the digestibility of the ration (as expressed as organic matter) nor the intake of digestible organic matter from the hay. The total digestible organic matter intake was higher, however, in those animals given molasses infusions. Urea and molasses infusions increased the level of total volatile fatty acids (VFAs) in the rumen frumen from 75.9 to 93.6 moles/litre but differences in molar proportions of VFAs were not consistent. The authors considered that rather than the N *per se* being of importance in the supplement, it was the molasses (with sufficient N for microbial fermentation) which was a useful energy supplement.

Asia

The only work to have been carried out in Asia appears to have been in Thailand.

Thailand

A large area of the highlands in northern Thailand are dominated by *Imperata cylindrica* resulting from shifting cultivation of agricultural crops (including opium) and maintained by the practice of annual fires. It was thought that the strategic use of supplementary urea would enable a more intensive use to be made of these areas with, at the same time, an increase in productivity of the cattle industry.

In an experiment carried out at the Highland Agronomy Project, Pa Kia, poor-quality pasture consisting mainly of *Imperata cylindrica* was supplemented with molasses/urea (3%) which was consumed at the rate of 1.2 kg/day by three-year-old female native cattle. A further group was allowed access to improved pasture (predominantly *Desmodium intortum*) for a period of 5 hours each day whilst unsupplemented animals were used as a control. The trial lasted ninety-eight days during the dry season when supplements were given and a further eighty-four days during the rainy season. Gains over the whole period were 29.1, 19.0 and 14.0 kg for molasses/urea, improved pasture and control animals respectively and it was concluded that such supplementation could usefully be incorporated into the traditional management pattern in Thailand where cattle are 'kraaled' at night (Mikled 1977, Falvey and Mikled 1978).

Discussion

The pathways of urea utilization in the rumen are well known and have been reviewed elsewhere (see, e.g. Schwartz 1967). However, it is convenient to outline them before further discussion.

On reaching the rumen, urea is hydrolysed by rumen urease to form ammonia. Urease activity is always relatively high and no period of acclimatization to urea is required. The ammonia produced may be used for bacterial synthesis, it may pass to the small intestine or it may be absorbed through the ruminal wall into the portal vein. Absorption is pH-dependent and is reduced when acid conditions prevail in the rumen. Although most of this absorbed ammonia is excreted, some is returned to the rumen via the saliva.

Various factors affect net bacterial synthesis from ammonia (there may be a loss as great as 30% because of fermentation of microbial protein), but the most important of these is the simultaneous availability of other nutrients. It was commonly believed that readily-fermentable carbohydrate was necessary as a source of carbon and energy for ammonia utilization. Work with rations high in cellulose (see, e.g. Pieterse 1961) have shown that this is not the case although ammonia uptake is more rapid when such carbohydrate is available and portal loss is minimized. It also appears that branched-chain and higher volatile fatty acids are essential and that sulphur is required for the incorporation of ammonia into methionine and cystine.

Certain micro-organisms can utilize ammonia and some of these (including some cellulolytic bacteria) are ammonia specific. Whilst ammonia is always present in the rumen (because of naturally occurring proteolysis, deamination and ureolysis), there appears to be a threshold below which insufficient ammonia is available for these bacteria. Various estimates have been made of this level but it appears to be about 80 mg N/litre rumen contents.

Positive responses to urea supplementation of natural pastures have occurred in various high rainfall 'sour' areas in Southern Africa (see, e.g. Barrie and Clark 1959; Kreft 1966; Altona et al. 1960; Bishop

1959; Verbeek and von la Chevallerie 1958; Pieterse 1961) either to prevent weight loss during winter (dry season) grazing or to support limited gains. Similar results have been obtained in Latin America (see, e.g. Carrera et al. 1969; Butterworth et al. 1973; Porres 1971; Chicco et al. 1975) and Australia (Beames 1963; Winks et al. 1972, 1976, 1979) and from Thailand (Mikled 1977). Various authors have associated this growth response with increased intake of dry matter by the supplemented animal (see, e.g. Barrie and Clark 1959; Pieterse et al. 1963; Beames 1963; Playne et al. 1973) and with higher digestibility (Playne et al, 1973), increased rate of passage through the rumen (Pieterse et al. 1963) and higher rumen fermentation rates (Carnevali et al. 1970). It has also been suggested that total rumen concentrations of VFAs may be increased by molasses and urea supplementation (Langlands and Bowles 1976; Hennessy et al. 1978). All these results are consistent with the role of urea in supplying ammonia to rumen cellulolytic bacteria when levels are below the threshold value.

On the other hand, several instances have been reported in which there has been no effect of urea supplementation on growth rate. Many of these have been from rainy season situations where neither forage nor protein was limiting (Kreft 1966; von la Chevallerie 1965; Veitia et al. 1972; Chicco et al. 1975) or where selective grazing permitted the animal to graze forage with an approximate N content of 1% which maintains the rumen ammonia level above the threshold already mentioned (O'Bryan 1960; Jones 1966).

The occurrence of such selective grazing would be indicated by a definite level of faecal protein, although estimates of the value at which this occurs vary; see, e.g. Winks and Laing (1972), Cohen (1974).

These results are also consistent with the suggestion of McDonald (1968) that supplementation with urea is not effective when pasture condition enables the animal to maintain body weight or, alternatively (Alexander et al. 1970), when weight gains are more than 0.15 kg/day.

It is likely that the availability of better-quality forage (including browse) to enable selective grazing to take place accounts for much of the observed differences between localities. One of the most noticeable differences between the South African highveld and the sweet lowveld (Skinner 1964), for example, is the widespread occurrence of protein-rich *Acacia* spp. etc. in the latter.

Various studies have investigated the most appropriate level of daily supplementation with urea. Langlands and Bowles (1976) showed that 40 g was better than 12 g. Winks et al. (1972, 1979) could not demonstrate any improvement above 28 g and the results of Butterworth et al. (1973) indicated value between 27 and 54 g/day. However, the most comprehensive experiment was carried out by

Grant (1979) who calculated regression equations indicating that the most efficient response was to 20 g/head/day (i.e. 500 g/head/day of the 4% mixture). In view of the wide range of conditions under which these experiments were carried out, there is little inconsistency among these results.

It has also been pointed that high levels of urea limit supplement consumption (Veitia et al. 1972).

Several workers have investigated the value of biuret as an NPN supplement and generally found it to be equivalent to urea (Winks et al. 1976, 1979; Mackenzie and Altona 1964a) with the exception of Chicco et al. (1975), who obtained overall gains of cattle supplemented with biuret no different than those of control cattle. This may have been associated with the lower intakes of biuret-containing licks as demonstrated by Mackenzie and Altona (1964a).

For production purposes (i.e. as opposed to maintenance or prevention of weight loss), NPN may not be sufficient. It has been suggested by Leng et al. (1977) that the presence of 'by-pass' protein (i.e. that which passes to the small intestine without deamination in the rumen) is an important limiting factor. This hypothesis recognizes that the ruminant has a requirement for certain essential amino acids *per se* rather than being able to rely on those synthesized by mirco-organisms in the rumen. The role of by-pass proteins as glucose precursors in the form of glucogenic amino acids is also important.

The frequently-encountered adverse effects on utilization of molasses alone has been attributed by Hennessy et al. (1978) to available nutrients being used up by rapid fermentation of the molasses with a consequent reduction of intake. They postulate that provided enough N is added to the molasses (i.e. 3% urea) to maintain microbial fermentation, the molasses may provide a useful energy supplement. Differences in N-content of molasses could thus explain other equivocal experimental results (e.g. Ernst et al. 1975).

As with all the supplementary studies mentioned in this review, responses also depend on the nutrient requirements and previous nutritional history of the class of stock concerned. These vary so widely that they would be expected to constitute an important source of variation in animal response.

8 *Energy and protein deficiencies: III. Their correction in breeding stock*

It has been shown that grasses in the tropics and subtropics are deficient in crude protein during relatively long periods of the year. These deficiencies may be corrected by feeding supplementary protein, resulting in an increased growth rate which can be attributed not only to the provision of protein *per se* but also to a consequent increase in the consumption of energy.

The pregnant cow requires protein for growth if she has not yet attained her mature body size. The growth of the foetus, placenta, foetal fluids and maternal uterus impose an extra demand for protein which reaches a level of more than 180 g/day during the last month of gestation. Similarly, in the case of the lactating cow, the output of protein in milk has been estimated to be 3.3 g/100 g milk (ARC 1965).

The pregnant and/or lactating cow is therefore subjected to a particular nutritional stress when exposed to the low levels of protein found in tropical forages.

The importance of nutrient provision for conception must also be considered. Age at first oestrus is affected by nutrient, specifically protein, status as is subsequent reconception following *post partum* oestrus. Similarly, protein plays a role in the hormonal relationships concerned with reconception, with normal fertilization of ovum and with changes in the endometrium favourable to implantation. Studies on supplementation of the breeding cow with protein are reviewed and discussed in this chapter.

Southern Africa

Republic of South Africa

The earliest trials with supplementary proteins were carried out on growing cattle, and these lagged behind the pioneer work of Theiler and his colleagues who concentrated their efforts on mineral deficiencies and thereafter on disease aspects.

It was not until 1948, after initial plans had been interrupted by the Second World War, that experiments were initiated on the supplementation of various breeds of heifers (Joubert 1954a). These studies were carried out over a period of several years in the Bankenveld (a region in the central Transvaal, bounded to the north by the Magaliesburg range). The area is referred to as 'scattered bush savannah' and the principal grasses of importance are *Themeda triandra* and *Setaria* and *Hyparrhenia* spp. The work was carried out with twin pairs of heifers of the breeds Shorthorn, Jersey, Friesian and Afrikaner (sic.). For the purposes of the present review, emphasis will be given to the beef breeds. Considerable loss of animals occurred and thus only four were available on the high plane and three on the low plane. The basic supplementation consisted of *Eragrostis tef* (Teff) hay and maize and cowpea silage and a small quantity of a standard concentrate (composition not given). This was administered during the winter seasons until the low-plane (unsupplemented) twins began to benefit appreciably from summer grazing. Animals were fed bonemeal and salt (2 : 1) licks at all times. After four years of investigation, it was concluded that the poor nutritional conditions significantly retarded development as well as growth *per se*. Skeletal development slowed down or stopped whilst muscular development actually decreased. However, low-plane individuals tended to make higher relative increases during the summer period than their high-plane mates. Growth during pregnancy differed appreciably between groups; the low-plane animals increased much less in weight and their calves weighed on average 3.4 kg less at birth. The low plane of nutrition delayed puberty on average by 221 days and supplemented animals could be served 9.6 months earlier than their low-plane counterparts.

In a final evaluation of the data, Joubert (1963, 1972) showed that the low plane in the study delayed puberty by as much as 271 days.

This finding was confirmed by Penzhorn and Meintjes (1968) who showed, also with Africander heifers, that a low plane of nutrition delayed puberty by 247 days. A tendency was also revealed for the whole of the subsequent reproductive cycle to be retarded by adverse nutrition. There was marked tendency for cows to go into anoestrus shortly after weight loss had become appreciable and this inactivity persisted into the following summer. Joubert (1954a) concluded that depleted body tissues were restored before reproductive function reverted to normal activity and that 'the problem seems primarily one of activating the normal sexual cycle'.

Finally, the growth rate of suckling calves was significantly influenced by the nutritional plane of their dams, the difference at one year of age amounting to approximately 69 kg both in Shorthorns and Afrikaners. Although small numbers of animals were used in this study, it has been reviewed in some detail because of its pioneer nature and because of the number of findings which have been duplicated in later work.

Several trials were carried out at the Agricultural College at Potchefstroom in the Transvaal. The grassland has been characterized as northern *Cymbopogon/Themeda* veld which merges easily into the western variation of the Bankenveld (Acocks 1975). Lesch et al. (1969) compared maize meal and urea fed with salt and dicalcium phosphate with salt and dicalcium phosphate fed alone to in-calf, grade Afrikaner cows. The total conception rate (over a three-year period) of the cows supplemented with urea and maize meal compared with those supplemented only with salt and dicalcium phosphate was 63 v. 46%. No differences in weight could be demonstrated between the two groups but it was concluded that this was because of compensatory growth on the part of the non-supplemented animals.

In a subsequent experiment, a similar maize meal/urea lick was compared with lucerne hay as a protein supplement (Penzhorn et al. 1970). Unfortunately, no composition was given for the lick, but average consumption was 38 g/day compared with 907 g lucerne hay. The trial was continued over a five-year period and the overall calving percentage was 72% for the lucerne hay-supplemented cows v. 63% for the maize meal/urea group. However, a large source of this variation was due to the first year of the trial when the lucerne group had a calving percentage of 83 v. 42% for the maize meal/urea group. If the result for this year are discarded, the corresponding overall means are 71 and 69%, an unimportant difference. There is no readily apparent explanation for this anomaly but it emphasizes the importance of carrying out this type of trial over a period of several years, before final conclusions are drawn.

Maree and Harwin (1971) carried out a trial on the effects of creep-feeding with or without subsequent supplementation on the onset of puberty in Bonsmara heifers grazing Transvaal bushveld. Calves were allocated to treatment at birth. Treatments were creep-feeding plus post-weaning supplementation, creep-feeding with no post-weaning supplementation, no creep-feeding plus post-weaning supplementation and neither creep-feeding nor post-weaning supplementation. Weight at puberty was 347, 294, 338 and 293 kg for the four groups of animals in the order stated above whilst age at puberty was 450, 484, 481 and 493 days. The creep-feeding plus post-weaning supplementation decreased the age at which puberty was attained but, in general, the difference in weight at puberty was more marked than the age differences. This is not so marked as the results of Joubert (1954a) but the heifers in Joubert's study were older at puberty. This occurred on average (beef breeds only) at 822 days, thus giving more opportunity for a difference between high and low planes. The advantage imparted by supplementation insofar as first conception was concerned was confirmed in this study where 73% of supplemented animals conceived v. 54% of the non-supplemented group.

Less beneficial results were obtained at Mara (near Pietersburg), by Lüdemann (1973). Cattle were supplemented with up to 2.7 kg of maize meal and 340 g of oil cake/day during conditions described as 'times of drought and scarcity of veld grazing'. Grazing was described as 'sweet bushveld' (although Acocks (1975) classifies this area as 'sourish mixed bushveld'). Supplemented cows had a calving rate of 21% compared with less than 10% of the unsupplemented controls. One can only conclude that grazing conditions were meagre. During more favourable seasons, supplementation did not improve the already reasonably high calving percentages of between 72 and 90% depending on breed group. The author concluded that factors other than nutritional prevented consistently high values for calving percentage from being attained. The intervention of certain (unspecified) physiological factors was implied although this assumption seems unnecessary as performance was adequate in favourable years presumably because of dietary supplementation with browse as has been noted in other bushveld areas.

Zimbabwe

Comprehensive trials were carried out in Zimbabwe which accumulated an impressive array of data to demonstrate the beneficial effects of protein supplementation on the reproductive performance of beef cows.

The first long-term trial was done by Bembridge and reported in 1963 after a three-year period of supplementation. This was carried out on the de Beer's ranch near Shongani. Rainfall was approximately 660 mm/year on granitic sand veld with predominantly *Hyparrhenia* spp. grasses. Cows started the trial at approximately 2.5 years of age and a stocking rate of 5.7 ha/head was maintained throughout. Supplemented animals were given from 450–900 g of cottonseed cake daily during the winter season. Although non-supplemented animals lost considerably more weight during the winter, this was almost entirely made up during the following summer (average overall weight gains 21 and 17 kg for supplemented and unsupplemented animals respectively). However, weaning weights were 176 and 153 kg for the two groups respectively whilst mortality was 0.6 and 9.1%. Unfortunately, no measures of statistical variation were given but the advantages of supplementation with protein were apparently clearly demonstrated.

Bembridge (1964) also carried out a preliminary trial on the effects of feeding a urea supplement to breeding cows. Three-year-old Sussex Africander-type heifers were fed protein cubes (sic.) containing 5% urea to supply 113 g digestible protein equivalent/day.

Unfortunately, there were no unsupplemented control animals but calving performance was 77.8% (76.2% weaned) over the three-year period. The author concluded that an increase of 20% in weaning percentage over non-supplemented animals could be expected and that the cost of the supplement would be more than compensated for by the expected increase in weaning weight. The increase in weaning percentage could therefore be safely considered as profit.

Concluding this series of experiments, Bembridge (1970) reported the results of a trial, also carried out on the de Beer's ranch, on the effects of two levels of supplementary feeding as well as three stocking rates on the lifetime performance of beef cows. The two levels of winter supplementary feeding employed were the accepted one of 227 g digestible crude protein equivalent and a lower level of 114 g. Protein was fed in the form of cubes containing 5% urea. The stocking rates used were equivalent to one Livestock Equivalent Unit per 4.4, 5.6 and 6.8 ha respectively. Differences in winter weight loss between the cows fed the two levels of feed was minimal (9 kg) although during one particularly prolonged dry season this increased to 17 kg. Similarly, differences in weaning percentage and birth weights of the animals of the two groups were minimal, amounting to some 8–11% and 2–3 kg respectively in favour of the high-plane animals.

As the most important economic measure was total weight of weaner produced, a comparison was made of gross return from the two groups. The gross return was greater in the higher feed level groups in all cases and the return increased progressively from the lightest to the heaviest stocking rates. However, after deducting feed costs, the gross margin in the low feed level groups was greater in all instances than in the higher.

The author concluded that relatively low levels of protein supplementation serve to achieve high calving percentages when these are combined with sound management. However, it was also pointed out that these lower levels are probably close to the minimum that can be expected to give these results, whilst the optimum is probably closer to the higher level (Bembridge 1963).

At about the same time as Bembridge started his supplementation studies at a commercial ranch near Shongani, Bauer (1965) started a similar trial near Enkeldoorn also on *Hyparrhenia* spp. veld. The cattle used were grade Herefords and they were studied over a five-year period from 1959 to 1964. It was concluded that the best results could be expected from first calf heifers where reconception was 83.4% in fed cattle as opposed to only 12.9% in unfed animals. The treatment consisted of 0.73 kg 50% protein cubes for a period of 140 days. Of the fed animals, 71% had a calving interval of less than twelve months in contrast to only 46% of the unfed controls. Older cows also benefited from supplementation but not so dramatically; the calving percentage of fed cows was 88 v. 77% for unfed controls.

Elliott (1964) supplemented heifers of mixed breeding with protein, minerals plus vitamin A or a combination of the two over a three-year period. Feeding started on 1 May and finished on 31 October (regardless of conditions at the time) and consisted of a 900 g decorticated cottonseed cake (approximately 42% crude protein)/head/day with bone meal and a commercial trace mineral mix where appropriate; salt was available at all times to all cows. The birth of full-term calves was used as the index of fertility. With the exception of the first year (where conception had already taken place before dietary stresses had become marked), conception rate of the cows on the protein supplement groups was higher than those supplemented with minerals alone or given no additional feed. Corresponding means were 39 and 76% respectively (calving rates for the first year were not included in this average).

The main effect on fertility of protein supplementation occurred in lactating cows. The reconception rate of non-protein-supplemented cows was 33% whilst that of their supplemented counterparts was 72%. Calves from protein-supplemented dams weighed significantly more at birth and at weaning (39 and 167 kg) than calves from non-supplemented cows (32 and 155 kg) for three out of the four seasons. Mortality was lower in cows fed protein supplements, as was calf mortality.

Elliott (1964) carried out a detailed analysis of factors effecting weight changes in cows and the relation of these with conception and reconception and concluded that an adverse weight balance was the main cause of infertility and that periodic 'rests' from reproduction enabled depleted body reserves to be restored. This was borne out by the marked 'cyclical' (i.e. tendency to calve in alternate years) reproductive behaviour of the unsupplemented cattle.

Later work carried out by Elliott and his co-workers (Elliott et al. 1966) elucidated the nutritional requirements of Afrikander cows fed for survival on three differing food levels. The two lower levels of feeding apparently supplied less energy than was theoretically required for maintenance of body weight over a 32-week period but, nevertheless, this was achieved; the most significant observation from the study appears to be somewhat self-evident to one having observed cattle in the tropics during the dry or winter seasons: 'cattle can survive for long periods on very little food' but, nevertheless, the study is important to quantify this often empirical observation.

Lactation stress has been mentioned elsewhere as an important factor inhibiting reconception. A novel approach to the correction of this was adapted by Rose et al. (1963) who attempted to measure the interactions between supplementation with high protein concentrates and early weaning. The probability of conceiving was 0 and 41.7% in non-fed and fed cattle respectively when calves were not early-weaned but 100% in both groups when calves were weaned early (prior to the breeding season).

Later work by Symington and Hale (1967) and Harwin and Venter (1970) failed to confirm this finding but Holness et al. (1976) commented that this might have been because the stress of separation was too great. These workers avoided this by the use of nose plates for a period from 50 to 58 days *post partum* and showed that this stimulated ovarian activity. At 60 days *post partum*, only 30% of nursing cows had shown oestrus compared with 55% of cows in which suckling had been temporarily prevented. They further observed that the use of nose plates (with creep feeding of the calves on this treatment) did not significantly reduce weaning weight (163 v. 170 kg for nose-plated and non-nose-plated calves respectively).

Ward (1964) made a preliminary report of the effects of feeding groundnut cake on the lifelong performance of cows. The calving rate of the supplemented animals over a three-year period was 81 whilst that of the unsupplemented controls was only 63%. A later report (Ward 1968) dealt with the long-term effects of various types of supplementation, including bone meal year-round, groundnut cake and vitamin A during the winter and various combinations of the three supplements. The trial was carried out at the Makaholi Experiment Station (near Fort Victoria) in the Rhodesian Midlands and the predominant veld grasses were *Hypharrhenia* spp. The cattle were Mashona heifers. The productivity of the cattle given groundnut cake at the rate of 900 g/day was 32% higher than those given no supplement measured as weight of calves weaned. The effect was particularly evident in terms of calves born to the supplemented animals and thus to the number of calves weaned. These were 83 and 71 for the two groups respectively (maximum possible being 120 in each case).

Ovarian palpation showed that 'the immediate physiological reason for this improvement in fertility was earlier resumption of *post partum* reproductive cycling'. The extent of this effect may be judged by the fact that the number of cows showing *post partum* anoestrus at the end of the breeding season was reduced by half. This effect was not observed in animals supplemented either with bone meal or vitamin A. Cows supplemented with groundnut meal gave birth to significantly heavier calves than those born to other cows; however, although the calves of the groundnut meal-supplemented dams increased more rapidly in weight during the suckling period than the calves of other cows, this difference was not significant.

Ward and Tiffin (1975) subsequently summarized the results of a series of trials on supplementary feeding and fertility of beef cows at Makaholi. They drew attention to the possible disadvantage of the traditional winter feeding in that a large proportion of the extra feed is given when the cow is non-productive. However, trials comparing supplementation during winter with that during summer and with both Mashona and Afrikander dry cows was designed to study the supplementation during both seasons indicated that reconception rates between the four treatments did not differ markedly (87.8, 84.8, 87.9 and 79.3% for the three treatments in the order given and the control respectively). They concluded that no biological or economic benefits could be expected from an extended period of supplementation. It was also pointed out that the relatively high reconception percentages for the control animals be attributed to the fact that they had been grown out with supplemental protein during the winter. The authors further comment that 'there is increasing circumstantial evidence to suggest that the need for protein-rich supplements in winter is much reduced or even eliminated as the cow attains a constant bodymass at successive matings'.

First priorities in the use of supplementary feed are the growing and in-calf heifer (see also Tiffin 1976). As had been pointed out earlier by Elliot (1964), Ward (1968) and Harwin et al. (1967), cow body mass, condition and/or body mass gain are critical at mating insofar as reconception is concerned. The authors presented data on weaning rate in relation to the mass distribution of lactating cows at mating and showed that not only did more cows conceive in the heavier groups but that supplementation of protein was effective in increasing weaning rate. Studies carried out in Uganda (Trail et al. 1971) and in Botswana (Buck et al. 1976) confirm these findings and emphasize that a carefully-controlled supplementary feeding programme was instrumental in increasing calving percentage. Cows which lost weight had a calving percentage of 68% whilst those that gained 20 kg had a calving percentage of 76%. The period between calving and reconception was also reduced by one week. However, a note of caution was sounded by the latter authors who noted that cows over 440 kg suffered a decline in fertility supporting the widely-held belief that over-fat cows may become infertile. Thus, whilst more cows in heavier weight groups conceive, when cows are of similar weight those in better condition are more likely to calve.

Richardson and his colleagues (Richardson et al. 1975a) studied various factors affecting the productivity of beef cows in Que Que, an area of Rhodesia with marginal rainfall. They estimated that the percentage body-weight changes from autumn peak to mean body weight at the beginning and end of the mating season which would be followed by successful conception rates of 50, 75 and 90% were − 14, − 6 and + 1% of autumn body weight respectively. Steenkamp et al. (1975) also discussed the relationship of weight and condition on reconception based on the results of a long-term trial carried out at Matopos. When fed and non-fed cows with a similar body weight of 343 kg were compared there was a 19% difference in reconception in favour of the former animals, indicating the importance of condition rather than weight *per se*.

Hale and Holness (1977) have investigated the relation between weight, condition and sexual activity in some detail. A trial using influence of chronic malnutrition on the ability to continue sexual activity, to determine the length of time necessary for the resumption of sexual activity after the start of realimentation and to determine the site of endocrine dysfunction during nutritional anoestrus. It was shown that the loss of body weight from maximum body weight to the point at which oestrus stopped was 19 and 33% in the case of the Afrikander and Mashona cows respectively. (Two Mashona cows did not cease regular cyclic activity despite losing 44 and 39% of maximum body weight.) Oestrus was resumed after realimentation at 99 and 82% of the maximum weight for the two breeds respectively. The greater the initial weight of the cow, the more weight could be lost before the cessation of oestrus or 'large cows had a greater ability than small cows to withstand the suppressive effect of a sub-maintenance plane of nutrition on sexual activity'. So far so good; however, the authors also included an estimate of body fat content based on the use of a radiosotope dilution technique after injection of tritiated water and found indications that there was no difference in body water (and thus, of fat) between cows at the start of underfeeding. They inferred that this indicated that the superiority of heavier cows to lighter cows insofar as reconception was concerned was attributable to the size of the cow *per se* and not to differences in condition. The resumption of sexual activity in the Afrikander cows after the injection of hypothalamic gonadotrophin releasing hormone implicated this hormone in anoestrus.

The beneficial effects of protein feeding on weaning weight has already been mentioned, and as Heyns (1960) has shown that this is closely related to the quantity of milk drunk by the young calf, it is to be supposed that feeding influences milk production. That this was indeed the case was demonstrated by Richardson et al. (1977d) who fed grazing cattle on three planes of nutrition. Cows on the high plane received 450 g/day of cottonseed cake together with maize meal as winter advanced. Medium-plane animals were given half this quantity of supplement and low-plane animals were given no additional feed. Supplementation was continued until eight weeks *post partum*. High-plane and medium-plane animals produced more milk than those fed the low plane but, although high-plane animals produced more than those on the medium, this difference was not significant. Mean 98-day lactation yield/cow/year was 746.8, 682.5 and 563.3 kg respectively.

Holness et al. (1978) at Henderson Research Station investigated the aspects of reconception in supplemented cows in more detail. Pregnant cows were allocated to four groups which were fed so as to gain 12 to 14% of their initial liveweight by mid-breeding season (HH), this level until 25 to 50 days *post partum* when they were cut to a level of feed calculated to provide 75% of maintenance requirement for the 25-day period (HL), a level calculated to allow for a 12 to 14% loss (LL) or this level until 25 days *post partum* and then a liberal supply of concentrates for 25 days. The occurrence of oestrus was observed using epididymectomized or entire bulls and by rectal palpation. Overall, oestrus occurred significantly earlier *post partum* in high-plane than in low-plane cows (66 v. 75 days), but within high- and low-plane groups the short-term changes in feed intake (HL and LH) had no significant effect on the occurrence of oestrus. An interesting feature of this work was that cows which lost liveweight from parturition until oestrus, exhibited oestrus earlier than cows that gained liveweight over this period. The authors surmize that this was because cows which lost more weight were in better condition than those that gained weight although this was not confirmed as no condition scores were recorded.

Botswana

Work was carried out in Botswana to determine the effectiveness of proprietary rumen-stimulatory licks in increasing reproductive performance of breeding cows (APRU 1975). Trials took place on five AI stations in various parts of the country and licks containing either 29 or 37 crude protein were compared with a lick containing bone meal and salt only. Provision of a rumen stimulatory lick significantly increased the liveweight of the cows in three of the five trials, but this did not result in a significant improvement in reproductive performance. Subsequent analysis showed that this was true for both lactating and dry cows, although the authors pointed out that the former group had all been in milk more than five months at the start of the insemination season and that by this time calves were consuming considerable quantities of herbage and therefore lactation stress would be reduced.

A subsequent trial showed clearly that supplementation benefited lactationally-stressed cows more than other groups. Overall, there was an improvement in conception of 7.1% in supplemented cows. However, when separate least squares analysis was carried out on stressed and non-stressed cows, it was shown that there was a 14.1% advantage of the former over the bone meal controls. It was concluded that there was no benefit in supplementing unstressed cows.

Subsequent studies therefore concentrated on 'stressed' breeding cows only (APRU 1977). Trials were carried out in nine ranches during September – December 1975. It was shown that supplementary feeding improved pregnancy rates (determined by palpation) by 3% and calving rate by 2.6%. The supplementary feed given consisted of maize meal (69%) and urea (6%) with molasses, salt, bone meal, vitamins and minerals. The trials were repeated in 1976 when there was a non-statistically significant improvement in pregnancy rate on the part of the supplemented cows of 87% as opposed to 82% of the non-supplemented animals. However, the overall pregnancy rate was high, reflecting the good grazing conditions during the year and the adequacy of the management system.

Swaziland
Similar marginal increases in calving rate were obtained in Swaziland when yearling cattle were supplemented with 300 g/day/head of whole cottonseed from from August to November (that is, from the worst of the dry season until sufficient green feed became available after the beginning of the rains). Supplemented animals gained more than unsupplemented animals during the supplementation period, gains being 116 v. 43 g daily. The control animals showed compensatory growth during a subsequent recovery period but not sufficient to overcome the initial advantage of the supplemented cattle. The heifers were bred in December and 52% of the supplemented animals calved v. 47% of those non-supplemented (FAO, 1978b).

Zambia
Finally, a similar study carried out in Zambia may be mentioned (Rakka and Igboeli 1972). Cattle were kept on veld hay alone or given a supplement of corn cob meal (75%), cottonseed meal (20%) and fish meal (5%) in sufficient quantities to maintain body weight. Supplemented cattle showed 85% of possible oestruses v. only 49% of their unsupplemented mates. The authors also concluded that climatic stress could be a cause of anoestrus but their results seemed to support this contention only tenuously.

It would appear that the benefits of supplementary feeding with protein or NPN under Southern African conditions have been demonstrated in most cases with respect to increased calving percentage, milk production and consequent weaning weight and reconception. The impact of this improved productivity on the economics of beef production have been studied by Steenkamp and van der Horst (1975), who compared the economics of feeding beef cows with cottonseed meal and silage. Projections were made over a twenty-three-year period and it was concluded that such supplementation was profitable when standards of management were high.

Australia

In Australia, several authors have drawn attention to the low reproductive efficiency of cows in Queensland and attributed this to nutritional deficiencies (Donaldson 1962; Churchward 1965; Lamond 1969).

Sparke and Lamond (1968) investigated the effect of a supplement of linseed cake on the growth and fertility of beef heifers in northern

New South Wales. The heifers were grazing pastures predominantly consisting of *Agrostis*, *Paspalum* and *Bothriochloa* spp. and were given the supplement at the rates of 3.6, 5 or 5.4 kg/week during the three years of the trial. Dramatic differences in conception rate were apparent between the two groups, values being 59.3 and 33.8% for supplemented and unsupplemented animals respectively. The authors also noted that all increase of liveweight up to 295 kg were accompanied by increases in fertility.

Results of field surveys reported by the Australian Meat Research Commission (1967) showed that the reproductive performance of lactating cows was closely dependent upon their body condition at mating rather than weight *per se*. This was confirmed by Barr (1971) who assessed body condition at the same time as pregnancy diagnosis was carried out at various farms and stations throughout Queensland. Effects of condition were not noted in dry, non-lactating mature cattle but contrasts were marked in heifers. Heifers selected on basis of weight and condition achieved a pregnancy rate of 95% whilst only about 60% of the remainder of the group became pregnant. The most critical effect of body condition is to be noted in lactating cows. Barr (loc. cit.) concludes 'There is a critical level of condition at which the reproductive function will commence in spite of the tendency to delay caused by the suckling calf, and the higher the level of *condition*, [my italics] the shorter the period from calving to the first activity of the reproductive tract, demonstrated by oestrus.'

It is interesting to compare these findings with those from Africa reviewed previously (see previous section).

The author continued to describe several trials on the feeding of supplements to breeding cows and has detailed the three approaches to feeding for fertility followed by the Beef Cattle Husbandry Branch of the Queensland Department of Primary Industries. These included correction of specific deficiencies, supplementation at strategic points in the breeding cycle and low-level supplementation of all breeding cattle. Barr (1971) cited a phosphorus supplementation trial carried out at Charters River as a typical example of the first approach. This has been referred to elsewhere.

Two examples were given of supplementation at strategic times. One of these was carried out in the Brisbane valley and supplemented the growing heifer from weaning until the second breeding period, thus theoretically correcting deficiencies at two critical points: the post-weaning check and the lactational anoestrus of the young cow with first calf at foot. Seventy-one weaner Hereford heifers were supplemented with a high-protein, energy and phosphorus block (proprietary) which it was recommended should be consumed at the rate of approximately 450 g/day (actual consumption was from 90–450 g/day). The author concluded that there appeared to be a slight but persistent advantage to be gained from supplementation as evinced by liveweight, pregnancy rate (67 and 60% for supplemented and unsupplemented cows respectively), conception pattern and weaning weights (155 and 141 kg for fed and control calves respectively). There was, however, a higher level of dystocia in the supplemented animals (4.2 v. 1.4%) and, although the author commented that this was unlikely to be the effect of supplementation, it must be considered as a possible undesirable consequence of this practice.

Barr (1971) goes on to detail further trials with mature animals. The effect of dry-season supplementation with either cottonseed meal or molasses/urea on Shorthorn cows was investigated at Swan's Lagoon. The cottonseed meal was fed at the rate of 450 and 900 g/day and molasses/urea was given so as to be isocaloric and isonitrogenous with the former. Only the higher level of cottonseed meal proved effective in reducing weight loss which was still approximately 200 g/day. An attack of vibriosis prevented the measurement of any differences in conception rate.

At Brian Pastures, supplementation of Hereford cows with 2 kg sorghum grain/day resulted in reduced weight loss than in unsupplemented cows (36 v. 77 kg). This was accompanied by a higher pregnancy rate (96 v. 80%).

In a further trial in the Bowen area, urea and biuret were compared as sources of non-protein nitrogen for breeding cows. There was little difference between groups insofar as weight change during the 218 days of the trial, these being approximately 6 kg and 90 g respectively. However, the results were clearly highly variable, as approximately one third of each group of cows lost weight.

Urea and urea/molasses were compared as supplements to three-year-old cows. Controls were superior insofar as pregnancy rates were concerned in the two years of the trial. It is concluded that 'a phenomenon similar to compensatory weight gain is a feature of reproductive performance with the result that initial advantages of supplementation tend to be more than compensated when the season breaks' and the question is raised as to whether low-level supplementation can be expected to achieve marked results under the pattern of seasonal mating practised.

It should be mentioned that in the Swan's Lagoon and Dululu trials there was insufficient roughage available and therefore the effects of supplementation may have been masked by other deficiencies.

Finally, Barr (1971) mentioned two trials to test the effectiveness of whole herd supplementation.

After demonstrating that an energy supplement alone would not enable breeders to survive under the prevailing grazing conditions, urea/molasses and urea/grain supplements were compared at Home Hill. The author stated that the trial had been plagued with drought conditions and various cattle showing advanced poverty had to be removed. Although a few more calves were born to the supplemented cows, this was not reflected in increased weaning percentages or weights and it would appear that there was no consistent advantage to supplementation.

A further trial was described which was carried out at Mackay and involved the comparison of lucerne and urea block supplements. The proprietary block group had the highest rate of conception but this difference was only significant during one season. However, the results obtained in three-year-old cows supported the view that this is the vulnerable group in the herd; advantages to block supplementation in these animals varied from 4 to 15%. The author also concludes that 'provided normal culling for age is practised, the view that fertility in mature cattle is not greatly influenced by low level supplementation . . . must also receive consideration'.

In a later publication (Barr and Burns 1972), cottonseed meal (1.36 kg) and molasses (220 g) plus urea (55 g) were compared as supplements for cross-bred Hereford Brahman cows at Swan's Lagoon in order to increase fertility during a drought. Ovarian activity was considerably better in the cows fed cottonseed meal, and pregnancy diagnosis revealed 77% pregnant contrasted with only 46% of those cows fed molasses and urea. This inferiority of the urea supplement is even more marked than that noted by Penthorn et al. (1970) mentioned above.

Studies initiated at Townsville by Little (1967) were designed to investigate the effects of several supplements including protein on the reproductive behaviour of mature lactating Droughtmaster cows kept in pens. Although it was stated that the greatest single response was to protein and there was undoubtedly a response to protein in the cows that showed oestrus (of the first six cows to show oestrus, five had been supplemented with protein), it was also mentioned that the most likely cause of failure to exhibit oestrus was a deficiency of digestible energy. This was supported by the higher incidence of cows to show oestrus in a group of grazing cattle.

A long-term grazing experiment was initiated at Lansdown Pasture Research Station 48 kg south of Townsville to examine the effects of specific nutrients on the reproductive performance of Droughtmaster cows grazing *Heteropogon contortus/Stylosanthes humilis* pastures. One of the treatments involved the administration of 22 g nitrogen, 5 g sulphur and 6 g sodium/cow through the drinking water during the first three years of the study. The mean conception rates over two years of those cows which were rearing calves when mated was 90% as compared with the control group where the comparable figure was 81%. However, it was stated that the significance of these differences was not evident because of the small number of cows per treatment (Playne et al, 1973). Later in the experiment, nitrogen was provided from three sources: urea, cottonseed meal or crushed maize. The effects of climate were emphasized in this study as there were not sufficient responses to justify supplementation in three out of five years of the study. Of the various supplements used, only cottonseed meal significantly improved pregnancy rate. Urea was ineffective and this was attributed to the digestibility of the forage and the frequency of urea ingestion. Pregnancy was 84% in the group fed cottonseed meal contrasted with an average of 36% in the other groups (Siebert et al. 1976).

Siebert and Field (1975) carried out a further trial at Lansdown to investigate the reproductive activity of beef heifers following post-weaning feeding on hay of *Heteropogen contortus* alone or with supplements consisting of either a small quantity of legume hay or legume hay plus grain. They were subsequently allowed to graze *Heteropogon contortus/Stylosanthes humilis* during the following wet season. During the pen-feeding part of the experiment, the control animals lost 211 g/day whilst legume and legume plus grain-fed animals gained 179 and 386 g/day respectively. Oestrus was observed during the grazing period and the addition of the legume hay was apparently responsible for a significant increase in activity as heifers in this group showed oestrus twenty-two times contrasted to the control animals for which the corresponding value was seven; those animals supplemented with both legume and grain showed oestrus forty-two times. It is concluded that even the addition of a small quantity of legume hay can maintain adequate rates of weight gain and stimulate oestrus activity. It is suggested that this is not only due to the effect of the protein *per se* but also because of the increased intake of energy which this promotes (see Siebert and Kennedy 1972).

Little (1975) investigated the effects of dry-season supplements of protein and phosphorus on the incidence of first *post partum* oestrus at the CSIRO Pasture Research Station at Samford in south-east Queensland.

Cows grazed pastures mainly composed of *Digitaria didactyla* and *Axonopus affinis* and were given 180 g crude protein/day (as peanut meal) as well as 10 g phosphorus/day as NaH_2PO_4. Positive effects of supplementation with protein occurred in both pregnant and empty cow weight gain and, notably, in the interval from calving to first oestrus, these being 74.9, 62.4 and 40.4 days for control phosphate- and protein-supplemented cows respectively. It is clear from this work that so-called 'lactational anoestrus' can be significantly curtailed by the provision of protein supplements with a consequent reduction of calving interval. Dramatic effects were also manifested in the proportion of cows in oestrus by eight weeks after calving, these being 38, 70 and 100% for animals of the three groups in the same order as above.

Apparently only one publication from Australia relates to the effect of supplementation on the milk produced by cows. Allan et al. (1972), working at Swan's Lagoon, found that milk production was increased from 3.2 to 3.6 kg/day by supplementation with

P.O. ABERYSTWYTH DYVERGELL

57 g/head/day of urea during the dry season. Both Shorthorn and Brahman cross-bred heifers grazed either native pasture stocked at the rate of 1 beast/4.86 ha or a mixed sward of native pasture and *Stylosanthes humilis* stocked at the rate of 1 beast/2.43 ha. A subsequent paper (Holroyd et al. 1977) reporting results of the same trial over a four-year period showed that urea compared with non-nitrogen containing supplements (molasses, phosphorus) produced increases in conception rates of cows grazing native pasture, particularly during a dry year. Values were 55% for unsupplemented animals and 75% for supplemented animals. During the last year, precipitation was only 604 mm compared with a thirty-nine year average of 867; precipitation during the first two years of the trial was 866 and 1458 mm respectively. On the pastures oversown with Townsville stylo, only marginal responses to urea supplementation occurred during the first two years; however, a significant response occurred during the third, dry year when supplementation with urea increased the conception rate from 55.8 to 74.1%. The urea supplement also significantly improved both the calving date (expressed as days from 1 January) and calving interval, these being 296 and 444 days for the unsupplemented control animals and 241 and 402 days for the supplemented animals respectively. Survival feeding (Burns 1971) was given to 27.5% of the non-urea treatment cows v. only 9.2% of the urea-supplemented group.

To conclude the review of Australian work, brief mention will be made of a trial carried out on Wallum pasture at the Coolum Research Station situated in the narrow belt of coastal lowlands of south-east Queensland. Although confounded by other factors (climatic, etc.), improved weaning percentages were obtained when cows grazed Pangola grass (*Digitaria decumbens*) with grazing oats drilled into it as compared with Pangola grass alone during the previous year. Weaning percentages were 64 and 87% respectively (Tierney 1973).

Briefly summarizing, it is clear that condition of the cow at breeding time is a critical factor in determining whether she will conceive or not. Supplementation which leads to improved condition may also be expected to have the same effect and this is most marked in cattle which are most stressed, i.e. first calf heifers and young cows with first calf at foot. In pastures consisting of both grasses and legumes, effects of supplementation are less pronounced (as would be expected) except during years with particularly low rainfall.

United States

Florida

There is some evidence on supplementary feeding of breeding females available from the subtropical United States. General recommendations for Florida conditions were given by Chapman et al. (1964b) although these were based apparently on textbook estimates of nutritional requirements and summarize mimeoed reports. It is recommended that heifers grazing pastures without legumes should be fed 1.35–2.25 kg of a high-energy feed containing 20% crude protein or 450 g of a 40% supplement plus 1.35–2.2 kg of a high-energy feed. A similar supplement was recommended for brood cows grazing grass pastures on mineral soils.

The influence of pasture and winter supplementation during growth on the productivity of beef cows was investigated by Warnick et al. (1969). Heifers were grazed year-round on two types of pasture: all grass pasture which consisted of Pangola grass (*Digitaria decumbens*) and Pensacola bahia grass (*Paspalum notatum*) and a similar pasture overplanted with Nolin's Louisiana white clover (*Trifolium repens*). The cross-bred heifers used in the experiment were divided into four groups, two of which were allocated to each pasture type. One of the groups on each pasture type was supplemented with 1.1 kg of protein supplement (cottonseed meal) whilst the other group received minimum requirements (450 g maximum) or no supplement. The heifers on the clover pastures were 47 kg heavier than those on the grass pasturage, but the supplemented heifers were only 18 kg heavier than the unsupplemented animals and this difference was not significant.

A similar advantage accrued to the clover-pastured females insofar as calving rate was concerned; the percentage of calves born was approximately 15% greater than to the unsupplemented animals.

Again, however, there was no advantage to extra supplementation. It is probable that the animals receiving the low level of supplement were protected from extreme nutritional stress and thus the effect of extra supplementation was not demonstrated.

In a later study, Kirk and his colleagues (Kirk et al 1974) studied the effects of various supplements on burnt and unburnt range. The use of cottonseed pellets for cows grazing native range at a stocking rate of 12 cows/64.8 ha of native range did not improve calving percentage over burning alone although burning improved calving percentage from 56 to 75%. Marginal improvements to weaning weight and production per cow and per ha were attributable to the cottonseed pellets, these weights being 150, 117 and 18 kg v. 142, 106 and 13.1 kg for supplemented v. non-supplemented cattle in the same order as mentioned above.

Martin et al. (1976) compared biuret with cottonseed meal cubes as a source of protein for cows and heifers, one of the few comparisons of this nature to be carried out in tropical or subtropical regions. Angus cattle grazed Argentine bahia grass (*Paspalum notatum*) pastures and were given cottonseed meal cubes containing 45.5% crude protein at the rate of 0.9 kg/day or citrus pulp cubes containing 13.75% feedgrade biuret (equivalent to 43.6% CP) at the rate of 1.0 kg/day. Control animals were given citrus pulp cubes without biuret at the rate of 0.86 kg/day. All cows lost weight during the winter period, losses being 49, 60 and 70 kg for the animals on the treatments in the same order as noted above. All values differed significantly. Conception rate, calf survival and time to reconception were not influenced by wintering treatment. Weaning weights of calves from mothers fed biuret were significantly lighter than calves from cottonseed-fed cows, weights being 162 and 166 kg respectively. However, this difference, although significant, was clearly marginal and the cottonseed-fed group weighed 1 kg less than the calves from mothers fed citrus pulp only. The authors comment that the capacity of cows to compensate for excessive winter weight loss during the summer months and the fact that the wintering treatment was terminated 120 days prior to weaning, suggest that the observed difference in weaning weight may have been due to factors other than wintering treatment. It would seem more likely, however, that differences, although significant, were not real.

This may have been due to the fact that differences were obscured by the feeding of Pangola grass hay to all groups.

Subsequently, Peacock and his colleagues (Peacock et al. 1976a) compared various grazing systems with native and improved pastures which also involved different levels of supplementation. However, since the level of supplementation was confounded with the pasture management used, this work will be dealt with elsewhere.

Louisiana

Finally, several studies have been carried out in Louisiana. As early as 1963, it was shown that supplemental feeding could substantially increase cattle production (Duvall and Whitaker 1963; Duvall and Hansard 1967), and subsequently Pearson (1974) compared two types of supplemention for cows grazing native range (predominantly of *Andropogon* and *Panicum* spp). A 32% crude protein supplement (based on urea and molasses) was dispensed with self-feeding roller tanks either year-long or during the winter only. Control cows were fed on a graduated schedule with cottonseed cake and hay from November to May inclusive. Performance of the year-round-fed cows was better than those fed on cottonseed cake, which was in its turn better than that of those fed only during the winter. Calving percentages for the three groups were 83, 93 and 54% for the three treatments in the order given above, whilst weaning weights were 192, 201 and 168 kg respectively. These values were a reflection of the weights of the cows; at the fall and spring weighings (unfortunately, no summer weighing were taken of the winter-fed group), winter-fed cows were substantially lighter than animals from other treatments. Although to a certain degree these results are to be expected, it is also clear that winter supplementation may not be sufficient on its own to increase performance if summer grazing is not good enough to maintain condition.

Subsequently, Duvall (1969) showed that cows fed cottonseed cake on alternate days had a higher calving percentage (78%) than those fed every day (70%); both these values were higher than those obtained when cows were allowed to consume cottonseed meal adulterated with one part of salt to three parts of meal (59%). The former observation is probably due to the fact that feeding every two days interfered less with the normal grazing habits of the cattle. Similar behaviour has been reported from Swaziland (Ffoulkes 1977). No explanation could be given for the inferiority of the salt-fed cows.

Several other studies from the United States, Mexico and elsewhere deal with supplementation of the growing female animal but they have been reviewed elsewhere as they are concerned with growth rather than reproductive functions.

Discussion

Before embarking on a discussion of the investigations reviewed above, it is probably as well to point out once more that very little information is given on the diet available to the grazing animals employed in the various studies. On only few occasions is even the proximate analysis of the herbage given and in any event even the most complete analysis gives very little indication of the amount of any given feed consumed by the animal. Similarly, deficiencies of any nutrient in the forage may well be corrected by browse which frequently contains higher levels of phosphorus and protein than wintering grasses. Thus, if the basal intake of nutrients above which supplementation is to be practised is so unclearly understood, it is hardly surprising that varied responses to supplementation have been encountered. Moreover, it is clear that the physiological state of the animal prior to and at the time of supplementation is of the greatest importance. These factors explain many of the differences in results obtained by various workers.

It is apparent that a lack of protein during the period before sexual maturity is attained, delays puberty and consequently age at first calving. There is evidence that supplementation with protein will correct this and also exert an effect on the birth weight of the calf.

However, there is also considerable evidence that calves deprived of protein during the winter will make up the weight difference significantly by compensatory growth during the following summer period given favourable conditions. There is clear evidence that lack of protein during the pre-pubertal period delays the onset of this phenomenon.

Puberty is reached not at a given age, but when age, weight and size have interacted in such a way as to reach a minimum level at which gonadotrophic activity commences. During periods of undernutrition, developmental growth may continue although weight remains constant. The comments on the role of the gonadotrophic hormones in ovulation given below is clearly also relevant to the onset of puberty.

The obvious consequence of a delay in reaching puberty is a corresponding increase in the age at calving. Supplementation with protein will correct this and also exert an effect on the birth weight of the calf although, given favourable conditions, this difference will probably be made up for by compensatory growth. The explanation for the cessation of oestrus during periods of prolonged protein deficiency is not entirely clear but may be associated with an insufficient synthesis of the gonadotropic hormones when the protein reserves of the body are depleted. The gonadotrophins, the follicle-stimulating hormone (FSH) and the luteinizing hormone (LH) are glycoproteins produced by the anterior lobe of the pituitary and their lack may be caused by the reduced size of this gland caused by protein deficiency. It may be relevant to the former point that Bedrak et al. (1964) have shown that serum albumin, the most labile of the body proteins, is significantly reduced in protein-deficient beef heifers.

Mulinos and Pomerantz (1940) suggested that 'pseudohypophysectomy', a condition resembling hypophysectomy, was induced in response to undernutrition and that this led to a reduction in LH. This was not confirmed by the findings of Gombe and Hansel (1973), however, who found that restricted intakes of energy led to increased levels of LH but that the ability of the ovary to respond to gonadotrophins was inhibited by undernutrition. This was accompanied by reduced progesterone levels said to be caused by an inhibition of cholesterol side-chain cleavage leading to reduced pregnenolone formation.

Alternatively, it has been suggested that environmental events acting through sensory extrareceptors may affect the activity of the hypothalamic-pituitary system; the effects of alternating light and dark and seasonal variations in day length on ovulation in various species are well-documented examples of this phenomenon (Johnson, Hatton and Goy 1969).

Finally, Yeates (1965) has cited work by Rinaldini (1949) indicating that during inanition, failure of release of the gonadotrophins rather than their synthesis plays the major role in the suppression of ovulation.

There is wide agreement that protein supplementation of deficient cows increases their calving percentage but the extent of this increase varies widely from one trial to another. Differences vary from those described as 'marginal' as in the Botswana and Swaziland work to the massive 48% reported by Siebert et al. (1976). It would seem likely, however, that improvements of some 15% can be expected. Many workers have drawn attention to the important part played by the 'condition' of the cow in conception and reconception and it is also generally accepted that this is associated with a shorter period of *post partum* anoestrus.

Despite the interesting and elegant work by Hale and Holness (1977) already referred to, it would seem that more evidence of this type must be accumulated before condition can be ruled out as an important factor influencing reconception. Whether the cow is lactating or not is clearly of importance as several workers have emphasized that the former animals should be favoured when given supplements. The additional vulnerability of the cow nursing her first calf is also recognized.

Supplementation of cows with calves at foot has been usually associated with increased weaning weight and this, in its turn, as would be expected, with increased milk yield although few critical studies have been carried out in this area. As would be expected, supplementation with protein reduces mortality. Presumably this occurs first in cows already stressed more than their companions.

The question of what level of supplementation to employ has been raised by several authors, particularly in Zimbabwe, and it appears that often only relatively small quantities are required to significantly increase calving percentage etc. Although higher levels may produce higher *gross* returns, these are not justified when the cost of the feed is taken into account. In this respect, it is interesting to note the anomalous effects caused by the use of urea; this has in some cases been shown to be ineffective although conventional protein supplements have caused large increases in conception rate. This could be due to the lack of a suitable energy source to enable the rumen microorganisms to use NPN for protein synthesis or because of the necessity for by-pass protein discussed elsewhere. Similarly, some of the beneficial effects of protein or urea have been attributed to a consequent increase in digestibility and energy consumption.

The effectiveness of small quantities of supplement are corroborated by the effects of small quantities of legume hay in improving performance and by the lessened effect in many cases of supplements when fed to cows already grazing grass/legume pastures when adequate rainfall has been experienced.

As a last point, two possible adverse effects of supplementation should be stressed: the lessened fertility of the obese animal is well known and, whilst clearly it is extremely inefficient to feed an animal to this level, it has been seen in cattle fitted for shows etc. The effect of supplementation in increasing the birth weight of the calf increases the incidence of dystocia, particularly in breeds already prone to this condition.

Summary

It appears that, under conditions of protein deficiency, supplementation of female stock with this nutrient causes the following effects:

1. Acceleration of growth, earlier puberty and consequently lower age at first calving.
2. Increased calving percentage (improvements from as low as 5 to as much as 48% have been quoted).
3. More rapid resumption of oestrus after parturition.
4. Improved milk production and consequent calf growth to weaning.
5. Decreased mortality.

However, although weight gains may be increased by supplementation, these advantages may be subsequently reduced by compensatory growth in non-supplemented stock. A 'compensatory' effect with regard to reproduction has also been mentioned.

Attention has also been drawn to the importance of body weight and 'condition' when reconception is considered. Logically, a cow in poor condition would be expected to be more responsive to supplementation. Although some difference of opinion exists as to which class of stock it is most important to supplement, it is clear that 'stressed' cows (i.e. those lactating) are as important in this respect as are in-calf heifers.

As would be expected, supplementation is less effective when cattle are grazing pastures containing legumes. The exception to this may be during hard years.

Finally, there is some evidence to support the contention that supplementation at low levels is more economically sound than that at higher levels.

9 The use of legumes in improving the productivity of tropical pastures

Legumes, in symbiotic association with bacteria of the genus *Rhizobium* (which establish themselves in root nodules), are able to reduce atmospheric nitrogen to ammonia and thence to protein. This series of reactions is known as nitrogen 'fixation' and requires the presence of pyruvate as an electron donor, ATP and the enzyme, nitrogenase. Possible reaction intermediates may be hydrazine and hydroxylamine but these have not been demonstrated.

In view of the importance of nitrogen in plant nutrition *per se*, nitrogen fixation has an important potential which has been recognized in agricultural systems in temperate regions for some time.

In tropical regions, however, there has been some doubt that nitrogen fixation actually takes place (Vine 1957; personal communication) or that it occurred in sufficient quantities to affect productivity significantly (see, e.g. Webster 1954; Bumpus 1957). This doubt was expressed verbally as recently as 1973 at the III International Congress of Livestock Production. The extent of nitrogen fixation may not have been recognized because of concurrent losses of nitrogen in experiments which were not designed to evaluate this aspect, but the work of Schofield (1945), one of the pioneers of legume use for tropical pastures in Queensland, and Bumpus (1957) in Kenya showed that nitrogen level of soil under legumes was significantly increased in comparison with that of soil under bare fallow.

Goldschmidt (1938), in a preliminary report on the nitrogen problem in pasture soils of the Natal sour veld, had already reported nodulation together with nitrogen fixation in the indigenous legume, *Trifolium africanum*.

Henzell (1977) stated that the role of legumes in the improvement of tropical pastures can be summarized by the following functions:

1. The ability to fix nitrogen from the air gives the legume a competitive advantage in association with grasses and weeds.
2. The presence of the legume in the forage provides a better diet for livestock.

3. The legume contributes available nitrogen to other non-legume plants in the pasture.
4. The legume contributes nitrogen to a subsequent cropping season.

In this work, we are particularly concerned with legumes insofar as improved diet is concerned, as well as the extent to which additional nitrogen availability contributes to the overall availability of herbage to the grazing animal.

The nutritional advantage of legumes is mainly due to their content of crude protein; Minson (1977) has quoted a mean of 17.2% crude protein for 166 legumes (presumably excluding edible trees and shrubs) v. the mean protein content of tropical grasses reviewed by the author of 7.7% (Butterworth 1967a).

Minson (1977) has also stated that the crude fibre content of legumes is slightly lower than the mean of the values for tropical grasses reviewed by Butterworth (1967a), i.e. 30.6 v. 33.4%. Parra *et al* (1972) showed that levels of both NDF and ADF were lower in tropical legumes than in grass.

Minson (1977) compared contents of phosphorus and differences are not conclusive; legumes appear to be consistently higher in calcium but this element is not usually deficient in the diet of the grazing animal.

There does not appear to be any reliable evidence to indicate that either the energy content (as measured by different methods) or the voluntary intake of tropical legumes is different from that of grasses under similar conditions.

Australia

Although occasional references to earlier work carried out in Africa, the Americas and the Pacific occur in the literature, by far the greatest amount of work on pasture legumes has been carried out in Australia. This was probably due to the fortuitous success of an accidentally-introduced legume *Stylosanthes humilis* (originally known as *Stylosanthes sundaica* or Townsville lucerne) or Townsville stylo. Townsville stylo was introduced to Queensland, possibly through the medium of cattle faeces, early in the century and rapidly established itself as a 'leguminous weed'. (An ability to establish itself in this manner is probably the most valuable attribute of any pasture legume!)

Nevertheless, it was not until the late 1950s that work on the evaluation of the productive capacity of Townsville stylo was started at Katherine in the Northern Territory and Rodds Bay, central coastal Queensland. Thereafter a large number of trials were carried out on various pasture legumes both in Australia and other tropical regions and the field has become one of the most intensively investigated aspects of tropical pasture research.

Beef production from legume-based pastures both in Australia and elsewhere has been reviewed by Norman (1974a and b) whilst other aspects of tropical forage legumes have been reviewed by Skerman and his colleagues (Skerman 1977).

Northern Territory

The Northern Territory (NT), the 'wet and dry' tropics, has a rainy season with precipitation ranging from 500 to 1500 mm which falls during a relatively short period, and a predictable dry season. Cattle production is based on the low productivity of extensive native grasslands. Research carried out in the Territory has concentrated on the role of Townsville stylo in both wet and dry season situations both in combination with natural pasture but also with such improved grasses as Buffel and Birdwood (*Cenchrus setigerus*) as well as Sabi and Pangola. A large proportion of the work has been carried out at Katherine but results have also been obtained from Adelaide River, Berrimah and Manbulloo.

Katherine

Katherine lies between latitudes 14 ° and 15 °8′S and the predominant vegetation is Eucalypt forest with an understory of *Themeda australis*, *Sorghum plumosum* and *Chrysopogon fallax*. Mean annual precipitation is 900 mm which falls during a four-month period.

In the first evaluation of the use of Townsville stylo as an adjunct to natural pasture, Shorthorn steers grazing on Townsville stylo during the dry season gained 42 kg more than similar steers grazing native pasture year-long; stocking rate on natural pasture was 0.14 head/ha year whilst the comparable figure for dry season legume stocking was 1.2 head/ha (Norman and Arndt 1959; cited by Norman 1974).

A subsequent trial used improved pastures to increase stocking rate and weight gain during the wet season. Pasture was composed of *Cenchrus setigerus* (Birdwood) intersown with Townsville stylo and this was compared with Bulrush millet (*Pennisetum americanum*). Cattle grazed native pasture the first month of the grazing season and then were transferred to the improved pasture at a stocking rate of 0.44 animal/ha (the stocking rate on natural pasture was 0.04 animal/ha). The trial was continued for three seasons. Cattle on native pasture alone gained 72 kg/head, those on Birdwood/stylo, 113 kg/head, and those on millet, 123 kg, an outstanding increase in terms of beef/ha. In addition to increased gains and carrying capacity attributable to the improved pastures, the grazing season was extended by four weeks in the case of the sown pasture and eight weeks for the millet (Norman 1963; Norman and Stewart 1964).

The use of Townsville grown in the wet season and left as standing hay for use in the dry season was investigated by Norman (1964). Paddocks were grazed by Shorthorn steers at a stocking rate of 2.2 ha/head for a period of twenty-two weeks during the dry season. Average daily gain during the period was 576 g/day (or 197 kg/ha). The author points out that the acceptability of Townsville stylo as standing hay is dramatically reduced on wetting and that the climate at Katherine, without winter rains and frosts and with only a short period of dew, is ideal for winter utilization.

Norman and Stewart (1964) compared rates of gain on Townsville stylo/Birdwood/volunteer pastures during the dry season and showed that the rate of gain was related to the percentage of legumes in the pasture. Results are summarized in Table 9.1. The stocking rate used was again 2.2 head/ha and production on the treatment with the highest proportion of Townsville stylo was equivalent to 198 kg/ha.

Table 9.1 Performance of cattle grazing pastures with different proportions of Townsville stylo during the dry season (Katherine) (adapted from Norman and Stewart 1964)

Townsville stylo (%)	*Liveweight gain (kg/head)*	*Period of gain (weeks)*
23	9	8
45	45	20
63	89	22

Norman and Stewart (1967) investigated the effects on liveweight gains of varying the length of time animals were allowed to graze Townsville stylo. The treatments used were the following:

1. Native pasture only.
2. Townsville stylo two days, native pasture five days.
3. Townsville stylo four days, native pasture three days.
4. Townsville stylo only.

The stocking rate on native pasture was approximately 6 ha/head (Shorthorn steers, 250–350 kg) whilst that on the Townsville stylo was 0.1, 0.2 and 0.3 ha/head for treatments 2, 3 and 4 respectively (this was reduced slightly from the first year).

Animals on native pasture alone and on treatment 2 lost weight slightly during the dry season, animals on treatment 3 gained about 18 kg and those on treatment 4, Townsville stylo alone, gained about 50 kg (gains have been estimated as the results were presented graphically). This is similar to the 179 kg/ha obtained in previous experiments. The authors likened the provision of legume pasture to supplementation with protein and calculated that 1.3 kg/day of

Townsville stylo was equivalent to 0.1 kg/day of protein as supplementation terms of weight gain.

In a later trial (Norman 1968), the performance of beef cattle on different sequences of Townsville stylo and native pasture was measured. Whilst previous trials had been of a short-term nature, either dry season or wet season, data were collected on cattle from weaning to slaughter. Shorthorn weaner steers were allocated to sown (S) and natural (N) pasture treatments as follows:

Group	*First dry season (phase 1)*	*First wet season (phase 2)*	*Second dry season (phase 3)*	*Second wet season (phase 4)*
A	S	S	S	S
B	N	N	S	S
C	S	S	N	N
D	S	N	S	N
E	N	N	S	N
F	S	N	N	N

Stocking rates ranged from approximately 4.4 to 6.2 ha/head for native pasture and 1.1 to 1.3 ha/head/year for sown pasture. Group A animals were ready for slaughter at the end of the second wet season and all other animals were continued on sown pasture (phase 5) until they reached the same mean liveweight. Rates of gain during the first part of the experiment (phases 1–4) and during phase 5 are given in Table 9.2, together with dates of slaughter of the various groups.

The author draws the practical conclusion that relation between the time spent on Townsville stylo and the total time from weaning to slaughter is linear; compensatory growth under the conditions of the experiment was quite limited.

A further trial was carried out to determine the relation between liveweight gain and the amount of Townsville stylo on offer. At a stocking rate of 0.61 animals/ha, there was a linear response to the amount of legume between 1000 and 7000 kg/ha. The author uses these results (Norman 1970), and those previously obtained (Norman and Stewart 1967; Norman 1968), to demonstrate that the progressive substitution of Townsville stylo for natural pastures at Katherine results in a linear improvement in liveweight gains. This is attributed to the extremely low level of production of the natural pasture combined with the fact that the introduced pasture is sufficient for beef cattle liveweight gain at a near maximum level for almost nine-tenths of the year. Presumably, the linear relationship would be altered by increasing the potential for beef production by either

Table 9.2 Liveweight changes, dates of slaughter and dressing percentage of steers fed natural and sown pastures according to different strategies (Katherine) (adapted from Norman 1968)

	Group					
	A	B	C	D	E	F
Liveweight change						
Phases 1–4 (kg/head)	285	227	151	141	106	85
Rate of gain (g/head/day)						
Phases 1–4	454	363	227	227	182	136
Phase 5	—	182	363	272	272	272
Date of slaughter	May'65	Apr'66	May'66	Sep'66	Apr'67	May'66
Dressing (%)	51	47	45	49	51	50

increasing the stocking rate or using animals capable of more rapid growth.

A subsequent experiment was carried out to investigate the effects of different stocking rates on Townsville stylo pastures during the wet season. Three stocking rates were used: 3.3, 2.5 and 1.7 Shorthorn steers (averaging 320 kg at the start of each year's test)/ha. Results are summarized in Table 9.3. The stocking rates used did not limit individual production and therefore total production/acre increased, as did stocking rate. The authors concluded that increasing grazing intensity reduced the proportion of invading grasses in the sward, and the Townsville stylo that resulted, supported higher gains per acre. It is perhaps unfortunate that stocking rates were insufficient to provoke declining productivity, either on a per head or per ha basis (Norman and Phillips 1970).

Table 9.3 Liveweight gains and Townsville stylo content of pasture grazed at three stocking rates (Katherine) (adapted from Norman and Phillips 1970)

	Stocking rate (head/ha)		
	3.3	*2.5*	*1.7*
Liveweight gain			
per head (kg)	117	118	120
per ha (kg)	383	291	196
% TS	75	67	58

Woods (1970) carried out a number of trials near Katherine to test the various recommendations made by Norman under semi-commercial conditions. Two types of improved grazing were used: native grasses (209 ha) supplemented with *Stylosanthes humilis* (5 ha) and *S. humilis* dominant pastures (these were either *S. humilis* alone or with *Cenchrus* spp. and/or *Urochloa mosambicensis*). When results of the twelve trials were combined, estimated total liveweight gain (kg/ha) were as follows:

Native pasture	7.1
Native pasture + *S. humilis*	11.3
S. humilis dominant	126

It was pointed out that one of the more important aspects of management was to maintain sufficient wet season grazing to keep down grass competition whilst ensuring that sufficient *S. humilis* was available for the dry season.

Later work at Katherine continued investigations into grazing systems on Townsville stylo pastures (Evans, McLean and Winter, 1979). Three grazing management systems using *S. humilis* and *Urochloa mosambicensis* included continuous grazing year round (stocking rate: one beast/0.59 ha); two paddock system: wet season stocking rate, one beast/0.91 ha; dry season stocking rate, one beast/ 0.3 ha with no grazing during the dry season. The average stocking rate over the three systems for the whole year was one beast/0.59 ha. Animal production for the three systems was similar averaging about 177 kg/ha and liveweight gain was linearly related to stocking rate during the wet season.

Although all the work reported here has concerned *S. humilis*, investigations have been initiated on *S. hamata* (Verano), *S. scabra* and *S. viscosa* (Winter, Mott, and Mclean, 1979) in association with various grasses.

Upper Adelaide River

The Upper Adelaide River Experiment Station is located 76 km south of Darwin. Average precipitation is 1270 mm, which falls mainly from December to March. The station is 180 km closer to the sea than Katherine and has a longer wet season and higher year-long relative humidity. Two types of country are included: upland and floodland, and the predominant grasses are *Chrysopogon, Themeda, Sorghum, Heteropogon* and *Eriachne*. More than forty short-term grazing trials were carried out to investigate the performance of cattle on native and improved pastures. Trials included evaluations of:

Native pasture
Townsville stylo
Legumes, mainly Phasey beans (*Macroptilium lathyroides*)
Perennial grasses (mainly Para grass. *Brachiaria mutica*)
Mixed improved grasses and legumes, mainly Para and Phasey bean

Cattle used were Shorthorn steers off range 'with a wide initial range of ages (2–5 years), weights (average 275 kg, range 209–356) and temperament' (this last important factor is rarely mentioned). Stocking rates were 1 beast/4–8 ha (native pasture) and 1 beast/ 0.4–1.2 ha for improved pastures.

The author (Wesley-Smith 1972) has expressed his results graphically, but some observations are made in Table 9.4.

Floodplain and upland grazing were compared in a long-term trial using Shorthorn and Brahman Shorthorn cross cattle. The upland pastures were Townsville stylo and floodplain were native pasture supplemented with blocks of Townsville stylo and Phasey bean. Gains on both systems averaged about 130 kg between December and July and losses between July and December averaged 44 kg. Gains of floodplain cattle were lower, and although compensatory gains occurred during the finishing period, these were insufficient to make up the difference in weight.

Table 9.4 General results of short-term grazing trials using native and improved pasture (Adelaide River) (adapted from Wesley-Smith 1972)

Pasture type	*Overall weight gain (kg/head)*	*Remarks*
Native pasture	18	Gain (1.3 kg/day) Nov–Dec Rapid loss July onwards
Townsville stylo	91	Gain until May Maintenance until Sep Loss Oct–Nov
Mixed improved pastures	85	Declining gain Mar–Aug Loss Sep–Nov
Perennial legumes	63	Gains Apr–Dec

Also working at Upper Adelaide River, Moran (1973a) carried out an experiment of the effects of meat-and-bone meal supplements on growth of weaner cattle grazing Townsville stylo (50–60%), *Pennisetum pedicellatum* (20–30%) pastures during the dry season. Animals used were Brahman (predominantly) Shorthorn heifers seven months of age with an initial liveweight of 156 kg. Meat-and-bone meal was offered at the rates of 0.21, 0.43, 0.85 and 1.71 kg oven-dried meat-and-bone meal/head/day. Actual consumption as well as liveweight changes are summarized in Table 9.5.

Overall results only are given in Table 9.5, but the paper indicates that the consumption of meat-and-bone meal increased as the quality of the pasture deteriorated and the effect on liveweight change also increased accordingly. However, overall response to the supplement was relatively small as would be expected on grazing of reasonably high quality (faecal nitrogen levels during the trial averaged 1.4% with a minimum value of 1.04%).

Table 9.5 Liveweight changes and meat-and-bone meal consumption of heifers grazing improved pastures (Adelaide River) (adapted from Moran 1973a)

	Meat-and-bone meal on offer (kg)				
	0	*0.21*	*0.43*	*0.85*	*1.71*
Weight change (kg/day)	–0.02	+0.06	+0.08	+0.16	+0.20
Meat-and-bone meal intake (kg/head/day)	0	0.21	0.42	0.78	1.05

Berrimah

Improved pastures composed of Pangola grass (*Digitaria decumbens*) and Townsville stylo were used for beef production at Berrimah, 14 km south of Darwin. The average precipitation at the station is 1524 mm which falls almost entirely in the five-month period from November to March (Moran 1973b). The experimental animals used were Brahman Shorthorn and Shorthorn steers, Buffaloes and Bantengs (*Bos banteng*). The stocking rate used was approximately 1 beast/0.35 ha. Average daily gains during the fifteen-month duration of the trial (March 1971–June 1972) were as follows:

Buffaloes:	0.40 kg/head
Brahman crosses	0.40 kg/head
Bantengs	0.21 kg/head
Shorthorns	0.20 kg/head

This production is representative of what may be obtained in the wetter parts of the Northern Territory.

Douglas Daly

A further trial comparing an introduced grass, Sabi (*Urochloa mosambicensis*), was carried out at the Douglas Daly Experiment Station 160 m south of Darwin where the average precipitation amounts to 1110 mm, which is still one of the wetter parts of the Territory. Falvey (1976) compared Townsville stylo alone with the legume sown with Sabi grass. Townsville stylo is frequently invaded by less desirable annual grasses and the introduction of a more palatable perennial grass could improve productivity. This did not appear to be the case as Townsville stylo pastures appeared to give consistently higher productivity. Mean dry season liveweight changes over a three-year period were 3.4 and –9.0 kg for the Townsville stylo and Townsville/Sabi pastures respectively, whilst the corresponding figures for the wet season were 99.8 and 87.3 kg. It would appear from this experiment that there is no appreciable difference between an 'improved' perennial grass and natural grasses insofar as beef production is concerned. The only advantage of the Townsville/Sabi grass pasture appeared to be that it sustained worse management for a longer period than the pure stylo and still responded to later improved management (a useful recommendation in view of the apparently universal prevalence of poor management!).

Tropical north-eastern Australia

Pasture research has been carried out in north and central Queensland north of latitude 22 °S where annual precipitation is between 500 and 1500 mm, and this description more or less defines the area also classed as 'wet and dry' by Norman (1974a), characterized by tropical woodland. As in the Northern Territory, most of the earlier work was based on Townsville stylo, both with natural and artificial pastures, but later work investigated the use of the perennial Oxley fine-stem stylo (*S. guianensis var intermedia*) and verano caribbean stylo (*S. haemata*). Work was initiated at Rockhampton and Townsville but later results came from the Cape York Peninsula.

Rodd's Bay

Rodd's Bay is a private property where legume introduction into the natural *Heteropogon contortus* (Spear grass) swards was examined as early as 1944 (Shaw 1961) and where experimental grazing started in 1955. Rodd's Bay is situated 32 km south of Gladstone at a latitude of 24 °S with an average annual precipitation of 856 mm.

In the first record of beef production from legume-based pastures in this part of Queensland, Shaw (1961) compared a pasture containing Townsville stylo with natural Spear grass pastures. Part of the Townsville stylo pasture had originally been sown to the legume and subsequent disc-harrowing caused it to spread and increase in density. The pasture received substantial fertilization with superphosphate whilst the control pasture was unfertilized. Grazing was continuous in both treatments, but the control was burnt occasionally to remove old grass whilst burning was avoided in the legume pasture. Results of liveweight gain/ha over a five-year period are summarized in Table 9.6. Shaw supported these results with observations from other commercial stylo paddocks where fat steers were sold earlier than comparable animals on natural pastures and heifers were mated at two, rather than three, years of age.

Table 9.6 Liveweight gain (kg/ha) of steers grazing Townsville stylo and natural pasture (Rodd's Bay) (adapted from Shaw 1961)

Year	*Grazing days*	*Liveweight gain (kg/ha)*	
		Townsville stylo	*Control*
1955–56	378	54	No control results first year
1956–57	361	49	13
1957–58	347	64	Grazed only part period
1958–59	350	102	21
1959–60	364	123	16

The author concludes that 'Beef production can be greatly increased by including Townsville Lucerne [sic.] in native spear grass pastures, the increase being a combination of higher carrying capacity, greater liveweight gain per head and quicker turn-over of stock'.

In a subsequent study (Shaw and t' Mannetje 1970), paddocks at Rodd's Bay were subjected to the following treatments:

Control: natural pasture stocked at 0.3 beast/ha
Natural pasture at high stocking rate: 0.6 beast/ha
Fertilized natural pasture at high stocking rate: 0.6 beast/ha
Natural pasture oversown with Townsville stylo and stocked at 0.8 beast/ha
Natural pasture oversown with Townsville stylo, fertilized and stocked at 0.25–0.50 beasts/ha.

Fertilization consisted of 530 kg superphosphate in four years and molybdenum. Livestock used were groups of Hereford steers either 2.5 or 1.5 years old which were introduced each June or July and grazing was continuous. Results for the seven-year period during which the experiment was conducted are summarized in Table 9.7. It was concluded that, because there was an excess of forage on offer in the natural pastures, the increase of animal performance due to

Table 9.7 Liveweight gains per head and per ha from natural pastures and natural pastures oversown with Stylosanthes humilis with and without fertilizer at different stocking rates (Rodd's Bay) (adapted from Shaw and 't Mannetje 1970)

	Liveweight gains (seven year means)	
	(kg/head)	*(kg/ha)*
Natural pasture	83	25
Natural pasture: high stocking rate	47	29
Natural pasture: high stocking rate + fertilizer	100	62
Oversown pasture	121	93
Oversown pasture + fertilizer, high stocking rate	149	148

fertilization and the presence of the legume was because of an improvement in the quality of feed on offer as well as the increased carrying capacity. The highest production was from the fertilized Townsville stylo paddock not only in terms of gain/acre but higher gains of individual animals. The authors also pointed out that although production in years of drought was reduced (as would be expected), the improved pasture proved to be 'safe' during drought and no deaths resulted. Effects of Townsville stylo and fertilizer were additive.

Shaw (1978a) initiated a complex trial at Rodd's Bay in 1965 which continued until 1973. The trial compared aerial with ground sowing, different fertilizer levels and different stocking rates. Fertilizer levels were F_0: nil, F_1: 125 kg/ha superphosphate annually, F_2: 500 kg/ha

Table 9.8 Percentage of Townsville stylo in differently treated pastures on two occasions (Rodd's Bay) (adapted from Shaw 1978a)

Treatment[1]	*Frequency Townsville stylo (%)*	
	June 1967	*June 1970*
Ground sown	24	44
Aerial sown	11	22
F_0	13	21
F_1	14	27
F_2	23	52
SR 1	17	30
SR 2	14	24
SR 3	13	31
SR 4	23	40

[1] For explanation of treatments, see text

initially followed by 250 kg/ha/year. Stocking rates (SR) started at 0.41 beasts/ha (SR 1) increasing to 0.82 beasts/ha (SR 4) in 1973. Increased stocking rates were possible because of the increase of Townsville stylo in the pastures during the course of the experiment (Table 9.8). Ground seeding produced a better stand than serial seeding and this effect persisted. Legume frequency increased with increasing levels of superphosphate application and (less conclusively) with stocking rate.

Data for animal production from successive groups of weaner Hereford and Brahman cross heifers is given by Shaw (1978b). This is summarized in Table 9.9.

Weight gain from both fertilizer treatments were considerably higher than those without fertilizer, but there was little difference between levels. When responses to fertilizer were further broken down, it could be seen that the response to the higher level was

Table 9.9 Mean liveweight gain (kg) per head of cattle grazing Townsville stylo/natural pastures with different fertilizer treatments and stocking rates (Rodd's Bay) (adapted from Shaw 1978b)

Fertilizer[1]	*Stocking rate*[1]				
	SR 1	*SR 2*	*SR 3*	*SR 4*	*Mean*
F_0	83.5	47.4	50.0	40.3	55.3
F_1	98.5	98.0	95.0	76.6	92.0
F_2	120.0	101.0	98.5	89.6	102.3
Mean	100.7	82.1	81.2	68.8	—

[1]For further details of fertilizer treatments and stocking rates, see text; for full details, see original reference

greater during the growing season. Fertilizer was the most important factor considered as it resulted in increased gains per head as well as enabling pastures to carry higher stocking rates. The detailed results also show that the superiority of the fertilized pastures continued to increase throughout the trial, indicating a successive improvement of feed quality attributable to additional superphosphate.

Millaroo

Swan's Lagoon is located at Millaroo, 112 km south of Townsville. The average rainfall is 780 mm, occurring mainly during the summer months and resulting in a growing season of four months' duration.

Work was carried out comparing different rates of fertilizer application and stocking rates on native pasture alone and on native pasture improved by sowing Townsville stylo (Winks and Lamberth 1968). The cattle used to evaluate the trial were Shorthorn steers and the various treatments used are given below:

	(steers/ha)
1. Native pasture	0.25
2. Native pasture	0.41
3. Townsville stylo–native pasture	0.41
4. Townsville stylo–native pasture	0.82
5. Townsville stylo–native pasture + superphosphate (375 kg/ha)	0.41
6. Townsville stylo–native pasture + supersphosphate (375 kg/ha)	0.82

Results are expressed in terms of maximum weight attained during the experiment and the time required to reach it. Values are also given of the rate of weight loss after maximum weight. Results are summarized in Table 9.10. The main effect of the Townsville stylo

Table 9.10 Effects of treatments on liveweight changes on native and Townsville stylo pastures (Swan's Lagoon) (adapted from Winks and Lamberth 1968)

Treatment[1]	*Max. lwt (kg)*	*Time to max lwt (days)*	*Daily gain max. lwt (kg)*	*Daily weight loss after max. lwt (kg)*
Native pasture	348.3	87	0.8	−0.6
Townsville stylo	386.7	189	0.6	−0.4
Townsville stylo + P	462.5	211	0.9	−0.4
Stocking rate:				
0.25 steers/ha	401.4	177	0.7	−0.2
0.41 steers/ha	415.9	180	0.8	−0.4
0.82 steers/ha	381.3	130	0.8	−0.9

[1]For further details of treatment, see text

was to extend the growing period and thus increase the maximum weight reached during the trial. This effect was enhanced by the application of superphosphate. The effect of the highest stocking rate was to reduce the length of the productive period but not the rate of gain during the growing period. Increase in stocking rate also increased the rate of weight loss, however (so much so that cattle had to be withdrawn from the higher stocking rate treatment), after the maximum weight had been reached.

Winks (1973) reviewed the work carried out at Swan's Lagoon and emphasized various aspects of management. As soil fertility improved as a result of fertilization with phosphorus as well as nitrogen fixation, there was an increased component of annual grasses in the sward (these increased from 10 to 90% during the years 1967–71). It was suggested that several methods might be effective in reducing this problem: these included no discing at establishment, less frequent fertilization, heavier stocking during the rainy season, and the introduction of an improved perennial grass (possibly *Urochloa* or *Cenchrus* spp.).

With regard to the quantity of phosphorus to be applied to the pasture, the author comments that whilst maximum production was promoted by 63 kg/ha of super, the application of 125 kg/ha markedly improved the nutritive value of the forage. The importance of molybdenum, copper and zinc was also stressed, particularly for dry years.

Liveweight change on both native and improved pastures follows a pattern of gain after the rains, followed by a period of loss; improved pastures are characterized by a longer growth period and a more rapid rate of gain.

Higher stocking rates result in a more rapid weight loss on both native and improved pastures. This is exacerbated by the tendency of the Townsville stylo as 'standing hay' to develop a heavy mould infestation because of winter rain, fogs and dews; this lowers palatability and the accompanying leaching effect reduces nutritive value (this is in marked contrast to the results from Katherine where winters are drier and no mould is encountered).

The supplementation aspects of an experiment carried out by Holroyd et al. (1977) have been mentioned earlier, but the comparison between natural pasture (mainly *Heteropogon contortus*, *H. triticeus* and *Chrysopogon fallax* on lighter soils and *Dicanthium teniculum* and *Bothriochloa bladhii* on heavier soils) and a similar area aerially seeded with Townsville stylo (and fertilized with superphosphate and trace elements) is of interest in the present context. The natural area was stocked at the rate of 1 breeding cow/4.9 ha and the oversown area at the rate of 1 breeding cow/2.4 ha. Calving rates for natural pasture and oversown pasture were 65.3 and 77.3% respectively, and thus the reproductive rate on the Townsville pasture was either equal to (in two years) or greater than (in the third year) the natural pasture at more than double the stocking rate.

Townsville

The first animal production data from the Townsville area appear to be those of Gillard (1970) who carried out a trial on improved pastures west of the Great Dividing Range of northern Queensland. Rainfall in the area is 630 mm and the evaporation rate is high; conditions are more arid than Rodd's Bay. The grass cover is predominantly *Heteropogon contortus*, *Bothriochloa intermedia* and *Chrysopogon fallax* and the carrying capacity is only 1 beast/15–20 ha. Soils are deficient in phosphorus, nitrogen and trace elements. The experiment was an unreplicated factorial design and the treatments were cleared or uncleared, fertilized (375 kg molybdenized superphosphate in the first year and 125 kg each year thereafter), or unfertilized, and stocking rates of 1 eighteen-month-old steer/4.8 ha or 2.43 ha. The pastures were aerially fertilized and seeded with *Stylosanthes humilis* pods and grazing was continuous. Results for liveweight gain are summarized in Table 9.11.

The most definite effect in this experiment was due to clearing and was because the Townsville stylo was able to establish more readily in cleared than in uncleared areas after aerial seeding. The effect of stocking rate on animal production was also significant and the effect was less so in years of above average rainfall.

It was surprising that no significant differences could be attributed to fertilizer in this experiment. The author explains this by the fact that site variability (the experiment covered 117 ha of undulating country) obscured the effects. Fertilized treatments were also located on less favourable sites than the unfertilized controls. However, there was an indication, which did not reach statistical significance, that there was a beneficial effect of the fertilizer by the third year of the experiment. It would be expected that this would increase with further fertilizer application. The initial lack of response to fertilizer was not because the phosphorus status of the soil was inadequate as

Table 9.11 Liveweight gains of steers grazing improved pastures under different conditions west of the Great Dividing Range (northern Queensland) (adapted from Gillard 1970)

Clearing	*Fertilizer*	*Stocking rate*	*Liveweight gain/annum (kg) three-year mean*
Cleared	Without	Low	166
		High	131
	With	Low	169
		High	127
Uncleared	Without	Low	70
		High	54
	With	Low	92
		High	24

fertilizer treatment of adjacent small plots considerably increased yield of dry matter.

Most of the work from tropical Australia that has been quoted has indicated that the use of improved pastures containing Townsville stylo increased carrying capacity, rate of gain or rate of turn-off of fattening cattle for slaughter. Trials carried out at Lansdown near Townsville showed that the fertility of breeding cows was increased by grazing fertilized pastures containing Townsville stylo.

The design used was randomized block with the following treatments:

Stocking rates: 1 cow/1.2 ha (high); 1 cow/2.4 ha (low)
Superphosphate: nil; 126 kg/ha/year; 377 kg/ha/year
Conservation of hay: not conserved; conserved (hay was fed during the dry season)

The ground flora of the unimproved area was dominated by *Heteropogon contortus* but invasion of *Stylosanthes humilis* made sowing unnecessary. The cattle used were Droughtmaster cows with calves at foot. Records were kept of mean annual conception and calving rate as well as changes of liveweight of the cows. Results are summarized in Table 9.12.

There was a highly significant effect of fertilization rate on conception and calving rate. The effect due to fodder conservation approached significance. Differences in liveweight due to the different treatments were significant in individual years, but cows grazing fertilized pastures were generally heavier than those on unfertilized pastures. The authors (Edye et al. 1971) concluded that grazing fertilized improved pastures was beneficial to cows as well as steers (as already demonstrated elsewhere).

Table 9.12 Conception and calving rate, and weight change of cows subjected to different management systems on improved pasture (Lansdown) (adapted from Edye et al. 1971)

Treatment	*Conception rate* (%)	*Calving rate* (%)	*Liveweight gain (kg/head/year)*
Stocking rate			
High	75	78	34
Low	75	80	37
Fodder conservation			
Not conserved	71	75	33
Conserved	79	83	38
Superphosphate			
Nil	66	68	29
126 kg/ha/year	73	76	42
377 kg/ha/year	84	90	36

In a subsequent paper (Ritson et al. 1971), it was shown that all these treatments significantly influenced forage yield and chemical composition. Under the lower stocking rate, pastures continued to be dominated by perennial grasses whilst the higher stocking rate caused the legume and annual grasses to predominate. Superphosphate increased the amount of legume at the lower stocking rate. The conception rate was significantly correlated with the phosphorus content of the pasture. The lower rate of fertilization resulted in phosphorus levels of less than 0.15% during the greater part of the year, and only the higher levels guaranteed enough phosphorus to ensure that cows would not be deficient. It was also stated that sulphur deficiency could have limited the utilization of Townsville stylo during much of the year (see Hunter et al. (1979) below).

Calf production was considered in a later paper (Edye et al. 1972). Birth and weaning weights of the get of the dams during the experiment were measured and the results are summarized in Table 9.13.

Superphosphate treatment significantly increased both weaning weight and mean annual liveweight of calf weaned/cow.

The response of cattle to supplementation with sodium and sulphur whilst grazing a predominantly *Stylosanthes guianensis* pasture in the Townsville area was studied by Hunter et al. (1979). They worked with eighteen-month-old Droughtmaster steers at the commercial beef property 'Meadowbank', 70 km south-west of Mt Garnet. The experimental treatments employed were:

Table 9.13 Birth weights and weaning weights of calves from dams subjected to different management systems on improved pasture (Lansdown) (adapted from Edye et al. 1972)

Treatment	*Birth weight (kg)*	*Weaning weight (kg)*	*Weaned calf weight (kg/cow/mated)*
Stocking rate			
High	28.6	188	[1]
Low	30.0	199	[1]
Fodder conservation			
Not conserved	29.1	191	[1]
Conserved	30.0	197	[1]
Superphosphate			
High	28.6	180	116
126 kg/ha/year	30.0	198	144
377 kg/ha/year	30.0	203	165

[1]No data given in original text

1. Control (no supplement).
2. Supplement of 35 g sulphur (as H_2SO_4)/week.
3. Supplement of 273 g sodium (as NaCl)/week.
4. Supplement of 273 g sodium and 21 g sulphur/week.

The experiment could be divided into two phases: from the start of the growing period in April or from the start of the supplementation period in May until the end of the experiment in October. Summarized liveweight changes are given in Table 9.14. Pasture analyses showed low levels of both sodium and sulphur. The authors point out that the fact that the animals given both supplements gained weight until August and then maintained it until October was unusual for the environment where the experiment was carried out.

Heavy stocking rates are required to maintain adequate amounts

Table 9.14 Liveweight changes of steers given sodium and/or sulphur supplements on *Stylosanthes guianensis* pastures (Meadowbank) (adapted from Hunter et al. 1979)

	Liveweight change (kg/head)	
	April–October	*May–October*
No supplement	4	−9
Sodium only	−4	−16
Sulphur only	22	6
Sodium + sulphur	46	26

of the legume and there is little growth in years of poor rainfall. However, perennial legumes have not always proved entirely satisfactory. Gardener (1980) carried out trials at Lansdown using native Spear grass pastures (*Heteropogon contortus, Bothriochloa decipiens*) reinforced with *Stylosanthes hamata* cv. Verano. Animal production was estimated using Droughtmaster steers between 160 and 220 kg and data over a four-year period were considered. Experimental treatments were two stocking rates (0.6 and 1.2 steers/ha) and two fertilizer rates (0 and 300 kg superphosphate/ha). Although *S. hamata* proved to be a stable and productive legume in terms of dry matter production, this was not reflected in terms of animal production during the dry season because of mouldiness caused by dry season rain and consequent lack of palatability of the forage. Liveweight gains reflected the green growth of the perennial grasses in the sward. Nevertheless, in two years of the four, the steers on the pasture which contained legume gained between 30 and 43 kg more than control animals.

Marlborough

Several trials were carried out on various aspects of Townsville stylo near Marlborough where the average annual precipitation is 813 mm. Cattle performance on natural Spear grass pasture fertilized with superphosphate at 125 kg/ha and supplemented with urea molasses was compared with an area intersown with Townsville stylo, fine stem stylo (*Stylosanthes guianensis*) and Siratro and a pasture comprising Townsville stylo and Buffel grass (*Cenchrus ciliaris*). The trial was carried out over a period of 250 days and stocked with ten-month-old steers and heifers at a rate of 1 beast/1.6 ha. The weight gains for the three treatments in the same order as given above were 97, 139 and 130 kg/head. On a further control area of Spear grass without supplement, the weight gain was 100 kg/head at a stocking rate of only 1 beast/3.2 ha (this confirms the pasture-sparing effects of supplementation to improve stocking rate as discussed in an earlier chapter). The author (Toon 1971) also mentioned that cattle grazed on native pasture oversown with stylo 'performed two-and-a-half times better' than those on unimproved native pastures.

Graham and Mayer (1972), cited by Norman (1974a), reported the results of a comparison of several methods of establishing Townsville stylo in association with the natural pasture and obtained the results summarized in Table 9.15.

Animal production in terms both of carrying capacity and liveweight gain were as high with the strip-sown treatment as with the overall sowing and costs were only one quarter. The application of superphosphate did not significantly increase animal production in this trial.

Table 9.15 Liveweight gains on natural pasture and pasture oversown with Townsville stylo (TS) (Marlborough) (adapted from Norman 1974a)

	Stocking rate (head/ha)	*Liveweight gain* (kg/head)
Natural pasture	0.31	35
+ strip-sown TS	0.62	64
+ TS sown overall	0.62	62
+ strip-sown TS + superphosphate	0.62	65

Cape York Pensinsula

The first animal production data to be obtained from improved pastures in the Cape York Peninsula appears to be from a trial carried out at Wrotham Park, west of Cairns (Anon 1973; cited by Robinson and Sing 1975). The weight gains of eighteen-month-old steers on Caribbean stylo (*Stylosanthes hamata*)/Sabi grass (*Urochloa mosambicensis*) pastures and Townsville stylo/Sabi grass pastures were 150 and

130 kg respectively at a stocking rate of 1 beast/1.3 ha.

Robinson and Sing (1975) used these data (among others) to estimate the profitability of investment in legume pastures in the far north of Queensland. The initial premise was that the level of private investment in the establishment of Townsville stylo pastures did not appear to be commensurate with the level of public investment in research into the role of Townsville stylo in the pastoral environment. Economic analyses were made for four locations at different distances from Cairns (as both a receiving point for cattle and a distribution point for the fertilizer necessary for pasture establishment). The centres chosen were Abingdon Downs, Highbury, Strathburn and Merluna which are 547, 419, 580 and 708 km from Cairns respectively. Internal rates of return were calculated and only for the location nearest to Cairns (Highbury) were they positive. It was suggested that fertilizer subsidies or investment in 'beef roads' would render investment in improved pastures attractive. These analyses were based on costs prevailing in 1972–3. It is likely that with large increases in costs (particularly of fuels) without corresponding increases in meat prices since that date, the investment would now (1981) be even less attractive.

Most of the early research in this part of tropical Australia was carried out to determine the effects of various treatments and management systems on *Stylosanthes humilis*. However, it became apparent that this legume was not ideal for all situations because the botanical composition of Townsville stylo-based pastures tends to be unstable. It is also insufficiently productive where annual precipitation is below 800 mm and may suffer badly from mould damage as standing hay during the winter. Subsequent work was carried out to evaluate other legumes and grasses which could combine with them.

Winter and his colleagues (Winter et al. 1977a) initiated a complex factorial experiment to determine the effects of various management variables on the productivity of either *Brachiaria decumbens* cv. Basilisk or *Panicum maximum* (common Guinea grass) each sown with *Stylosanthes guianensis* cv. Endeavour and *Macroptilium atropurpureum* cv. Siratro. Other treatments included stocking rate: 0.7, 1.2, 1.7, 2.2 beasts (Droughtmaster steers or heifers)/ha; fertilization with phosphorus: 10, 20, 40 kg/ha/year; supplement: nil, 10 g phosphorus/head/day (as Biophos, 21% P).

On a yearly basis (results are presented for a three-year period) there was no differences between the grasses. There was, however, an interaction between grass and stocking rate in that, in June, cattle on the 2.2 beasts/ha stocking rate lost weight on the Panicum pastures but still continued to gain weight on the Brachiaria pastures. Response to fertilization ceased during the third year after a positive response in years 1 and 2. Mean daily liveweight gain per head and per ha are shown in Table 9.16.

Table 9.16 Daily liveweight gains of steers subjected to different management systems on improved pastures (Heathlands) (adapted from Winter et al. 1977a)

	Liveweight gain (kg/day)	
	Per head	Per ha
Stocking rate		
0.7 beasts/ha	0.38	0.27
1.2 beasts/ha	0.35	0.42
1.7 beasts/ha	0.32	0.54
2.2 beasts/ha	0.29	0.56
Phosphorus rate		
10 kg/ha	0.32	0.42
20 kg/ha	0.33	0.44
40 kg/ha	0.36	0.49

The Panicum pastures did not support the 2.2 beasts/ha stocking rate after the first year and this was reduced to 1.9 beasts/ha for all treatments. The higher applications of fertilizer were required to support the higher stocking rates. Response of liveweight gain to stocking rate was not linear as suggested by Jones and Sandland (1974), and this disparity was attributed to effects of stocking rate on pasture quality and higher compensatory growth at higher stocking rates. The authors commented that *Stylosanthes guianensis* was more persistent than Siratro, but the production of the latter was similar to stylo during the early part of the wet season.

A further analysis of the data from this experiment was carried out to explain the absence of any significant effect of stocking rate on animal gain (Edye et al. 1978). It was suggested that this was due to the existence of an antagonistic wet and dry season relationship and, whilst compensatory growth undoubtedly was involved in the effect, grazing behaviour played the major role. It was observed that at the higher stocking rates, pastures were uniformly grazed but that at lower stocking rates, the animals developed 'mosaics' of closely grazed and ungrazed areas. The ungrazed areas became rank and of a lower nutritional value. It was estimated from aerial photographs that amount of grazed pasture was 29% at 0.7 beast/ha and 45% at 1.2 beast/ha. Thus the 'effective' stocking rate was 2.4 and 2.7 beasts/ha respectively. This observation will be referred to again in the chapter on management.

Gillard et al. (1980) carried out a series of trials which compared the perennial *Stylosanthes hamata* with the annals *S. subsericea* and Townsville stylo at four different sites representative of this area and central Queensland. The sites (with average rainfall in brackets)were Wrotham Park (937 mm), Lansdown (871 mm), Kangaroo Hills (634 mm) and Westwood (720 mm). At all sites except Westwood, the legumes were sown with Sabi grass (*Urochloa mosambicensis*) and the management systems employed were representative of management thought suitable for the particular station. All sites were fertilized with superphosphate at various rates. At all the sites, *S. subsericea* performed poorly because of its susceptibility to anthracnose (the better results obtained from this pasture at Lansdown are because the pasture was invaded by volunteer Townsville stylo). Stocking rates varied between 0.62 and 0.99 beast/ha during the two years of the trial. A summary of results is given in Table 9.17.

Table 9.17 Liveweight gains[1] in g/day/head from *Stylosanthes humilis*, *S. hamata* and *S. subsericea* pastures in various parts of the Queensland dry tropics (adapted from Gillard et al. 1980)

	Station			
	Wrotham Park	*Lansdown*	*Kangaroo Hills*	*Westwood*
S. hamata	429	410	452	456
S. humilis	394	396	377	341
S. subsericea	—	421	289	399

[1]All data (except Westwood) averages of two years' date; Westwood, one year only

The authors concluded that Verano persisted well, improved soil fertility and increased liveweight gains of cattle in sites where Townsville stylo was poorly adapted. The latter factor implies that 'legume-based pasture improvement technology' is available for a greater area of northern Queensland.

Subtropical eastern Australia

This large area lies approximately between latitudes 22° and 30 °S and has an annual precipitation of between 500 and 1500 mm, some of which falls during the winter period with consequent problems of mould formation etc. Research has not been dominated by Townsville stylo, as has been the case further north, but has included various legumes including Siratro, finestem, Desmodium spp. and Neonotonia amongst others in combination with introduced grasses such as Rhodes, Buffel, Setaria and various Panics. Work has been centred on Gayndah, the Brigalow and the Wallum (where beef production has been based on combinations of a wide variety of introduced grasses and legumes supported by complete fertilization).

Gayndah

A considerable amount of early work was carried out at the 'Brian Pastures' Pasture Research Station at Gayndah which was operated by the Queensland Department of Agriculture and livestock, later the Queensland Department of Primary Industries. The climate is subtropical with an average annual rainfall of 750 mm of which some 230 mm is recorded in the six winter months. The natural pasture consists mainly of *Bothriochloa intermedia*, *Dichanthium* spp. and *Heteropogon contortus*.

The first report of legume use in the area was by Christian and Shaw (1952), who showed the benefit of the incorporation of *Medicago sativa* (lucerne) into an experimental Rhodes grass (*Chloris gayana*) pasture at Lawes near Brisbane, and this resulted in the first long-term study which was carried out by Young et al. (1959).

Commercial Rhodes grass, commercial green Panic (*Panicum maximum* var. *trichoglume*) and Buffel grass (*Cenchrus ciliaris*, Gayndah strain) were each sown with Hunter River lucerne and Phasey bean *(Macroptilium lathyroides*). Store and weaner Hereford steers at stocking rates of 1 beast/1.6 ha (for the older animals) and 1 beast/ha (for weaners) were used to measure weight changes and these are summarized in Table 9.18. The performance of Rhodes grass was inferior to that of the other two grasses. The authors commented that the critical factor in animal production was the yield of lucerne and that this legume persisted well over the first five-year period. It was also suggested that the improved pastures might usefully be used for bridging the nutritional gap between winter and early spring (Young et al. 1959).

In a subsequent paper (Young and Daly 1967), results for a further five-year period were reported. These are also summarized in Table 9.18. During the second period, pasture mixtures did not regain their

Table 9.18 Beef production from different grass/legume mixtures over a ten-year period (Gayndah)

	Rhodes grass mixture	*Green panic mixture*	*Buffel grass mixture*	*Native[1] pasture*
Liveweight gain (kg/ha)				
First period[2]				
1955–8	309	382	399	121
Second period[3]				
1958–63	367	400	330	161
Overall (total)	676	781	729	282

[1]The native pasture was grazed at only half the stocking rate of the other treatments
[2]Young et al. (1959)
[3]Young and Daly (1967)

Table 9.19 Effects of summer pasture on body weight gains of grazing animals (Gayndah) (adapted from Addison 1970)

	Gain (kg/head)	
	1966/7	1967/8
Green Panic with nitrogen	89	78
Green Panic with lucerne	75	76
Green Panic with Siratro	73	65
Unimproved pasture	73	66

productivity after a severe drought and the recovery of the Rhodes grass mixture was inferior to the other two grasses. During the latter part of this phase, the amount of lucerne in the Rhodes grass pastures was superior to that in the other pastures, which presumably accounts for the more favourable results from this treatment.

Addison (1970) carried out another experiment at Gayndah to compare Green Panic alone sowed with lucerne or with Siratro (*Macroptilium atropurpureum*) to be grazed during the months of December to May.

The experiment compared cattle on these pastures during the summer and on unimproved native pasture from June to November. The effect of the summer pasture on bodyweight gains is summarized in Table 9.19.

The stocking rates were 1.85 and 0.62 steers/ha for unimproved and improved pastures respectively. The author states that performance was not satisfactory because of lower gains during the latter part of the period, probably due to lower protein intake (as judged by faecal protein). Higher stocking rates would have improved gains/ha.

Siratro was evaluated in association with the natural Spear grass-dominated pastures near Gayndah by Bisset and Marlow (1974). High (1.24), medium (0.82) and low (0.62 beasts/ha) stocking rates were compared at two sites, Lowmead and Brooweena, where the

Table 9.20 Liveweight gains[1] from Siratro/native pastures at different stocking rates (adapted from Bisset and Marlow 1974)

	Stocking rate (head/ha)			*Length of period* (days)
	1.24	*0.82*	*0.62*	
Lowmead	173 (213)	183 (150)	183 (113)	314
Brooweena	93 (115)	130 (106)	182 (112)	329

[1]Five-year means; kg/beast, kg/ha in brackets

rainfall was 1090 and 1008 mm respectively. Animal production was evaluated using Hereford and cross-bred weaners. Results are summarized in Table 9.20.

The better performance at Lowmead was associated with a greater proportion of Siratro; this reflected better land preparation than at Brooweena but also probably a more suitable environment. At the latter site, only a trace of Siratro remained at the heaviest stocking rate.

In a later trial at Gayndah, fine-stem stylo (FSS) was investigated as an associate of Spear grass and *Rhyncheletrum repens* natural pasture. The legume was *Stylosanthes guianensis* var. *intermedia* Oxley cultivar. Experiments were designed to investigate:

1. The ability of native pasture sowed with FSS to provide satisfactory liveweight gains.
2. The effect of protection from grazing on the botanical composition of the pasture.
3. The comercial usefulness and effect of fertilization with superphosphate.

Three experiments were carried out, the first two at 'Brian Pastures' and the third at 'Tecoma' Station 116 km northwest of Gayndah. Experiment 1 compared three grazing systems: heavy put-and-take, light put-and-take and set-stocked using weaner Hereford steers over a five-year period. Results are summarized in Table 9.21.

Differences in liveweight gain/beast were not significant. Liveweight gains/ha were dependent on stocking rate and represented a 90% increase on those from unimproved natural pasture.

The results of the second experiment showed that the highest densities of FSS occurred with constant grazing and that protection caused the legume to decline.

The third trial compared the productivity of pastures fertilized with superphosphate (250 kg/ha in 1968 (establishment), 1972, 1973 and 1974) with non-fertilized pastures on a commercial property.

Table 9.21 Liveweight gains and equivalent stocking rates of Fine-Stem Stylo/natural pastures (Gayndah) (adapted from Bowen and Rickert 1979)

	Treatment			
	LPT	*HPT*	*PTT*	*SST*[1]
Mean annual liveweight gain (kg/head)	167	167	167	147
Equivalent stocking rate (head/ha)	1.25	1.71	1.47	1.39

[1]LPT, HPT, PPT: light, heavy and all put-and-take treatments respectively; SST: set stocked

Table 9.22 Performance of weaner cattle on fertilized and unfertilized FSS/natural pasture (Tecoma) (adapted from Bowen and Rickert 1979)

	Liveweight gain[1]	
	kg/head	*kg/ha*
Fertilized	154	123
Unfertilized	143	104

[1]Four-year means

Results for liveweight gain and stocking rate are summarized in Table 9.22.

There was a marked response to the application of superphosphate both in gain/beast and gain/ha. The authors suggest three alternative methods of management for FSS/natural pastures. The most practical of these was the use of a variable stocking rate to maintain FSS density with a target liveweight gain of 180 kg/head (Bowen and Rickert 1979).

A further trial in this area was carried out by Bewg et al. (1970) at Oakwood Station, 40 km south of Gympie. Pastures were established after clearing and ploughing and a mixture consisting of *Desmodium uncinatum, D. intortum, Macroptilium atropurpureum* and *Neonotonia javanica* was sown; no details of the natural grass cover are given but the carrying capacity of the unimproved pasture was about 0.25 beast/ha. The pastures were stocked at the rate of 1.23 beasts/ha during the first year of the trial and 1.23 and 1.82 beasts/ha during the second year. Daily gain during the first year averaged 0.52 kg/ha and the overall gain was 222.6 kg/ha. During the second year, the mean daily gain for the animals was 0.45 and 0.39 kg for low and high stocking rates respectively, whilst performance expressed on a unit area basis amounted to 209.8 and 269.0 kg/ha. It was suggested that there was no shortage of protein but that productivity could have been limited by the low energy value of the pasture.

The use of sod-seeded Siratro to improve natural Spear grass pasture near Ipswich was investigated by Lowe (1974). The stocking rate was 1 beast/1.6 ha during the first year of the experiment, but this was increased to 1 beast/1.2 ha for both improved and unimproved control pastures and finally to 1 beast/0.8 ha on the more productive Siratro pasture.

Increased liveweight gain/ha from the improved pasture was 22 and 34 kg for the two years of the experiment. Pasture improvement increased productivity in autumn and perhaps in spring, increased stocking rates and was a low-cost method to develop property. In a later paper (Lowe et al. 1977), the greater carrying capacity of the improved pasture was confirmed and liveweight gains per head and per ha were also superior in the case of the pasture containing Siratro; these were 0.7 v. 0.9 beasts/ha, 0.51 v. 0.65 kg/head/day and 0.36 and 0.60 kg/ha/day for the three parameters respectively.

Factorial trials on associations of *Setaria sphacelata* cv. Nandi with either *Desmodium intortum* or *Macroptilium atropurpureum* were carried out at the CSIRO Pasture Research Station, Samford near Brisbane (Jones, 1974). The associations were compared with the grass alone fertilized at the rate of 336 kg N/ha/annum and stocking rates of 1.11, 1.73, 2.35 and 2.96 beasts/ha for the grass/legume associations 2.96, 3.58, 4.20, 4.82 and 5.44 beasts/ha for the fertilized grass treatment. Stock used were yearling Hereford heifers. Long-term average annual precipitation at Samford is 1218 mm. Treatments were compared by using linear regressions of gain/animal on stocking rate. The slopes were significantly different and there was an linear decrease in gain/beast with increased stocking rate. The equations were used to derive the stocking rate from which maximum weight gain/ha would be produced. These were 2.12, 2.42 and 5.58 beasts/ha for the Desmodium, Siratro and fertilized pastures respectively, with corresponding maximum gains/ha of 256, 256 and 491 kg/ha/annum respectively.

Increased stocking rate reduced the legume component of the sward, and this was associated with the more rapid decline in weight gain with stocking rate in the legume treatments than in the fertilized treatment. The highest production/ha was reached at a stocking rate which did not permit as high gains as the legume systems. Thus, although stocking rate of the fertilized pastures was considerably higher than that of the legume associations, turn-off rate would be lower (i.e. yearling cattle would not attain slaughter weight at 2 years but would require 2.6 to 3.2 years).

A further trial on Setaria/Siratro pastures at Samford was carried out by Jones, (1979). Stocking rates ranged from 0.8 to 2.8 beasts/ha and grazing frequencies were four days every three, six or nine weeks. The experiment was designed to measure the effect of these variables on the persistence of the legume, and no animal production data are presented. With a three-week grazing cycle, both total productivity and Siratro yield declined markedly at the higher stocking rates and this was less noticeable in the nine-week cycle. This effect was associated with a decline in soil seed reserves and it was predicted that if the heavily stocked treatments had been continued, the pastures would have reached a stage where the Siratro would not have recovered during the deferred period. It was suggested that the effectiveness of the long period of deferment was limited to high stocking rates where overgrazing occurred. This statement is of interest when compared with that of Bowen and Rickert (1979) in which a variable stocking rate was used to maintain the legume component (in that case, fine-stem stylo).

The Brigalow

This represents an area of approximately 4.85 million ha of moderate to high fertility lying between the 500–750 mm and rainfall isohyets where Brigalow (*Acacia harpophylla*) is normally dominant. However, Belah (*Casuarina cristata*) or *Eucalyptus crebra* is dominant in some areas which have been included in this section. Ground cover in the area is sparse and includes such grasses as *Chloris*, *Paspalidium*, *Dichanthium*, *Heteropogon* and *Eragrostis* spp. Rainfall is variable throughout the area with winter as well as summer rainfall. Drought can, however, be a hazard. Diurnal variations in temperature are wide and frosts occur. Because of frosts, early legume use was limited to lucerne, although establishment of this was sometimes difficult. An early trial carried out by Yates and his colleagues at Lawes was intended to compare *Medicago sativa* (var. Hunter River) with or without the summer-growing legumes *Neonotonia javanica*, *Macroptilium atropurpureum* and *M. lathyroides* as associates for *Sorghum alum* but failed because of poor establishment of all the legumes leading to a virtually monospecific sward (Yates et al. 1964). Similar results were obtained by Coaldrake and Smith (1967), who failed to establish lucerne and Siratro at the 'New Banana' Station, Banana. A further unsuccessful trial with lucerne was carried out by Coaldrake and his colleague (Coaldrake et al. 1969) at the 'Tarrewinnabar' Station, 53 km north of Goondiwindi. However, despite the failure of the lucerne to establish two naturalized legume introductions, *Medicago denticulata* and *M. minima*, appeared to make a positive contribution to the productivity of native pastures which were mainly composed of *Chloris, Paspalidium Dichanthium* and *Eragrostis* spp. Although it was not possible to quantify this effect, the authors commented that cattle either gained or maintained weight during the dry season rather than losing weight as might have been expected from native pasutres and attributed this effect to the legumes.

The first effective demonstration of the value of pasture legumes in the Brigalow country was provided by 't Mannetje (1972) Experiments were carried out on native (predominantly Spear grass, *Heteropogon contortus*) pastures and introduced grasses with Townsville stylo and Siratro at the CSIRO Research Station at Narayen, Mundubbera, where average annual precipitation is 713 mm with a range of 315–1368 mm. Comparisons included Buffel grass (*Cenchrus ciliaris*), either fertilized with 168 kg N/ha/year or sown with Siratro. Liveweight gains were the same from both treatments (162 kg/head) at a stocking rate of 1.1 beast/ha. Liveweight gain from fertilized Buffel grass was 173 kg/heat at a stocking rate of 0.7 heast/ha, whilst non-fertilized Buffel grass only produced 96 kg/head at the same stocking rate.

A comparison was also made of pasture species which could be associated with Siratro. The various grasses were sown in 0.4 ha Siratro-based plots and stocked with two steers each for 115 days

Table 9.23 Liveweight gains from different grasses associated with Siratro (Narayen) (adapted from 't Mannetje 1972)

Grass grown with Siratro	*Liveweight gain*	
	kg/head	*kg/ha*
Sabi Panic (*Panicum maximum*)	55	272
Petrie green Panic (*P. maximum* var. *trichoglume*)	60	296
Nunbank Buffel (*Cenchrus ciliaris*)	59	292
Gayndah Buffel (*C. ciliaris*)	33	163
Pioneer Rhodes grass (*Chloris gayana*)	33	163

from January 1952. Results are summarized in Table 9.23.

The productivity of the Panics and the Nunbank Buffel was clearly superior to the other grasses. The author also commented that liveweight gain was linearly related to the Siratro content of the pasture to 1000 kg/ha but that gain was not increased with Siratro above this level. The importance of heavy stocking to remove forage before deterioration set in was also emphasized.

Further work on the use of Siratro to improve pastures at Narayen was carried out by Tothill (1974). Siratro introduction to the native Spear grass pasture was compared with the establishment of a Siratro/Buffel grass pasture. Different stocking rates were used on each treatment. Preliminary results indicate that there was a considerable increase in productivity by the incorporation of Siratro amounting to some 4 to 10 times in terms of production/ha or 1.4 to 2.0 times in terms of production/head.

't Mannetje and Nicholls (1975) collected production data over a five-year period for liveweight gains of Hereford steers grazing Siratro associated with *Cenchrus ciliaris* cv. Biloela at stocking rates of 1.1, 0.7 and 0.5 beasts/ha. These were compared with data from Buffel grass, both fertilized and unfertilized, and with fertilized and unfertilized native pastures. Results are summarized in Table 9.24.

The experiment originally included lucerne but this did not persist for more than two years. The author does not comment on the equivalence of the fertilized and unfertilized native pastures, but this was presumably because the fertilizer produced more forage when this was already underutilized and did not extend the grazing season.

't Mannetje (1974) used the results from this experiment to derive relations between various attributes and liveweight gains. Liveweight gains were related to total green material, green grass or green

Table 9.24 Productivity of improved and unimproved pastures (Narayen) (adapted from 't Mannetje and Nicholls 1975)

Pasture	*Stocking rate* (beasts/ha)	*Mean animal liveweight gain* (kg/head)	(kg/ha)
Buffel/Siratro	0.54	172	93
	0.74	156	115
	1.09	142	155
Unfertilized native	0.27	124	33
	0.54	92	50
Fertilized native	0.54	92	50
Buffel only	0.74	100	74
Buffel + N	1.09	156	170

legume. The animals selected a better-quality diet and the nitrogen content of the green material was always sufficient to keep the animal in positive nitrogen balance. Although pen-feeding trials have demonstrated that the nutritive value of mature tropical forage is generally low, the animals' capacity to graze selectively clearly enables it to narrow the gap between the herbage on offer and its nutrient requirements. This point has already been made in the chapters dealing with protein and mineral supplementation.

Later work at Narayen investigated the effect of improved pastures on the breeding cow. Four systems were used in the comparison, as follows:

1. Native pasture (mainly *Heteropogon contortus*) at a stocking rate of 0.17 cow/ha.
2. Green Panic (*Panicum maximum*) var. *trichoglume* at a stocking rate of 0.51 cow/ha.
3. As above but stocking rate 0.68 cow/ha.
4. Native pasture during the summer (November until April) at a stocking rate of 0.34 cow/ha and access to sown pasture (as above) for the rest of the year at a stocking rate of 1.00 cow/ha.

Productivity was measured in terms of reproductive efficiency (conception and calving rates) growth rate to weaning and reproductive index (adjusted kg of calf weaned/cow exposed to the bull). Results are summarized in Table 9.25. Thus, on a production/ha basis, the improved pasture at the high stocking rate had a productivity 5.4 times that of the native pasture ('t Mannetje and Coates, 1976).

A further confirmation of the value of improved pastures for calf production in the Brigalow was obtained by Silvey and his colleagues (Silvey et al. 1978), also working at Narayen. Grasses used were *Panicum maximum* var. *trichoglume* and *Chloris gayana* cv. Pioneer

Table 9.25 Reproductive characteristics and calf weaning weights of cows on different pasture treatments (Narayen) (adapted from 't Mannetje and Coates 1976)

	Calving %	*Weaning weight (kg)*		*Reproductive index (kg)*
		Female	*Male*	
Native pasture	78	186	198	166
Panic/Siratro – low	82	233	244	210
Panic/Siratro – high	94	223	248	221
Native/sown	82	197	212	193

grown alone or with Siratro, *Neonotonia wightii* cv. Cooper and *Medicago sativa* cv. Hunter River. Stocking rates were 1.31, 0.93 and 0.66 cows/ha, and fertilizer was applied at the rate of 18 kg P/ha plus 52 kg N/ha for the pasture of grass alone.

Adequate body weights were generally maintained throughout the trial although some hay feeding was necessary. Annual conception rates for all treatments for the three years of the trial were 90, 100 and 90%, whilst calving rates were 88, 101 and 89% for the same years. The authors conclude that both liveweight and reproductive performance were excellent but that 'the destructive effects of drought can nullify the high production potential of the soil–plant–animal system'.

In view of the importance of *Macroptilium atropurpureum*, Siratro, for this area, it is appropriate to cite the review of Jones and Jones (1978) on the ecology of Siratro-based pastures as well as that of Walker (1977) on their productivity.

The Wallum

The undeveloped vegetation of the Wallum, which covers some 800 000 ha in the Central Lowlands of southern Queensland, 'is incapable of carrying productive animals' (Evans 1970). The native vegetation is largely composed of various forms of eucalypt/*Melaleuca* forest and woodland with a sparse understory of native shrubs and grasses. The rainfall ranges from 1070–1900 mm with the northern part being the drier region. Effective pasture establishment depends on fertilization with phosphate, lime and potassium as well as applications of copper, zinc and molybdenum. Grazing trials have been carried out at the CSIRO Pasture Research Station, Beerwah, and at the Research Station, Coolum (DPI) (Department of Primary Industry).

Beerwah

Experiments were initiated in 1959 to investigate pasture mixtures, management and possible methods of improvement of pastures in the Wallum (Bryan 1968a). The pastures used were:

Paspalum dilatatum sown with *Centrosema pubescens*, *Macroptilium lathyroides* and *Trifolium repens*
Digitaria decumbens with legumes as above
Paspalum scrobiculatum sown with *Desmodium uncinatum*
Paspalum plicatulum sown with *D. uncinatum*

The stocking rate was 1.6 beasts/ha (cattle were Herefords with an average age of fourteen months and weight of 250 kg) for the first three years and 2.5 beasts/ha thereafter. Overall gains are summarized in Table 9.26.

Table 9.26 Annual gains on different pastures (Beerwah) (adapted from Bryan 1968a)

	Annual liveweight gain (kg/ha)	
	Period 1 (3-year mean)	*Period 2 (4-year mean)*
Paspalum dilatatum	311	243
Digitaria decumbens	273	302
P. scrobiculatum	237	342
P. plicatulum	250	220

The legume component on average was only 13.5%. The *Macroptilium lathyroides* was not self-regenerating and therefore made little contribution to the pastures, the *Centroscema pubescens* was not suited to the conditions of southern Queensland and the presence of white clover was insignificant. Only *Desmodium uncinatum* persisted throughout the trial in any quantity (although treatments containing this legume were not noticeably more productive) and it was thought that this would have been better had a more appropriate *Rhizobium* sp. been used at establishment. Nevertheless, as this represented some of the first results on animal productivity from improved pastures in this area, it was considered that results had been reasonable. The author also noted the importance of selective grazing in filling the nutrient requirements of the stock (Bryan 1968a also quotes the unpublished work at Beerwah of Villarreal, who obtained liveweight gains of 500 kg/ha on Pangola pasture containing 40% *Lotononis bainesii* and *Desmodium intortum*).

In a complementary and more or less contemporaneous trial, Bryan (1968b) carried out grazing trials on complex mixture of grasses and legumes under common grazing. Every mixture contained two grasses (except one which contained one only) and species included *Paspalum dilatatum*, *P. notatum*, *P. scrobiculatum*, *Chloris gayana*, *Brachiaria mutica*, *Digitaria decumbens* and *Panicum maximum* var. *trichoglume*. Twelve of the mixtures contained three legumes and the remainder four; the legumes used included *Trifolium repens*, cv. New Zealand and cv. Ladino, *T. incarnatum* cv. Dirie (crimson Clover), *Desmodium uncinatum*, *D. sandwicense*, *Lotononis bainesii*, *Centrosema pubescens* and *Medicago sativa* cv. Hunter River. Hereford cattle aged fourteen months were used and stocking rate was increased throughout the experiments as the density of the pasture improved so that the constancy of pressure on the feed available was maintained more or less constant. This increase in carrying capacity was reflected in increased liveweight gains/ha: these were 157 kg in year 1, 224 kg in years 2 and 3 and 336 kg in years 4 to 7 (although, obviously, gains in individual years varied with rainfall). Selectivity was maximized by common grazing and this vitiated particularly against Para grass (*Brachiaria mutica*) which was already at a disadvantage because of the relatively dry conditions of the experiment, whilst *Paspalum notatum* and *P. plicatulum* prospered because of their relative unpalatability. The nitrogen content of the pasture was normally above 1.15%, considered to be the minimum amount required to maintain nitrogen balance during the period from spring to autumn, and this was confirmed by the liveweight gains of the cattle.

Evans (1970) summarized these early experiments and also briefly mentioned an experiment carried out on a grass/legume pasture which indicated that liveweight gain (kg/ha) increased as did stocking rate even though the amount of legume in the sward tended to decrease. A summary of these results is given in Table 9.27.

Table 9.27 Liveweight gains and legume of an improved pasture grazed at different stocking rates (Beerwah) (adapted from Evans 1970)

Stocking rate (beasts/ha)	*Liveweight gain (kg/ha)*	*Legume* (%)
1.23	295	30
1.65	326	28
2.47	384	25

The author mentioned another trial which had been carried out on nitrogen fertilization of Pangola grass and, in a comparison of the results, concluded that 'the ceiling for animal production from grass/legume pastures is determined by the legume content of the pasture' (presumably, in view of the above, at equivalent stocking rates). Thus the supply of nitrogen to the plant/animal system is determined either from fixation from legumes or by fertilization and is the main factor which affects livestock production.

In a experiment specifically designed to compare different levels of nitrogen application with legume incorporation, Bryan and Evans (1971) compared low and high nitrogen fertilization (112 kg N/ha/year in 1966 and 1967 and 168 kg N/ha/year in 1968–70; and

Table 9.28 Liveweight gains (kg/ha) from a mixed legume/grass pasture at different superphosphate levels and stocking rates (Beerwah) (adapted from Evans and Bryan 1973)

Stocking rate (beasts/ha)	*Superphosphate application*		*Days to turn-off* (*all treatments*)
	125 kg/ha	*250 kg/ha*	
2.47	271	348	393
1.65	271	303	317
1.23	208	282	283

449 kg N/ha/year throughout) with *Macroptilium lathyroides, Desmodium uncinatum, Trifolium repens, Desmodium intortum* and *Lotononis bainesii* sown with appropriate strains of *Rhizobium*. Over the entire four-year period, liveweight gains for the three systems were 533, 892 and 527 kg/ha respectively.

Evans and Bryan (1973) have given a fuller description of the stocking rate and fertilizer trial briefly mentioned by Evans (1970). Pastures consisted of the grasses: *Chloris gayana, Digitaria decumbens, Paspalum dilatatum* and *P. scrobiculatum*; and the legumes: *Desmodium intortum, D. uncinatum, Lotononis bainesii, Macroptilium lathyroides* and *Trifolium repens.*

The stocking rates were as above and the cattle were commercial Hereford steers. Applications of 125 and 250 kg single superphosphate were made annually (potassium levels were also tested but there was no response above 63 kg/ha). Summarized results for the six years of the trial are given in Table 9.28. The application of the higher rate of phosphorus significantly improved performance and the highest rate of liveweight gain/ha was from the highest stocking rate, although individual animal gains and hence rate of turn-off were higher at the lower stocking rates. There was a significant positive correlation between liveweight gain and the amount of legume in the sward. Despite the higher productivity at the highest stocking rates, these pastures tended to be unstable and legume content declined over the course of the experiment with a concomitant decline in animal production suggesting that, for long-term optimum production, lighter rates might be more advisable.

In a study of the steers and the pasture material made by Thornton and Minson (1973) during the concluding phase of the trial, it was shown that higher superphosphate applications were associated with higher mean plasma concentrations of both phosphorus and nitrogen. It was suggested that increased weight gain was not because of increased forage production but because the animals grazing these pastures had greater intake of forage or higher nitrogen content due to the stimulation of legume growth by the phosphorus application. A further point was made that, although the pastures could not be

Table 9.29 Performance of cows on nitrogen-fertilized grass and grass/legume pastures (Beerwah) (adapted from Evans and Biggs 1979)

	Nitrogen-fertilized grass	*Grass/legume*
Calving rate (%)	94	95
Weaning rate (%)	89	81
Weaning weight (kg at 200 days)	205.5	192.0
Calf growth rate (kg/day to 200 days)	0.85	0.79

considered deficient in phosphorus, additional intake of this element on the fertilized pastures could not be ruled out as a cause of the increased weight gains. This point is of interest in view of the frequent lack of response to supplementary phosphorus in deficient or marginal pastures mentioned in an earlier chapter.

Evans and Biggs (1979) compared heavily-fertilized (280 kg N/ha/year) Pangola grass pastures with Pangola grass/legume pastures in terms of breeding cow performance. Stocking rates were 2.5 cows/ha on the former and 1.0 cow/ha on the latter, which was similar to the grass/legume pastures previously described. There were no significant differences between the two systems and performance was consistently high. Average values for the various performance characteristics measured over a five-year period are given in Table 9.29.

Coolum

Young and Chippendale (1970) have described experiments carried out at the Research Station, Coolum. This represents a different ecological system from that of Beerwah whilst still classified as Wallum. Heath plains are interspersed between dunes, low hills and swamps, and the total area of this lowland ecosystem is more than 33 000 ha. Soils are of poor fertility and the area is subject to flooding. Precipitation during the period of the experiments described by Young and Chippendale (1970) averaged 1778 mm between 1963 and 1968.

In their first experiment, the combination of Pangola grass (*Digitaria decumbens*) with *Lotononis bainesii*/white clover was investigated at different stocking rates and fertilizer applications. However, *L. bainesii* did not constitute an appreciable part of the pasture and the white clover failed to persist. Another trial was a carried out to investigate the association of *Setaria sphacelata* cv. *Nandi* with *Desmodium uncinatum* and white clover; however, the legumes also constituted an insignificant part of the sward and these experiments will not be considered further in this section except to note that the authors suggest that the periodic inundations and the fluctuating water table were probably factors which contributed to the failure of the legumes; similar findings have been reported from the 'dambo' areas of Central Africa (Savory, personal communication)

New South Wales

One study carried out at Grafton, NSW will be included in this section (trials carried out at Glen Innes by Hennessy and Robinson 1978 have not been dealt with here as they represent species which are more typical of temperate than subtropical areas).

Cohen and O'Brien (1974) investigated various stocking rates on natural pasture as well as two improved pastures which had been originally seeded for demonstration purposes. The treatments are summarized in Table 9.30.

Table 9.30 Treatments and stocking rates at Grafton, NSW (adapted from Cohen and O'Brien 1974)

Pasture type	*Stocking rates (cows/ha)*
Pasture A unfertilized natural pasture mainly consisting of *Axonopus affinis, Bothriochloa decipiens*	0.5, 0.65 and 0.8
Pasture B natural pasture fertilized with 625 kg/ha superphosphate and oversown with naturalized white clover, *Trifolium repens*	0.65, 0.8, 1.1 and 1.4
Pasture C natural pasture fertilized as above oversown with ladino white clover, *Lotononis bainesii, Macroptilium atropurpureum* and *Lespedeza striata*	0.8, 1.1, 1.4 and 2.0

The establishment of the clovers was good, as was that of the Lotononis after a second oversowing; the Siratro died out at the seedling stage as rainfall was poor. The Lespedeza persisted throughout the experiment. The pastures were evaluated using cows and a summary of the effects on fertility is given in Table 9.31. Clearly, the improved pastures not only supported an increased stocking rate of cows but this was accompanied by increased fertility (although this also declined at the higher stocking rates). Thus the difference between the rate of calving on the natural pasture v. the best rate of calving on the improved pasture (both at a stocking rate of 0.65 cow/ha) was 0.3 v 0.5 calf/ha. Calves at weaning (185 days) were suitable for slaughter from both improved treatments whilst those from natural pastures were only in store condition. It was also suggested that with the increased rate of gain of the calves from improved pastures, it would be possible to join heifers at an earlier age than the three years usual for the area. Overall, there was little

Table 9.31 The effects of pasture type and stocking rate on calving percentage (Grafton) (adapted from Cohen and O'Brien 1974)

	Stocking rate (cows/ha)		
	0.65	*0.8*	*1.1*
Unfertilized natural pasture	47	33	—
Fertilized natural pasture with naturalized white clover	80	73	60
Fertilized natural pasture with Ladino, Lotononis, Lespedeza	—	67	58

difference in beef production between the two improved pastures, and the naturalized white clover was recommended as the least expensive alternative. It was noted that the use of this legume in the district had already (1972) reached significant proportions.

The Queensland wet tropics

The last area to be considered in the development of legumes in tropical and subtropical Australia is that part of the north coast where moisture is not a limiting factor. The precipitation at the stations where experiments have been carried out in this area ranges from 1220 mm at Kairi to 3200 mm at Utchee Creek. For convenience, those studies carried out under irrigation have been included in this section. Some of the highest levels of productivity recorded from the tropics and subtropics have been reported from this area.

Ayr

Impressive results were obtained at the Ayr Regional Experiment Station in the Burdekin delta from trials which were initiated as early as 1950. Pastures studied were Para grass (*Brachiaria mutica*) and Centro (*Centrosema pubescens*), Guinea grass (*Panicum maximum*) and Centro, Guinea grass and Stylo (*Stylosanthes humilis*), and Rhodes

Table 9.32 Average stocking rate and cattle liveweight gains on improved pasture (Ayr) (adapted from Allen and Cowdry 1961)

Pasture	*Average stocking rate (beasts/ha)*	*Liveweight gains*	
		Per day	*Per year*
Rhodes + Stylo	3.7	1.5	543
Para + Centro	4.0	1.8	662
Guinea + Centro	4.0	1.7	614
Guinea + Stylo	3.6	1.6	577

grass (*Chloris gayana*) and Stylo. Pastures were irrigated fortnightly with 60–75 mm in periods when this had not been supplied by rainfall. Pastures were rotationally grazed (one week on and three weeks off) and eight groups of steers were fattened during the period 1950–7 (Allen and Cowdry 1961). Average stocking rates and cattle gains/ha for this period are summarized in Table 9.32.

Total yearly gains/ha from August 1950 at the start of the trial to December 1957 averaged 656 kg and ranged from 237 to 1066 kg. The authors estimate that the carrying capacity of the unimproved natural pasture was 1 beast/4 ha.

Utchee Creek

Research has been carried out at South Johnstone Research Station which is located some 32 km west of Innisfail, north Queensland. The climate is humid tropical with an average annual precipitation of 3200 mm. Investigation has been directed towards the utilization of Guinea grass (*Panicum maximum*) in association with different legumes. Pastures consisting of Guinea grass alone, with or without nitrogen (168 kg/ha), or Guinea grass in association with *Centrosema pubescens* were grazed with a rotation of two weeks on, two weeks off at a stocking rate of 4.2 beasts/ha. Liveweight gains for non-fertilized, grass/legume and fertilized pastures during the two years of the trial were 344, 468 and 660 kg/ha respectively, indicating a substantial increase due to the legume but not so much as that obtained by the use of nitrogenous fertilizer (Grof and Harding 1970).

Later work compared different Guinea grass cultivars in association with either *Centrosema pubescens* alone or a mixture of Centro, *Neonotonia wightii* cv. Tinaroo, *Vigna luteola* cv. Dalrymple and *Desmodium intortum* cv. Greenleaf (Mellor et al. 1973). The cultivars of Guinea grass were: common, Hamil, Coloniao and Embu. The rotation used was the same as that in the previous experiment and the stocking rate was 4.94 beasts/ha during the rainy season and 2.47 beasts/ha during the drier period from April to mid-December. Mean liveweight gain (kg/ha) over a three-year period from the first three cultivars in the order given were: 756, 698 and 633 kg (Embu died out after the second season of full grazing). Although all the legumes established well, *Desmodium* and *Vigna* had died out by the end of the second year. Differences between legume treatments were not apparent but, in association with Hamil and Coloniao, the proportion of Centro increased when compared with the legume mixture in the former and decreased in the latter. There was no overall advantage of including a variety of legumes over the use of Centro alone, although the inclusion of Glycine increased autumn and winter gains. According to Miller and van der List (1977), Teitzel has obtained even higher rates of gain from adequately fertilized Guinea grass/Centro pastures.

Atherton Tablelands

Research on associations of Guinea grass and *Neonotonia wightii* cv. Tinaroo were also carried out at the Kairi Research Station, 8 km north-east of Atherton, where the mean annual precipitation is 1230 mm. Green Panic (*Panicum maximum* var. *trichoglume*) was introduced to a stand of Tinaroo glycine on two soil types: a deep red basaltic clay loam and a light-coloured granitic soil. Pastures were stocked with Brahman cross-bred and Shorthorn steers at rates of 2.5 and 1.25 steers/ha. Results of weight gains and carcass weights are summarized in Table 9.33.

Table 9.33 Weight gains and carcass weights of steers grazing green Panic/glycine pastures (Kairi) (adapted from Winks et al. 1970)

	Red soil	*Light soil*	*Stocking rate*	
			1.25 steers/ha	*2.50 steers/ha*
Body weight gain (kg/day)	0.66	0.75	0.63	0.79
Hot carcass weight (kg)	224	227	219	232

Values for liveweight gain obtained in this study overall were 578 kg/ha at 2.5 steers/ha and 348 kg/ha at 1.25 steers/ha. This was considered to be unsatisfactory, as grazing pressure during the summer months was inadequate to utilize all the forage available. Thus stocking rates should be flexible if optimum utilization of the forage on offer is to be realized (Winks et al. 1970).

In a further study (Winks et al. 1975), the use of different stocking rates on Green Panic/Neonotonia pastures was investigated. Stocking rates of up to 8 head/ha could be achieved during the early part of the rains, whilst during the later rains a stocking rate of 4 beasts/ha could be maintained without depressing animal gains unduly. This dropped to 2.5 beasts/ha later during the winter and spring. Liveweight gains of up to 1 kg/head/day were achieved for short periods and total liveweight gains of more than 650 kg/ha/year were obtained, a noteworthy performance.

Parada

Norman (1974a) has quoted values for beef production from Parada (J. Evans and J. Pulsford, cited by Norman; I have not been able to trace any other reference) which exceed those already quoted (even higher values will be mentioned in a later chapter in connection with fertilization). Liveweight gains of 692, 636 and 622 kg/ha were obtained from Pangola plus Centro, Para plus Centro, and Paspalum plus white clover respectively, whilst unfertilized Para grass plus Centro in a later trial produced an average liveweight gain of 870 kg/ha over a three-year period.

Murray Upper

Murray Upper is situated between Cardwell and Tully and has a mean annual rainfall of 2400 mm. Trials were carried out with Hamil Guinea grass associated with either Schofield or Endeavour stylo (*Stylosanthes guianensis*). The Schofield stylo supported a beast to 0.8 ha with a daily liveweight gain of 0.68 kg/day for at least six months. Both carrying capacity and weight gain were higher for Endeavour at 1.5 beasts/0.8 ha, gaining at the same rate for a similar period of time (Mortiss 1974).

It would appear that Guinea grass in association with various legumes will form a pasture which will support very high rates of gain in high rainfall areas.

Mareeba-Dimbulah Irrigation Scheme

This scheme is located in the moist (1072 mm/year during the three-year trial described here) Queensland tropics west of Cairns, and investigations of beef production from grass/legume pastures have been carried out at the Walkamin Research Station. In the first experiment, Para grass (*Brachiaria mutica*) was established with *Neonotonia wightii* cv. Tinaroo, *Stylosanthes guianensis* cv. Schofield or *Macroptilium atropurpureum* cv. Siratro and rotationally grazed with Brahman cross steers at a rate which was varied seasonally so as to achieve uniform utilization throughout the year. Superphosphate was applied at the rate of 250 or 500 kg/ha and flood irrigation was applied at the rate of 40–50 mm every twenty-eight days (complementary to rainfall). Annual liveweight gains (three-year means) were 686, 521 and 428 kg/ha for Tinaroo, Schofield and Siratro associations respectively; the superiority of the first was due to 'much higher rates of gain in winter' (these approached summer gains, i.e. 0.54 v. 0.69 kg/day).

A further experiment compared Para grass and *Setaria sphacelata* cv. Kazungula in association with *Neonotonia wightii* cv. Tinaroo or *Desmodium intortum* cv. Greenleaf at two stocking rates, 3.4 5.1 steers/ha. Pastures were top-dressed with 250 kg superphosphate in November of each year and irrigation was supplied as in the first trial. Results for the two-year period of the experiment are summarized in Table 9.34.

Legumes did not persist in the Kazungula pastures. The authors comment that at high stocking rates, only extremely well adapted legumes could be expected to compete with a relatively non-aggressive grass to remain a significant part of the association. Only in the Para/Tinaroo mixture was beef production substantially increased by increased stocking rate, indicating that pasture availability was not limiting.

Despite the high rates of livestock production from these pastures, the authors indicate that the high fixed and variable costs associated

Table 9.34 Liveweight production (kg/ha) from grass/legume pastures at two different stocking rates (Walkamin) (adapted from Miller and van der List 1977)

Pasture	*Stocking rate* (steers/ha)	
	5.1	*3.4*
Para/Tinaroo	724	670
Kazungala/Tinaroo	647	674
Kazungula/Greenleaf	597	574

with irrigation made the returns only marginal (Miller and van der List 1977).

Ord Valley
Although the Ord Valley is in north-western Australia, the following trial has been included in this section because of the use of irrigation and the high yields obtained.

Blunt (1976) planted *Leucaena leucocephala* cv. Peru in rows 2.3 m apart interplanted with Pangola grass (*Digitaria decumbens*). This was grazed by cross-bred Brahman and Kimberley Shorthorn steers at different stocking rates. However, although liveweight gains/ha were as high as 830 kg at a stocking rate of 6.2 beasts/ha, individual liveweight gains were not satisfactory and it was suspected that this was connected with the occurrence of mimosine in the *Leucaena* as symptoms of hair loss were also encountered.

The results presented in this section indicate that very high gains/ha can be obtained from grass/legume associations in the humid tropics but that these are associated with high inputs, especially when irrigation is involved, and problems of persistence of pasture legume often occur.

As would be expected, technology developed in tropical and subtropical Australia has been adapted to suit conditions in parts of Asia and some of the Pacific Islands.

Malaysia

In 1973, the Pasture Unit of the Malaysian Agricultural Research and Development Institute (MARDI) instituted a joint programme in collaboration with the CSIRO Division of Tropical Crops and Pastures in Brisbane. A grazing trial was set up at the MARDI Research Station at Kluang, Johore. The station is situated at 2 °N and the mean annual precipitation over the period of the experiment was 1764 mm. This was more or less evenly distributed with the exception of brief dry spells; there were no seasonal variations in temperature, mean maximum and minimum being 31 °C and 21 °C respectively. Conditions were therefore very favourable to pasture growth.

The experimental design used has been detailed by Eng et al. (1978a). Guinea grass and *Stylosanthes guianensis* were sown into existing *Pueraria phaseoloides* and *Centrosema pubescens* cover. Four fertilizer treatments and three stocking rates were applied: the former were 20, 40 and 80 kg P/ha as Christmas Island rock phosphate and 20 kg P/ha as triple superphosphate. Stocking rates were 2, 4 and 6 Kedah-Kelantan bull calves/ha. Animals were between eight and twelve months of age at the start of the experiment. Pasture availability declined over the three-year period at the high stocking rate and the proportion of the dominant Guinea grass declined with stocking rate during the third year. Centro and Stylo were relatively less affected by stocking rate but Puero disappeared from all treatments after two years.

Liveweight gains on the various treatments are summarized in Table 9.35 (Eng et al. 1978). There were significant effects of phosphorus level at the two lower stocking rates during the third year of the experiment (low pasture availability probably masked this effect at the highest stocking rate) and this was evident in both gain/head and gain/ha. Liveweight gains/ha were significantly higher for the heaviest stocking rate during the first year but subsequent differences between the high and the medium rate were not significant; this effect was presumably due to the decline in pasture availability. It was calculated that the optimum stocking rate was about 4 beasts/ha. There was no effective relationship between liveweight

Table 9.35 Effect of stocking rate and rate and kind of phosphorus fertilization on liveweight gains (kg/ha) of grass/legume pastures (Johore) (adapted from Eng et al. 1978)

	Liveweight gains (kg/ha)		
	1974–5	*1975–6*	*1976–7*
Stocking rate (beasts/ha)			
2	128	190	249
4	226	355	465
6	310	311	485
Phosphorus application (kg P/ha/year)			
20[1]	220	294	362
40[1]	194	281	436
60[1]	222	302	416
20[2]	248	269	384

[1]As rock phosphate [2]As triple superphosphate

gain and the legume content of the pasture (which ranged from 12 to 25%, averaging almost 21%). This was attributed to the fact that the more or less even growth habit of the grass ensured that protein would rarely become deficient and thus intake would not be limited by low protein levels; similarly, the relatively high level of legume in the pasture was outside the normal levels at which the relationship was critical. The authors commented that the main effect of the legume was likely to be on the growth of grass and consequent availability of forage, rather than in correction of protein deficiencies. These trials clearly demonstrate the capacity for high-level beef production in tropical areas, when factors which limit beef production in more extreme climates are not critical.

Thailand

Thailand has also been able to use technology of pasture improvement developed in Australia. The use of limited grazing on pastures improved with *Desmodium intortum* as a protein supplement (Falvey and Mikled 1978) has been mentioned elsewhere. Members of the Thai–Australian Highland Agricultural Project have also carried out preliminary studies on improved pastures in Chiang Mai province where the mean annual rainfall is 1630 mm, most of which falls in the six-month period between May and October.

Pastures were established based on *Setaria sphacelata* cv. Nandi or *Brachiaria decumbens* cv. Basilisk with various legumes including *Desmodium intortum* cv. Greenleaf, *Stylosanthes guianensis* cv. Schofield, *Macrotyloma axillaris* and *Trifolium repens* cv. Ladino. They were rotationally grazed on a six- to eight-week rotation by Brahman cross heifers at a stocking rate of 1.04 beasts/ha for approximately 8.5 hrs/day. Control animals were grazed on native pasture at a stocking rate of 0.07 beast/ha for a similar period of time each day, simulating traditional management. The persistence of the legumes was not good throughout the two-year period of the trial: neither Trifolium nor Stylosanthes persisted despite good establishment. Macrotyloma persisted reasonably well, possibly because of its low palatability. Grass yield also declined due to weed invasion. Nevertheless, the productivity of the improved pasture was 122 kg/ha over the approx. twenty months of the experiment, whilst the corresponding value for the unimproved pasture was only 34 kg/ha.

Not many results of the use of improved pastures integrated into a traditional system are available and this experiment is valuable in that it indicates how improved productivity may be obtained without abandoning altogether the custom of night 'kraaling' of cattle which is common in many parts of the world (this will be discussed in more detail in a later section).

The Philippines

Large areas of grassland in the Philippines are predominantly *Imperata cylindrica* on steep or undulating slopes. Pastures are particularly subject to overgrazing at high stocking rates, as they do not withstand trampling or frequent defoliation. Various attempts have been made to improve the carrying capacity of the natural pasture by the incorporation of legumes, or by replacement with an improved grass species, either with fertilization or legumes.

The latter approach was investigated by Magadan and his colleagues (Magadan et al. 1974) in Bukidnon province in the southern Philippines. The site is at an elevation of 610 m and the precipitation during the year of the experiment was 1685 mm, which was lower than average. The treatments used were unimproved *Imperata cylindrica* pasture, Para (*Brachiaria mutica*) sown with *Centroscema pubescens* or fertilized with either 100 kg N and 50 kg P, or 200 kg N and 75 kg P/year. Animal production was measured using one- to two-year-old Nellore-Brahman steers. A summary of the results is given in Table 9.36. The superiority of the improved

Table 9.36 Liveweight gains and weed incursion on native Imperata and improved Para pastures (Bukidnon) (adapted from Magadan et al. 1974)

Treatment	*Liveweight gain*		*Stocking rate (beast/ha)*	*Weeds (%)*
	kg/head/day	*kg/ha/year*		
Imperata	0.27	100.0	1	59.0
Para/Centro	0.42	305.4	2	18.0
Para + 100 N	0.36	260.0	2	25.8
Para + 200 N	0.28	310.8	3	47.4

pastures over the unimproved Imperata is clear, as is the lack of persistence of the Imperata at even relatively low stocking rates. The inclusion of legume is equivalent to a level of fertilization between 100 and 200 kg of N/year.

An experiment was carried out at Masbate to determine the productivity of native 'Cogon' (*Imperata cylindrica*) pasture unimproved at two stocking rates, and improved by the incorporation of *Stylosanthes guianensis* cv. Schofield with the application of 50 kg P_2O_5/ha/year. Cattle used were Brahman × Ongole heifers and the stocking rates were 0.5, 1.0 and 1.0 beast/ha for the three treatments. Liveweight gains during the experimental period of 400 days are summarized in Table 9.37.

The authors (Siota et al. 1977) state that 'this experiment provides overwhelming evidence for the soundness of incorporating legumes in native pastures'.

Table 9.37 Liveweight gains of heifers on improved and unimproved 'native pasture' (Masbate) (adapted from Siota et al. 1977)

	Stocking rate (beast/ha)	*Liveweight gain (400 days)*	
		kg/head	*kg/ha*
Native pasture	0.5	55.7	27.9
Native pasture	1.0	17.4	17.4
Native pasture + *Stylosanthes guianensis* + 50 kg P_2O_5	1.0	125.7	125.7

This evidence was corroborated in a subsequent experiment (Siota et al. 1978) in which native pasture was compared with associations which included *Stylosanthes guianensis* or *Centrosema pubescens*. Results are summarized in Table 9.38. Both legume associations were superior to native pasture at either stocking rate and returns from improved pastures were significantly better on both a per head and per hectare basis.

It seems clear therefore that improvement of the typical *Imperata cylindrica* grassland is quite feasible either by using Stylo or Centro, or by the more radical method of resowing with Para grass and legumes.

Table 9.38 Liveweight gains of heifers on native pastures unimproved and improved with Stylo or Centro (Masbate) (adapted from Siota et al. 1978)

	Stocking rate (beasts/ha)	*Liveweight gain (380 days)*	
		kg/head	kg/ha
Native pasture	0.5	43.3	21.6
Native pasture	1.0	26.6	26.6
Native pasture + Schofield	1.0	116.6	116.6
Native pasture + Centro	1.0	91.8	91.8

Papua and New Guinea

Some empirical observations were made on the possible beneficial effect of *Leucaena leucocephala* on fertility of beef cows in Papua and New Guinea in 1963 (Territory of Papua and New Guinea 1965), but subsequent work (Holmes 1976) showed that hair loss and goitre occurred together with growth checks, unpalatability and poor reproduction. The author concluded that this legume 'cannot be recommended for extensive use'.

Disappointing results were also obtained by Chadhokar (1977), who endeavoured to improve the indigenous *Imperata cylindrica* (Kunai) pastures with oversown *Stylosanthes guianensis*. The experiments were carried out in the Beef Cattle Research Centre near Lae in the Markham valley, where the annual precipitation is about 1120 mm. Pastures were grazed at a common stocking rate of 1 beast/0.8 ha and liveweight gains averaged 78 and 106 kg/ha/year for unimproved and improved pastures respectively. The author attributed these relatively low gains to possible deficiencies of phosphorus or cobalt (although no evidence was presented to support this) or the use of a common stocking rate. The proportion of Stylo in the pasture was reported to be high, however, and this exerted an effect on the total dry matter available.

Fiji

Contrary to experiences in Papua and New Guinea, *Leucaena leucocephala* has been successfully used in Fiji. Pastures comprising basically *Dichanthium caricosum* (Nadi, pronounced Nandi in Fiji; Partridge 1979) were supplemented by sowing 10 or 20% to *Leucaena* at the Sigatoka Research Station (precipitation 1630 mm approx. during the year of the trial). Liveweight gains/ha/year for the three treatments in the order given were 215, 300 and 500 kg. It was suggested that the overall stocking rate of 1.5 beasts/ha was too heavy for the unimproved Nadi grass. *Leucaena* persisted well under grazing conditions and no problems of hair or weight loss were experienced (Partridge and Ranacou 1974).

Partridge (1979) carried out further studies with Nadi blue grass in association with Siratro (*Macroptilium atropurpureum*) at Nawaicoba Research Station. Pastures were fertilized with super-phosphate and stocking rates of 5, 2.5 and 1.25, and 2.5, 1.25 and 0.83 beasts/ha were used on improved and unimproved pasture respectively.

Annual liveweight gains are summarized in Table 9.39. Steers used to evaluate the experiment were part Zebu or Santa Gertrudis × Hereford with an initial weight of about 200 kg.

It was calculated that the optimum stocking rate in this high rainfall area (mean 2657 mm over the four years of the experiment) was 3.5 beasts/ha at which the productivity was 380 kg/year. This represented a 100 to 200% increase attributable to the improvement.

Siratro was unusually persistent under the high rainfall conditions experienced at the site, and infection with *Rhizoctonia* fungus (as has

Table 9.39 Weight gains of steers grazing improved and unimproved pastures at different stocking rates (Nawaicoba) (adapted from Partridge 1979)

Stocking rate (beasts/ha)	*Liveweight gain (kg)*			
	Improved		*Unimproved*	
	kg/head	*kg/ha*	*kg/head*	*kg/ha*
5	61	305	—	—
2.5	141	352	57	143
1.25	180	225	93	116
0.83	—	—	108	90

been reported elsewhere) did not occur. The use of superphosphate encouraged other naturally occurring legumes, some of which, particularly *Desmodium heterophyllum* and *Mimosa pudica*, became important at the higher stocking rates where the Siratro component was considerably reduced.

Africa

Although the use of Beggar weed (*Desmodium tortuosum*) to improve the veld in Rhodesia was suggested as long ago as 1919 (Walters 1919), it was some time before work on legumes was started in Africa. Trials were initiated in various countries in the late 1950s but the first work to be reported appears to be that from Nigeria.

West Africa

Nigeria

Moore (1962) quoted unpublished work by Hill at Ibadan indicating that White Fulani cattle maintained average daily gains of 370 g on pastures consisting of Stargrass (*Cynodon dactylon* or *plectostachyus*) and Centro (*Centrosema pubescens*) whilst comparable gains on Stargrass alone were 310 g. Using the same mixtures, Oyenuga (1958) (also quoted by Moore (1962)) showed that the advantages due to Centro were greater in the semi-dry season than during either the wet or dry seasons and that, contrary to experience with legumes elsewhere, Centro afforded no advantage during the dry season. These results are summarized in Table 9.40.

A further trial involving grass/legume mixtures was carried out at

Table 9.40 Gains of White Fulani steers grazing Stargrass and Stargrass/Centro pastures during wet and dry seasons (Ibadan) (adapted from Moore 1962)

	Daily weight gains (kg)		
	Stargrass alone	*Stargrass/ Centro*	*Difference*
Rainy season (49 days)	0.51	0.61	0.10
Semi-dry season (70 days)	0.30	0.45	0.15
Dry season (98 days)	0.05	0.03	0.02

Ibadan using Ndama cattle (Okorie et al. 1965). The pastures used were composed of the following mixtures:

1. *Cynodon plectostachyus* plus *Centrosema pubescens*.
2. As above plus *Chloris gayana* plus *Digitaria decumbens*.
3. As 2 above plus *Stylosanthes guianensis*.

The trial was continued for two years and the stocking rate was 4.2 steers/ha during the first year and 6.7 steers/ha during the second. Pastures were rotationally grazed. Gains/ha/year for treatments 1, 2 and 3 respectively were 303, 350 and 386 kg and 372, 297 and 261 kg for the first and second years respectively indicating that the inclusion of the Stylo improved in the first but not during the second year. The authors reported that the complex mixtures tended to revert to the simple Stargrass/Centro mixture.

In a subsequent study (Oyenuga and Olubajo 1966), the Stargrass/Centro pasture was compared with two other grass/legume mixtures: Stargrass/Centro plus *Stylosanthes guianensis* and Pangola grass (*Digitaria decumbens*)/Centro plus Stylo. This experiment was also continued for a two-year period and the stocking rate was 4.2 Zebu steers (170 kg mean liveweight)/ha in year 1 and 3.2 steers/ha in year 2. Yields of herbage dry matter before and after grazing, intake and digestibility were determined. The results are summarized in Table 9.41 (differences in digestibility among treatments were small and inconsistent and results have not been included).

Although no statistical analyses of the results is presented to support their conclusions, the authors claim that the addition of *Stylosanthes guianensis* to the Stargrass/Centro pastures promoted a higher rate of gain although it did not result in higher dry matter yields or intake. Botanical analyses showed that in the Stargrass associations, the Stylo and Centro were present in equal proportions

Table 9.41 Weight gain and herbage intake of steers grazing grass/legume mixtures (Ibadan) (adapted from Oyenuga and Olnbajo 1966)

	Grass/legume mixture		
	Stargrass/ Centro	*Stargrass/ Centro + Stylo*	*Pangola/ Centro + Stylo*
Average daily gain (kg)	0.22	0.24	0.17
Gain/ha/year (kg)	526	567	412
Consumption kg DM/head/day)	8.2	8.1	7.3
Consumption (kg DM/100/kg liveweight day)	2.6	2.5	2.1

DM: dry matter

(14% each) whilst Stargrass dominated the pasture (56%). In the simple Stargrass/Centro pasture, the proportion was 66 and 20% respectively. When the Pangola grass replaced Stargrass, however, the legumes proved more dominant, the composition being as follows:

	(%)
Centrosema	39
Stylosanthes	20
Digitaria	16
Weeds	25

The authors commented that, besides being a late starter, Digitaria did not withstand heavy grazing as well as Cynodon and its slow fine shooting caused it to be dominated by the legumes.

In a subsequent paper, Olubajo and Oyenuga (1971) reported results of the same experiment for a further two years. Estimates of herbage intake were made by clipping and by the nitrogen-chromic oxide faecal index method. The results obtained indicated the superiority of the Cynodon/Centro/Stylo mixture, but the differences were almost certainly not statistically significant (no statistical analysis of weight gain data is presented). A summary of the results obtained is given in Table 9.42. Finally, Oyenuga (1971) reported gains of 0.36 kg/day during the wet season and 0.14 kg/day during the dry season (equivalent to an annual liveweight gain of 308 kg/ha) from a combination of *Pennisetum purpureum* and Centro.

In this series of experiments, the benefits resulting from incorporation of legumes into the pastures appeared to be minimal and absolute rates of gain from intensively-managed pastures to be low. The severe dry season was mentioned as a factor limiting rate of gain as was also an outbreak of trypanosomiasis. It was pointed out that

Table 9.42 Liveweight gain and herbage intake of steers grazing grass/legume mixture (Ibadan) (adapted from Olubajo and Oyenuga 1971)

	Grass/legume mixture		
	Cynodon/ Centro	*Cynodon/ Centro/Stylo*	*Digitaria/ Centro/Stylo*
1965–6			
Mean liveweight (kg)	363	386	285
Av. daily gain (kg)	0.27	0.29	0.20
1966–7			
Mean liveweight (kg)	193	209	200
Av. daily gain (kg)	0.14	0.15	0.09
Dry matter consumption (Overall av. kg/day)	9.1	9.0	7.8

the greater part of the food consumed was used for maintenance and it was suggested that rates of gain could be maintained by supplementation with hay or silage during the dry season. However, it would appear that some other factor was involved in limitation of performance.

de Leeuw (1969–71) initiated a series of trials on *Stylosanthes guianensis* at Ahmadu Bello University, Zaria. The experiments were carried out at the Shika Station in the Northern Guinea Savannah zone. Stylo was used as a supplement to range grazing nightly, or every second, or every third night. Weight changes for the three groups were 0.14, 0.02 and −0.03 kg/day respectively, whilst animals grazing pure stands of Stylo gained 0.17 kg/day. This is clearly a useful method of incorporating legumes into a management system, especially under African conditions where cattle are herded and usually brought in from pasture every night to avoid losses from predators. de Leeuw (1969–71) also reported gains of more than 150 kg/ha during the wet season from pastures of *Chloris gayana, Digitaria smutsii, Brachiaria brizantha* and *Stylosanthes guianensis* compared with only 18 kg/ha from unimproved shrub savannah. This is characterized by *Andropogon gayanus, A. pseudapricus, Hyparrhenia rufa, Hyperthelia dissoluta* and *Isoberlinia* spp. (Rattray 1960).

More details of the work carried out in mixed sown pastures are given by Haggar (1971). *Digitaria smutsii* oversown with stylo was compared with *D. smutsii* alone and without fertilization, *D. smutsii* fertilized at the rate of 84, 168 and 252 kg N/ha. All treatments were compared at stocking rates of 10 Fulani heifers (average weight 190 kg)/ha (normal grazing pressure), 8 head/ha (lenient grazing) and 13 head/ha (restricted grazing). This rate was maintained from August to September when the stocking rate was halved, and all

Table 9.43 Mean liveweight gains (kg/ha) of cattle grazing *D. smutsii* alone, with *Stylosanthes gracilis* and with various levels of fertilizer (Shika) (adapted from Haggar 1971)

	Stocking rate (see text)			
	Lenient grazing	*Normal grazing*	*Restricted grazing*	*Mean*
Grass + stylo	134	152	69	118
Grass + no N	150	101	– 3	82
Grass + 84 kg N/ha	216	299	289	268
Grass + 168 kg N/ha	271	342	344	318
Grass + 252 kg N/ha	291	276	420	328
Mean	212	233	223	—

cattle were removed from the experiment in November. The experiment was repeated the following year with some modifications. The results obtained in the grazing trial are summarized in Table 9.43.

Although the results were derived from a very short grazing period, there are some interesting indications. Fertilization with 84 kg N/ha produced a highly significant increase in liveweight gain and increasing levels resulted in progressively smaller increments. When the liveweight gain on the plots containing stylo plots was compared with the smoothed curve of the liveweight gains from fertilized plots, it was calculated that the incorporation of stylo was equivalent to fertilization with only 15 kg of nitrogen/ha. However, the inclusion of stylo reversed losses as grazing intensity increased. Over the two-year period, stylo became dominant over the grass component of the sward and the grass did not therefore benefit from nitrogen fixation by the legume.

The role of stylo in veld improvement was also investigated at Shika (Haggar et al. 1971). Stylo was oversown into savannah and used as a supplement to savannah grazing. In the first experiment, stylo was sown into cultivated strips at the rate of 4.5 kg seed/ha. Weight gains of Fulani and Fulani × Friesian bulls grazing improved and unimproved savannah were compared. Animals were given high (0.5–1 kg) or low (0.25–0.5 kg/day) supplements of whole cottonseed. Mean values for liveweight gain over the two years of the experiment are summarized in Table 9.44.

Differences between improved and unimproved savannah were significant at the 5% level and were exceeded by those attributable to supplementation which were significant at the 1% level. The authors mentioned several ways in which the establishment of the legume could have been improved, including phosphate application at the time of oversowing and heat treatment or scarification of the

Table 9.44 Weight gains of bulls (kg) grazing improved and unimproved savannah with high and low supplements of whole cottonseed (average of two years) (Shika) (adapted from Haggar et al. 1971)

	Supplement (for levels see text)	
	Low cottonseed	*High cottonseed*
Improved	58	85
Unimproved	43	72

seed to improve germination. It was also pointed out that further investigation was required with regard to management of the sward so as to maintain the presence of stylo in competition with native grasses.

The same paper (Haggar et al. 1971) also described complementary grazing of stylo in association with unimproved savannah, as mentioned earlier (de Leeuw 1969–71). Cattle grazed stylo for 100, 50 or 25% of the time, or were kraaled on stylo at night. Mean weight changes (over a period of ten weeks only) were 12, 2.4, −2.3 and 9.6 kg respectively. The difference between weight gains of cattle grazing stylo continuously and during the night only were not significant. There was a significant relationship between weight gain and time spent grazing stylo, and animals lost weight if the time was less than approximately 37%. It was pointed out that some form of rationing was necessary to avoid luxury consumption and to ensure that the stylo lasted throughout the dry season. A ratio of approximately 0.5 ha of mature stylo to 1.8 ha unimproved savannah was suggested to meet the requirements of growing cattle throughout the dry season. A further advantage of the system would be the widespread distribution of stylo seed in the faeces of grazing cattle.

The value of stylo as a supplement for the two best treatments (i.e. 100% and night stylo grazing) was equivalent to feeding 800 g whole cottonseed daily during the dry season. An economic analysis was made using prices operative in northern Nigeria at the time and it was concluded that the replacement value of the stylo was considerable. Compensatory growth was no greater in animals fed low levels of cottonseed than in those fed high levels (this might have been different had the comparison been with unsupplemented cattle.

Ivory Coast

One trial carried out at the experimental station at Bouake was indicative of the extent to which stocking rate can be improved by the incorporation of legumes. Carrying capacities were estimated using Ndama steers of approximately 300 kg and were as follows for the various treatments (which are self-explanatory):

Bush savannah, 0.25 head/ha
Cleared savannah, 0.5–1.0 head/ha
Cleared savannah, oversown with stylo, 1.5 head/ha
Clear Savannah with stylo sown in seedbed, 3.0 head/ha

Maximum production was indicated by the stocking rate of fertilized, irrigated *Panicum maximum*: 13.7 head/ha (Letenneur, L. cited by Crowder and Chheda 1977).

East Africa

Zambia

An early trial on grass/legume pastures was carried out by Smith (1962b) at Mazabuka. Four treatments were duplicated in eight 0.4 ha paddocks. These were:

1. Giant Rhodes grass (*Chloris gayana*).
2. Rhodes grass mixed with Velvet beans (*Stizolobium deeringianum*).
3. Rhodes grass mixed with *Stylosanthes guianensis*.
4. Rhodes grass mixed with pigeon pea (*Cajanus cajan*).

Pastures were used as hay or forage during the dry season. There were no significant differences in cattle weight gain among treatments (these may have been obscured by differences in management). Digestibility trials carried out on Rhodes grass/Stylosanthes hay and Rhodes grass hay showed a marked superiority of the former in terms of digestibility of crude protein, crude fibre and ether extract; the voluntary intake of the former was also greater (6.7 v. 5.9 kg/day). Animals fed the two hays whilst grazing the aftermath gained 4.8 and 8.7 kg on the Rhodes and Rhodes/stylo hays during a twenty-three-day period. This difference was not significant. These results do not indicate any marked advantage of legume incorporation insofar as cattle weight gains are concerned, and in a parallel trial the extra yield of N attributable to legumes was only 24 kg N/ha/year. It was concluded that this relatively low rate of nitrogen fixation was the main reason why legumes did not benefit the associated grasses. It was stated that, although not inoculated, the legume nodulated satisfactorily and it was assumed that the pink colour of the nodule indicated satisfactory nitrogen fixation. It can only be concluded that the low rate of nitrogen fixation was due to some other factor; perhaps a deficiency of some specific element, e.g. molybdenum.

Uganda

Stobbs and Joblin initiated a series of trials on intensive grass/legume pasture utilization at the Serere Research Station in Uganda. These trials were largely concerned with the techniques of measuring animal production on tropical pastures but also provide some interesting information on the productivity of grass/legume associations.

In the first trial (Stobbs and Joblin 1966a), mixtures of *Stylosanthes guianensis* with Rhodes grass or Guinea grass were compared. Treatments were as follows:

1. *Chloris gayana* Giant strain (SR 5316/Rhodes).
2. *Chloris gayana* Giant strain plus *Stylosanthes guianensis* H.B.K. (Stylo).
3. *Panicum maximum* Jacq. Local Teso strain (Guinea).
4. *P. maximum* plus stylo.

Treatments occupied 0.2 ha plots and were rotationally grazed by groups of five small (about 150 kg liveweight) East African Zebu steers according to a Latin Square design. Differences attributable to the legume were significant at the 5% level and amounted to some 13% over the two years of the experiment. Liveweight increases are summarized in Table 9.45. The authors comment that differences due to the legume were greater in the second year than the first (although this does not appear to be true for the Rhodes grass association) and also in the third (although no data are presented to support this).

Table 9.45 Liveweight gains from Rhodes and Guinea grass pastures with and without stylo (Serere) (adapted from Stobbs and Joblin 1966a)

	Gains (kg/ha)	
	1960–1	*1961–2*
1. *Chloris gayana*	208	211
2. *Chloris gayana* + stylo	231	235
3. *Panicum maximum*	230	189
4. *Panicum maximum* + stylo	233	251

In a further trial, stylo and Centro were compared in association with local strains of *Hyparrhenia rufa* and *Panicum maximum* (Stobbs and Joblin 1966b) and 'natural regeneration'. (No details are given of natural regeneration but, presumably, veld was composed of mainly *Hyparrhenia* spp.; Rattray 1960.) Animal production was measured using small East African Zebu heifers grazing according to a 'put-and-take' method during the first year and a common stocking rate during subsequent years. This stocking rate was based on the productivity of the best pasture and no details are given except that it was built up to 4.3 beasts/ha.

The results for the first nine months of the trial and total liveweight gains from the second and third years of the trial when a constant stocking rate was used are shown in Table 9.46. The authors comment that the Hyparrhenia and the Centro combined well as the

Table 9.46 Weight gains on various grass/legume mixtures (Serere) (adapted from Stobbs and Joblin 1966b)

	First year (different stocking rates) (kg/ha)	*Second + third year (fixed stocking rate) (kg/ha)*
Hyparrhenia	244	704
Hyparrhenia + Centro	259	830
Panicum + Stylo	278	736
Panicum + Centro	250	700
Natural regeneration + Stylo	223	712

legumes were able to climb the tall fibrous inflorescences of the grass. The Hyparrhenia also persisted better than the Panicum despite the greater palatability of the former. The Centro was of greatest value during the dry season, the advantage attributable to the legume being 10 kg/month during the dry and only 6 kg/month during the wet seasons.

A further trial evaluated the effects of inorganic fertilizers on grass and grass/legume pastures. A mixture of grasses was used comprising *Panicum maximum*, *Hyparrhenia rufa* and *Chloris gayana* sown with a mixture of *Stylosanthes guianensis* and *Centrosema pubescens*. The following treatments were established:

1. Grass with no fertilizer.
2. Grass plus N.
3. Grass plus N, P, S, K and lime.
4. Grass/legume.
5. Grass/legume plus P, S, K and lime.
6. Grass/legume plus N, P, S, K and lime.

Small East African Zebu steers were used in a 'put-and-take' method of evaluation. Overall average daily gain was 280 g. Production was significantly (at the 1% level) greater from the treatments that included legumes and the grass treatments fertilized with nitrogen. Gains from the grass/legume pasture fertilized with P, S and K were similar to the grass pasture fertilized with all the nutrients. It was concluded that although fertilization with nitrogen increased beef production, it was unlikely to do so economically. The application of nitrogen had the effect of suppressing the amount of legume in the pasture and could therefore be a possible method of influencing botanical composition. A summary of the results obtained is given in Table 9.47.

Investigation into grass/legume pastures continued with a comparison of either *Centrosema pubescens* or *Stylosanthes guianensis* in association with *Hyparrhenia rufa*. Animal productivity was measured

Table 9.47 Animal production from grass and grass/legume pastures with and without fertilization (*Serere*) (adapted from Stobbs 1969a)

	Average daily gain (kg)	*Animal days*	*Total liveweight gain (kg/ha)*
Grass	0.25	1 078	653
Grass + N	0.30	1 444	1 048
Grass + N, P, S, K	0.28	1 646	1 127
Grass/legume	0.29	1 384	973
Grass/legume + P, S, K	0.27	1 595	1 058
Grass/legume + N, P, S, K	0.31	1 735	1 319

using small East African Zebu heifers grazed at a common stocking rate which was varied according to seasonal forage availability. Cumulative gains over the two years of the experiment were about 1000 kg/ha (results are expressed graphically) and differences between treatments were not significant (Stobbs 1969b).

Stylosanthes guianensis was compared with *Desmodium intortum* (contaminated with *Desmodium sandwicense*) as part of an association with *Panicum maximum* (Stobbs 1969c). Evaluation of animal production was similar to the previous trial except that cross-bred (Hereford × Zebu, Angus × Zebu) heifers as well as East African Zebus were used. There was a significant difference between breeds but not between pastures. The interaction between pastures and animals was significant, cross-bred cattle producing more on the Stylo pasture and Zebu cattle on the *Desmodium* pasture. It was suggested that the grazing potential was greater than the genetic potential for gain of the local stock and that improved pastures should be grazed by improved stock. Results are summarized in Table 9.48.

In the last publication of this series, the effects of varying stocking rate on the productivity of *Hyparrhenia rufa/Stylosanthes guianensis* and *Panicum maximum/Macroptilium atropurpureum* (Siratro) pastures was

Table 9.48 Animal production from pastures of Panicum maximum with Stylo or Desmodium intortum (Serere) (adapted from Stobbs 1969c)

Liveweight gain (kg/ha)	*Replicate 1*		*Replicate 2*	
	Desmodium	*Stylo*	*Desmodium*	*Stylo*
First 105 days	218	177	212	176
First 245 days	390	354	379	366
Following 525 days	861	778	1 084[1]	1 191[1]

[1]Cross-bred animals used during this period

Table 9.49 Liveweight gains from grass/legume pastures at different stocking rates (Serere) (adapted from Stobbs 1970)

	Hyparrhenia/Stylo			*Panicum/Siratro*		
	Low	*Medium*	*High*	*Low*	*Medium*	*High*
Mean liveweight gain (kg/ha)	449	816	1 179	534	685	944
Mean liveweight gain (kg/head)	114	138	119	136	118	96

[1]For explanation, see text

studied (Stobbs 1970). The cattle were small East African steers and the stocking rates used were 2, 3 and 5 steers/ha (low, medium and high).

Significant differences between pastures could not be detected but increasing stocking rate improved liveweight gains/ha in both cases. No indication is given of the persistence or otherwise of the legume at the various stocking rates. A summary of the results is given in Table 9.49.

The principal objective of this series of trials was to investigate the use of different experimental designs for the evaluation of tropical pastures using liveweight gains. The subject is not pertinent to the present discussion but the interested reader is referred to the original papers. Some of the results from these trials have been reported elsewhere (Stobbs 1965a and b). In a further series of trials (Stobbs 1969d, e, f and g), the effects of different management systems on grass/legume associations were investigated. Although these will be discussed in more detail in the relevant chapter, some comments are appropriate here.

Different stocking rates were employed on a *Hyparrhenia rufa*/*Stylosanthes guianensis* pasture (Stobbs 1969d). Although the level of the legume was maintained at more or less the same level, there was a marked increase in less desirable species, notably *Sporobolus pyram-*

Table 9.50 Botanical composition of Hyparrhenia/Stylo pasture at various stocking rates (Serere) (adapted from Stobbs 1969d)

	Stocking rate		
	1 beast/ha	*2 beasts/ha*	*5 beasts/ha*
Hyparrhenia rufa (%)	71	68	26
Stylo (%)	19	22	20
Sporobolus (%)	1	2	39
Other	10	8	15

idalis as stocking became more intensive. A summary of botanical composition is given in Table 9.50.

The heavy stocking rate was clearly overgrazing the pasture as judged by the high levels of an 'increaser' species (*Sporobolus pyramidalis*), but it is interesting that despite this, the legume was able to maintain itself.

In a subsequent trial, Stobbs (1969f) showed that *Macroptilium atropurpureum* became dominant when an association with *Panicum maximum* was grazed, but that continuous grazing also allowed a build-up of *Sporobolus pyramidalis*. This was presumably because cattle preferred the more palatable *P. maximum* to either the legume or the Sporobolus. A further trial shed light on this grazing behaviour (Stobbs 1969g). Samples were taken from the oesophageal fistula of a steer grazing a *Hyparrhenia rufa/Stylosanthes gracilis* sward. Consumption of the legume on the first day was only about 10% of the total dry matter intake, but this stabilized to an estimated level of approximately 27.8% of the total diet over the six-week period during which measurements were made. No intake studies were made on *Panicum maximum*/stylo associations but observation indicates that *P. maximum* is more palatable than *Hyparrhenia* and it might be expected to suffer more from overgrazing.

In a further trial *S. guianensis* and *Centrosema pubescens* were compared in association with *Hyparrhenia rufa* at different grazing frequencies (Stobbs 1969e). During the 420 days of the trial, gains off the *Hyparrhenia/Centrosema* pasture were 773 kg compared with 754 kg/ha from the Hyparrhenia/Stylosanthes pasture; these differences were not significant. (Stobbs 1969e). The previous trials described by Stobbs have concerned the use of legumes in leys under Ugandan conditions. However, Stobbs (1969h) also investigated the use of *S. guianensis* to oversow a natural *Hyparrhenia* spp. pasture with or without a top dressing of locally-produced single superphosphate.

The small East African Zebu steers used to evaluate the effects of fertilizer gained 474 kg liveweight gain/ha on the top dressed pasture and only 279 kg/ha where no superphosphate was applied. Fertilization increased the stylo in the sward and reduced the weed, *Imperata cylindrica*. Mineral analyses of blood did not suggest that improved growth rate of the animals on the fertilized areas was due to correction of mineral deficiencies.

This study of oversowing natural pasture was continued by Otim (1975) who compared its productivity with that of a ley composed of: *Chloris gayana*, *Panicum maximum*, *Hyparrhenia rufa*, *Stylosanthes guianensis*, *Centrosema pubescens* and *Macroptilion atropurpureus*. Details of the initial botanical composition of the natural pasture are not given but, judging by subsequent botanical analyses, it contained a predominance of *Imperata cylindrica*; this was markedly reduced after three years whilst the *P. maximum* and *C. gayana* increased. (Rattray 1960 has noted that *I. cylindrica* in Uganda forms part of the succession towards *Pennisetum purpureum* which takes place after clearing.)

The animal production from the improved natural pasture was 357 kg/year as compared with 462 kg/ha from the ley (measured with small East African Zebu steers) but the improved natural pasture offered a better profit margin. The plots were 0.2 ha in area (the total experiment was 2.4 ha) and the stocking rate was 1 beast/0.3 ha in the wet season and 1 beast/0.4 ha in the dry season.

Zaire

Risopoulus (1966) demonstrated the advantages of improving the savannahs in the Gandajika area of southern Zaire with *Stylosanthes guianensis*. Although the effects of legumes are confounded by methods of management (i.e. day grazing only and day and night grazing on improved paddocks) and supplementation, there is no doubt of their role in increased animal production. Husbandry methods and liveweight gains are given in Table 9.51. Skerman (1977) has mentioned this as a case history justifying the introduction of legumes on economic grounds and states that a profit is made over expenditure.

Table 9.51 Liveweight gains from different management systems (Gandajika) (adapted from Skerman 1977)

Husbandry system	*Liveweight gains (kg/ha)*
1. Traditional: burned day pasture only	10–15
2. Improved: Day pastures burned every 4 years, night pastures oversown with *Stylosanthes guianensis*[1]	46
3. Improved: all pastures oversown with *S. guianensis*[1]	76

[1]10 kg/day silage during last 3 months of the dry season

Southern Africa

Despite (or perhaps because of) the advances made in this part of Africa on other means of improving the nutrition of beef cattle, attention was given to legumes at a comparatively late stage. Investigations have been carried out in Zimbabwe, Malawi, Swaziland, Madagascar and South Africa.

Zimbabwe

Apart from the observations of Walters (1919) mentioned earlier, the first data to be collected on the incorporation of legumes into pasture in Zimbabwe were from Melsetter in the Eastern Highlands at an altitude of between 1000 and 1500 m. The legumes used were clovers, *Trifolium ruepellianum* and *T. semipilosum*, which are usually found in temperate rather than tropical regions, but the climate in Melsetter is modified by altitude and this work will be included for completeness. Steers grazing Kikuyu (*Pennisetum clandestinum*)/clover pastures at a stocking rate of about 5 head/ha gained 47 kg each or a total of 246 kg/ha in sixty-nine days (December–February), i.e. 3.6 kg/ha/day. Subsequent production during the summer season (February–May) amounted to 1.8 kg/ha/day. Although no comparable data are given for unimproved pasture, there is no doubt that the Kikuyu/clover mixture produced considerably more liveweight gain/ha than would be expected from the *Themeda triandra* association that forms the natural veld (Rattray 1960). The author makes the point that intensification of livestock production by this and other means would allow smaller holdings to be farmed, which could be a means of providing more land resources for the indigenous population (Atkinson 1963).

More recent work on legumes has been carried out at the Grasslands Research Station at Marandallas and the Henderson Research Station. Clatworthy (1975a) sowed a reverted (from arable) land with a mixture of five strains of stylo, fine-stem stylo and Siratro and found that animal production as measured by the liveweight gains of weaners (192 kg) amounted to 103 kg/ha from the unimproved reverted land and 192 kg/ha from the legume-improved area (a difference of 86%) during a 352-day period. The steers grazing the improved pasture took five weeks less to finishing (in feed lot) than the other group.

Gains from improved and unimproved land in the following year were 205 and 139 kg respectively, an advantage of 48% for the improved grazing; the decreased advantage of the legumes pasture was due to the better performance of the reverted land. The content of the fine-stemmed stylo in the pasture continued to increase whilst the other legumes decreased (Clatworthy 1976).

In 1976–7, the experiment was modified and one replicate of each treatment grazed until the end of the summer season (June). During the dry period, responses were similar to previous years in that animals grazing improved land lost only 2 kg between July and November whilst those grazing reverted land lost 28 kg. Gains to January were (approximately) 80 kg (reverted) and 99 kg (improved), and until June 120 kg (reverted) and 167 kg (improved). Stocking rate on the improved land was also higher at 0.80 ha/head v. 1.22/ha/head (Clatworthy and Holland 1977).

In 1977–8, steer gains until June were appreciably lower than in the previous year, being 111 and 124 kg for reverted and improved respectively. The cattle on unimproved reverted pasture showed marked compensatory growth during the early growing season and became nearly as heavy as the other group. However, the cattle on the improved grazing gained more rapidly in the later part of the growing season. The results of this experiment were compared with those from a pasture which has been sown with a similar mixture of legumes without grass (Although invasion of *Eleusine indica* had occurred). This pasture contained more legume than those in the previous trial and this is reflected by the better performance of the steers, i.e. 192 kg/head for a comparable period (Clatworthy and Holland 1978).

This experiment is interesting in view of the large amount of previously-cultivated land which has reverted and the need to exploit if profitably. In many parts of Africa, land has been abandoned after repeated cropping and the incorporation of legumes would improve soil fertility at the same time as providing a means of intensifying livestock husbandry (see also Clatworthy and Holland 1979).

In a trial to assess animal production from a dryland pasture of *Desmodium intortum* and varying proportions of grass, steers maintained body weight during the first winter and gained 145 kg/head (236 kg/ha) during the following summer (until the end of June). However, there was a downward trend in winter performance in succeeding years associated with a decline in the legume content of the sward (Clatworthy 1975b).

Clatworthy (1975c) carried out an exploratory trial to determine the carrying capacity of fertilized grass v. grass/clover pastures. Both pastures were irrigated. The grass was *Pennisetum clandestinum* (Kikuyu) fertilized at the rate of 500 kg N/ha/year and this was compared with a pasture intersown with a locally-selected strain of Kenya white clover (no species given). Mean daily gain of steers on the two treatments respectively was 0.71 and 0.62 kg/head/day respectively. Problems of bloat were encountered when clover pastures were grazed without a supplement of corn-and-cob meal.

A trial was carried out comparing different areas sown with Silverlea desmodium (*Desmodium uncinatum*) at the rate of 3 kg/ha on disced strips in veld covering 0, ⅓, ⅔ and the whole area, and at two stocking rates (0.5 and 0.75 ha/weaner). Results in this experiment were disappointing and are summarized in Table 9.52. It was supposed that the Silverleaf plants were largely defoliated during the

Table 9.52 Weight changes (kg) of weaners grazing Silverleaf-oversown veld at different stocking rates (Grasslands) (adapted from Clatworthy and Boultwood 1978)

Proportion sowed with legume (%)	*Stocking rate* (ha/weaner)	
	0.5	*0.75*
0	−20	−24
33	−16	−16
66	−3	−15
100	−17	−14

dry season and therefore the grazing was of poor quality (Clatworthy and Boultwood 1978).

Malagasy Republic

Delhaye (1967) carried out a trial to investigate the potential for beef production of various improvement systems on reverted land on both high and low ground.

The treatments were as follows only most important species represented):

(a) Permanent artificial pasture (Ambatofotsy – high ground)
Chloris gayana (more than 75% surface cover)
Melinis minutiflora (at least 5% surface cover)
Stylosanthes guianensis (at least 5% surface cover)

(b) Improved natural pasture (Ambatofotsy)
Spontaneous grasses:
Cynodon dactylon (50–75% surface cover)
Digitaria humbertii (25–50% surface cover)
introduced spp.:
Melinis minutiflora (at least 5% surface cover)
Stylosanthes guianensis (at least 5% surface cover)

(c) Natural pasture (Ambatofotsy)
Cynodon dactylon (50–75% surface cover)
Digitaria humbertii (25–50% surface cover)

(d) Permanent artificial pasture (Ambahikely – low ground)
Chloris gayana (more than 75% surface cover)
Stylosanthes guianensis (at least 5% surface cover)

(e) Improved natural pasture (Ambahikely)
Spontaneous grasses:
Brachiaria madecassa (at least 5% surface cover)
Cynodon dectylon (25–50% surface cover)
Digitaria humbertii (25–50% surface cover)
Paspalum scrobiculatum (at least 5% surface cover)
introduced spp.:
Stylosanthes guianensis (at least 5% surface cover)

(f) Natural pasture (Ambohikely)
Brachiaria madecassa (at least 5% surface cover)
Cynodon dactylon (25–50% surface cover)
Digitaria humbertii (25–50% surface cover)
Paspalum scrobiculatum (at least 5% surface cover)

The cattle used to measure productivity of the various treatments were 1–2-year-old Brahman and Sahiwal cross-bred heifers.

A summary of the values obtained for stocking rate and meat (sic) production/ha are given in Table 9.53. Although the results cannot be attributed to the introduction of a legume alone, this experiment is interesting because of the applicability of the results to the intensification of livestock production on reverted lands (cf. Clatworthy 1975a etc.).

Table 9.53 Stocking rates and meat production from different pasture treatments (Madagascar) (adapted from Delhaye 1967)

	Ambatofotsy (high ground)		*Ambohikely (low ground)*	
	Stocking rate (kg/ha)	*Meat production (kg/ha/year)*	*Stocking rate (kg/ha)*	*Meat production (kg/ha/year)*
Permanent artificial	443	286	449	438
Natural improved	161	115	288	179
Natural unimproved	82	−19	160	105

Swaziland

I'ons (1968) carried out preliminary grazing trials at the Malkerns Research Station in the Middleveld. The pastures employed were follows:

Kikuyu grass (*Pennisetum clandestinum*)/*Desmodium intortum* (Greenleaf).

Weeping Lovegrass (*Eragrostis curvula*)/*Desmodium sandwicense*/stylo (*Stylosanthes guyanensis*).

Paspalum dilatatum (This had originally been seeded with the same legumes as above but these were frosted out).

Nandi Sataria (*Setaria sphacelata*)/Silverleaf *Desmodium*.

Camps were rotationally grazed by a single group of five to seven steers which started the season with a mean weight of 250 kg and reached a maximum of 410 kg. Because of the rotational grazing, it was not possible to determine liveweight gains on the individual pastures, but results were based on grazing days and carrying capacity and are summarized in Table 9.54.

The *Paspalum dilatatum* pasture provided the most grazing days during summer and early autumn and the Kikuyu/Desmodium and Setaria/Desmodium pastures provided useful forage during May and early June, but after this there was considerable leaf loss in the legume due to trampling and frost.

In a subsequent (unpublished) experiment (Peter Jones 1973, personal communication), the effects of introducing *Stylosanthes guianensis* into the natural veld (Lowveld sour bushveld, Acocks 1975) consisting largely of *Hyparrhenia dissoluta* and other *Hyparrhenia* spp. were investigated. Over seven seasons, starting in 1966–7, the average gain from the control (unimproved) pastures was 115 kg/ha whilst that from the veld with stylo was 175 kg/ha, i.e. an increase in production of 54% over the control.

It seems clear that legume introduction under these conditions would be an important method of increasing beef production.

Table 9.54 Grazing days and carrying capacity of different pastures (Swaziland) (adapted from I'ons 1968)

	Spring (70 days)		*Summer and autumn (150 days)*		*Early winter (75 days)*		*Late winter (70 days)*	
	GD	*CC*	*GD*	*CC*	*GD*	*CC*	*GD*	*CC*
Kikuyu/Desmodium	—	—	—	—	119	0.65	—	—
Lovegrass/legume	64	1.1	201	0.75	—	—	—	—
Paspalum	67	1.0	404	0.37	—	—	—	—
Setaria/Desmodium	47	1.5	—	—	98	0.79	127	0.56

GD = grazing days (450 kg/steer) CC = carrying capacity (acres/animal unit)

Malawi

Addy and Thomas (1976) carried out an experiment to evaluate the effect of sowing one third of the total area of a Rhodes grass (*Chloris gayana*) pasture with legumes. Legumes used were Stylo (*Stylosanthes guianensis*), Silverleaf Desmodium (*Desmodium uncinatum*) and Siratro (*Macroptilium atropurpureum*). Cattle either grazed the sowed pastures or were allowed access to legumes for 25% of the total grazing time. Cattle grazing the grass alone gained 72 kg in 144 days (0.5 kg/day) whilst those grazing the improved pasture gained 80 kg in the same period (0.56 kg/day). The stocking rates used were 1–1.3 livestock units /ha (1 livestock unit – 340 kg) and it was said that adequate grazing was always available. The experiment was carried out at the Chitedze Agricultural Research Station on the Lilongwe plain in the central region of Malawi.

Apparently there was a small response to legumes in this experiment. However, no statistical analysis of the data is presented and there is no indication of the number of animals used. Results are presented as means and there is no indication of whether animals on the two management systems differed in weight gain. In a subsequent publication (Thomas and Addy 1977), the authors state that the response to legumes was only apparent after March.

Addy and Thomas (1976) also investigated the effect of legumes in winter grazing. Rhodes grass was sown with Siratro or Schofield and Endeavour Stylosanthes. Cattle (no indication of number or weight given) on the Rhodes grass alone lost weight (1.4 kg) from July to November whilst those on the legume-improved pasture gained 24.5 kg. This gain was similar to that of another group of steers grazing the Rhodes grass forage and supplemented with cottonseed cake. There is some disparity between the results in the two references cited and no statistical analysis is presented. The authors commented that no responses were obtained when levels of legume of 12 and 17% of DM and that the legumes were grazed out by August. No data are given to support this, however.

Attempts at veld improvement using Cook stylo were also made by the UNDP/FAO Project MLW/75/020 in the West Mzimba Land Utilization Scheme in the northern region, and it was shown that when this legume was incorporated into the predominantly *Hyparrhenia* grasslands, dry season weight losses could be reversed. In a subsequent trial it was shown that the mean daily gain/hectare from the stylo-based pasture was 0.51 kg/day compared with a loss of 0.1 kg/day from natural unimproved veld (Dzowela 1980). The legume established well and it only remains to see how it stands up to village management conditions. This particular part of Malawi is important for the provision of feeder steers for small-scale fattening schemes in the central region, and the West Mzimba scheme, if successful, would be an interesting development for other parts of Central Africa to follow in the improvement of grassland productivity and intensification of presently uncontrolled grazing practices.

Republic of South Africa

Little work on legumes in association with improved or natural grasses has been carried out in South Africa, and both the studies which are cited here are somewhat atypical of the subtropics in that the environment at Ntambamhlope was modified by altitude and a temperate legume was used, whilst the other dealt with a tropical legume at 32° S in Cape Province. However, both studies will be included for completeness.

Theron et al. (1974) investigated various methods of Radical Veld Improvement (RVI) of the highland sour veld of Natal at Ntabamhlope. The authors defined RVI as 'the improvement of herbage production (and so the carrying capacity) by the replacement of the native grasses with improved herbage species'. RVI can be either *by replacement*, when the natural veld is ploughed out and improved species are established by conventional techniques, or *reinforcement*, when they are established by minimum tillage or over-sowing accompanied by fertilizer application. The treatment relevant to this discussion was the reinforcement of veld using ladino white clover introduced using the Cedara Sodseeder (a minimum tillage implement where seed and fertilizer are introduced after a tine has cut the sod of the veld). Control pastures were unfertilized, unreinforced (or unimproved) veld and the same veld fertilized with 92 kg N, 25 kg P and 100 kg K/ha/year. The performance of beef steers is summarized in Table 9.55.

Grazing days were counted until there was no remaining available forage or animals had stopped gaining weight. It is perhaps surprising that there was such a marked difference between the length of the grazing season for unimproved veld and that which had been reinforced, but no doubt this was caused by the difference in stocking rate between the two treatments. The authors note that the cattle were removed from the reinforced veld earlier, in order to enable the legume to establish more effectively. The authors also noted the

Table 9.55 Performance of beef steers on reinforced fertilized or unimproved veld (Ntabamhlope) (adapted from Theron et al. 1974)

	Unimproved	*Reinforced*	*Fertilized*
Mean stocking rate (mlu/ha)[1]	0.86	2.33	2.22
Days grazing	206	117	189
Total liveweight gain (kg/ha)	154.4	310.5	407.2

[1]Mature livestock unit

improved carcass quality of cattle grazed on the legume-reinforced pasture. This was a major factor contributing significantly to the profitability of the system which, in terms of return for capital invested, was comparable with more intensive pastures (in this case, *Eragrostis curvula* and *Pennisetum clandestinum*).

Apparently the only other study which has been carried out on the use of legumes was to investigate the use of Greenleaf Desmodium (*Desmodium intortum*) for improving beef production potential in the coastal belt to the north of East London in the Eastern Cape. The actual location was the Dohne Agricultural Research Institute at about 32° S. The legume was established with Rhodes grass (*Chloris gayana* cv. Katambora) and the pasture was evaluated using young cross-bred (predominantly) Friesland (q.v. Friesian) steers. Siratro was also established but this died out due to mismanagement. The number of steers was adjusted annually according to the availability of feed under rotational grazing conditions.

Liveweight gains (kg/ha) on a yearly basis were:

1973/4	*1974/5*	*1975/6*	*1976/7*	*1977/8*	*Average*
276	508	494	624	548	544

The authors (Nel and Bosman 1979) conclude that, provided the legume is established, the potential of *Desmodium intortum* appears to be 'unlimited' for the frost-free coastal areas. This is interesting in view of the comment by Skerman (1977) that *Desmodium intortum* extends to 25° S and indicates that the legume could well be more widely used.

Tropical and subtropical America

United States of America

Tropical and subtropical areas of the United States include the widely different ecological zones of mainland Florida, Caribbean Puerto Rico and the Hawaiian islands in the Pacific Ocean. In Florida, the use of legumes was largely confined to traditionally temperate species established with applications of lime at a high rate. Early Hawaiian workers evaluated Koa hoale (*Leucaena leucocephala*), a leguminous browse, as cattle feed but subsequent studies incorporated tropical pasture legumes. Puerto Rican work investigated the use of tropical Kudzu (*Pueraria phaseloides*) in association with various tall grasses, but more recent work has concentrated on heavy fertilization of grass pastures under conditions of abundant moisture.

Florida

Work was started in 1940 at Gainesville by Blaser et al. (1952) who used a combination of eight clovers to improve Carpet grass (*Axonopus compressus*) pastures. However, after the first year more than

95% of the legume present was Louisiana white clover. This was compared with grass plus lime plus complete fertilizer (no details given) and Lespedeza sown into the grass sward. Liveweight gains were 84, 166, 245, 219 and 693 kg/ha/year for the control, fertilized, Lespedeza and white clover treatments respectively.

Subsequent experiments were carried out at the Range Cattle Station at Ona (Hodges et al. 1953). Average beef cattle liveweight gains (1943–51) from Carpet grass alone and Carpet grass interseeded with Louisiana white clover were 78 and 277 kg/ha respectively. Gains from irrigated and non-irrigated grass/clover pasture were 329 and 907 kg/ha respectively.

Jones et al. (1954 and 1960) used native and improved pastures for year-long grazing. The native range consisted mainly of wiregrass (*stricta*) *Aristida* and the improved pastures were composed of common and Pensacola Bahia (*Paspalum notatum*), with hairy Indigo (*Indigofera hirsuta*) Pangola grass, Coastal Bermuda and Pensacola Bahia with Hubam clover (*Melilotus alba*). Cattle grazed both native and improved pastures intermittently and there is no indication of the proportion of time spent on the legume-containing pastures. However, it was stated that one cow could be supported by 2.5 ha of native range and 0.5 ha of improved pasture without any supplemental feeding.

In a later study at the Range Cattle Experiment Station, the beef production from fertilized Pensacola Bahia was compared with that from Pensacola Bahia pastures, 25% of which had been oversown with a mixture of hairy Indigo and *Aeschynomene americana*. The grass pastures were fertilized thrice yearly with a total of N: 168 kg, P: 37 kg, K: 70 kg. Animal performance was measured using cross-bred Brahman × European weaners. Over a three-year period, liveweight gains averaged 288 and 311 kg on the grass and grass/legume systems respectively; the differences were not significant (Hodges et al. 1977a).

Koger and his colleagues (Koger et al. 1961) established a long-term study in north-central Florida near Gainesville to evaluate different pasture programmes on flatwoods soils. The Beef Production Research Unit represents some 15 million acres which could be used for beef production under favourable economic conditions. In an evaluation of the the first five years of the trial, it was concluded that clover/grass programmes produced beef at approximately 60% of the cost of the all-grass programmes. The advantages of the former included a higher weaning percentage (83 v. 64%), slightly heavier calves and a lower total cost/cow. Although the fertilization of the grass pastures improved carrying capacity, this was not economically justifiable at prevailing prices (Koger et al. 1970).

The second stage of the study (1958–64) included five pasture programmes. Grass composed of one third Pensacola Bahia, one third Pangola grass, and one-third Coastal Bermuda (*Cynodon dactylon*) was fertilized at the rate of 500 kg 0–20 and 180 kg N/ha. Clover/grass pastures included the grasses previously mentioned as well as white clover (*Trifolium repens*). These were fertilized at three rates, 336, 560 and 780 kg 0–10–20 and 67, 134 and 200 kg N/acre; the fertilization with N was only carried out during the first two years as it did not appear to improve productivity. A final grass/clover treatment was irrigated and fertilized with 100 kg 0–10–20 and 200 kg N/ha/acre. No further details are given of establishment and it is not clear how the clover was incorporated into the pastures.

Botanical analyses indicated that Pangola grass did not survive but that the other two grasses persisted well.

Animal performance from pasture was evaluated from cow and calf performance and management was designed to ensure that all utilizable forage was grazed. Stocking rates on the different programmes were similar, varying from 0.5 to 0.6 ha/cow and production/ha (kg calf weight) was 390, 430, 395, 410 and 412 kg for programmes 1 to 5 respectively. However, heavy supplemental feeding was necessary on programme 1 which led to poor economic performance. Net returns per cow were calculated for the various programmes and a summary of results is given in Table 9.56.

In the third phase of the project, the number of pasture programmes was reduced to three. Programme 1 was a standard grass/clover pasture and programme 2 involved renovation of old grass/clover pastures infested with *Sporobolus poiretii* (Smut grass). Programme 3 consisted of grass/clover pastures, half of which were under seepage irrigation, with an increased level of fertilization. Individual animal response evaluated by cows and calves showed no significant differences attributable to pasture programme. Net returns

Table 9.56 Net return per cow for various pasture programmes (north-central Florida) (adapted from Koger et al. 1970)

Programme	*Net return/cow (US $)*	*Comment*
1. Grass	−22	High fertilizer and wintering costs, lower production
2. Grass/clover; low NPK	43	
3. Grass/clover; med. NPK	3	Higher fertilizer costs than 2 without increased production
4. Grass/clover; high NPK	22	Higher fertilizer costs than 2 without increased production
5. Grass/clover (irrigated)	−12	High costs of sprinkler irrigation

High (4.95)

per cow were US$48.12, 43.32 and 37.08 for programmes 1, 2 and 3 respectively (Koger et al. 1977).

This long-term systems approach to beef production from pasture showed the utility of white clover in pasture under subtropical conditions. The objective of the heavy liming programme used was to maximize the growth of white clover by maintaining surface soil pH at 6.0 and providing an optimum balance between calcium and magnesium. The results have been reported in some detail because of the long-term nature of the trial (twenty-one years in total) and because of the lack of results on traditionally temperate legumes in the tropicas or subtropics.

Performance on three different management programmes was also compared by Peacock and his colleagues (Peacock et al. 1976b) at Gainesville. Pasture programmes were: native grasses (mainly *Aristida* and *Andropogon* spp.); a combination of 30 ha of improved pasture with 127 ha native grasses (the improved pasture consisted of Pensacola Bahia, Coastal Bermuda and Pangola grass; part of the Pensacola Bahia was grown in conjunction with hairy Indigo and part with white clover and Hubam clover); and improved pastures only (Pangola, white clover, Pensacola Bahia). Least square means for production per cow on the three pasture programmes in the same order as given were 102, 145 and 160 kg respectively, and differences were significant.

Although this trial and those from Ona previously quoted are not critical examinations of the productivity of legume pastures as such, they illustrate effectively how legumes can be fitted into practical management programmes.

Hawaii

Early work was carried out on 'Koa haole', *Leucaena glauca* (now known as *Leucaena leucocephala*) which was known in the territory prior to 1888. Although cattle had grazed the plant extensively, the first trials to determine its feeding value were carried out with dairy cows in 1928. It appears that the first trials with beef cattle were carried out by Henke and his colleagues (Henke 1936; Henke et al. 1939; Henke and Work 1940) when the weight gain of Aberdeen Angus steers grazing Leucaena with and without molasses was measured.

The trial was repeated for three years and the results were summarized by Henke (1958). A summary is presented in Table 9.57.

The stocking rate was 0.8 ha/280 kg steer and molasses was provided *ad libitum*. Weight gains were poor when rainfall was low and molasses consumption was markedly increased to compensate for low availability of forage. Molasses supplementation increased weight gain, indicating that protein was adequate even though forage production might have been limited by low precipitation.

After the Second World War, intensive beef production from

Table 9.57 Weight gains of steers on Leucaena pastures with (+) and without (-) molasses supplementation (Hawaii) (adapted from Henke 1958)

	1936–7		*1937–8*		*1938–9*	
	−	+	−	+	−	+
Days on test	246		444		368	
Rainfall (mm/year)	1 800		860		700	
Average daily gain	0.52	0.68	0.38	0.48	0.20	0.34
Molasses consumption (kg/day)	—	1.6	—	2.2	—	2.9

heavily limed or fertilized Pangola grass/legume pastures was investigated. Pangola grass (*Digitaria decumbens*) and *Desmodium intortum* pastures were established on land previously under pineapples (abandoned because of low yields) by discing in chopped material (Younge and Plucknett 1966). The authors state that the legume established badly and therefore nitrogen fertilizer was applied (no details are given of the botanical composition of the pasture during the experiment). Bad establishment from cuttings has also been reported by Risopoulus (1966), who noted that this method gave only a 30–40% strike in Zaire. The pastures were limed at a rate of 2300 kg/ha and phosphorus was applied at rates of 280, 560, 1120 and 1680 kg/ha. Liveweight gain was evaluated by using yearling steers. A curvilinear response curve to phosphorus application was reported with yields reaching 1300 kg/ha at the level of 1680 kg P/ha. Stocking rates at this level of phosphorus application were 5.9 animal units/ha whilst the corresponding figure for the 280 kg/ha treatment was only 2.9.

Results for further years have been given elsewhere (Plucknett 1970) and are summarized in Table 9.58.

The trial was carried out at Kauai Branch Station where the soils are acidic aluminous-ferruginous latosols with high fixation capacities for both phosphorus and potassium and with an annual mean precipitation of 2300 mm.

Table 9.58 Liveweight gains on Pangola grass/*Desmodium intortum* pastures fertilized with different levels of phosphorus (Kauai) (adapted from Plucknett 1970)

Phosphorus treatment (kg/ha)	*Beef liveweight gain* (kg/ha: 4-year average)
280	561
560	637
1 120	674
1 680	650

Plucknett (1970) also gives results of a trial carried out at Kauai to evaluate beef production from lands which had been cleared of bush and seeded to stylo, Desmodium, Siratro and green Panic. Liveweight gain of grade yearling Charolais-Hereford steers was 218 kg/ha/year.

The effects of lime treatment on a Pangola grass/*Desmodium intortum* pasture were studied on the island of Maui by Younge et al. (1964). Pasture productivity was markedly increased by an application of fertilizer at the beginning of the trial and the legume content of the pasture was increased by liming to a rate of about 3 tonnes/ha with a consequent increase in liveweight gain. Results are summarized in Table 9.59. The location was on humic ferruginous latosol formerly cropped with pineapple and sugarcane with an annual average rainfall of approximately 2500 mm.

Table 9.59 Beef liveweight gains on Pangola grass/*Desmodium intortum* pastures with different levels of lime (Maui) (adapted from Plucknett 1970)

Treatment lime (kg/ha)	*Soil pH*	*Pasture legume (%)*	*Liveweight gain (kg/ha/year)*
0 +	5.0	34	623
1 450 +	5.6	27	717
2 900 +	5.5	40	586
4 350 +	5.9	40	762
5 800 +	6.3	36	780
5 800	6.2	33	320

Treatments marked + were fertilized at start with 49 kg N, 94 kg P, 116 kg K, 3.9 kg B and 2.8 kg Mo

The islands lie between 19° and 23° N approximately. This experiment was not replicated and no details of stocking rates are given (although there are indications that these were between 2.5 and 3.5 beasts/ha). It will be noted that there was a greater response to fertilizer than to lime, even at the relatively low level of P used in this trial.

Younge et al. (1964) also gave details of animal performance in an experiment laid down to investigate different grasses in association with *Desmodium intortum*. The authors state that 4 to 8 ha of native pasture were required to support one (454 kg) livestock unit. Although it was admitted that the rates of P and K application were grossly inadequate to maintain maximum productivity, the results are of interest in that they illustrate the superiority of Pangola grass when combined with the legume. Results are summarized in Table 9.60.

Again, no indications as to the stocking rate used are given. It is interesting to note that the performance of fertilized and limed native grasses in this experiment approached that of the introduced species.

Table 9.60 Liveweight gains from various limed and fertilized grasses in association with *Desmodium intortum* (Maui) (adapted from Younge et al. 1964)

Grass	*Liveweight gain (kg/ha; four-year mean)*
Digitaria decumbens	806
Paspalum dilatatum	706
Pennisetum clandestinum	644
Native grasses	588

Puerto Rico

Pueraria phaseoloides (tropical Kudzu), a vigorous climbing legume native to Malaya, was introduced into Puerto Rico in 1940. Rivera Brenes (1947) evaluated the livestock production of the legume in association with Para grass (*Brachiaria mutica*), and from Para grass to Guinea grass (*Panicum maximum*) alone. Lime was used to correct the soil pH to 6.5 but no other fertilizer was applied. Animals used were yearling native heifers which were withdrawn from the experiment when they reached breeding age of about 18 to 19 months. The actual stocking rate used was 2.5 heifer/ha but the theoretical carrying capacity was calculated by a complicated equation which included a measument of the residual forage which was clipped after the animals were removed.

The interpretation of the experiment is difficult because groups of heifers on the various treatments had different starting weights; nevertheless, the animals grazing the grass/legume combination gained significantly more (after covariance analysis) than other groups. However, the results for carrying capacity were less equivocal:

Mixture of Para grass and Kudzu	1.02
Para grass alone	0.41
Guinea grass alone	0.42

Results were expressed as standard cow days/ha/days in grazing season. The author suggested that 'the mixture benefited the nitrogen balance of the soil'. This conclusion was drawn as the Para grass grown in the absence of Kudzu contained 4.9% crude protein whilst that grown with Kudzu contained 9.0%, an interesting comment in view of the subsequent controversy on the nitrogen-fixing ability of tropical legumes.

A further trial was carried out by Rivera Brenes et al. (1949); this evaluated Guinea grass alone fertilized with relatively high (1120 kg 10–10–5 + 224 kg $(NH_4)S_2O_4$ ha) levels of nitrogen whilst Kudzu alone and Kudzu plus Para were fertilized with superphosphate

(560 kg/ha). The plots were evaluated using native, Holstein and cross-bred heifers.

Weight gains and stocking rates (calculated as before) are summarized in Table 9.61. These results are difficult to interpret because no statistical analysis of gain/ha was done and there was a considerable difference in initial weights of the three groups of animals. However, it is clear from the approximate similarity in weight gains of groups from the fertilized grass and grass/legume pastures that the legume was exerting a significant fertilizer-sparing effect. The authors explained the poor results of the Kudzu alone as a result of damage to the plants caused by trampling.

A third trial was carried out to compare fertilized Guinea grass with Kudzu/Guinea grass and Kudzu/Para grass/pasture (Rivera Brenes et al. 1950). The fertilizer application was 112 kg/ha ammonium nitrate at the beginning of the trial and 224 kg/ha three months later. Results are difficult to evaluate because of apparent discrepancies between the text and the tables as to the number of cattle used during the various periods. There were also large differences between the initial weight of animals on the various treatments as in previous trials. However, it seems likely that the grass/legume mixtures were superior to the Guinea grass alone. Carrying capacity was 1.9, 2.7 and 1.7 cows/ha for the Para/Kudzu, Guinea/Kudzu and Guinea grass. Guinea grass pasture was insufficient to maintain high yields.

Despite this apparently promising early work on legumes, emphasis was subsequently placed on the use of highly-fertilized grass pastures. Vicente Chandler et al. (1964) observed that weight gains of unfertilized Kudzu/molasses grass pasture were about 560 kg/ha when grazed by young cattle. This was similar to that produced from an adjacent lightly-fertilized grass pasture (638 kg/ha). Application of higher fertilizer rates to the grass pastures, however, produced more than 1100 kg of gain/ha. Vicente Chandler et al. (1974) raise the following points with regard to the comparison between grass/legume pastures and highly-fertilized grass pastures:

1. It is difficult to maintain the desired proportion between grass and

Table 9.61 Weight gains and stocking rates for fertilized Guinea grass, Kudzu/Para and Kudzu pastures (Puerto Rico) (adapted from Rivera Brenes et al. 1949)

	Weight gain (kg/ha in 273 days)	*Stocking rate (cows/ha; calculated)*
Kudzu/Para grass	549	2.1
Kudzu alone	370	1.3
Fertilized Guinea grass	573	2.9

legume; after several years, one predominates and the pasture must be replanted.
2. The productivity of grass/legume pastures is limited by the ability of the legumes to fix nitrogen; fertilization with nitrogen results in the grass crowding out the legume.
3. Legumes are more exacting than grasses with respect to their nutrient requirements and more care is required with respect to liming and the provision of trace elements.
4. The control of broad-leaved weeds is difficult in grass/legume pastures whereas these may be readily controlled in grass pastures by using the appropriate herbicides.

These authors conclude that productivity of grass pastures exceeds that of grass/legume pastures under conditions where moisture is abundant and intensive management, including heavy fertilization, is possible.

Mexico

Garza et al. (1973a) carried out trials on grass/legume associations at the Livestock Experimental Centre of Paso del Toro in the state of Vera Cruz where the annual mean precipitation is 1208 mm. The treatments used were: *Clitoria ternatea* alone, *Macroptilium atropurpureum* with Guinea grass (*Panicum maximum*) or Pangola grass (*Digitaria decumbens*), or Pangola grass alone. Holstein and Brown Swiss heifers aged thirteen months and weighing 216 mg were used at a stocking rate of 4 beasts/ha. Pastures were rotationally grazed every fourteen days from August to September. Liveweight gain during this period was 180 kg/ha for Clitoria, 148 kg/ha for Siratro/Guinea, 125 kg/ha for Siratro/Pangola and 106 kg/ha for Pangola alone. Notwithstanding this relatively high level of production, the Clitoria was resistant to trampling and recovered rapidly after grazing. The Siratro was adversely affected by grazing because of its high palatability.

Although this trial was carried out on plots of only 1 hectare for each treatment and with few animals, it is a useful indication that the productivity of Pangola grass can be improved by the incorporation of a legume and also confirms the observation made in Brazil that upright grasses combine better with legumes than prostrate ones.

Brazil

The first animal production data obtained from legume-containing pastures appears to have been published by von Schaffhausen (1965) as a possible solution to the problem of seasonality of forage production and quality in tropical areas.

Pigeon pea (*Cajanus cajan*) was seeded in double rows 20–30 m apart on contours in an already established Pangola grass (*Digitaria decumbens*) pasture whilst the same pasture without Pigeon pea served as a control. The trial was carried out during the dry season for approximately ninety-three days and bulls grazing the legume-containing pasture gained 29 kg whilst animals on the control pasture lost 6 kg.

Observations were made on animals given access to an area seeded with *Lablab purpureum* and Pigeon pea over corn stubble after having been on range (no details given). In twenty-two days on range, bulls gained 4 kg (180 g/day) whereas on the legume arable area, gains ranged from 680 to 970 g/day. It was also observed that lactating cows also did well on a pasture which included perennial soya beans (*Neonotania wightii*). It was concluded that 'Legumes give more weight in a shorter time and improve health and fertility', an emphatic endorsement of the advantages of legumes! The English summary of the paper suggested that legumes were already (1965) used on many ranches in the state of Sao Paulo, but this is not reflected by either the main body of the paper or by the summary in Portuguese.

The following paper (Buller et al. 1970) confirmed that the use of legumes in pasture improvement was insignificant in the savannahs and 'campos cerrados' of the state of Sao Paulo. These authors carried out trials at Matao with Pangola grass in association with *Neonotonia wightii* or *Stylosanthes guianensis* or fertilized with 100 kg N/ha. (*Digitaria pentzii* was also included but was extensively damaged by Rhodes grass scale (*Antonina graminis*) and this treatment was abandoned.). Results are summarized in Table 9.62.

Despite the apparently good performance of stylo, however, it had died out almost entirely by the end of the second year of grazing. This was attributed to year-round grazing at stocking rates of 1.6 and 2.3 steers/ha for the winter dry and summer wet seasons respectively. Nevertheless, it was concluded that stylo had several advantages, which included good establishment and production, retention of leaves and growth during the dry season, good acceptability by animals and resistance to disease and insect problems, and further work was suggested.

Legumes did not compare so favourably with nitrogen fertilization

Table 9.62 Liveweight gains from fertilized and unfertilized Pangola grass alone and with legumes (Matao) (adapted from Buller et al. 1970)

	Liveweight gain		
	Winter (140 days)	*Summer (224 days)*	*Total (364 days)*
No nitrogen	35	104	139
100 kg N/ha	65	174	239
with *Neonotonia javanica*	47	161	208
with *Stylosanthes guianensis*	54	205	259

Table 9.63 Liveweight gains from fertilized, unfertilized and legume-reinforced Pangola grass pastures (near Rio de Janeiro) (adapted from Aronovich et al. 1970)

	Winter	*Summer*	*Total*
No N	85	264	349
100 kg N/ha	155	376	531
with Centro	111	299	410

in an experiment carried out at the Institute de Pesquisas e Experimentaçao do Centro Sul near Rio de Janeiro (Aronovich et al. 1970). Beef production from unfertilized Pangola grass was compared with Pangola grass fertilized with 100 kg N/ha/year or in association with *Centrosema pubescens*. Results are summarized in Table 9.63.

The authors attributed the good response to nitrogen to favourable soil moisture conditions (annual average precipitation of 1304 mm) and the relatively poor response to legume to bad nodulation because of inappropriate *Rhizobium* strains.

Zuniga et al. (1975), cited by Vilela et al. (1976), investigated the association of various legumes with Jaragua (*Hyparrhenia rufa*), Gordura (*Melinis minutiflora*) or Pangola (*Digitaria pentzii* – sic.). The legumes used were *Macroptilium atropurpureum*, *Neonotonia wightii*, *Stylosanthes guianensis* and *S. humilis*. The legumes accounted for 13% of the association and dry season weight gains for associations and pure pasture stands were 279 and 138 g/day respectively. Dry season gains were also investigated by Sartini et al. (1974), quoted by Vilela et al. (1976); gains of 30.7 kg/ha from an association of *Melinis minutiflora* and *Centrosema pubescens* were obtained during a period of 196 days.

Sartini and his colleagues (Sartini et al. 1975) evaluated the productivity of *Melinis minutiflora* associated with *Centrosema pubescens* at the Nova Odessa Experimental Station at various stocking rates. The stocking rates were 0.6, 1.2 and 1.8 beasts (steers of the 'Pitangueira' breed about 16 months old)/ha. It was originally intended that this trial would last three years but it was destroyed by fire before completion (the susceptibility of *Melinis minutiflora* to fire was noted). Gains/ha for the three stocking rates are summarized in Table 9.64.

The weight loss in the third period on the low stocking rate was explained by poor acceptability of the available forage. The results clearly indicate the high productive capacity of an association which is said to account for 13.5% of the area of the state of Sao Paulo, a considerable advance on the 'insignificant' use already quoted.

Also working at Nova Odessa, Silveira and his colleagues (Silveira et al. 1975) investigated the carrying capacity of an impressive number of grasses in association with a mixture of legumes. The grasses used are noted in Table 9.65 whilst the legumes were *Stylosanthes guianensis*, *Neonotanio wightii*, *Macroptilium atropurpureum*,

Table 9.64 Weight gains (kg/ha) from *Melinis minutiflora/Centrosema pubescens* pastures at different stocking rates (Nova Odessa) (adapted from Sartini et al. 1975)

	Stocking rate (beasts/ha)		
	0.6	*1.2*	*1.8*
Winter (196 days)	39.6	58.0	57.3
Summer (216 days)	98.4	167.9	257.3
Winter (168 days)	–1.6	16.4	15.0

Centrosema pubescens and *Galactia striata*. Stocking rate was varied after observation of pasture condition and expressed as animal days/month from which the carrying capacity was calculated for summer, autumn

Table 9.65 Stocking rates of grass/legume mixtures during summer, winter and year-long (Nova Odessa) (adapted from Silveira et al. 1975)

	Stocking rate (beasts/ha)		
	Summer	*Winter*	*Year-long*
Pennisetum purpureum cultivars			
Taiwan A-143	3.0	1.7	2.5
Taiwan A-144	3.9	2.0	3.2
Taiwan A-148	2.9	1.6	2.4
Taiwan A-241	2.6	1.3	2.2
Mineiro	3.6	1.8	2.9
Cameron	2.9	1.8	2.5
Vrukwona	3.0	1.3	2.4
Digitaria decumbens	3.3	1.6	2.7
D. sp. 'slender'	3.5	1.6	2.8
D. sp. Florida X 1594	2.0	0.6	1.5
D. valida	1.9	0.7	1.5
Cynodon dactylon cv. Coast cross	3.9	1.8	3.1
C. dactylon cv. Estrela (Star)	3.4	2.0	2.9
Brachiaria decumbens	2.5	1.6	2.2
B. sp.	2.6	0.8	2.0
Panicum maximum cv. Gatton	2.1	0.9	1.7
Setaria anceps cv. Nandi	2.3	1.3	2.0
Melinis minutiflora	1.8	1.1	1.6
Hyparrhenia rufa	3.1	1.2	2.4
Paspalum guencarum cv. Ramirez	2.8	1.8	2.4
Panicum maximum cv. trichoglume	2.8	1.1	2.2
Chloris gayana cv. Callide	2.9	1.2	2.3
Paspalum plicatulum	3.0	1.2	2.4
Setaria anceps cv. Kazungula	3.0	1.4	2.4

and winter as well as year-round. Apologies are due for this catalogue of stocking rates but few comparative figures are available elsewhere.

Although no data for liveweight gain/ha are given for individual cultivars, average values were for the *Pennisetum purpureum* varieties (449 kg/ha/year), the prostrate grasses (364 kg/ha/year) and the upright grasses (396 kg/ha/year). Two applications of fertilizer were applied during the experiment totalling 200 kg P_2O_5 and 120 kg K_2O/ha. Although these liveweight gains are not comparable with some of those obtained in parts of Australia and elsewhere, they nevertheless demonstrate that high yields of beef may be obtained from grass/legume associations in the humid tropics of Sao Paulo. It was observed that Elephant grass varieties and grasses with a tussock growth habit formed better associations with legumes than the grasses with a prostrate growth habit and that *Centrosema pubescens* and *Macroptilium atropurpureum* had the highest populations followed by *Galactia striata*. *Antonina graminis* is again mentioned, this time as a pest of Pangola grass. The presence of Stunt virus disease of Pangola is also remarked upon.

Experiments were carried out in the state of Minas Gerais by Vilela and his colleagues (Vilela et al. 1977a). Dry season production of grass and grass/legume pastures was studied in the municipality of Luz where the annual rainfall during the year of the experiment was 1063 mm distributed over eight months. Pastures were composed of *Panicum maximum* alone or associated with equal quantities of *Macroptilium atropurpureum*, *Neonotonia wightii* and *Stylosanthes* (no species given). Dolomitic lime (2500 kg/ha) was applied before planting and 110 kg of P_2O_5 with micronutrients after planting. The animals used were Gir-type Zebu steers of about 2.5 years old. The gain from Guinea grass alone was 11 kg/ha at a stocking rate of 1.21 animal units/ha whilst that from the legume association was 38 kg/ha at a stocking rate of 1.30 animals units/ha over the eight-four days of the experiment. The legume content of the pasture varied between 13% (September) and 18% (July). These results were also published elsewhere (Vilela et al. 1977b).

Vilela et al. (1977a) carried out a similar experiment at Fernandes Tourinho. *Panicum maximum* (Coloniao) seeded with *Macroptilium atropurpureum* and *Neonotonia wightii* was compared with a native pasture mainly comprising Coloniao, *Brachiaria mutica* and *Paspalum notatum* (but also containing some 5–6% of the native legume *Calapogonium muconoides*). Cattle used were Nelore-type steers about two years old, weighing approximately 310 kg. The results obtained during the 243 days of the experiment are summarized in Table 9.66. The improved pasture contained about 20% legume and the advantages that this conferred were highly significant.

A further trial was carried out at Iguatama (rainfall 952 mm) comparing native pasture (principally *Melinis minutiflora*, *Hyparrhenia rufa*

Table 9.66 Stocking rates and production from native and improved pastures (Fernandes Tourinho) (adapted from Vilela et al. 1977a)

	Stocking rate (beast/ha)		*Liveweight gain (kg/ha)*		
	Wet season	*Dry season*	*Wet season*	*Dry season*	*Overall*
Improved pasture	2.3	1.7	190	70	260
Native pasture	1.3	0.9	101	9	110

and *Aristida palens* with small amounts of native *Desmodium* and *Stylosanthes* spp.). Pasture improvement was carried out by mechanical bush clearing, fertilization with 80 kg P_2O_5 and interseeding with *Stylosanthes guianensis* and *Macroptilium atropurpureum*; the average occurrence of the former in the pasture was 27.5% and of the latter 5.5%. Liveweight gain from the improved pasture was 240 kg/ha/year and from unimproved pasture 64 kg/ha/year from cross-bred Friesian (Hollands) × Zebu steers (Vilela et al. 1978).

An experiment to compare the effects of different stocking rates and levels of phosphorus fertilization was carried out at Brasilia (CIAT 1978). Pastures of *Brachiaria ruziziensis* and *Stylosanthes guianensis* cv. Endeavour were fertilized with 120 or 240 kg P_2O_5/ha. Animal gains were inversely related to stocking rate and performance was not affected by phosphorus application. Daily gains were 878, 923, 1182 and 929 g/ha for stocking rates of 1.5, 1.9, 2.3 and 2.7 steers/ha respectively. The two intermediate stocking rates were superior to both high and low rates in terms of total weight gain during the experiment which were 209, 219, 265 and 208 kg for the various stocking rates in the same order as given above.

The legume in this trial proved to be susceptible to anthracnose, however, and it is emphasized that such cultivars should be avoided.

After early trials with crop legumes in association with Pangola grass, it has been shown that tropical forage legumes effectively increase liveweight gain/hectare in association with grasses, particularly those with an upright growth habit. Emphasis has been given to improvement of liveweight gain during the dry season in many of the studies carried out in Brazil whilst poor persistence has been noted as a problem in connection with some of the pasture legumes. It is interesting that native legumes have been mentioned, but their insignificant role has been emphasized. The importance of various pests and diseases of improved pastures was also noted.

Peru

Santhirasegaram (1974) has reviewed work with forage legumes carried out at Pucallpa. It was shown that the productivity of *Pueraria phaseoloides/Hyparrhenia rufa* pastures could be increased from 120 kg liveweight gain/year to 159 kg by the application of 100 kg superphosphate, but that supplementation of cattle with minerals could increase this to 350 kg. It was estimated that the application of 200 kg of superphosphate together with mineral supplementation could produce more than 400 kg. The animals used were Nellore heifers at a stocking rate of 2/ha. It was also shown that the fertility of breeding cows was higher when they were grazed on legume-based pastures than when grass-only pastures were used.

Discussion

The various advantages which may be attributed to legumes when incorporated into a grass sward are discussed in this section, together with some aspects of management which are of particular importance insofar as livestock production is concerned.

Beneficial effects of legumes

In most of the examples quoted the incorporation of legumes into a grass sward has had a beneficial effect. This has either been manifested as increased gain/head or increased stocking rate or both. In any event, the most valid way of comparing the results is as total liveweight gain/hectare (as individual gains and stocking rates are related). Examples of increased liveweight gain/hectare are given in Table 9.67. Although values are subject to wide variation, the average increase in liveweight gain/hectare attributable to legume incorporation is more than 120 kg (121.5 ± 90.72 kg) with a range of from 23 to 396 kg. Clearly, the extent of production from grass alone is influenced by the species (native or improved), rainfall and soil fertility, but it seems that, with very few exceptions, legume incorporation improves productivity.

In the relatively small number of cases where no improvement have been demonstrated, this may usually be attributed to poor establishment, e.g. Yates et al. (1964), Coaldrake and Smith (1967), to inappropriate management, e.g. Addison (1970), or to some other unexplained factor, perhaps a deficiency of some specific element, e.g. Smith (1962b).

Other beneficial effects have included extension of the grazing season (Norman 1965, 1970; Wesley-Smith 1972; Winks and Lamberth 1968; von Schaffhausen 1965); and earlier maturity or readiness for slaughter for males (Shaw 1961; Norman 1968; Cohen and O'Brien 1974).

In addition to advantages for the growing animal, improved reproduction and calf production have been mentioned by Holroyd et al. (1977), Edye et al. (1971, 1972), 't Mannetje and Coates (1976), Sil-

Table 9.67 Some examples of legume incorporation improving liveweight gains per hectare

Country	*Beef production* (kg/ha)		*Reference*
	Grass/alone	*Grass + legume*	
Australia	18	304	Norman (1963)
	7	126	Woods (1970)
	17	92	Shaw (1961)
	25	93	Shaw and 't Mannetje (1970)
	31	86	Toom (1971)
	11	40	Norman (1974)
	283	679	Young and Daly (1967)
	45	128	Addison (1970)
	131	219	Lowe et al. (1977)
	33	155	't Mannetje and Nicholls (1974)
Thailand	34	122	Falvey and Mikled (1978)
The Philippines	28	125	Siota et al. (1977)
	22	117	Siota et al. (1978)
Papua and New Guinea	78	106	Chadhokar (1977)
Fiji	143	352	Partridge (1979)
Nigeria	18	150	de Leeuw (1969–71)
	82	119	Haggar (1971)
Uganda	210	233	Stobbs and Joblin (1966a)
	706	833	Stobbs and Joblin (1966b)
	655	975	Stobbs (1969a)
Zaire	10	76	Risopoulus (1966)
Madagascar	–19	115	Delhaye (1967)
Swaziland	115	176	Jones (1973)
United States of America	79	277	Hodges et al. (1953)
Mexico	106	180	Garza et al. (1973)
Brazil	139	259	Buller et al. (1970)
	110	260	Vilela et al. (1977)
Mean	116.8	263.9	
SD	±175.87	±230.10	
	Difference = 121.5 ± 90.72 with a range of 23–396		

very et al. (1978), Cohen and O'Brien (1974), Evans and Biggs (1979) and workers in Papua New Guinea (1965).

Thus, earlier reservations on the effectiveness of pasture legumes to improve livestock production in the tropics and subtropics were unfounded.

Various authors have related the beneficial effects of legumes to its occurrence in the sward (Norman and Stewart 1964, 1967; Norman 1968, 1970; Evans 1970; Partridge 1979; Clatworthy 1974–5; Clatworthy and Holland 1977, 1978; Clatworthy and Boultwood, 1977–8; Shaw 1978a), the general conclusion being that higher levels of legume in the pasture are associated with better livestock production. However, 't Mannetje (1972) suggested that this was true until the legume (in this case Siratro) reached 1000 kg/ha but that liveweight gain was not increased above this level. Eng et al. (1978b) could demonstrate no relation between legume content and rate of gain; it was suggested that this was because the more or less even growth habit of the grass under the conditions of the experiment ensured that protein would not become deficient nor would intake be limited by low protein levels. Inconclusive results were reported from Hawaii (Plucknett 1970) but this was probably because legume levels were relatively high (ranging from 27 to 40%) ensuring that protein was not limiting. In fact, Bewg et al. (1970) suggest that energy could become limiting at higher legume levels.

In view of the above, it is logical that improved animal performance would be related to increased time on legume pastures and this has been reported by Norman (1968), de Leeuw (1971), Haggar et al. (1971) and Norman and Stewart (1967).

Effects of phosphorus fertilization of grass/legume pastures

'Deficiency of phosphorus, more than any other element, limits the realization of the production potential of the tropical legume' (Skerman 1977). It is not surprising, therefore, that several authors have demonstrated a response to fertilization with this element both in growth rate/head and/hectare (Shaw and 't Mannetje 1970; Shaw 1978a; Winks and Lamberth 1968; Winter et al. 1977b; Bowen and Rickert 1979; Evans and Bryan 1973; Eng et al. 1978b; Stobbs 1969a and h; Plucknett 1970; Santhirasegaram 1974.) Similarly, positive effects of phosphorus have been demonstrated in the breeding cow in terms of improved conception rate (Edye et al. 1971, Santhirasegaram 1974) and weight at weaning (Edye et al. 1972).

Shaw (1978b) has shown that the amount of legume in an association is increased with increasing levels of phosphorus and have suggested that improved performance is a result of improved quality as well as increased quantity of feed. Similarly, Evans and Bryan (1973) have suggested that dry matter intake is increased at higher levels of phosphorus fertilization because of increased legume content and more protein in the forage.

Plucknett (1970) demonstrated responses to levels of phosphorus as high as 1120 kg/ha but Koger et al. (1970) and Robinson and Sing (1975) have pointed out that such levels may not be economically justified because of reduced yield per unit of fertilizer applied as well as high costs of transportation etc.

Winks (1973) also pointed out that with an increase in soil fertility attributable both to phosphorus fertilization and to nitrogen fixation, invasion of annual grasses could become a problem and suggested that less frequent fertilization might be advisable.

However, whilst economic and management limitations must always be borne in mind, there is little doubt that phosphorus fertilization plays an important role in the maintenance of the productivity of grass/legume pastures.

Effects of different stocking rates on grass/legume pastures

General effects of different stocking rates on productivity will be discussed in a later chapter and only aspects specifically related to grass/legume associations will be dealt with here.

Vicente Chandler et al. (1974) have referred to, *inter alia*, the difficulty of maintaining the desired proportion between grass and legume in an association, and several of the works which have been quoted in this chapter have shown that this balance is closely related to stocking rate.

Observation, and a limited number of studies with an oesophageal fistula (Stobbs 1969d), indicate that legumes are less palatable than grass forages. Using range management terminology, we may refer to them as 'increasers' i.e. they are not so sought-after by the grazing animal and prosper under moderate to heavy grazing. An increase in legume component of the sward under these conditions has been noted by Stobbs (1969d) and Edye et al. (1971).

At a higher stocking rate, the more palatable 'decreasers' having been reduced by severe grazing, the legume is more heavily grazed and also declines (Bisset and Marlow 1974; Evans and Bryan 1973; Stobbs 1969d), to be replaced by the unpalatable annual grass 'invader' species such as *Sporobolus pyramidalis* (Stobbs 1969d), etc.

Good grazing management therefore involves a delicate balance between enough grazing pressure to moderate grass competition but not so much that the legume disappears.

Performance of grass/legume pastures v. N-fertilized grass pastures

Henzell (1977) has reviewed the total quantity of nitrogen in soil profiles under unimproved tropical grasslands and has estimated that nitrogen available for plant growth by mineralization amounts to between 10 and 100 kg/ha/year. To this may be added some 5 kg from dust and rain, variable amounts from the dung and urine of grazing animals, and nitrogen fixation by indigenous legumes. A further important potential source of nitrogen is fixation by *Azotobacter* and *Beijerinckia* spp. which have been shown to be associated with tropical grasses (recent results have indicated that fixation from

Rhyncheletrum repens, a common colonizer of arable lands, may amount to more than 120 kg/ha/year; Weir et al. 1979).

Thus native soil nitrogen is only sufficient to support extensive exploitation, and the supply of additional nitrogen to the plant/animal system either from fixation by legumes or by fertilization is the main factor which affects livestock production. It has been shown that tropical grasses respond to both legume incorporation and nitrogenous fertilization by increased animal productivity as a result of increased quantity and improved quality of the herbage on offer.

Most of the trials which have been carried out to compare the productivity of grass/legume pastures with grass pastures to which nitrogen has been applied indicate that the nitrogen fixation of the legume is equivalent to a moderate level of nitrogen application of between 100 and 170 kg N/ha/year (Addison 1970; 't Mannetje 1972; 't Mannetje and Nicholls 1974; Grof and Harding 1970; Stobbs 1969h; Hodges et al. 1977a; Vicente Chandler et al. 1974; Buller et al. 1970; Magadan et al. 1974) although Haggar (1971) and Smith (1962b) only attributed benefits to legume incorporation of the equivalent of some 15 kg N/ha/year. However, the benefits of higher levels of fertilization exceeded those of legumes (Jones 1974; Haggar 1971; Vicente Chandler et al. 1974), particularly where moisture is abundant and intensive management is practised. This was particularly evident in terms of greater carrying capacity and, similarly, benefits were more in evidence at higher stocking rates (Haggar 1971).

This may be offset to a certain extent by lower individual rates of gain/head and a consequent slower turn-off rate so that the animal is a longer time on the inventory (Jones 1974).

It is clear then that incorporation of legumes into a grazing system increases productivity but not as much as fertilization with nitrogen 'from the bag'. The decision as to which system to follow must depend on an economic assessment of all the factors involved. In addition to establishment and fertilization costs, these include land price and availability as well as the opportunity benefits of other enterprises. Similarly, the disadvantages of lower turn-off rate might be overcome by the exploitation of breeding stock (Evans and Biggs 1979).

Different grazing systems

Little critical work has been carried out on the use of different grazing systems for grass/legume pastures and the results that have been obtained have not been consistent. Evans et al. (1979) could not demonstrate any difference in liveweight gains between a system of set-stocking and two paddock systems grazed during both wet and dry seasons or during the wet season only.

Similarly, Bowen and Rickert (1979) could demonstrate little difference in terms of liveweight gain/head between set and put-and-take stocking, major differences being attributable to changes in stocking rate. In a subsequent experiment, high levels of the legume were maintained with constant stocking, and protection caused a decrease in legume content. Jones (1979) suggested that long periods of deferment would only benefit the legume component of the sward when high stocking rates had caused a significant reduction in soil seed level.

It would appear that the dominant factor in the persistence of the legume in a grass/legume pasture is the overall stocking rate and, provided this is carefully adjusted, different methods of rotation or deferment do not alter animal production or sward composition.

Summary

There is no doubt that the successful incorporation of a legume into a grass sward has a beneficial effect on livestock production in terms of either increased gain/head or stocking rate, or both. This is often accompanied by extension of the grazing season and earlier readiness for slaughter. Increased breeding efficiency may also accrue.

These improvements in livestock production normally increase with the quantity of legume in the sward until a level of approximately 1000 kg/ha is reached. This may not be the case, however, when the protein content of the grass alone does not limit growth. Similarly, increased production is related to length of time grazing legume pastures when these are used as a supplement.

Phosphorus is generally the first element to become limiting to legume growth and its application results in increased quality as well as quantity of feed and consequent improved livestock performance. The effectiveness of response to phosphorus supplementation, however, must be judged in terms of prevailing economic conditions.

Maintenance of the correct stocking rate of grass/legume pastures requires considerable judgement, as legumes usually behave as 'increasers' under such conditions. Continual heavy grazing, however, may well result in a decline of the legume accompanied by an increase in undesirable grass species. Stocking rate is more important the type of rotation or deferment system followed.

Legume incorporation normally gives benefits equivalent to rather more than 100 kg N/ha 'from the bag'. These benefits are particularly evident at high stocking rates and may be offset by the greater interest charges of animals on inventory. Higher rates of gain can only be achieved by increased nitrogen fertilization of grass pastures and all relevant factors must be taken into account when deciding on the particular grazing system to be used.

10 *Improvement of beef production through pasture fertilization*

Pastures produce an annual crop of herbage which is normally removed by the grazing animal for the production of meat, milk and hides. This annual removal of plant nutrients must be restored in some way if productivity is to be maintained. Release (and return) of nutrients by natural processes is not adequate for sustained high productivity and fertilization of grasslands has been practised for many years in various parts of the subtropics such as South Africa and the southern part of the United States. More recently, fertilizers have been extensively employed in more tropical environments, notably the Caribbean and Brazil.

The nutrient normally most frequently limiting grassland productivity is nitrogen (N). This is supplied to the soil naturally by symbiotic and non-symbiotic fixation, in rain and dust, and to the plant by mineralization of soil organic nitrogen. The extent of these various contributions has been discussed by Henzell (1977) and Sanchez (1976), and the former has stated that 'the quantity of nitrogen available for plant growth is usually less than 100 kg/ha/year and frequently only 10 to 20 kg under established undisturbed grassland'. Sanchez (1976) has indicated the extent of nutrient removal by cultivated grasses and representative values are given in Table 10.1.

The values for nitrogen removal (not taking into account leaching or other losses) would correspond to a crude protein content of the grasses concerned of between 6.7 and 11.4% (mean 8.31 ± 1.58) which is higher than would be generally expected from unimproved grasses. Thus the production of grass that would be expected under natural conditions, assuming a crude protein content of (say) 5% of dry matter, would amount to between 1.25 and 12.5 tonnes from soils with 10 and 100 kg of available N respectively. It is clear that only rarely under natural conditions can high yields be achieved without additional inputs of nitrogen from one source or another.

The phosphorus status of a soil depends less on the total phosphorus present than on the forms in which it occurs. The mineralization of organic phosphorus is the most important factor in situations where no fertilizer is applied, whilst inorganic phosphorus may be present in active or inactive forms. Fixation, associated with high contents of iron and aluminium 'oxides' and low pH values, reduces the availability of fertilizer phosphorus. Many improved grass species are tolerant of low levels of available phosphorus. Typical extraction levels are given in Table 10.1. Generally, phosphorus applications have been useful when high extraction levels have been promoted by high forage production due to application of N. Similarly, replacement of the very large quantities of potassium(K) removed is also necessary at these high levels of forage production. Micronutrients may be required under specific deficiency conditions which are identified by soil and plant analysis and subsequent practical confirmation. It must also be borne in mind that multiple deficiencies of nutrients are likely to be found in pastures and that, under such circumstances, the provision of one element alone will not increase production.

Table 10.1 Nutrient removal by some tropical grasses (adapted from Sanchez 1976)

	Yield (tons/ha)	*Nutrient (kg/ha)*		
		N	*P*	*K*
Panicum maximum	10	107	27	180
	23	288	44	363
	35	560	77	600
Digitaria decumbens	10	120	22	180
	23	299	47	358
	31	400	53	558
Pennisetum purpureum	10	144	24	180
	25	302	64	504
	46	800	92	900
Brachiaria mutica	8	80	17	160
	24	307	43	383
	33	600	69	660

This chapter will review results of trials which have been carried out with grazing beef cattle. Many other trials have been done in pots and on plots but the results may not necessarily be confirmed by response in the grazing animal.

Similarly, improved pastures have been dealt with separately from natural pastures because of the different levels of production expected as well as the different capital and maintenance costs involved. Fertilization of grass/legume pastures has been reviewed in some detail in the previous chapter.

Fertilization of natural pastures

Although fertilization of natural pastures was carried out in the southern United States from a relatively early date, most early work was carried out in South Africa. Hall and his colleagues from the South African Explosive and Chemical Industries laid down a wide-reaching series of trials to measure productivity. Experiments carried out elsewhere have given priority to the fertilization of introduced pastures and studies on natural veld have been occasional and unsystematic.

Africa

The first trials to investigate the use of fertilizers to improve the productive capacity of natural veld were started more than fifty years ago by Staples and Taylor at Cedara in Natal following the successful experiences of Falks and Warmbold in developing the 'Hohenheim' system in Germany (Hall et al. 1955). However, it was the Agricultural Advisory Department of African Explosives and Chemical Industries Ltd, led by Dr T. D. Hall, which was responsible for laying down a comprehensive series of twelve co-operative experiments 'well distributed over the main pastoral areas of the Union', including the Orange Free State, Cape Province, Transvaal and Natal. This programme expanded until, eventually, there were twenty-four experiments on 280 ha in South Africa and twenty-one on 200 ha in Zimbabwe (then Rhodesia), as well as one experiment in Kenya. Farmers were given fencing materials, gates, etc. to enable the veld to be fenced off into one or two morgen camps (1 morgen = 0.86 ha) and fertilization was supervised. Thereafter, periodical visits were made, whilst farmers recorded number of animals and grazing days. The improvement to carrying capacity in some cases was dramatic (up to 300%) although responses varied markedly according to year and season as well as location. The greatest benefit was obtained from the application of nitrogen which gave 'the camps a darker green colour, more leafy growth and a greater bulk of herbage'. The application of phosphate did not increase the carrying capacity but apparently made the grass more palatable. Fertiliser treatments comprised P, PN, PNK, Ca, PCa, PNCa, and PNKCa where P was 210 kg/rock phosphate/ha, N was 105 kg sulphate of ammonia/ha, K was 40 kg potassium chloride/ha and Ca was 1060 kg calcium carbonate/ha (Hall 1932). These experiments have been summarized by Hall et al. (1955) and further details are to be found in the *South African Journal of Science* between the years 1931 and 1941. They have not been dealt with in more detail here as data are in most cases limited to grazing days and little information as to beef production is given. However, it was noted that in most cases, the chemical composition of the grass was improved by up to 20% in the case of crude protein, and 15 to 58% in the case of phosphorus(P), the latter being the most marked and consistent. It was also noted that the palatability of the fertilized grass was improved, as was its ability to recover after drought. Similarly, fertilizers not only reduced the percentage of bare ground but also the number of undesirable grasses. Hall and his colleagues (Hall et al. 1950) subsequently laid down a more permanent experiment from which more detailed information could be obtained. This was done at 'Frankenwald', the Botanical Research Station of the University of Witwatersrand, between Johannesburg and Pretoria in the Bankenveld (Acocks 1975). The experiment used N (at two levels), P and K and was established on poorly-covered (18.5%) veld of which the more important grasses were *Elyonurus argenteus, Eragrostis racemosa, Digitaria tricholaenoides* and *Heteropogon contortus*, none of which are particularly noteworthy grazing species. The experiment continued from 1933 until 1949 (although there was a four-year period when no cattle were used and only hay was harvested due to wartime conditions). Four camps were fenced for each fertilizer treatment and these were grazed rotationally by steers during the summer grazing season and were then sold, enabling economic evaluation of the results. Fertilizer levels, stocking rates and weight gains are given in Table 10.2.

The greatest response was to the two levels of N, which produced 2.54 and 2.27 kg extra liveweight gain for each kg of N applied respectively. A botanical analysis of the veld in 1948–9 showed that the total basal cover had increased in all treatments (that in the control was attributed to the improved management), but particularly in the high-nitrogen treatment (to 37%), and that there has been a considerable increase in the more palatable *Eragrostis* spp. (*E. curvula, E. planiculmis* and *E. chloromelas*) and *Cynodon dactylon*, which amounted to as much as 17.7% in the high-nitrogen treatment. The N_3PK treatment was also replicated on abandoned cultivated land ('old fallows') of which the original botanical composition was only

Table 10.2 Fertilizer levels, stocking rates and liveweight gains from fertilized pastures (Frankenwald) (adapted from Hall et al. 1950)

Treatment	*Fertilizer level (kg/ha)*			*Stocking rate (head/ha)*[1]	*Liveweight gain (kg/ha)*
	N	P_2O_5	K_2O		
O	—	—	—	0.87	65.7
P	—	34.7	—	0.87	69.8
NP	26.8	34.7	—	1.16	133.7
NPK	26.5	33.8	16.4	1.16	130.6
N_3PK	70.6	33.8	16.4	1.74	225.7

[1] Some adjustments to stocking rate occurred from year to year

9.7% total basal cover of which 3.3% was *Cynodon dactylon*. The liveweight gain from this treatment was 218 kg and, after only one season, basal cover had increased to 25.3% (although presumably part of this was due to natural colonization).

The authors note that that 'nitrogen has given (economic) returns despite very high prices for sulphate of ammonia (£25/ton)' and conclude that with lower fertilizer prices and higher beef prices 'there is a promising future for the use of nitrogen on veld in livestock maintenance and production', a statement which sounds most familiar to the present-day ear!

Hall et al. (1955) have also mentioned an experiment carried out at Athole (now part of the Nooitgedacht Station) near Ermelo in the eastern Transvaal close to the Swaziland escarpment where the veld type is North-Eastern Sandy Highveld (not unlike the Bankenveld) and the predominant species are *Tristachya hispida*, *Trachypogon spicatus* and *Heteropogon contortus* (Acocks 1975). Experiments started in the 1936–7 season compared rotational and continuous grazing of fertilized and unfertilized pastures. Fertilizer treatments comprised 211, 254, and 42 kg/ha/year of superphosphate, ammonium sulphate and potassium chloride during the first four years followed by 106 kg/ha/year of both superphosphate and ammonium sulphate, whilst the potash treatment was discontinued. A brief summary of the results obtained are given in Table 10.3.

Table 10.3 Liveweight production from rotationally or continually-grazed fertilized or non-fertilized pastures (Athole) (adapted from Hall et al. 1955)

Treatment	*Continuous grazing (kg/ha)*	*Rotational grazing (kg/ha)*	*Mean*
No fertilizer	93.6	139.5	116.6
+ NPK	146.9	187.9	167.4
Mean	120.3	163.7	—

Thus the incremental liveweight gain attributable (mainly) to nitrogen application was about 50 kg or approximately 1.6 kg/kg of nitrogen applied (assuming that the average application was 151 kg $(NH_4)_2SO_4$/year).

Edwards (1974), in an attempt to estimate the carrying capacity of veld from herbage yields and grazing days, set up a complicated (3 × 3 factorial) experiment with three levels of nitrogenous fertilizer (0, 40 and 80 kg N/ha), three levels of other fertilizers (0, 21 kg P/ha, and 21 kgP/ha plus 53 kg K/ha) and three stocking rates (1.92, 1.28 and 0.64 ha/mature livestock unit). The experiment was carried out at Zwartrand Research Station in the western variant of the Bankenveld (Acocks 1975). The liveweight increases for the three nitrogen treatments were 97.1, 147.4 and 162.2 kg/ha respectively. There was also a significant response to phosphatic fertilizer but not to the potassium treatment. The rates of liveweight gain were 1.26 and 0.81 kg/kg N applied (Edwards and Nel 1973b). The authors also made some interesting botanical observations over the four years of the trial in that the 'decreasers' tended to decrease more rapidly at the higher levels of N than the 'increasers' increased with a net drop in total cover (from 20.5 at 0N to 19.7% at 80N). Increasing species were *Setaria flabellata, Erogrostis gummiflua, Tragus berteronianus* and *Eragrostis curvula* whilst decreasers were *Themeda triandra, Andropogon amplectens, Tristachya rehmannii, Cymbopogon excavatus* and *Aristida congesta*. Botanical responses to fertilization with P and K were minor (Edwards and Nel 1973a).

The work by Theron et al. (1974) on radical veld improvement has already been mentioned in connection with reinforcement with legumes, but the same study compared fertilized with unfertilized veld. The control treatment was unfertilized veld grazed at a stocking rate of 1 mature livestock unit/1.52 ha, whilst the treatment was fertilized with 92 kg N, 25 kg P and 100 kg K/ha each season with 180 kg Ca being applied every fourth season, and stocked at a rate of 1 mature livestock unit/0.6 ha. The total liveweight gain was 154 kg/ha for the control as opposed to 407 kg/ha for the fertilized area. The study was carried out on Highland Sourveld at Ntabamhlope where the predominant grass species are *Themeda triandra, Tristachya hispida* and *Trachypogon spicatus* (Acocks 1975) and not dissimilar from the Bankenveld. The increment in liveweight gain attributable to the application of (mainly) nitrogen is most impressive in this case, amounting as it does to some 2.75 kg liveweight gain/kg N applied. However, despite the increase in productivity, there was little difference in return/Rand invested/ha, which was 0.38 for the control and only 0.40 for the fertilized veld. Returns in terms of cattle were even less, being 0.38 and 0.39 respectively (the returns from the reinforced veld were 0.59 in both cases).

Fertilization is only one part of the process of radical veld improvement; it has been more fully described by Theron and Harwin (1976) as a process of fertilizing followed by oversowing with improved grasses when the veld is in a suitably receptive condition (the fertilizer has acted 'as a tool with which to open up the sward thus making it more receptive to the oversown grass'). It is also recommended that the process should be accompanied by legume sod-seeding wherever possible. The authors give some distinctive features of pastures established in this way and quote examples of successful adoption by farmers. In one of these, there was 167% increase in stocking rate and a 33% increase in duration of grazing at a cost of R0.11/grazing day. The authors conclude that such veld improvement can bring about a highly significant improvement in weight of weaned calf/unit area due to both increased carrying capacity and

U.C.W. ABERYSTWYTH LIBRARY

daily gain of the calves. This is coupled with similar increases in marginal returns/hectare.

Such a co-ordinated approach to veld improvement, which makes the best use of all the various techniques to hand such as fertilization, interseeding, legume introduction as well as a system of camps which enables rotational grazing to be carried out, is obviously a formula for the intensification of veld use without going so far as to change the land-use pattern by cultivation and subsequent conventional establishment of pastures.

Grunow et al. (1970) have reviewed much of the early work on nitrogen fertilization in South Africa and made several interesting observations. It was generally shown (Pietermaritzburg, Pretoria, Warmbaths, Athole, Pretoria, Glen, Johannesburg) that the crude protein percentage and yield increased with increasing levels of nitrogen to an apparent maximum of between 300 and 450 kg/ha N, but it was also pointed out that changes of species composition were also important in this respect. It was also concluded that overfertilization in years with below average rainfall led to inefficient use of nitrogen and therefore that the overall water economy of the site should be taken into account when considering fertilizer application and when evaluating the response. In this connection, it has also been noted that high levels of nitrogen application in areas with rainfall of less than about 560 mm may cause sward deterioration in that the typical cover changes from small-tufted grasses to large-tufted species with a consequently higher susceptibility to erosion (this is an example of the advantages of long-term experimentation as carried out in South Africa where the effects of succession may be observed) and possible weed encroachment.

Grunow et al. (1970) compiled a list of 'increasers' and 'decreasers' in veld subject to nitrogen fertilization in various localities. *Themeda triandra* and *Heteropogon contortus* were decreasers in all eight sites. whilst *Elyonurus argenteus* was mentioned in six and *Tristachya hispida* in four. Common increasers included *Eragrostis curvula*, *E. chloromelas* and *Cynodon dactylon*.

Compared with the large volume of work carried out in South Africa over a period of more than forty years, little information on animal production data from fertilized natural pastures is available for other parts of Africa (or indeed elsewhere).

After Hall carried out one of his early experiments in Rhodesia, West (1956) reported the effects on natural veld fertilized with different levels of fertilizer at Grasslands Research Station. The predominantly *Hyparrhenia* spp. veld was dressed with 0, 70, 140, 210 kg/ha nitrogen from Nitro lime, together with 170 kg rock phosphate plus 56 kg potassium chloride/ha. Liveweight gains for the four treatments in the order given above were 103, 156, 180 and 190 kg/ha. The author concludes that although 'a great increase in the productivity of cleared veld can be brought about by fertilizing, these results fall far short of the improvement that can be effected by replacing the veld grassland by fertilized pastures of selected cultivated grasses'. This was the philosophy which was to direct future research on the effectiveness of fertilization in Zimbabwe.

Walker (1969b), working at Ukiriguru in Lake Province, Tanzania, applied different levels of sulphate of ammonia to natural pastures largely composed of *Bothriochloa insculpta*, *Panicum infestum*, *Hyparrhenia* spp., *Cynodon dactylon* and *Sporobolus pyramidalis*. Levels of sulphate of ammonia used were equivalent to 0, 80 and 160 kg N/ha approximately. Results are difficult to interpret as they are given graphically, but the authors state that animals grazing non-fertilized veld gained only 14.6 kg during the period December 1966–November 1967 whilst the animals on the low level of nitrogen gained 25 kg and those on the higher level, 34 kg. The stocking rate used was about 2 Indo-African Zebu steers/ha and thus the performance can hardly be considered outstanding. However, the effects of nitrogen application on growth rate were clearly positive whilst the high level of nitrogen exerted an overall positive effect on the protein level of the grasses (8.6 v. 7.5% for the control v. unfertilized pasture). It was concluded that although the responses to nitrogen fertilizer were positive, its use was unlikely to be economic given the low productivity and cash value of the local cattle.

In West Africa, Adegbola et al. (1968) investigated the effect of the application of nitrogen on natural *Andropogon gayanus* pastures with Ndama and Keteku steers at Fashola, Nigeria. Steers grazing pastures fertilized with 110 kg N gained 245 kg during the period June–December (0.76 kg/day) whilst animals grazing unfertilized pastures only gained 110 kg (0.38 kg/day) during the same period. At the same time there was sufficient production from the fertilized area to make grass silage which could be used during the dry season. The increase in liveweight amounts to 1.23 kg/ha for every kg of nitrogen applied.

The Americas and the Caribbean

The effects of fertilization of flatwoods pastures was investigated at Ona in Florida from 1943 (Hodges et al. 1949). Native pasture (no further description given) fertilized with 2245 kg/ha rock phosphate produced 20.8 kg/ha whilst the equivalent non-fertilized area produced 18.5 kg/ha. Apart from this experiment, pasture research in Florida has been largely concerned with the fertilization of improved grasses.

Jamaica was in the forefront of pasture development in the Caribbean for many years and an early trial was carried out to investigate the effect of fertilization on 'Commons' pasture (*Stenotaphrum secundatum*). Fertilization was at the rate of about 80 kg/ha N/year. The experiment was evaluated using Aberdeen Angus steers. The carrying capacity was 1.25 steers/ha except during the growing period when it became as high as 2 steers/ha. Liveweight gains were 265 and 164 kg/ha for fertilized and non-fertilized plots respectively, or about 1.3 kg liveweight/kg N applied. However, the increased production did not offset the cost of the fertilizers applied. In a subsequent year, the experiment was split to include the effects of mowing twice a year or after each grazing; the liveweight gains for fertilized and unfertilized pastures were 164 and 143 kg/ha and, again, the increase in liveweight gain was not sufficient to cover the cost of the fertilizer used (Jamaica 1950, 1953).

Isolated trials have been carried out in parts of Latin America. Pallares and Mufarrege (1970), working at the Livestock Experimental Station at Mercedes in the Province of Corrientes in the subtropical north of Argentina, compared fertilized with non-fertilized native pastures. The more important species in the natural pasture were *Paspalum alum, Sporobolus poiretii, Paspalum notatum, Bothriochloa laguroides, Schizachyrium intermedium* and *Axonopus argentinus*. Fertilization was 67 kg N/ha (as sulphate of ammonia), 45 kg P_2O_5/ha (as basic slag) and 30 kg K_2O (as potassium chloride). Over the four-year period of the experiment, the difference in weight gains between the animals grazing fertilized pasture and the controls was equivalent to 81 kg/ha or 1.21 kg LWG(Liveweight gain)/kg N applied. As has been frequently noted elsewhere, the use of fertilizers improved the content of crude protein and total phosphorus of the forage. The interesting observation was also made that the animals on the fertilized pastures consumed significantly less of the mixture of equal parts of salt and bone meal which was on offer free-choice throughout the experiment (101 v. 137 g/head/day). It also appeared that the fertilizer treatment might have encouraged the *Paspalum alum* in the sward and discouraged *P. notatum*.

Treviño et al. (1976) investigated the effects of fertilizing native pasture at Hueytamalco in the state of Puebla in Mexico. The grasses were *Axonopus* and *Paspalum* spp. and the rate of fertilizer application was 150 kg N/ha. Experimental animals were Santa Gertrudis and Charbray steers weighing 219 kg on average at the start of the experiment. The annual mean precipitation at the station is 2300 mm. The liveweight gains from fertilized and non-fertilized pastures were 322 and 177 kg/ha respectively, or an advantage attributable to fertilization of 144 kg or 0.96 kg/kg N applied. The fertilizer treatment gave a return on investment of 7.8%, compared with 1.2% in the case of the control (although the costs of production of the unfertilized pasture seem to be inordinately high and it is not clear what fixed costs were included). The authors point out the poor carrying capacity of the native grasses in comparison with introduced species such as *Cynodon dactylon* (Ferrer), *Brachiaria brizantha* or *Cynodon plectostachyus* as well as the relatively small response to fertilization (see also Garza et al. 1973b).

Fertilization of improved (or introduced) pastures

Although Moses (1934) recorded the improved productivity of fertilized Kikuyu grass at Umbogintwini in Natal, greatest emphasis in South Africa was given to the fertilization of natural veld and little work was carried out on improved pastures. Thus the Caribbean area provided the focus for intensive research on the fertilization of pastures which started after the Second World War. The Caribbean islands and surrounding countries on the mainland provide an ideal environment for the successful exploitation of fertilized pastures as favourable precipitation and temperature usually ensure that the application of nitrogen is effective. There was also a necessity to intensify agricultural production because of limited land availability in some of the islands and, at the same time, use land which was unsuitable for arable cultivation because of steep slopes. In other islands, the emphasis was on the use of lands which were surplus to the requirements of sugar or bauxite production. As this coincided with the availability of highly productive introduced grasses, notably Pangola, and the development of new beef breeds (Jamaica Black, Jamaica Red, Nelthrop) or Holstein steers and heifers surplus to the requirements of a dairy industry made possible by tick control, the use of fertilizers was an attractive proposition.

North and Central America and the Caribbean

Puerto Rico

Puerto Rico is in many ways ideally suited for tropical beef production from grass. It has warm weather, well-distributed rainfall of from 1500–2000 mm/year and deep porous soils. Large areas of land with relatively steep slopes are not suitable for mechanical cultivation but can be fertilized from the air, and vigorous grass cover forms the ideal protection against soil loss. The use of fertilized, well-managed pasture enables 5 steers/ha to produce in excess of 1100 kg liveweight gain/year from grass alone.

Experiments using fertilizers started as early as 1945 when an experiment was initiated at Rio Piedras (rainfall, 1500 mm) to compare Guinea grass (*Panicum maximum*), Para grass (*Brachiaria mutica*) and a mixture of Para and tropical Kudzu (*Pueraria phaseoloides*). During the first phase of the experiment, no fertilizer was applied to the Guinea grass pasture (Rivera Brenes 1947) but during the subsequent phase, 1100 kg of 10–10–5 fertilizer plus 190 kg muriate of potash were applied per ha. Although the trial did not constitute a true comparison, the authors (Rivera Brenes et al. 1949) comment that the carrying capacity of the pasture was more than twice that noted in the previous experiment (1.0 v. 3.0 beasts/ha).

This superiority was not so marked in a subsequent phase, however (Rivera Brenes et al. 1950).

Rivera Brenes carried out an experiment on the fertilization of Guinea grass with different levels of nitrogen at the Lajas sub-station in the south-west of the island, where rainfall fluctuates between 1000 and 1500 mm and where there are 'rather prolonged' dry spells (although mean monthly rainfall during the period of the experiment was 110 mm). The application rates were 0, 180 and 360 kg/ha N/year applied at three-monthly intervals. Liveweight gains for the three treatments in the same order were 383, 595 and 892 kg/ha. The authors concluded that fertilization was economically beneficial 'even in this relatively dry section of the Island' (Rivera Brenes et al. 1958). The increments are equivalent to 1.18 and 1.41 kg/liveweight gain/kg N applied.

A further experiment in the semi-arid south coast area was carried out by Caro et al. (1961) at Santa Isabel where annual rainfall is 900 mm with a marked dry season from December to April. Pastures were Guinea, Napier and Pangola fertilized at the rate of 1360 kg/ha 14–4–10 and grazed at a stocking rate of at least 6.25 steers/ha. Sprinkler irrigation supplied the equivalent of 40 mm weekly (although it sometimes broke down) and pastures were mowed twice-yearly. Yearly liveweight gains amounted to 1450, 1490 and 1195 kg/ha for the three pastures in the order given. Although this experiment does not compare the production of fertilized and unfertilized pastures, it does show the extremely high potential of well-fertilized, irrigated pastures in the tropics and for this reason it has been included.

Later work concentrated on the beef-producing capacity of higher land with an average slope of 45% and precipitation of between 1500 and 2500 mm.

Caro and Vicente (1961) initiated a series of experiments employing heavy fertilizer rates with a study on the effect on carrying capacity and beef production of Napier grass pastures. Three levels of fertilizer were used equivalent to 100, 300 and 500 kg/ha as 15–4–10. Pastures were evaluated using dairy heifers (initially weighing about 160 kg) and the basic stocking rate for the three fertilizer treatments was 3.3, 5 and 6.6 head/ha respectively. The gains in liveweight were increased from 640 to 1200 kg by the application of the additional 1350 kg of fertilizer, but little additional advantage was conferred by the further application.

In a later experiment, Caro et al. (1965) compared the productivity of five intensively-managed grasses in the same area. Although the same fertilizer treatment was applied to all grasses, this work has been included here to show the consistent high productivity obtained in the various experiments. Pangola, Guinea, Napier, Para and Molasses grass were fertilised with 1100 kg/ha/year of 14–4–10 and pastures were stocked with Holstein heifers or bulls and the grazing capacity was equivalent to six steers of 270 kg/ha over the four-year period of the experiment. Gains in liveweight for the five grasses in this trial averaged 1125, 1320, 1110, 780 and 645 kg/ha respectively. These gains compared favourable with those obtained in the irrigated coastal area mentioned earlier, and the maximum liveweight gain, obtained from Guinea grass in 1962, 1560 kg/ha, ranks as one of the highest values for liveweight gain from tropical pastures obtained anywhere in the world.

In a later trial, a similar fertilizer rate was applied to Pangola and Star grass pastures at the same station and the liveweight gains for the two grasses were 1517 and 1064 kg/ha respectively. Star grass was superior in terms of both carrying capacity and average daily gains/head (Caro et al. 1972a). These results appear to be the same as those reported by Caro et al. (1973).

Caro et al. (1972b), also working at Orocovis, measured the effect of different levels of fertilizer on Pangola grass grazed by young Holstein heifers. Rates of 450, 1345, 2690 and 3810 kg/ha/year of 14–4–10 were applied over a two-year period. Average daily gains of 545 g/head/day were obtained for all treatments, and results for carrying capacity and liveweight gains are given in Table 10.4.

Table 10.4 Stocking rates and liveweight gains from Pangola grass fertilized at various rates (Orocovis) (adapted from Caro et al. 1972b)

Fertiliser rate (kg/ha N)	*Stocking rate (270 kg steers/ha)*	*Liveweight gain (kg/ha)*
63	1.9	398
188	3.2	633
376	5.0	976
533	5.9	1 088

The pastures responded in terms of carrying capacity and liveweight gain to all levels of fertilization, although this response was not significant nor economically justified above the 2690 kg level. The response to fertilizer was 1.88, 1.84 and 1.40 kg liveweight gain/kg N applied for the three fertilizer treatments above the 63 kg/ha N level. Although the levels of beef production at the higher levels of fertilizer application may be considered good by more general standards, the authors comment that the lower results than those obtained in other studies could be explained by an attack of aphids and invasion by Para grass (*Brachiaria mutica*).

Finally, high levels of fertilizer were applied to Napier grass at Orocovis (Caro and Vicente 1972). Treatments were applied at the rates of 1790, 3140 and 4480 kg/ha of 14–4–10 in four equal applications per year. The grazing animals used were Holstein heifers for the first three years of the trial and Holstein bulls implanted with diethylstilboestrol thereafter. Pastures were mown once a year and

Table 10.5 Stocking rates and liveweight gains from Napier grass fertilized at various rates (Orocovis) (adapted from Caro and Vicente 1972)

Fertilizer rate (kg/ha)	*Stocking rate (270 kg steers/ha)*	*Liveweight gain (kg/ha)*
250	5.5	1 060
440	7.1	1 435
625	8.8	1 772

a rotation of approximately one week grazing followed by three weeks rest (cattle were moved when all leaves had been removed from the Napier grass stalks). The average rainfall during the five years of the experiment was 2000 mm with the 'normal' dry period from January to March. Results for stocking rates and liveweight gains are shown in Table 10.5 (average daily gains were similar for all treatments and averaged 595 g/day).

The response to nitrogen application was 1.99 and 1.90 kg liveweight gain/kg N applied. Heavy fertilization was economically justified except in 1967 when rainfall was only 1580 mm and which was apparently the only year when rainfall was limiting. It would appear, therefore, that heavy rates of fertilization would be justified in years with more than about 1800 mm of rainfall.

These results from Puerto Rico are interesting in that they show the high productive potential of well-managed, highly-fertilized introduced grasses in areas with high rainfall distributed relatively evenly throughout the year. These levels were generally economically to be recommended as the local market for beef was strong. They had the additional advantage of using land which was topographically unsuitable for cropping. The systems are probably the most intensive reported for beef production from unsupplemented pastures in the tropics, although some of the experiments from irrigated and high rainfall areas in Western Australia and Queensland are comparable insofar as cumulative liveweight gains/unit area are concerned.

Florida

Early work in the south-central part of Florida demonstrated the advantages of both compound and phosphorus fertilizers on pastures grown on acid soils. These frequently require lime to increase the pH and are deficient in all major nutrients as well as trace elements such as copper, manganese and zinc. Experiments were carried out at the Range Cattle Station at Ona by Kirk, Hodges and their colleagues and represent some of the first significant work to be carried out on fertilizers in the Americas. The pastures studied included Carpet grass (*Axonopus compressus*) followed by Pensacola Bahia (*Paspalum notatum*), various strains of Bermuda grass together with Bermuda crosses (*Cynodon dactylon*), Pangola grass (*Digitaria decumbens*) and experimental Star grass (*Cynodon nlemfuensis*) varieties.

Results from experiments started in 1943 were reported by Hodges et al. (1949). Carpet grass fertilized with rock phosphate (500 kg/ha) or 6–6–6 (560 kg/ha) performed slightly better than the unfertilized grass (16, 92 and 6 kg liveweight gain respectively) but these gains were insignificant when compared with Carpet grass/clover pastures fertilized with 0–14–10 and trace elements, where the gain/ha was 800 kg. These results are difficult to interpret, however, as no details are given of stocking rates, grazing days, etc.

Later experiments evaluated the performance of different grasses at two levels of fertilizer application. This was not a direct comparison, however, as the production was measured in different years. All grasses gave improved performance in the year that more fertilizer was applied, but this effect may well have been confounded with rainfall or management differences. Results are summarized in Table 10.6.

Table 10.6 Liveweight gains (kg/ha) of steers grazing different grasses at two fertilizer levels (Ona) (adapted from Hodges et al. 1958)

Grass variety	*Fertilizer level (kg/ha/year)*	
	225 kg 6–6–6	*410 kg 9–6–6*
Carpet	68	185
Common Bahia	83	167
Pensacola Bahia	170	241
Argentina Bahia	114	242
Coastal Bermuda	145	224
Torpedo	140	227
Pangola	226	379
Native	65	136

Nevertheless, the effect of the fertilizer, although not clear-cut, was equivalent to about 2 kg liveweight gain/kg N applied, a figure consistent with other results (Hodges et al. 1958).

Later work by Hodges and his colleagues was directed towards the use of different sources of phosphorus in cow/calf programmes on Pangola grass pastures. A complex fertilizer application programme was employed which favoured phosphorus at the expense of nitrogen during the first few years of the trial; however, in later years the importance of nitrogen was eventually recognized and application of this element was made at the rate of 112 kg/ha, 'intensifying the utilization of phosphorus'. Phosphorus application appeared to increase beef production in terms of weaned calf crop percentages as well as weaning weight, and the most effective source was basic slag (Hodges et al. 1964).

The superiority of basic slag was confirmed in an analysis of later results (Hodges et al. 1968) as it averaged 12% more beef production than other treatments: that the natural phosphorus supply of the flatwoods soils where the experiment was carried out was too low was indicated by the fact that all phosphorus treatments, when taken together, averaged 92% more beef production than the control. However, it was concluded that pasture productivity 'requires a balanced fertilizer treatment which supplies all major and minor elements and limestone'.

Finally, a trial which included the application of two levels of nitrogen to a mixture of perennial pastures (Pangola, common Bermuda, Pensacola Bahia and Star grass) was evaluated in a cow/calf programme. Nitrogen applications was 143 v. 215 kg/ha N. Both calving percentage and weaning weight were improved on the system employing a higher level of nitrogen, and the net results in terms of annual calf production for the two treatments were 266 v. 312 kg respectively. Many experiments have evaluated the effects of fertilization or liveweight gains of slaughter stock, but few have tackled the longer-term effects of fertilization on cow and calf performance and these results from Ona supply valuable complementary information on this aspect (Hodges et al. 1977b).

A long series of multidisciplinary trials was initiated at the Beef Research Unit near Gainesville, north-central Florida, in 1952. These were designed to evaluate beef production from different pasture programmes and, at the same time, collect data on soil and forage analyses as well as carry out studies of economic returns. The study started with eight pasture programmes which through the years were reduced to three. Unfortunately, only three of the programmes were relevant to this discussion and two of these were discontinued at the end of phase I of the trial. However, they provide useful information, particularly on aspects of the cow and calf exploitation so lacking from elsewhere. The grasses used were Pangola, coastal Bermuda and Pensacola Bahia, each forming a third of each treatment. The three levels of fertilizer used were (in kg/ha/year) 38, 20, 20; 65, 40, 40; and 134, 81 and 81 of NPK respectively. As would be expected, there were clear differences in stocking rate among the three treatments, these being 1.1, 0.8 and 0.5 ha/cow in the same order as given above. There was a slight increase in weaning percentage for the cows on the high fertilizer treatment but no difference in weaning weight of calves. It was concluded that despite the improved animal performance on the highest fertilizer levels, all-grass (as opposed to grass/legume) programmes were not profitable for cow and calf operations (Koger et al. 1961).

Jamaica

The first work to be carried out on the fertilization of improved grasses in Jamaica was initiated by Motta in 1955 and compared unfertilized Pangola with the same grass dressed with 60 kg/ha N together with 60 kg/ha KCl. Motta summarized the production obtained during the first two years of the trial. This was 1026 and 860 kg/ha liveweight gain for fertilized and unfertilized Pangola grass pastures respectively (Jamaica 1957). Incidentally, the production from the fertilized pasture during the second year of the trial was 1173 kg/ha, an outstanding performance (even the performance from the unfertilized pastures in this trial was noteworthy). The stocking rates (adjusted so that animals on both treatments gained approximately the same amount, an unusual approach) were 5.2 and 4.4 beasts/ha for fertilized and unfertilized pastures respectively, a very high stocking rate considering that this was the average of wet and dry season rates. The average gains for the five years of the trial were fertilized, 1069 kg/ha and unfertilized, 847 kg/ha (Motta, cited by Nestel and Creek 1962) or 1.72 kg liveweight gain/kg additional N applied.

A further trial was carried out on poorer soil and the rates of application of nitrogen were 0, 72, 144 and 217 kg N/ha. Liveweight gains for the treatments in the same order were 582, 863, 1013 and 1236 kg/ha which compare with some of the highest liveweight gains recorded from tropical pastures (loc. cit.).

Despite the high levels of productivity obtained by fertilization of Pangola grass, Nestel (1964) reported that 'in practice, very low levels of fertilizer application are commonly found', this despite that the responses were most favourable in comparison with studies carried out elsewhere (the responses in terms of liveweight gain in the trial mentioned above being 3.90, 2.99 and 3.01 kg liveweight gain/ha/kg N applied). Nestel (1964) recorded gross outputs of meat of over £125/ha and concluded that input costs (i.e. on fertilizer) could rise considerably before they reached such a high level.

Richards (1965) has reported a further trial carried out at the Government Livestock Station at Bodles applying different rates of nitrogen to Guinea grass pastures. Levels of nitrogen used were 0, 52, 103 and 155 kg/ha and the pastures were irrigated when rainfall was under 50 mm at three- to four-weekly intervals. Total liveweight gain for the four treatments were 461, 616, 713 and 811 kg/ha in the same order as above, representing a response of 2.98, 2.45 and 2.26 kg liveweight gain/kg N applied. Although these weight gains and responses to fertilizer were not as dramatic as those noted in the case of Pangola (irrigation applied was not sufficient to maintain full pasture growth), they still represent a considerable beef production/unit area.

These results from Jamaica illustrate the high productive potential of some of the improved tropical grasses under favourable management conditions and high levels of fertilizer application in regions where rainfall is reasonably favourable or where supplementary irrigation is applied.

St Croix

Oakes (1960) described the effects of fertilization on Pangola and Guinea grasses in the island of St Croix, one of the drier parts of the Caribbean. The average precipitation during the years 1956 to 1958 was 800 mm and the year in which the trial was started (1957) was the second driest year recorded on the island. The fertilized pastures were top-dressed with 336 kg/ha/year N and the test animals were St Croix (Nelthrop) steers and heifers. Results for stocking rate and liveweight gains are summarized in Table 10.7.

Table 10.7 Stocking rate and liveweight gains of fertilized and unfertilized grasses (Kingshill) (adapted from Oakes 1960)

	Carrying capacity (Au/ha)	*Liveweight gains (kg/ha/year)*
Pangola grass		
fertilized	2.3	1 178
unfertilized	1.5	364
Guinea grass		
fertilized	1.5	507
unfertilized	1.1	262

Considering the relatively low rainfall, the response to fertilizer by Pangola grass is particularly impressive, amounting to some 2.42 kg/liveweight gain/kg N applied. Indeed, the beef production of the fertilized Pangola grass pastures compares most favourably with values from other parts of the tropics. This was largely attributable to extremely high liveweight gains per animal (1.4 kg/day). There was a significant superiority in liveweight gain of the animals grazing fertilized pastures but stocking rates of these pastures, although better, were not significantly so. This is not generally true of most of the experiments quoted in this review, but no details are given as to the decisions on stocking rates and the finding is consistent with other experiments quoted elsewhere on the relation between liveweight gain and stocking rate.

El Salvador

Apparently the first trial in Central America was carried out by Dyer and his colleagues (Dyer et al. 1954) who reported an early trial carried out at the Centro Nacional de Agronomia to compare fertilized and unfertilized Elephant, Para, Pangola and Guinea grass pastures for Criollo and Brahman × Criollo bulls. Unfortunately, no details of the fertilization programme used are given, but daily gains from the fertilized pasture were 250 v. 160 g/day during the dry season. Only an abstract has been published which accounts for the lack of detail.

Panama

An early trial was carried out on the fertilization of Jaragua grass (*Hyparrhenia rufa*) at different levels at the National Institute of Agriculture in Divisa, Panama. Previously established pastures consisting mainly of Jaragua were fertilized with 0, 200, 500, 1000, 1500 and 2000 lb/ha (sic.) of 10–30–10 (the use of lb/ha apparently stems from the use of 'quintales', one quintal being 100 lb). The average yearly rainfall at the station where the experiment was carried out was 1550 mm, with a relatively dry period from January to March inclusive. The soils were said to be relatively infertile. Animal production was measured using Brown Swiss and Brown Swiss × Red Sindhi heifers. The experiment was grazed intermittently for a total period of ninety-three days between May and October 1956. Summarized results are given in Table 10.8.

Table 10.8 Grazing days and liveweight gains of heifers grazing fertilized and unfertilized Jaragua grass pastures (Divisa) (adapted from Noland et al. 1959)

Fertiliser level (kg/ha N)	*Grazing days*	*Total liveweight gain (kg/ha)*
0	567	75
9	567	134
23	688	139
46	618	136
68	715	248
91	749	336

The results are difficult to compare with those of other experiments because of the lack of information, but it is clear that there was a response to fertilizer application although this was very uneven. The authors also demonstrated that the protein content of the forage produced increased with fertilizer application (Noland et al. 1959).

Brazil

Work carried out on the fertilization of improved pastures in Brazil at first concentrated on increasing the productivity of the 'resident' variety of *Panicum maximum* known as 'Coloniao', and thereafter comparing the productivity of the resident grasses, Coloniao, Jaragua (*Hyparrhenia rufa*) and Molasses (*Melinis minutiflora*), with the more recent introductions, Pangola (*Digitaria decumbens*), Tanganyika (*Panicum maximum*) and coastal Bermuda (*Cynodon dactylon*). The comparison between winter and summer fertilization of Coloniao was also made and the effects of residual nitrogen after several years' fertilization were studied.

Quinn and his colleagues of the Ibec Research Institute obtained the first animal production data at the Fazenda Jangada, Municipio de Aracatuba, state of San Paulo (Quinn et al. 1960a). Three levels of nitrogen were used, 0, 100 and 200 kg/ha (as ammonium nitrate), to fertilize well-established (eleven-year-old) *Panicum maximum* pastures. The animals used were Nellore steers. Weight gains/ha on the three levels of nitrogen were 308, 494 and 703 kg/ha respectively, representing liveweight gains of 1.86 and 2.02 kg/kg N applied. These increments were a result of higher stocking rates. In addition, all pastures received 200 kg/ha P_2O_5 and 40 kg/ha S. When 200 kg/ha N was applied in the summer or the dry winter period, it appeared that the winter application was more effective as the liveweight gains from the two treatments were 618 and 682 kg/ha respectively. When three levels of P_2O_5 (0, 100 and 200 kg/ha) were compared in the presence of 200 kg/ha N, increasing levels of phosphorus gave small but significant increments in liveweight gain, values for the three treatments being 627, 675 and 703 kg/ha respectively. Fertilization with sulphur in the presence of nitrogen was economically advantageous, but there was no economic advantage to fertilization with phosphorus or sulphur in the absence of nitrogen. Economically, the most advisable levels appeared to be 200 kg/ha N, 100 kg/ha P_2O_5 and 60 kg/ha S. In a further experiment at the same Fazenda (Mott et al. 1967), molasses was used as an energy supplement to fertilized (100 kg/ha N) or non-fertilized *Panicum maximum* pastures; the liveweight gain for the two unsupplemented treatments was 253 and 472 kg/ha respectively or 2.2 kg liveweight gain/kg N applied. Increments were associated with both increased individual gains and higher stocking rates. The effects of application of nitrogen during the winter or summer seasons were summarized by Quinn et al. (1970). Yearly precipitation averaged 1150 mm over the seven-year period of the experiment, 80% of which falls during the summer (November–May). Applications of 200 kg/ha N were made either at the beginning of winter or at the start of summer. Although the earlier results (referred to above, Quinn et al. 1960a) indicated that there might be a favourable effect of winter application, this was not confirmed in the final analysis, and although winter fertilization increased the daily gain of steers during the dry winter by 145 g/day over those grazing the summer-fertilized pastures, this advantage was lost by the end of the following wet summer period. Thus the practice of winter fertilization would only be recommendable under a management system that aimed at marketing animals during the early summer months of November–February when beef prices are 15–30% higher than at the end of the summer, the normal peak marketing period.

Nitrogen was applied to the above-mentioned pastures from 1957–64, and a subsequent estimation of total digestible nutrients indicated that substantial recycling of nutrients occurred in later years

(1964–8), leading to significantly positive effects of residual nitrogen (Mott et al. 1970).

The Ibec workers also reported the results of an experiment in which the effect of fertilization on six tropical grasses was measured. The grasses were all dressed with 75 kg/ha P_2O_5 and 100 kg/ha N prior to the start of the experiment and the fertilized plots were dressed with 100 kg P_2O_5 once thereafter and 100 kg/ha N yearly. The grasses were *Panicum maximum* (var Coloniao and Tanganyika), *Hyparrhenia rufa* (Jaragua), *Digitaria decumbens* (Pangola), *Melinis minutiflora* (Molasses) and *Cynodon dactylon* (Coastal Bermuda). Initial results were reported by Quinn et al. (1962) and in more detail by Quinn et al. (1964). Results from the three years of the trial are summarized in Table 10.9.

The advantage due to fertilization for the six grasses in the order given was equivalent to 1.26, 0.85, 0.90, 0.29, 1.89 and 0.84 kg liveweight gain/kg N applied. In most cases, there was little difference between the daily weight gains on fertilized and unfertilized pastures but the number of grazing days was always substantially

Table 10.9 Daily gains, carrying capacity and liveweight gain from six fertilized and unfertilized tropical grasses (Matto) (adapted from Quinn et al. 1964)

Grass	*Daily gain (g/day)*	*Carrying capacity (steer days)*	*Liveweight gain (kg/ha)*
Coloniao			
fertilized	435	680	330
unfertilized	446	472	204
Jaragua			
fertilized	477	589	320
unfertilized	494	453	235
Pangola			
fertilized	375	766	330
unfertilized	365	607	240
Molasses			
fertilized	366	347	120
unfertilized	350	287	91
Tanganyika			
fertilized	385	632	296
unfertilized	348	313	107
Coastal Bermuda			
fertilized	264	495	140
unfertilized	234	297	56

greater in the fertilized pastures. Quinn et al. (1962) noted that the crude protein content of the fertilized pastures was consistently higher than that in the non-fertilized pastures (7.4 v. 6.2%).

Quinn and his colleagues (Buller et al. 1970) also briefly commented on the effects of 100 kg/ha N on Pangola grass. Daily gains were somewhat higher in the fertilized than the non-fertilized pasture (294 v. 231 g/day) whilst stocking rate was substantially higher in the former (2.21 v. 1.63 steers/ha), as was liveweight gain/ha (239 v. 139 kg/ha). Differences between the latter figures correspond to an increment of 1.0 kg liveweight gain/kg N applied.

Lima and his colleagues carried out an experiment to determine the effects of fertilization with nitrogen on four tropical grasses at Sertãozinho, Sao Paulo. The grasses were Guinea, Napier, Pangola and Suwannee Bermuda and the level of fertilization was 200 kg N/ha (as nitrocalcio, 20–22% N). Pastures were either continuously or rotationally grazed. The weight gains for the fertilized grasses in the order as above were as follows: 303, 406, 296 and 416 kg/ha and 274, 328, 244 and 342 kg/ha without fertilization. The responses were equivalent to 0.15, 0.39, 0.26 and 0.37 kg liveweight gain/ha/kg N applied. There were no significant differences between rotational or continuous grazing (Lima et al. 1968). In view of the limited responses to nitrogen alone, a further experiment was carried out to determine whether phosphorus and potassium were limiting forage production. In this case, all pastures (the same as above) were fertilized with 100 kg/ha P_2O_5 and 60 kg/ha K_2O every third year, and the nitrogen treatment was reduced to 100 kg/ha. The experimental animals were Nellore steers eighteen months old. A summary of the results obtained is given in Table 10.10.

There was a significant difference in liveweight gain/ha in favour of the fertilized pastures amounting to some 0.77 kg liveweight gain/kg N applied, indicating that phosphorus and potassium were limiting in the previous experiment (Lima et al. 1969).

Table 10.10 Stocking rates, daily gains and liveweight gains/ha for four tropical grasses either with or without fertilizer (Sertãozinho) (adapted from Lima et al. 1969)

Grass	*Stocking rate (steers/ha)*	*Daily gain (g/day)*	*Liveweight gain (kg/ha)*
Napier	3.64	462	343
Coloniao	2.40	456	281
Pangola	2.04	486	268
S. Bermuda	3.20	375	307
Fertilizer			
with N	2.82	435	338
without N	2.07	454	261

In what was apparently a second phase of the same experiment (only the breed of the animals used was different, being cross-bred Santa Gertudis × Zebu) carried out during the following year, the difference between fertilized and unfertilized pasture was somewhat higher, amounting to some 0.89 kg additional liveweight gain/kg N applied (Lima et al. 1969).

Thus all the experiments on fertilization with nitrogen in Brazil have shown positive effects in terms of liveweight gain/ha, largely associated with increased grazing capacity of fertilized pastures. The response varied from 0.29–2.2 kg liveweight gain/kg N applied. In several cases, improvements in crude protein content of the fertilized pastures were also noted.

Argentina

Pallares and his colleagues have carried out experiments on the fertilization of improved pastures in the subtropical area of Argentina at the INTA station of Mercedes, Corrientes. In the first experiment, Pangola, Pasto Ramirez (*Paspalum gueroarum*) and Pasto Rojas (*Paspalum rojasii*) were compared with levels of 67 kg/ha N, 45 kg/ha P_2O_5 and 30 kg/ha K_2O or without fertilizer. In the second experiment, the productivity of Pangola grass was investigated with or without 200 kg/ha N at four stocking rates.

In the first experiment, it was unfortunate that the Ramirez and Rojas grasses were invaded by natural grasses and only limited data were available on animal days/ha. However, the effects of fertilizer were marked in the case of Pangola grass where production was increased from 176 to 290 kg liveweight gain/ha or 1.70 kg increment in liveweight gain/kg N applied (Pallares and Mufarrege 1970).

Pallares and Mufarrege (1974) also investigated the effect of four stocking rates on Pangola grass pastures fertilized with 200 kg N/ha/year (as urea) and two stocking rates on unfertilized Pangola grass pastures. The stocking rates for the fertilized and non-fertilized treatments were 1.25. 2.50. 3.75 and 5.00 heifers (Santa Gertrudis, 13–14 months old) and 1.25 and 2.50 heifers/ha. The average production from the fertilized pastures at the first three stocking rates was 267 kg/ha liveweight gain (the highest stocking rate obviously 'broke down' as the production was only 87 kg/ha) whilst the average production from the non-fertilized pastures was 116 kg/ha. This difference is equivalent to 0.76 increment in liveweight gain (kg/ha) for every kg of nitrogen applied.

Mexico

Garza and his colleagues of the Instituto Nacional de Investigaciones Pecuarias have carried out an interesting series of experiments in the tropical grasslands on the gulf coast of Mexico. Whilst the natural or induced pastures consist mainly of *Axonopus compressus*, *A. affinis* and *Paspalum notatum*, these are increasingly giving way to such introduced but now resident species as *Panicum maximum*, *Brachiaria mutica* and *Pennisetum purpureum* as well as the more recent arrivals, Pangola, African Star and 'Ferrer' (apparently coast cross 1 *Cynodon*). Trials have therefore been concentrated on the latter grasses, although a trial on the fertilization of native pastures has been mentioned earlier.

The first trial to be carried out was at the Centro de Investigaciones Agricolas del Sureste, Cotaxtla, Veracruz, 'Paso del Toro'. The comparison was between Pangola and Jaragua with or without 100 kg/ha N and the experimental animals were Criollo × Zebu steers of about two years old and 250 kg. The experiment was carried out during the rainy season (mean precipitation during the two years of the trial was 1130 mm) which lasted between five and six months. The grazing capacity of the fertilized and non-fertilized pastures were 4.58 and 3.39 steers/ha (720 v. 528 animal days) respectively, whilst liveweight gain was 275 and 161 kg/ha respectively, equivalent to 1.14 kg liveweight gain/kg N applied. On the other hand, average daily gain on the fertilized pastures was similar to the control and there was no significant difference between pasture species although there was a strong indication of the superiority of Pangola grass (Garza et al. 1970).

A similar result was obtained by Garza and Arroyo (1971) when Guinea grass was fertilized with 100 kg/ha N at Playa Vicente, Veracruz, from August to December. The response was 205.5 v. 91.5 kg/liveweight gain/ha in favour of the fertilized treatment, equivalent to 1.15 kg liveweight gain/kg N applied. Garza (1973), in another experiment carried out at Playa Vicente during a full year, fertilized Guinea grass with 150 kg/ha N and obtained 834 kg liveweight gain v. 425 kg/ha from the unfertilized control. Daily weight gains were the same but the fertilized pasture supported a higher stocking rate than the unfertilized (2 v. 4 beasts/ha). This represents 2.7 kg liveweight gain/ha for each kg additional nitrogen.

Garza et al. (1973b) compared six tropical grasses both fertilized and unfertilized for liveweight gain during 168 days of the ten-month rainy season in Hueytamalco, Puebla (mean annual precipitation 2500 mm). The grasses were Ferrer, Pangola, Giant Pangola (*Digitaria valida*), African Star (said to be *Cynodon plectostachyus* but probably *Cynodon nlemfuensis*), Gordura (*Melinis minutiflora*) and the native grasses, *Paspalum notatum* and *Axonopus affinis*. The trial was evaluated using cross-bred heifers aged thirteen months with an average weight of 205 kg. Nitrogen was applied at the rates of 0 or 100 kg/ha. The results are summarized in Table 10.11.

These results are interesting in that they illustrate the wide range in response to fertilizers that can be expected from different grass species under similar soil and climatic conditions.

In a further experiment carried out at Hueytamalco, animal production was compared from Ferrer, African Star and Signal

Table 10.11 Liveweight gain from fertilized and unfertilized tropical pastures (Hueytamalco) (adapted from Garza et al. 1973b)

Grass	*Liveweight gain* (kg/ha)		
	Fertilized	*Unfertilized*	*Difference*
Ferrer	394	211	183
Pangola	316	238	78
Star	330	139	191
Giant Pangola	295	124	171
Gordura	267	135	132
Native	143	86	57

(*Brachiaria brizantha*) fertilized at the rate of 150 kg/ha N. The animals used were Indubrasil heifers aged thirteen months with an average weight of 203 kg, and the stocking rates were 2 and 4 beasts/ha for unfertilized and fertilized treatments respectively. The liveweight gains during the year in which the experiment was carried out were 636, 583 and 637 kg/ha and 334, 287 and 341 kg/ha for fertilized and non-fertilized pastures respectively. In contrast to the experiment quoted above, the homogeneity of response to fertilization was noteworthy, being equivalent to 2.01, 1.97 and 1.97 kg liveweight gain/kg N applied for the three pastures respectively. The experiment also illustrates the high productive potential of improved grasses in the humid subtropics (Treviño et al. 1975).

Molina et al. (1977) carried out an experiment to evaluate the effects of fertilization with 100–100–0 kg/ha on Guinea and Buffel (*Cenchrus ciliaris*) grass pastures at Tizimín in the east of Yucatan. The animals used were Zebu steers aged fourteen months with an average weight of 330 kg (which seems high). Pastures were grazed for a period of 364 days but, unfortunately, no rainfall data are given. The liveweight gain for fertilized Guinea and Buffel grass was 591 and 465 kg/ha respectively whilst that for the unfertilized controls was 168 and 216 kg/ha. The response to fertilization was higher in the case of the Guinea grass, amounting to 438 kg/ha or 4.23 kg/ha liveweight gain/kg N applied, whilst the corresponding value for the Buffel grass was 2.49 kg/ha, the former being one of the highest recorded responses. This was attributed by the authors to the fact that the pasture was well established and that it had not previously been fertilized. It was also noted that an attack of 'mosca pinta' (*Aeneolamia postica*) was better resisted by the fertilized than by the unfertilized pastures.

Colombia

The first work on nitrogen fertilization of pastures in Colombia appears to have been carried out by Vélez et al. (1977) at the ICA station at Palmira near Cali to the west of the Cordillera Central, whilst the CIAT workers have carried out experiments on fertilization of grasses introduced to the Ilanos (or plains) at Carimagua.

At Palmira, the pasture was *Brachiaria mutica*, which was fertilized with urea (46% N) at the rates of 0, 50, 75 and 100 kg/ha N. The cattle used to evaluate the trial were Romosinuano steers implanted with stilboestrol, which continuously grazed the pastures for 252 days, including part of both wet and dry seasons. Liveweight gains for the four treatments in the order given were 14.8, 41.8, 44.1 and 50.1 kg/ha for this period. The protein content of the forage was 6.6 and 7.5% for non-fertilized and all fertilized treatments respectively (this might have been confounded by the fact that the treatment with nitrogen reduced the legume content of the pasture; the Para content was also reduced in the 0 and 50 kg/ha N treatments). The total liveweight gains for the experiment are relatively low for all fertilizer treatments and it appears that some other nutrient might have been limiting.

Paladines and his colleagues (Paladines et al. 1974) have briefly reported on the use of fertilized Molasses grass (*Melinis minutiflora*), the only introduced grass available to replace the low-producing native savannah in the llanos region south of the Meta river. The savannah largely consists of *Trachypogon*, *Andropogon*, *Paspalum* and *Leptocoryphium* spp. corresponding to the *Trachypogon vestitus* type described by Blydenstein (1965). Unfortunately, results are presented graphically and are difficult to interpret, but it would appear that fertilization with 75 kg/ha P_2O_5 and 75 kg/ha P_2O_5 plus 40 kg/ha K only made an appreciable difference to liveweight gains during one season only out of approximately a three-year period. In view of the low levels of crude protein encountered in the forage, it would seem likely that, as in other experiments, nitrogen was the most limiting nutrient affecting pasture production.

East Africa

Few data are available from this part of Africa, but Stobbs has reported beneficial effects of fertilizers to mixed grass plots at Serere whilst Tiharuhondi and his colleagues reported impressive gains attributable to the application of high levels of nitrogen to both irrigated and non-irrigated mixed grass pastures at Makerere University Farm, near Kampala, Uganda.

Stobbs (1969a) fertilized mixed plots of *Panicum maximum*, *Hyparrhenia rufa* and *Chloris gayana* with levels of 0 or 560 kg/ha N or with or without additional P, K, S and lime. Treatments were evaluated using small East African Zebu steers. Total liveweight gain for the three pastures was 217, 350 and 375 kg/ha/year whilst stocking rate increased from 2.6 on the control treatment to 4.2 beasts/ha on the fully-fertilized treatment. There were therefore clear advantages to fertilization for both liveweight gains/ha and stocking rate (differences in individual liveweight gains were insignificant) but the author doubted that these would be economically warranted at the prices prevailing at the time. In any event, the increments attributable to the application of nitrogen were small, amounting to only 0.28 kg liveweight gain/kg N applied and, whilst the author states that relatively high rates were chosen to 'investigate production limits', lower levels might have been more effective.

Tiharuhondi et al. (1973) must be credited with the highest beef production obtained per hectare closest to the Equator! In an experiment carried out at Kabanyolo (0° 28′N), the productivity of split plots of a mixed ley consisting of *Panicum maximum* and *Setaria sphacelata* was compared at different levels of nitrogen with or without irrigation. The levels of nitrogen used were 0, 224, 448 and 672 kg/ha. Phosphorus and potassium were applied uniformly at the rates of 112 kg/ha P_2O_5 and 224 kg/ha K_2O. The mean rainfall at the station was 1410 mm with two main rainy seasons but with rarely less than 50 mm/month. The results are given in Table 10.12.

Table 10.12 Liveweight gains and stocking rates of irrigated and non-irrigated mixed grass pastures with different levels of nitrogen (Kabanyolo) (adapted from Tiharuhondi et al. 1973)

Nitrogen level (kg/ha)	*Liveweight gain (kg/ha)*	*Stocking rate (beasts/ha)*
Non-irrigated		
0	340	3.2
224	558	4.9
448	843	5.9
672	990	7.4
Irrigated		
0	279	3.5
224	889	5.4
448	1 195	7.2
672	1 360	8.9

These results indicate the extremely high yields that can be obtained when high levels of nitrogen are combined with adequate moisture levels in tropical conditions. As would be expected, the responses to additional nitrogen decreased as the levels increased in the case of the irrigated pasture. These were 2.7, 2.0 and 1.6 kg liveweight gain/kg additional N. In the case of the non-irrigated pasture, however, all responses were lower but more or less linear (1.0, 1.1 and 1.0 kg liveweight gain/kg additional N). High levels of fertilization were economically sound both with and without irrigation but the authors suggest that utilization might have been limited by light intensity because of the generally overcast conditions prevailing at the station. The cattle used in the trial were Nganda × Jersey cross-bred steers or heifers and it was also suggested that cattle with better growth rates could make better use of the forage available.

West Africa

Very little work has been carried out in West Africa on the fertilization of improved pastures. Haggar (1971), working at Zaria in Nigeria, applied four levels of nitrogen (0, 84, 168 and 252 kg/ha N) to *Digitaria smutsii* pastures. Pastures were grazed by Fulani heifers at different stocking rates (lenient, normal and restricted) from August to November. The mean liveweight gains for the four levels of nitrogen application were 82, 268, 318 and 328 kg/ha respectively. Corresponding responses were 2.2, 1.4 and 1.0 kg liveweight gain kg N applied. It is interesting, however, that the response to nitrogen was greater in the heavily-stocked treatment than in the others. Liveweight gains were −3, 289, 344 and 423 kg/ha, corresponding to responses of 3.5, 2.1 and 1.7 kg liveweight gain/kg N applied. The heaviest stocking rate corresponded to 0.10 ha/190 kg heifer. This interaction between fertilizer and stocking rate is important as it is clear that the full benefits of nitrogen application may only be realized at high stocking rates.

Zimbabwe

West (1956), working at the Grasslands Research Station at Marandellas, recognized that much greater returns would be expected by fertilizing improved cultivated grasses than by fertilizing veld. This was supported by the fact that liveweight gains from improved pastures fertilized with up to 120 kg N/ha produced more beef than application of up to 207 kg N/ha to natural veld (515 v. 199 kg liveweight gain/ha).

Most later work on fertilization of cultivated grasses was carried out at Henderson Research Station which serves a maize-farming area. For the maize to be effectively used in fattening cattle it was necessary to increase the summer grazing capacity by intensification of pasture utilization. Rodel and Boultwood (1967) reported that by applying 270 kg N/ha, pastures could be stocked by twelve 250 kg steers during the period from December to the end of March with a liveweight gain of 810 kg/ha. Later work resulted in recommendations for management of No. 2 Star grass (*Cynodon aethiopicus*) although high levels of cyanogenetic glycosides were apt to cause problems in breeding cattle.

Later work with No. 2 Star grass (Rodel et al. 1975) indicated that carcass gains of up to 529 kg/ha (equivalent to more than 1000 kg liveweight/ha) could be achieved by a stocking rate of 14.7 steers/ha combined with application of nitrogen at the rate of 367 kg/ha (or 25 kg/steer). In view of the rise in prices of nitrogenous fertilizer, however, a further trial was started to determine the effects on production of various stocking rates and fertilizer levels. The experimental design is given below:

Nitrogen level (kg/ha)	*Stocking rate (yearling steers/ha)*
100	5 6 7 8
200	7 8 9 10
300	9 10 11 12
400	11 12 13 14

The cattle used were Hereford × Afrikander steers which were grazed from December until they started to lose weight after April.

At the time of writing, a final analysis of the results of the four years of the trial had not been made, but it was clear that increased nitrogen levels resulted in increased liveweight gains/ha although this relationship was curvilinear as would be expected. The role of stocking rate on liveweight gain/ha, however, was less clear and varied from year to year (Parkin et al. 1979). Carcass yields of more than 450 kg were obtained from several fertilizer stocking rate combinations.

This work on grazing trials with No. 2 Star grass is backed up by studies of accessions of *Cynodon* spp. as well as breeding work in which yield and livestock preferences are studied.

Australia

The great majority of Australian soils in the tropics and subtropics are deficient in either nitrogen or phosphorus or both. Deficiencies of several trace minerals also impose constraints on both plant and animal production. Although emphasis has been given to nitrogen fixation by legumes, nitrogen fertilizer use became widespread in the 1960s when it was encouraged by excess production by fertilizer plants and by governmental subsidies. However, price increases during the 1970s, combined with low beef prices in Australia, have reversed this trend. On the other hand, phosphorus, chiefly in the form of single superphosphates, is widely used but largely in fertilization of legume pastures. Potassium is not normally deficient in Australian soils.

Compared with the volume of research that has been carried out on legumes, relatively few data on animal production from nitrogen fertilizer experiments are available. These are either from the wet tropics of the coast of north Queensland (or from irrigated areas) where soil moisture is not the principal limiting factor, or from the subtropics of southern Queensland and New South Wales. Only fertilization of grass pastures will be dealt with in this chapter as phosphorus application to grass/legume pastures has been dealt with elsewhere.

The first livestock production data on fertilized v. non-fertilized *Sorghum alum* were derived from an experiment which attempted unsuccessfully to establish the grass with legumes (Yates et al. 1964). The experiment was carried out at Lawes near Townsville and the intended treatment were *S. alum* top dressed with 504 kg/ha N, *S. alum* sown with *Medicago sativa* (Hunter River lucerne) and *S. alum* sown with summer-growing legumes (*Glycine javanica, Macroptilium atropurpureum* and *M. lathyroides*). However, the legumes failed to establish and 'from the point of view of grazing, all pastures remained virtually monospecific (*S. alum*)'. The second and third treatment results were therefore combined and compared with the fertilized treatment. The experiment was grazed at three stocking rates (3.75, 2.50 and 1.25 beasts/ha) and the effect of nitrogen was most marked on the highest stocking rate where liveweight gains were 570 and 302 kg/ha for fertilized and unfertilized pastures respectively. The authors pointed out that the marked decline in forage production on the non-fertilized pasture indicated a need for additional nitrogen. It was also suggested that the grazing method used (constant stocking) was inefficient and that alternative management systems ('put-and-take' grazing, fodder conservation, etc.) might have permitted a better assessment of the treatments.

Also near Brisbane, Moir and his colleagues (Moir et al. 1969) investigated the effect of nitrogen on (predominantly) *Paspalum dilatatum* pastures at Rocklea. Despite the application of 522 kg/ha N over the four years of the experiment, the differences in liveweight gain between fertilized and non-fertilized treatments was negligible (a precise value is difficult to arrive at because of the way the results are presented). This lack of response was due to inefficient utilization of the pasture because of lack of flexibility in the stocking rate as well as the suppression by the nitrogen of the white clover present in the sward. Analysis of faecal samples indicated that the ability of cattle to select a diet of high quality, whatever the pasture treatment, was obviously a factor in obscuring treatment differences and also supports the view that higher stocking rates would have been more informative.

't Mannetje (1972) has decribed an experiment at Narayen in the Brigalow to compare pasture management systems. Liveweight gain from Buffel grass (*Cenchrus ciliaris*) fertilized with 168 kg/ha N was 282, 237 and 178 kg/ha for stocking rates of 2.2, 1.5 and 1.1 steers/ha respectively. This represented responses of 1.28, 1.01 and 0.66 kg/liveweight gain/kg N applied (a control pasture of Buffel grass to which no fertilizer was applied produced 67 kg/ha at a stocking rate of 0.7 steers/ha). These results clearly indicate the very important interactions between fertilizer application and stocking rate. Later results were summarized by 't Mannetje and Nicholls (1973–4) and the increment due to 168 kg/ha N year was 96 kg liveweight gain/year. Although application of this quantity of nitrogen increased animal production, it was costly and the experiment was modified so as to allow the use of half the quantity of nitrogen. This treatment showed lower liveweight gains after two years ('t Mannetje and Nicholls 1974–5).

Mears and Humphreys (1974a and b) carried out an experiment at Wollongbar (18° 38′S) to investigate the effects of different stocking rates and nitrogen treatments on pastures of Kikuyu grass (*Pennisetum clandestinum*). The experimental design was as follows:

N applied (kg/ha)	*Stocking rate (beasts/ha)*					
0	2.2	3.3	4.9			
134		3.3	4.9	7.4		
336			4.9	7.4	11.1	
672				7.4	11.1	16.6

The cattle used were Aberdeen Angus weaners. The mean annual rainfall at Wollongbar is 1660 mm which is reasonably well distributed (average monthly precipitation = 138 ± 53 mm). All treatments were also dressed with 80 kg P and 50 kg S/ha.

Liveweight gains/ha were positively related to annual rate of nitrogen application. Liveweight gains at successive nitrogen levels were 380, 502, 871 and 1056 kg/ha in 1970–1 and 637, 711, 1141 and 1477 kg/ha during 1971–2 (there was more rain during the latter year). Liveweight response to nitrogen application was highest (3.4 kg/ha liveweight gain/kg additional N) at 7.4 beasts/ha and 134 kg N/ha in the 1971–2 season, but the overall average was 1.2 kg/ha liveweight gain/kg N applied). Botanical measurements indicated that Kikuyu became more dominant in response to nitrogen application each year of the trial until, in the final year, it comprised 100% cover of the 336 kg/ha N treatment compared with only 62% on the 0 kg/ha N treatment. This was at the expense of *Paspalum dilatatum* and *Axonopus affinis* which occupied 26 and 27% respectively of the original sward.

The authors emphasized that great changes in output can be caused by manipulation of nitrogen application and stocking rate and that maximum profit can only be calculated by taking both factors into account.

The feasibility of producing large quantities of beef from fertilized Pangola pastures was demonstrated by Evans (1969) at Beerwah in the Wallum. Nitrogen was applied at the rate of 450 or 900 kg/ha in addition to 48 kg/ha P plus 126 kg K/ha. Animal production was measured using yearling Hereford heifers which were grazed according to a 'put-and-take' system. Liveweight gains for the two treatments were 1280 and 1365 kg/ha at stocking rates of 7.4 heifers/ha. The small difference between nitrogen application rates was noteworthy, but the grazing system employed did not apparently differentiate between treatments and the stocking rate was probably insufficient to utilize available forage on the pastures fertilized at the higher rate. The author supposes that the increased quantity of forage on offer should have allowed more selective grazing which would lead to higher *per capita* weight gains. However, Mears and Humphreys (1974b), in the study on Kikuyu grass already referred to, showed that the performance of individual animals was not dependent on the quantity of forage available for the major part of the year. The different levels of nitrogen application in this Beerwah study led to a clear difference in protein content of the Pangola grass fertilized at high and low levels (10.7 and 13.9% respectively): these levels are so high that selective grazing would not be expected to make a significant difference to the diet of the animals. The results of this experiment are also discussed by Evans (1970).

Bryan and Evans (1971) subsequently carried out another study to compare beef production from nitrogen-fertilized Pangola grass and from Pangola/legume pastures. The levels of nitrogen used were either 112 or 168 kg/ha/year and 449 kg/ha/year together with a complete fertilizer ('the Wallum fertilizer mixture') and average liveweight gains from the two treatments over the five years of the experiment were 556 and 923 kg/ha at stocking rates of 3.6 and 4.6 beasts/ha respectively. Protein levels were again affected by the fertilizer levels; these were 7.4 and 3.1% in the winter and 12.0 and 7.8% in the summer, substantially lower than in the previous experiment. This probably was one of the causes of the lower liveweight gains, but the author also made the point that the pasture used earlier was older and denser and supported a higher stocking rate.

Evans (1973) commented on an experiment comparing high nitrogen levels with Pangola and Nandi Setaria pastures and stated that serious problems in pasture growth can occur at high nitrogen levels (above 476 kg/ha/year). Evans also stressed the sensitivity of pasture fertilizer budgets to changes in beef prices and fertilizer costs but reaffirmed that internal rates of return obtained from grass/N systems were satisfactory in comparison with legume-based pastures. The Australian work described so far in this chapter has been from the cooler subtropical regions of southern Queensland and New South Wales. However, work on fertilization of grass pastures has also been carried out on the tropical humid north coast region of Queensland.

Grof and Harding (1970) reported that fertilization of Guinea grass (*Panicum maximum*) with 168 kg/ha N increased liveweight gains form 345 kg/ha from the unfertilized pasture to 589 kg/ha, an increment of 1.9 kg liveweight gain/kg N applied. The experiment was carried out at Utchee Creek with cattle rotationally grazed at a stocking rate of 4.2 beasts/ha.

Norman (1974a) has reported beef production studies from Parada which must count as the highest liveweight gain recorded from grass alone in the tropics or subtropics. The experiments, which were carried out by J. Evans and J. Pulsford, involved three heavy applications of nitrogen (336, 532 and 673 kg/ha/year N) and three stocking rates on recently-established Pangola grass. Results are summarized in Table 10.13.

Table 10.13 Liveweight gains (kg/ha) from Pangola grass pastures fertilized with different levels of nitrogen at different stocking rates (Parada) (adapted from Norman 1974a)

Nitrogen level (kg/ha/year)	*Stocking rate* (beasts/ha)		
	7.4	*9.9*	*12.4*
336	1 888	2 150	—
532	1 908	2 178	2 614
673	—	2 372	2 760

The experiment lasted for three years and gains were maintained during this period except for the treatment with the highest level of nitrogen which was invaded by Couch grass in the third year causing gains to decline. Individual gains varied from 0.58 to 0.71 kg/day and thus the high stocking rates used did not affect the gain/head.

Evans and Pulsford (Norman 1974a) also compared levels of 112 and 336 kg/ha/N year and obtained liveweight gains of 1150 and 2051 kg/ha/year. This corresponds to a very effective use of nitrogen as 4.0 additional kg of liveweight gain were produced for each additional kg N applied.

Ebersohn and Lee (1972) have stated that 3001 kg/ha liveweight gain has been recorded at Parada from Pangola grass pastures fertilized at the rate of 674 kg/ha N. Unfortunately, no reference is made to the workers who achieved what must be the highest liveweight gain recorded from a tropical pasture (perhaps a single value from the previous study).

The Ord Valley Irrigation Scheme in the Kimberley region of Western Australia is another tropical area where soil moisture is not limiting. An experiment was carried out at the Kimberley Research Station on established Pangola grass pastures. Levels of nitrogen used ranged from 311–800 kg/ha and stocking rates from 5.1 to 12.6 weaners/ha using a central composite design. In this case, increased levels of nitrogen did not improve liveweight gain/ha, indicating that the level of 311 kg/ha was adequate. Production was probably therefore limited by some other factor. The author (Blunt 1978) suggests either low temperatures inhibiting grass growth during the cooler months or high temperatures depressing intake and therefore weight gains during the warmer months. The maximum gain obtained was 1330 kg/ha at a stocking rate of 11.5 steers/ha and 311 kg/ha N.

Discussion

In the majority of work quoted in this review, the application of nitrogen, with or without phosphorus and potassium. has increased pasture productivity, resulting in more forage being available to the grazing animal. Clearly, the degree to which such fertilization is effective depends on many factors. These include soil humidity, soil nitrogen and the form it is in, time of application of fertilizer, the presence or absence of other nutrients etc., and information given in publications is frequently inadequate to enable a complete interpretation of the results to be made. Similarly, although it is well known that time and frequency of application, for example, are critical in efficient utilization, no studies involving the grazing animal have been designed to measure this effect under tropical conditions. Excellent responses are obtained where rainfall is relatively high and well distributed throughout the year, as in Jamaica (Jamaica 1957) and Puerto Rico (Caro and Vicente 1972), or when supplementary irrigation is applied (Richards 1965).

Too little rainfall is a limiting factor to plant growth, whilst too much causes leaching. There is some discussion as to what is the minimum useful rainfall and this must depend on many factors, including soil type, distribution, rate of plant growth etc. Nuthall and Whiteman (1972) consider that an economic response to continuous nitrogen use would be unlikely at less than 45 in (1130 mm) under tropical conditions, whilst Caro and Vicente (1972) considered heavy (629 kg/ha N) fertilization only justified when there was more than 70 in (1750 mm) rainfall.

Some pastures respond more efficiently to application of high levels of nitrogen than others. The liveweight gains from highly-fertilized Napier grass in Puerto Rico have already been commented on but various strains of *Panicum maximum* have also been mentioned in this respect. Other grasses which respond well to high levels of nitrogen include Pangola grass, Star and Coastal Bermuda (Quinn et al. 1964; Garza et al. 1973).

In the relatively few cases where little or no increase in pasture productivity has occurred in response to nitrogen application, e.g. Lima et al. (1968), subsequent experimentation has shown other elements to be limiting (Lima et al. 1969).

Application of N also increases the amount of crude protein in the forage. The extent to which this occurs depends on the amount of nitrogen applied. From the response curve for a number of experiments reviewed in South Africa, Grunow et al. (1970) showed that the range was from less than 6% CP at 0 kg/ha N to about 10% CP at 400 kg/ha N with some indication that the curve was flattening at this level.

The response may be greater with improved grasses as Mears and

Humphreys (1974b) showed that the nitrogen content of *Pennisetum clandestinum* leaf varied from 1.5% at 0 kg/ha N to 4.0% at 672 kg/ha N (equivalent to 9.4–25.0% CP) on a heavily-stocked pasture, whilst the increase in the stem fraction was comparable.

This increased supply of nutrients is reflected in increased grazing capacity, either measured in grazing days (Quinn et al. 1964) or stocking rate (Rodel et al. 1975), and clearly increased liveweight gain per unit area must be an expression of both individual gain and stocking rate imposed (as will be discussed in greater detail elsewhere).

Despite all the variables involved, an attempt has been made throughout this review to express the response to nitrogen (only) as additional liveweight gain/kg N applied. The average value for this ratio in native pastures was 1.46 ± 0.72 (n = 11) and for improved grasses, 1.69 ± 0.87 kg liveweight gain/kg N applied (n = 61). It is somewhat surprising that the value for native grasses is so close to that for improved grasses in view of the theory that nitrogen is better applied to the latter (e.g. West 1956). Native grasses may be generally well adapted to low soil fertility and recovery of applied nitrogen is inefficient. However, applications of nitrogen to natural pastures have generally been small and it is unlikely that in most cases there were enough grass species able to respond to these quantities, whereas higher levels would not have been efficiently utilized. In any event, the most important aspect to be considered is that of cost effectiveness. Fertilization of veld may be recommended if the returns are sufficiently high.

The economics of nitrogen fertilization have been discussed by several authors (e.g. Ibach and Adams 1965; Nuthall and Whiteman 1972) and it would be presumptuous of the author to venture into this field. Nevertheless, there are several important management factors which should be taken into account when the economic viability of fertilizer application is considered. It has been stated already that fertilization enables higher stocking rates to be maintained. However, high stocking rates are frequently associated with lower *per capita* gains (this point is discussed more fully elsewhere) and thus animals are finished more slowly. It has also been noted that they may be deficient in fat cover (Winks et al. 1970) and thus command a lower market price. In addition to increased liveweight gain per unit area therefore, rate of turnover, price obtained for the finished beast and the interest charges involved in the larger number of animals on inventory at higher stocking rates must all be considered. Increasing stocking rate also presupposes that suitable animals are available to purchase when extra forage is available. This is often not the case in developing countries where markets are not well organized.

As would be expected, increased production of forage is also reflected in a higher potential stocking rate for beef cows and their calves and consequent improvement in weaning percentage as well as weaning weight. This therefore increases both per cow and per hectare performance (Hodges et al. 1977a). Unfortunately, little work has been done on this aspect of beef production from fertilized pastures.

Apart from the relatively large body of work from South Africa, little information is available on the effects of nitrogen fertilization on the botanical composition of natural pasture. However, it seems clear that there are species which increase and decrease in response to fertilization as there are species which react to grazing pressure in a similar manner. An isolated observation indicated that decreasers decrease faster than increasers increase at high levels of nitrogen application, resulting in a net decrease in ground cover (Edwards and Nel 1973a). It has also been suggested that large bunch species increase at the expense of small bunch species at high levels of nitrogen over a considerable period (although *Cynodon dactylon* is the stoloniferous exception), thus leaving the veld more liable to erosion. It has also been observed that this may lead to invasion by undesirable forb species (Grunow et al. 1970). It should perhaps also be mentioned in this context that high levels of nitrogen application will generally be accompanied by a higher stocking rate and it is likely that the effects of both management systems will be confounded. Grunow and his colleagues (Grunow et al. 1970) have also compiled a useful list of grass species which increase or decrease as a response to nitrogen fertilization under South African conditions.

Some improved species also become more dominant when fertilized with high levels of nitrogen, whilst others become less dominant. Mears and Humphreys (1974b) clearly demonstrated that *Pennisetum clandestinum* was in the former category whilst *Axonopus affinis* and *Paspalum notatum* were in the latter.

Summary

Work on fertilization of both native and improved pastures with nitrogen (with or without additional phosphorus or potassium) under tropical and subtropical conditions has been reviewed. Although the response to fertilization depends on many factors, it has normally resulted in increased forage becoming available for the grazing animal.

It has also been demonstrated that some grasses respond more effectively than others to nitrogen fertilization. The response to nitrogen has not only been noted in terms of increased forage production *per se*, but as an increased supply of nutrients, particularly crude protein, which result in improved grazing capacity of the pastures as measured as either grazing days or stocking rate.

Responses have been evaluated in terms of beef produced per unit of nitrogen applied and this ratio was approximately 1.5 : 1 and 1.7 : 1 in native and improved grasses respectively. The economic advisability of nitrogen application must depend on the prevailing prices of both fertilizer and beef, but it is also emphasized that management must be such that full advantage can be taken of the increased forage availability by increasing stocking rates. This involves more animals on inventory for a greater length of time (as gains are normally inversely related to stocking rate) and may also result in deficient fat cover of slaughter stock.

Application of nitrogen also results in improved performance of breeding stock in terms of both weaning percentage and weaning weight. However, little information is available on this point.

Finally, work on the effects of nitrogen fertilization on botanical composition of pastures has been reviewed and it has been noted that species may be 'increasers' or 'decreasers' in response to fertilizer application.

11 Stocking rate and the improvement of beef production

Defoliation of grasses by the grazing animal involves a number of morphological and physiological changes. These include a reduction of the leaf area and thus the photosynthetic capacity of the plant as well as the carbohydrate held in reserve. All aspects of growth are controlled by the shoot apex, i.e. the apical meristem and the sub-apical region where the leaf primordia are to be found. Removal of the shoot apex by grazing during either the vegetative or reproductive phase exerts an effect on the subsequent tillering and regeneration of the plant and consequently on its productivity. The seasonal development of the shoot apex varies from one grass to another and therefore the season of defoliation also exerts an important effect on productivity.

Defoliation can be harmful when so much vegetative material is removed that the vigour of the plant is lessened. This reduces competitiveness and initiates pasture deterioration. The more desirable and palatable species decrease and less palatable and nutritious species increase; reduced vigour and reproductive performance allow the invasion of less desirable 'invaders', bush encroachment may occur or soil may be removed by the erosion caused by reduced water-holding capacity. Overgrazing is a phenomenon increasingly obvious in many parts of the tropics and subtropics and is one of the major constraints to livestock production in many parts of Africa and Latin America where uncontrolled grazing is practised.

Undergrazing can also be harmful; a build-up of material of low nutritive value occurs, shading reduces photosynthetic activity, conditions prevail which encourage the proliferation of pests and diseases, and forage production and animal productivity per unit area are reduced.

It is clear, therefore, that both intensity and timing of defoliation is of the greatest importance in pasture productivity. However, as the timing of defoliation is largely determined by the grazing system used and is generally less important than the extent of defoliation which is determined by the stocking rate, the latter will be dealt with first. Trials on the productivity of pasture at different stocking rates have been carried out in both Africa and Australia, but surprisingly little information is available from elsewhere. In most cases, treatments have been limited to three or less stocking rates, thus making estimation of the optimal carrying capacity of the pasture hazardous. Similarly, differences in rainfall in longer-term trials make results difficult to interpret.

Data on productivity of pastures at different stocking rates carried out in the tropics and subtropics will be briefly reviewed and the relationships between pasture productivity and stocking rate will be discussed. Data on liveweight gains both per head and per hectare from trials involving three or more stocking rates have been grouped together in Table 11.1 for ease of comparison and derivation of the various relationships discussed. Many interesting botanical data are given, especially from Africa, illustrating the effect of grazing pressure on different species. These are discussed in connection with currently accepted theories of ecology and range management derived from experience in other countries.

Africa

Early work was carried out on the thornveld at Matopos in Zimbabwe, and the predominantly *Hyparrhenia* veld in Zambia. In Uganda, a series of trials was established on natural pastures at Makyesasa and Ankole as well as on improved pastures at Serere. Isolated studies have been carried out in Nigeria, South Africa and Malawi.

Zimbabwe

The earlier work carried out at the Matopos Experimental Station near Bulawayo has been reviewed by Kennan (1969). The grass cover of the thornveld where the trials were located was largely composed of *Themeda triandra* and *Heteropogon contortus*. Stocking rates and weight gains are given in Table 11.2. The results over a twelve-year period confirm the finding that heavy stocking usually provides the highest return in animal gains per unit area, but in all the experiments quoted by Kennan (1969) heavy rates of stocking caused palatable perennial grass species to be replaced by pioneer annual grasses whilst the relatively unpalatable perennial species remained vigorous. Similarly, bush encroachment was noted on several sites as a result of less vigorous competition from grasses enabling seedlings of woody species to become established whilst, at the same time, resulting in less grass being produced as fuel for fires which would normally control bush.

Some results from trials using different stocking rates on predominantly *Hyparrhenia/Hyperthelia* spp. veld are available from the Grass-

Table 11.1 Weight gains of cattle grazing tropical pastures at different stocking rates

Stocking rate (beasts/ha)[1]	*Weight gain* (kg/ha)	(kg/head)[2]	*Comments*	*Reference*
Africa				
0.7	38.4	58.2	Mazabuka, Zambia; *Hyparrhenia* veld and dambo	Smith (1966)
1.0	56.8	56.8	rainy season	
2.0	100.0	50.0		
0.8	76.8	116.4	dry season	
1.3	113.4	98.6		
2.0	197.2	98.6		
0.7	70.8	107.3	Marandellas, Zimbabwe; natural veld summer rainy	*Grasslands Res. Sta.*
0.5	59.9	119.8	season only; mean of two years' results	*Ann. Rep* 67/68,
0.4	53.5	133.9		68/69
1.6	92	147	Ukiriguru, Tanzania; natural veld; total gain during	Walker and Scott
0.8	132	105	93-week period	(1968a)
0.4	152	61		
0.31	21.8	70.2	Matopos, Zimbabwe; natural veld; mean of two	Kennan (1969)
0.23	16.8	72.6	years' results	
0.19	13.9	75.0		
0.14	12.0	87.7		
4.9	1270.8	0.25	Serere, Uganda; *Hyparrhenia*/Stylo (1050 days)	Stobbs (1969d)
2.5	752.8	0.25		
1.6	495.5	0.29		

2	449	114	Serere, Uganda; *Hyparrhenia*/Stylo 672 days	Stobbs (1970)
3	816	138		
5	1 179	119		
2	534	136	Serere, Uganda; *Panicum*/Siratro	
3	685	118		
5	944	96		
7.9	135	17.1	Shika, Nigeria; *Digitaria smutsii*/Stylo (August–November, 1966 and 1967)	Haggar (1971)
9.9	153	15.5		
13.2	70	5.3		
7.9	151	19.1	As above, *D. smutsii*, no N	
9.9	101	10.2		
13.2	–3	–0.2		
7.9	217	27.5	As above, *D. smutzii*, 84 kg/ha N	
9.9	300	30.3		
13.2	290	22.0		
7.9	272	34.4	As above, *D. smutsii*, 168 kg/ha N	
9.9	343	34.6		
13.2	345	26.1		
7.9	292	37.0	As above, *D. smutsii*, 253 kg/ha N	
9.9	276	27.9		
13.2	421	31.9		
0.5	134.3	258.9	Zwartrand, South Africa; natural veld, total of three grazing seasons	Edwards and Nel (1973a)
0.8	166.6	214.1		
1.6	105.8	68.2		
12.0	354.8	27.9	Henderson, Zimbabawe; fertilized rotationally-grazed, No. 2 Star grass (carcase gain/ha)	Rodel (1974)
14.7	367.5	25.0		
17.3	254.3	14.7		
20.0	122.0	6.1		

Table 11.1 (contd)

Stocking rate (beasts/ha)[1]	*Weight gain* (kg/ha)	(kg/head)[2]	*Comments*	*Reference*
0.4	55.8	136	Ankole, Uganda: *Cymbopogon/Themeda/Hyparrhenia* veld (differing periods from 1961 to 1972)	Harrington and Pratchett (1974a)
0.8	86.9	106		
1.6	117.2	71		
2.5	370	148	Lilongwe, Malawi; fertilized Rhodes grass (mean of three seasons, average length 297 days)	Addy and Thomas (1978)
5	524	105		
7.5	539	72		
Australia				
1.7	289	175	Lawes; *Sorghum almum* (mean of two years)	Yates et al. (1964)
2.5	361	146		
5.0	391	79		
2.5	384	155.5	Beerwah; grass legume mixtures (mean of three years, 1965–8)	Evans (1970)
1.7	326	197.6		
1.2	295	239.8		
5.0	590	118.0	Coolum; Pangola grass/white clover/*Lotononis bainesii* (summer gains)	Young and Chippendale (1970)
2.5	419	167.8		
1.7	294	176.2		
3.3	339	101.8	*Setaria anceps* with legumes as above	
2.5	311	124.5		
2.0	267	133.8		
2.2	282	128	Narayen; Buffel grass with 168 kg/ha N (one year's data)	' tMannetje (1972)
1.5	237	158		
1.1	178	162		

1.2	264	0.62	Beerwah; grass/legume mixtures (mean of six years: 1966–71)	Evans and Bryan (1973)
1.6	295	0.54		
2.5	328	0.41		
2.2	171.3	0.44	Wollongbar; *Pennisetum clandestinum*, summer 1971, 177 days, no N	Mears and Humphreys (1974b)
3.3	257.0	0.44		
4.9	156.1	0.18		
3.3	344.6	0.59	134 kg/ha N	
4.9	338.3	0.39		
7.4	327.5	0.25		
4.9	399.0	0.46	336 kg/ha N	
7.4	576.3	0.44		
11.1	805.5	0.41		
7.4	641.8	0.49	672 kg/ha N	
11.1	844.9	0.43		
16.6	940.2	0.32		
2.2	252.6	0.56	Summer 1972, 205 days, no N	
3.3	297.7	0.44		
4.9	401.8	0.40		
3.3	405.9	0.60	134 kg/ha N	
4.9	431.9	0.43		
7.4	515.8	0.34		
4.9	562.5	0.56	336 kg/ha N	
7.4	773.7	0.51		
11.1	1 115.0	0.49		
7.4	804.0	0.53	672 kg/ha N	
11.1	1 069.5	0.47		
16.6	1 497	0.44		

Table 11.1 (contd)

Stocking rate (beasts/ha)[1]	*Weight gain* (kg/ha)	*Weight gain* (kg/head)[2]	*Comments*	*Reference*
2.2	115.8	0.28	Winter 1971, 188 days, no N	
3.3	130.3	0.21		
4.9	101.3	0.11		
3.3	117.9	0.19	134 kg/ha N	
4.9	165.8	0.18		
7.4	13.9	0.01		
4.9	414.5	0.45	336 kg/ha N	
7.4	306.1	0.22		
11.1	–41.7	–0.02		
7.4	417.4	0.30	672 kg/ha N	
11.1	250.4	0.12		
16.6	–249.7	–0.08		
2.2	147.8	0.42	Winter 1972, 160 days, no N	
3.3	323.3	0.44		
4.9	243.0	0.31		
3.3	190.1	0.36	134 kg/ha N	
4.9	282.2	0.36		
7.4	153.9	0.13		
4.9	501.8	0.64	336 kg/ha N	
7.4	414.4	0.35		
11.1	319.7	0.18		
7.4	532.8	0.45	672 kg/ha N	
11.1	479.5	0.27		
16.6	239.0	–0.09		

0.8	150	183	years	(1974)
1.2	213	173		
0.6	112	182	Broweena; *Heteropogon contortus*/Siratro, mean of five years	
0.8	106	130		
1.24	115	93		
3.0	383.0	129.4	Samford; Nandi Setaria; 336 kg/ha N, mean of three years	Jones (1974)
3.6	428.2	119.6		
4.2	461.2	109.8		
4.8	484.0	100.0		
5.4	491.2	90.3		
1.1	197.8	178.2	Setaria/*Desmodium*	
1.7	247.4	143.0		
2.4	253.6	107.9		
3.0	217.0	73.3		
1.1	181.0	163.1	Setaria/Siratro	
1.7	235.3	136.0		
2.4	255.9	108.9		
3.0	243.6	82.3		
4.9	580	0.33	Kimberley; irrigated Pangola/Leucaena, 375 days	Blunt (1976)
6.2	830	0.37		
7.4	710	0.26		
8.6	490	0.16		
7.4	1 282	0.56	Ayr; irrigated Pangola grass, mean of three years	Deans et al. (1976)
9.9	1 536	0.50		
12.4	1 697	0.44		
0.7	97.6	0.38	Cape York Peninsula; grass/legume, mean of three years	Winter et al. (1977b)
1.2	155.1	0.35		
1.7	197.1	0.32		
2.2	204.4	0.29		

Table 11.1 (contd)

Stocking rate (beasts/ha)[1]	*Weight gain* (kg/ha)	(kg/head)[2]	*Comments*	*Reference*
5.1	178.8	Kimberley;	Blunt (1978)	
6.3	167.0	irrigated		
8.9	141.4	Pangola		
11.5	115.8	grass,		
12.6	105.0	mean of		
		two years		
911.9				
1 052.1				
1 258.5				
1 331.7				
1 323.0				
Elsewhere	2.6	0.63	Matanzas, Cuba; fertilized Pangola grass, wet season of about six months, daily gains	Veitia et al. (1974a)
4.2	2.8	0.49		
5.7	3.2	0.45		
7.1				
2	189.0	0.33	Johore, Malaysia; Guinea grass/legume, mean of 2.5 years	Eng et al. (1978b)
4	348.7	0.30		
6	368.7	0.22		
5	305	0.17	Nawaicoba, Fiji; Nadi blue grass/Siratro fertilized with superphosphate	Partridge (1979)
2.5	352	0.39		
1.3	225	0.49		
2.5	143	0.16	Nadi blue grass, unimproved, mean of four years	
1.3	116	0.26		
0.8	90	0.30		

[1] Stocking rates have been rounded to the first decimal place except when this would have obscured a treatment difference; e.g. Kennan (1969)

Table 11.2 Productivity of three grasses stocked at two different stocking rates (Henderson) (adapted from Rodel and Boultwood 1967)

Grass	*Liveweight gain* (kg/ha)	
	7.5 steers/ha	*12.5 steers/ha*
Giant Rhodes	644	596
No. 2 Star	415	652
Panicum repens	637	972

lands Research Station for summer grazing periods. Results for two summer seasons are given in Table 11.1.

Later work in Zimbabwe was concerned with improved pastures. The differences in resistance to grazing among various introduced grasses was pointed out by Rodel and Boultwood (1967). Giant Rhodes gave results as good as did *Panicum repens* when stocked at a rate of 7.5 steers/ha but this was not the case at 12.5 steers/ha, and the Giant Rhodes was rapidly invaded by *Cynodon dactylon* varieties and Star grass. The authors point out that there was 'a marked difference between grasses in ability to withstand intensive grazing'. Production of the various grasses at the two stocking rates is given in Table 11.2.

In a further trial at Henderson, Rodel (1974) measured the carcass gain of cross-bred steers grazing a pasture of No. 2 Star at four different stocking rates (results are given in Table 11.1). Each group of steers grazed fifteen paddocks in rotation. Although pastures had been fertilized, low gains were explained by the presence of numerous biting flies and waterlogged pastures.

The complex trial, involving four levels of nitrogen application each at four stocking rates, has already been mentioned in connection with fertilization.

Zambia

Both trials on the effects of different stocking rates which have been reported from Zambia were carried out on natural grazing at Mazabuka. This was tall *Hyparrhenia* veld with some browse, the main grasses being *Hyperthelia dissoluta, Hyparrhenia filipendula* and *Heteropogon contortus*. The first trial investigated the grazing behaviour of indigenous cattle at light (0.2 beast/ha) and heavy (0.73 beast/ha) stocking rates (Smith 1959). Observations indicated that the limiting factor for the heavily-stocked animals was the ability to graze selectively so as to obtain a better diet than that afforded by Hyparrhenia stems and senescent leaves. Total herbage dry matter *per se* was not limiting. There was little difference in grazing times between the two groups in the dry season during which the heavily-stocked group lost more weight than the other cattle, but it was interesting that the heavily-stocked group grazed up to 43% longer when there was forage available. This was associated with a higher rate of daily weight gain (1.45 v. 0.95 kg/head). It was also noted that both groups increased their grazing time (to about 13 hours during the dry season v. about 8 hours during periods of feed abundance) and it was suggested that the lack of difference between groups was because grazing had reached its practical maximum as both groups were losing weight. Pasture differences were noted between the two grazing regimes in that, in the undergrazed area, the pasture formed a fairly uniform cover (although towards the end of the season, circular closely-grazed areas became apparent; cf. the 'mosaic' of Edye et al. 1978) whilst in the heavily-grazed areas the *Hyparrhenia* spp. appeared less vigorous and there was an increase in low growing annual grass species as well as bare ground amounting to some 30% of the paddock (no comparable figure for bare ground in the other pasture is given).

In later trials at Mazabuka, Smith (1966) compared different stocking rates during wet and dry seasons on upland and 'dambo' pastures (dambos are waterlogged riverine areas). It was noted that stocking rates exerted a significant negative result on liveweight gain/head although liveweight gain/ha was higher at heavier stocking rates. The greater efficiency of these was associated with more efficient use of the nutritious pasture produced in the dry season on the dambo areas. Stocking rates and liveweight data are presented in Table 11.1.

Some interesting observations on the botanical composition of the dambo margin was given which shed light as to which of the common grasses to be found in Central Africa may be considered decreasing and increasing grass species. These are summarized in Table 11.3.

Table 11.3 The effects of different stocking rates on botanical composition and forage yields of a dambo margin (Mazabuka) (adapted from Smith 1966)

	Botanical composition (% dry weight)	
	Heavily grazed (*1.23 beasts/ha*)	*Lightly grazed* (*0.31 beasts/ha*)
Hyparrhenia spp.	31	74
Heteropogon contortus	22	15
Cynodon dactylon	15	3
Setaria sphacelata	15	6
Brachiaria spp.	14	—
Other spp.	3	2
Yield (kg DM/ha)	2 247	2 742

These results show very clearly that *Hyparrhenia* spp. decrease when grazing pressure increases whilst the other species mentioned show the reverse tendency or are 'increasers'. It is also clear that the forage yield is reduced by overgrazing. However, the author states that the grass cover recuperated after one year's rest following four years' heavy grazing, but it is not stated whether the recuperation was qualitative as well as quantitative.

Uganda

A series of trials was carried out on different stocking rates in Uganda over a period of more than ten years. These were on natural grazing on the *Pennisetum purpureum/Imperata cylindrica* tall grass areas of Buganda and the fire-induced *Acacia/Cymbopogon/Themeda* savannah in Ankole. Trials were also carried out on sown, intensively-managed grass/legume pastures at the Serere Research Station in eastern Uganda.

The trial in Buganda was carried out at the Livestock Husbandry Experimental Unit, Nakyesasa, and the stocking rates used were 1.23, 2.5 and 5 beasts/ha. The results of the first year of the trial were apparently described by Harker and McKay (1962), but I have not been able to trace this reference. However, in the second year of the trial, gains were 0.20, 0.14 and 0.01 kg/day/head for the three treatments in the same order as above (Harker 1963).

Thornton (1970) summarized the results of the last two years of the trial when the heaviest stocking treatment had been alleviated to 3.29 beasts/ha. These are given in Table 11.4.

Table 11.4 Liveweight gains at different stocking rates from natural pasture (Buganda) (adapted from Thornton 1970)

Stocking rate (beasts/ha)	*Liveweight gain*	
	(kg/head)	*(kg/ha)*
1.23	85	105
2.5	92	230
3.29	93	306

These results show the usual relationship between gains per beast which decline with increased stocking rate and gains per hectare which increase. Possibly the most interesting observation from this trial was that both *Pennisetum purpureum* and *Imperata cylindrica* decreased under heavy grazing whilst *Cynodon dactylon, Setaria sphacelata, Brachiaria* spp. and *Panicum maximum* increased. It was commented that the increase in lower growing spp. (*Cynodon, Brachiaria*) would result in lower total production of dry matter but that this would have a higher nutritive value.

A series of trials was started at Serere by Stobbs (1969b). Pastures of *Hyparrhenia rufa* and *Stylosanthes guianensis* were grazed by small East African Zebu steers at different stocking rates and details of these as well as liveweight gains are given in Table 11.1. Interesting information on botanical composition is also given which confirms earlier findings. These are summarized in Table 11.5.

Table 11.5 Pasture botanical composition at different stocking rates (Serere) (adapted from Stobbs 1969b)

Species	*Stocking rate (beasts/ha)*		
	Low (1.64)	*Medium (2.47)*	*High (4.94)*
Hyparrhenia rufa (%)	70.8	68.5	25.7
Stylosanthes guianensis (%)	18.9	22.0	20.2
Sporobolus pyramidalis (%)	0.5	1.6	38.7

The above confirms the results of Smith (1966) that *Hyparrhenia* spp. do not withstand heavy grazing and also show the resistance of the unpalatable *Sporobolus pyramidalis* to high grazing pressures.

A further trial (Stobbs 1970) confirmed this finding for *Hyparrhenia rufa* at stocking rates of 2, 3 and 5 beasts/ha and also showed that *Pancium maximum* was a decreaser when in association with Siratro. There was again confirmation of the role of *Sporobolus pyramidalis* as an increaser. Results are summarized in Table 11.6.

Table 11.6 Summary of herbage yields and botanical composition of pastures at different stocking rates (Serere) (adapted from Stobbs 1970)

	Stocking rate (beasts/ha)		
	2	3	5
Hyparrhenia/Stylo pasture			
Hyparrhenia rufa (%)	57.9	43.5	16.9
Stylosanthes guianensis (%)	32.2	43.6	63.9
Weeds (monocots.) (%)	3.1	7.9	17.0
Weeds (dicots.) (%)	6.8	5.2	2.3
Total yield (kg DM/ha/100)	67.2	43.7	29.0
Panicum/Siratro pasture			
Panicum maximum	73.0	68.4	31.6
Siratro	3.3	5.7	49.5
Weeds (monocots.) (%)	3.2	3.7	15.4
Weeds (dicots.) (%)	20.5	22.1	3.5
Total yield (kg DM/ha/100)	45.1	26.0	6.3

The monocotyledenous weeds referred to largely comprised *S. pyramidalis*. It is interesting that the dicotyledenous weeds present in the low stocking rate treatment were reduced at the heaviest stocking rates; this is contrary to usual thinking with regard to bush encroachment but, unfortunately, no details as to species found are presented. It is also notable that both legumes flourished at heavy stocking rates; this would tend to keep individual animal gains higher than would otherwise have been expected. As would be expected, yields of dry matter were considerably decreased at high stocking rates. Results on liveweight gains obtained in this experiment are summarized in Table 11.1. The work of Stobbs and Whitting (1970) has not been included here because stocking rates were adjusted according to herbage availability.

The effect of stocking rate on the range land of western Uganda was investigated by Thornton and Harrington (1971). Range was dominated by *Themeda triandra* or *Cymbopogon afronardus* with 'abundant' *Brachiaria decumbens* (no quantitative data on botanical composition are given). Stocking rates of 0.82, 0.41 and 0.20 beast/ha were used. It was clear that the heavier stocking rates favoured the procumbent grasses, particularly *B. decumbens*, rather than the taller, fire-climax grasses, and it was noted that these were generally of a higher nutritive value. Lower grazing pressures resulted in uneven grazing patterns and in closely-grazed areas, which were normally also dominated by *B. decumbens*. It was suggested by earlier work (Harrington and Thornton 1969) that even heavier stocking rates would result in an overall dominance of this grass.

Subsequent results from the same trial have been reported by Harrington and Pratchett (1974a) and these are summarized in Table 11.1. Attention was drawn once more to the increase in frequency of occurrence (% of total quadrats in which species is recorded) of *B. decumbens* with higher stocking rates. This observation was coupled with analysis of samples from a steer with an oesophageal fistula which showed that animals grazing highly-stocked areas could select a diet of higher protein content (10.0 v. 8.8% CP for stocking rates of 0.82 and 0.41 beast/ha respectively) because of the presence of more *B. decumbens*. This also accounted for the fact that gains from individual animals grazed at high stocking rates were higher than would have been expected.

These trials from various parts of Uganda add considerably to our knowledge of the reaction of various grasses to different grazing pressures as well as to the productivity of both natural and improved pastures at different stocking rates.

Tanzania

Walker and Scott (1968a) compared three stocking rates on the cemented hardpan soils of Ukiriguru Experimental Farm near Mwanza in Sukumaland (mean annual precipitation 825 mm). The stocking rates used were 1.6, 0.8 and 0.4 steers/ha and liveweight gain data are given in Table 11.1. Greatest gains/head resulted from the lowest stocking rate but more liveweight gain/ha was obtained from the heaviest stocking rate and it was concluded that its use was to be recommended, particularly since botanical observations did not reveal any deterioration of the predominantly *Bothriochloa insculpta* veld (although some increase in *Cynodon dactylon* was observed).

Republic of South Africa

Surprisingly, less attention has been given to stocking rate than to other aspects of grassland productivity in South Africa. Edwards and Nel (1973a) carried out studies on the Bankenveld at Zwartrand where the predominant species in the extremely sour grazing are *Andropogon amplectens, Digitaria eriantha, Setaria flabellata* and *Themeda triandra.* Stocking rates of 0.52, 0.78 and 1.56 mature livestock units/ha were used and production data are given in Table 11.1. Increasing stocking rate caused a significant decrease in *Themeda triandra* (generally regarded as a palatable grass of good nutritive value under sour veld conditions although it may not be so considered elsewhere; see Bogdan 1977) from 7.2 to 5.3% as well as decreases in various other grasses including *Andropogon amplectens* and *Loudetia simplex*. There was also a reduction in overall cover from 21.5 on the low to 19.1% on the high-stocking rate. Clearly such differences may be of considerable importance in grasslands where bunch grasses predominate and cover is poor even under light stocking rates.

Malawi

The Chitedze workers (Addy and Thomas 1978) carried out a trial using three stocking rates (2.5, 5 and 7.5 beasts/ha) on fertilized Rhodes grass pastures (productivity results are given in Table 11.1). They reported that little weed encroachment had occurred on the higher stocking rate treatments, contrary to the evidence of other workers. They attributed this to the fact that the stoloniferous Rhodes grass was more aggressive and competes better with weeds and is more tolerant of heavy grazing than the bunch grasses used elsewhere. This is particularly interesting as *Sporobolus pyramidalis* is a prevalent weed of overgrazed pastures on the Lilongwe plains.

Australia

As would be expected, most of the work carried out on different stocking rates in Australia has been concerned with pastures containing legumes, either reinforcing natural pastures or sown with improved grasses. Trials have been reported from throughout

Queensland, parts of New South Wales and an irrigation project in Western Australia.

North Queensland

Workers of the Queensland Department of Primary Inudstries carried out a trial employing two stocking rates (2.5 and 1.25 steers/ha) on mixed *Neonotonia wightii/Panicum maximum* pastures at Kairi near Atherton. Body weight gains were 0.63 and 0.79 kg/head/day for the two treatments respectively, resulting in 578 and 348 kg/ha. The authors (Winks et al. 1970) also reported a lower killing-out percentage of those animals on the high stocking rate (54.7 v. 55.3%) and point out that this was associated with an excessively fat carcass at the low stocking rate and too little finish at the high stocking rate. It was also commented that there was insufficient stocking pressure during the summer months (the rainfall at Kairi is 1230 mm, which falls mainly during the summer) and that undergrazing occurred at this time. The high stocking rate appeared to provide a near to optimum grazing pressure during the winter months, however.

Gillard (1970) carried out a trial on improved native pastures in a contrasting part of north Queensland, west of the Great Dividing Range where rainfall is only 630 mm annually with potential evaporation of about 2000 mm. Normal stocking rate on commercial properties is from 15 to 20 ha/head. By a combination of fertilization, bush clearing and reinforcement with *Stylosanthes humilis*, it was possible to increase stocking rates to 0.21 and 0.41 beast/ha. Productivity was higher on the higher stocking rate (34 v. 26 kg/ha) but individual liveweight gains were 124 and 84 kg for low and high stocking rates respectively; differences were smaller in high rainfall years. Herbage production was generally less on the highly-stocked treatments.

Shaw initiated a series of stocking rate trials at Todds Bay in 1959 which were to continue until 1973. Native pasture, whose most important grass components were *Heteropogon contortus, Chrysopogon fallax* and *Eragrostis* spp., was stocked at 0.27 and 0.62 beast/ha. Average liveweight gains for the six-year period from 1959–66 were 25 and 29 kg for low and high stocking rates respectively. The botanical composition of the low stocking rate treatment remained remarkably constant throughout the period but pronounced changes occurred in the higher stocking rate pastures. *Heteropogon contortus* decreased so much that it was no longer the dominant species, the decrease representing a combination of both fewer and smaller plants. There were associated increases of *Eremochloa bimaculata, Chrysopogon fallax, Tripogon loliiformis* and *Brachiaria miliiformis*. However, the major change was that vegetative cover was severely decreased; bare ground increased, total grass cover decreased and forb spp. increased. Similar trends were reflected by results for percentage dry matter yield of main species. The contribution of *Heteropogon contortus* was

59 v. 17% in low and high stocking rates respectively, *H. contortus* being largely replaced by *Eremochloa bimaculata* (Shaw and 'tMannetje 1970).

A further trial at Todds Bay clarified some of the results of the previous trial, started in 1966 (Shaw 1978a). Treatments included fertilizer rates, sowing method and stocking rates. Stocking rates ranged from 0.41 to 1.37 beasts/ha and relevant livestock production data are listed in Table 11.1 (Shaw 1978b). Botanical analyses of the various pastures showed that Townsville stylo (which was oversown in all treatments), *Sida acuta* and *Bothriochlea decipiens* increased at high stocking rates whilst *Themeda australis* and *Neptunia gracilis* declined. *Heteropogon contortus* remained dominant in all treatments and botanical changes were small. This was in contrast to the results reported for the earlier experiment where the grass was no longer dominant after heavy stocking without reinforcement with Townsville stylo (although the period of the first trial included two dry years). The author comments that the relevance of the work 'concerns the evidence for botanical stability in the presence of treatments that had large effects on pasture yield and animal production' (Shaw and Dale 1978).

Cape York Peninsula

Winter et al. (1977b) investigated the effects of stocking rate on beef production on grass/legume pastures at Heathlands in the northern Cape York Peninsula. The pastures used were *Brachiaria decumbens* cv. Basilisk or *Panicum maximum*, each sown with *Stylosanthes guianensis* cv. Endeavour and *Macroptilium atropurpureum* cv. Siratro. The stocking rates used were originally 0.7, 1.2, 1.7 and 2.2 beasts/ha but two of the Panicum pastures 'broke down' at the highest stocking rate and this was subsequently reduced to 1.9 beasts/ha. Results for liveweight gain both per beast and per hectare are summarized in Table 11.1. There was little difference between gains/head at the various stocking rates and this was non-significant (although there was a clear trend indicating a lower gain/head at higher stocking rates when the average values for the three-year period were considered, which was attributable to the results of the first year of the experiment).

Winter et al. (1977a) reported on the botanical and chemical composition of the pastures. Stocking rate had little effect on botanical composition except that there was more green Siratro at higher stocking rates, particularly during the first two years of the experiment. It was also noted that *Brachiaria decumbens* was highly resistant to grazing. This is in agreement with the African work which noted that *Brachiaria* spp. tend to be 'increasers' (Smith 1966; Thornton 1970; Thornton and Harrington 1971; Harrington and Pratchett 1974) and that *Panicum maximum* tends to be a 'decreaser' in association with Siratro (Stobbs 1970).

There was an interesting effect of stocking rate on chemical composition in that it was noted that stocking rates of 2.2 or 1.9 beasts/ha gave slightly but consistently more nitrogen, phosphorus and sulphur in the forage than the lower stocking rate. This is also in accord with the oesophageal fistula work carried out by Harrington and Pratchett 1974b. In the latter case, this was due to the presence of the 'increaser' *Brachiaria decumbens*, but probably because of the stimulation of tillering by grazing and consequent new growth in the former.

Edye et al. (1978) carried out a further analysis of the Heathlands data to explore the reduced effect of stocking rate on gain of individual animals. They postulated that this was due to an antagonistic effect between wet and dry season performance. During the dry season, the commonly-observed effect of increased stocking rate, i.e. reduced gain per animal, was observed. However, during the succeeding wet season, the animals from the higher stocking rate treatments grew faster when ample herbage was available because of compensatory growth. At the same time, the gain of animals on the lower stocking rate treatments tended to be reduced because of uneven grazing, the accumulation of rank herbage and the consequent formation of 'mosaics' of closely-grazed and ungrazed pastures. It was estimated that 'effective' stocking rate amounted to 45% of the 'imposed' stocking rate at 1.2 beasts/ha and only 29% at 0.7 beast/ha. The authors concluded that grazing behaviour was an important variable in estimation of the optimum stocking rate and also that adequate stocking rates for the wet season would be much higher than those commonly considered suitable for year-long or particularly dry season grazing. This blinding glimpse of the obvious emphasizes one of the most salient problems of tropical grassland management of which more later.

South-east Queensland and New South Wales

A substantial amount of work has been carried out on stocking rate in this area, both from the Brigalow and similar country, as well as in coastal high rainfall areas. The first trial to be reported appears to be that by Yates and his colleagues who failed to establish legumes with *Sorghum alum* at the Queensland Agricultural College at Lawes where conditions are similar to those on the margin of the Brigalow region. Stocking rates used were 1.67, 2.5 and 5.0 beasts/ha and liveweight gains are given in Table 11.1. The authors were the first to confirm in Australia the close relation between stocking rate and total animal production on a basis of unit area. They also considered the adverse effect of lowered liveweight gains on carcass quality and market value (Yates et al. 1964).

Different stocking rates on fertilized Buffel grass were compared by 't Mannetje (1972) at Narayen. Stocking rates were 2.2, 1.5 and 1.1 beasts/ha and liveweight data are given in Table 11.1. The author emphasizes the profound effect of stocking rate on profitability. Low stocking rates lead to under-utilization and consequent low returns on investment whilst overstocking leads to poor carcass quality and over-investment.

Bewg et al. (1970) carried out a stocking rate trial on the coastal ranges at Oakwood, 40 km south of Gympie. Pastures were sown with a mixture of *Neonotonia wightii, Desmodium intortum, D. uncinatum* and *Macroptilium atropurpureum*. Stocking rates were 1.23 and 1.82 beasts/ha. Mean daily gains for the whole period of the trial (379 days) were 0.45 and 0.39 kg/head/day for low and high stocking rates respectively, whilst the liveweight gains on a unit area basis were 209.8 and 269.0 kg/ha. The legume was present at approximately 40% at the lower stocking rate and 25% at the higher respectively.

Bisset and Marlow (1974) carried out research on the coastal foothills at Burnett. Trials were located at two sites, Lowmead and Broweena, and the stocking rates used were 1.24, 0.82 and 0.62 beasts/ha. Liveweight gains are given in Table 11.1. It was noted that only animals from the low stocking rate were fit for slaughter during most of the trial. The Siratro population was reduced at high stocking rates.

A number of trials were carried out in the Brisbane area by both CSIRO and DPI workers. Moir and his colleagues compared stocking rates of 2.5 and 5 beasts/ha on *Paspalum dilatatum*-dominant pasture at Rocklea (Moir et al. 1969). Mean changes in fasted body weight totalled 719 in the light v. 412 kg/head in the heavy stocking rate treatment. There was thus more production per animal but less production per hectare at low than at high stocking rates. The higher stocking rate resulted in a significant decrease in *P. dilatatum* and a marked, though non-significant, increase in Couch grass (*Cynodon dactylon*).

Working at the CSIRO Pasture Research Station at Samford, Jones (1974) carried out a comprehensive study of the effects of different stocking rates on pastures of *Setaria sphacelata* cv. Nandi plus greenleaf Desmodium, Nandi Setaria plus Siratro, or Nandi Setaria plus 336 kg/ha N. A wide range of stocking rates were used and liveweight gain results are reported in Table 11.1. Liveweight gains/head decreased with increasing stocking rate as would be expected, but the slope for the fertilized grass pasture was much less than for the grass/legume pastures. There was also a lower gain/animal on the fertilized than on the grass/legume pasture, at optimum stocking rates. The turn-off rate from the former would therefore be lower and animals would not be finished before 2.6 to 3.2 years. It was also noted that the legume declined at high stocking rates (for example, the percentage contribution to the total yield by Siratro fell from 41 to 9%). This would cause a marked decline in nitrogen input and therefore the legume pastures would be more likely to break down at high stocking rates than the nitrogen-fertilized pastures where the input of nitrogen was constant. At the same time, there was a considerable increase in weeds, mainly Queensland Blue Couch (*Digitaria didactyla*), at the higher stocking rates.

The Wallum is an area in central lowlands of southern Queensland where the mean annual precipitation is 1650 mm but where the native vegetation is incapable of carrying productive livestock. Evans and Bryan carried out stocking rate trials on mixed pastures at the CSIRO Pasture Research Station at Beerwah. Mixed pastures of *Paspalum dilatatum, P. commersonii, P. plicatulum, Chloris gayana* and *Digitaria decumbens*, and *Desmodium intortum, D. uncinatum, Lotononis bainesii, Macroptilium lathyroides* and *Trifolium repens*. Early results were reported by Evans (1970) and relevant details on liveweight gain and stocking rates are given in Table 11.1. There was a tendency for the legume to decline at higher stocking rates, as noted elsewhere, and this was accompanied by a decrease in animal production. The decrease in legume content of the sward was attributed to grazing pressure on the legume being greatest in winter and early summer when growth was least.

Bryan and Evans (1973) gave botanical details of the completed experiment and some of the more interesting details are summarized in Table 11.7.

It is evident from Table 11.7 that the legume was drastically reduced at the high stocking rate, the only legume to survive as a significant contributor to the pasture being *Trifolium repens*. Similarly, the taller growing grasses were badly affected by the heavy stocking as only *Paspalum dilatatum* was a major contributor at the end of the experiment. On the other hand, the lower growing grasses, notably *Digitaria decumbens* and *Axonopus affinis* (which was an 'invader' in that it was not present at the beginning of the experiment), made a considerable contribution to the total production of dry matter. Similarly, the total weed component increased considerably at the high stocking rate (this was also true of the amount of bare ground). The deterioration of sward occurred rapidly towards the end of the experiment and this was associated with a decline in liveweight gains/head (Evans and Bryan 1973) which are summarized in Table 11.1. This would be expected because of the reduced amount of legume in the pasture.

Young and Chippendale (1970) carried out stocking rate trials at the Department of Primary Industries Research Station, Coolum, also in the Wallum. Pastures used comprised Pangola grass (*Digitaria decumbens*), white clover and *Lotononis bainesii* or *Setaria sphacelata* cv. Nandi with the same legumes. Stocking rates and liveweight gains are given in Table 11.1. The establishment of legumes was poor and animals on the highest stocking rate required supplementary feeding during the first experiment. These unpromising results on coastal

Table 11.7 Important changes in botanical composition of the high stocking rate pasture from 1965–71 (Beerwah) (adapted from Bryan and Evans 1973)

	1[1]	2	3	4	5	6	7	8	9	10	11	12	13	14	15
								(% dry weight)							
Dec 1965	25.7	6.5	16.6	3.7	2.2	3.6	22.0	6.0	3.1	0	1.2	0.1	52.5	33.7	13.7
Dec. 1971	7.4	0.4	0.3	27.9	0	0	1.3	8.4	0.6	33.4	9.1	5.6	35.9	9.7	54.3

[1] 1 = *Paspalum dilatatum* 2 = *P. Scrobiculatum* 3 = *Chloris gayana* 4 = *Digitaria decumbens* 5 = *Demosdium uncinatum* 6 = *D. intortum* 7 = *Lotononis bainesii* 8 = *Trifolium repens* 9 = *Paspalum urvillei* 10 = *Axonopus affinis* 11 = *Juncus prismatocarpus* 12 = *Eragrostis parviflora* 13 = Grass 14 = Legume 15 = Weed

lowland soils were attributed to inundation and water table movement not experienced on the neighbouring lateritic podzolic soils at Beerwah.

Kikuyu grass (*Pennisetum clandestinum*) withstands high stocking rates and high levels of nitrogen. Mears and Humphreys (1974a and b) carried out a comprehensive trial to investigate the effect of these two management factors at the New South Wales Department of Agriculture Research Station at Wollongbar. Stocking rates varied for different rates of nitrogen application and full details of these and associated liveweight gains are given in Table 11.1. Detailed measurements were made of tiller growth as well as nitrogen distribution within the plant. It was shown that the number of tillers increased with increasing stocking rate but that their size decreased. The nitrogen concentration in leaf and stem fractions was higher with increasing stocking rate and less in the stolons. At the same time, the relative proportion of leaf plus stem nitrogen located in the leaf was higher at higher stocking rates. Concentration of nitrogen in the rhizome and total available nitrogen in the plant were reduced at high stocking rates. Thus high-quality grazing can be maintained at high stocking rate provided nitrogen is applied (cf. results obtained with legume/grass pastures). The authors also pointed out that the rate of recycling of nitrogen by animals was high at higher stocking rates at the expense of nitrogen returning to the soil as litter. Kikuyu grass was resistant to all grazing pressures applied and eventually became dominant despite the presence in the original sward of Carpet grass, *Axonopus affinis*, which has been shown in other trials to be most resistant to heavy grazing pressures (Bryan and Evans 1973). The data obtained for liveweight gains suggested that optimum gain/ha occurred where *per capita* gain was reduced by about 25% and again showed the adverse effect on individual carcasses. It was suggested that nitrogen-fertilized pastures could be used to 'store' cattle at a high stocking rate which could later be finished in lot or on high-quality pastures.

Trials on irrigated pasture

Several trials have been carried out under irrigation to investigate optimum stocking rates. *Leucaena leucocephala* was interplanted with Pangola grass at the Ord Irrigation Scheme. Gains were not satisfactory at higher stocking rates, however, and although initially reasonable on lower stocking rates, collapsed later due to supposed mimosine toxicity or 'induced deficiency'. Liveweight data and stocking rate used are given in Table 11.1. (Blunt 1976).

In a later trial, Blunt (1978) investigated the effect of different stocking rates on Pangola grass pastures, also at Kimberley Research Station. Different fertilizer levels as well as stocking rates were used, and the latter ranged from 5.1 to 12.6 steers/ha in a 'central composite' design. As usual, liveweight gain was linearly related to stocking rate and the effect varied seasonally, but was always negative. Maximum gain/ha was 1330 kg at 11.5 steers/ha. Temperature was said to limit pasture growth during the winter and high temperatures limit intake in the summer. Also, weaners were used rather than older steers and these factors might have accounted for gains not being as high as those reported by Ebersohn and Lee (1972) from Parada.

In northern Queensland, stocking rate trials on irrigated pastures were carried out at the Mereeba-Dimbulah irrigation scheme. Some of the soils are not suitable for annual cropping and there is a market for beef at the seasonally-operated meatworks in Cairns. The experiment was carried out at the Walkamin Research Station and the pastures were composed of Para grass (*Brachiaria mutica*) plus *Neonotonia wightii* cv. Tinaroo, *Setaria sphacelata* cv. Kazungula plus N. *wightii*, or *S. sphacelata* plus *Desmodium intortum* cv. Greenleaf. Stocking rates used were 3.4 and 5.1 steers/ha. Results for liveweight gains are summarized in Table 11.8.

Table 11.8 Liveweight gains from grass/legume pastures grazed at different stocking rates (Walkamin) (adapted from Miller and van der List 1977)

	Stocking rate (steers ha)			
	5.1		*3.4*	
	kg/head	*kg/ha*	*kg/head*	*kg/ha*
Para/Tinaroo	142	724	197	669
Kazungula/Tinaroo	127	647	198	673
Kazungula/Greenleaf	117	597	169	574

Legumes did not persist in the *S. sphacelata* pastures at either stocking rate but *N. wightii* persisted at 25 and 40% of lightly and heavily stocked treatments respectively when in association with *B. mutica.* The authors commented that only a well-adapted legume can remain a significant component of a pasture mixture and only when associated with a non-aggressive grass (Miller and van der List 1977).

Finally, Deans and his colleagues carried out an experiment on the effects of varying stocking rate on the weight gains of dairy heifers grazing irrigated Pangola grass at the Department of Primary Industries Research Station at Ayr (Deans et al. 1976). As has been the case in other experiments, increasing stocking rate decreased *per capita* gain on a year-long basis. There was no effect of stocking rate on gain during spring and summer, however, suggesting that greater grazing pressure could have been maintained at this time or that forage could have been conserved. Liveweight gains and stocking rates are summarized in Table 11.1.

The brief summary of experiments carried out on the effect of stocking rate on livestock productivity and botanical composition on both grass and grass/legume pastures in Australia indicates the importance of this factor in overall pasture management. Australian workers have emphasized the necessity of maintaining the maximum stocking rate concomitant with both liveweight gain and carcass value. It has also been shown that overstocking can have a marked effect on sward characteristics, particularly insofar as legume and weed content is concerned.

Experiments carried out elsewhere

In comparison with the large amount of work to come out of Africa and Australia, remarkably few trials on the effects of stocking rate on the productivity of beef cattle on pasture have been carried out in other countries. One of the reasons for this may be that some workers (see Mott 1960, for example) do not regard experiments involving fixed stocking rates as a valid method of assessing pasture productivity and prefer a method where the number of animals per unit of forage is maintained at an equivalent level for all treatments. This certainly appears to have been the case in Brazil, for example, where almost all pasture experiments were designed on the 'put-and-take' system.

However, stocking rate trials have been reported from Cuba, Malaysia and Fiji and these will be briefly summarized below.

Cuba

Working at the Pasture and Forage Experimental Station at Indio Hatuey, Veitia et al. (1974a) investigated the effects of stocking rate and supplemental feeding on bulls grazing fertilized Pangola grass pastures. The stocking rates used were 4.8, 5.7 and 7.1 bulls/ha and the customary relationship between *per capita* gain and gain per unit area was found to hold true. Further details of liveweight changes are given in Table 11.1. It was concluded that reduced gain was due to reduced availability of dry matter. This appears to be the same work as that reported by Veitia et al. (1974b).

Malaysia

Guinea grass/legume pastures at Serdang, Malaysia, were grazed by Kedah-Kelantan bulls at stocking rates of 2, 4 and 6 beasts/ha. The associated legumes included *Stylosanthes guianensis, Centrosema pubescens* and *Pueraria phaseoloides*. Liveweight gain *per capita* was related to forage yield and allowance rather than to sward legume content, and it was suggested that this was because of the better conditions for even grass growth in a humid tropical climate and also because the mean legume content of the pastures was fairly high (21%) and the nitrogen supply to the animals was not therefore limiting their intake. Liveweight data are given in Table 11.1.

Fiji

Nadi blue grass (*Dichanthium caricosum*) has been widely used in Fiji to improve drier ranching land, but it responds poorly to application of nitrogen and *Macroptilium atropurpureum* has been used to improve pastures, together with the application of phosphorus. Partridge (1979) carried out an experiment at Nawaicoba Research Station to determine the effects of stocking rate on productivity of natural and improved pastures and the persistence of Siratro.

The customary negative relationship between stocking rate and *per capita* gain was demonstrated and there was little difference between the slope of the regression line for improved or unimproved pastures. Results for liveweight gains are given in Table 11.1. Determination of botanical composition showed that Siratro disappeared at higher stocking rates but that the native legumes, being lower growing and more resistant to heavier grazing, increased markedly on both improved and unimproved pastures, thus presumably minimizing differences between the two. Native legumes included *Desmodium heterophyllum, D. triflorum, Alisicarpus vaginalis, Atylosia scarabaeoides* and *Mimosa pudica*. In some cases, increases were dramatic: for example, *D. heterophyllum* increased in frequency from 1 to 70 on one of the improved sites, with an increase of stocking rate from 1.25 to 5 beasts/ha. It was also confirmed that *D. caricosum* was highly resistant to abuse. The author noted the importance of dung beetles in removing dung pats at high stocking rates so that animals were free to select herbage. In view of the obvious importance of the native legumes in pasture productivity, the effects of phosphorus application to natural pasture would be interesting.

Discussion

It is appropriate at this stage to define some of the terms used when discussing the relationship between stocking rate and pasture productivity. Mott (1960) has given useful clarifications: stocking rate: the number of animals/unit area of land (in this case, head/ha). This bears no relation to the amount of forage available.
Grazing pressure: the number of animals per unit of available forage.
Carrying capacity: the stocking rate at the optimum grazing pressure.
It should be emphasized that no attempt has been made by Mott

(1960) or other workers to define 'optimum' as other than 'maximum' in this content; i.e. that grazing pressure at which maximum animal gains are obtained regardless of the long-term effects on the pasture. Mott (1960) also quotes 'grazing capacity' as a synonym of carrying capacity whereas the definitions of the American Society of Range Management are preferable in that they take into account the interaction between livestock productivity and pasture trend (ASRM 1964) i.e.

Grazing capacity: the maximum stocking rate possible without inducing damage to vegetation or a related resource, and

Carrying capacity: the maximum number of individual animals that can survive the greatest period of stress each year on a given land area. This does not infer sustained production and has become 'erroneously' synonymous with grazing capacity.

Having made this important distinction, the two terms will be used interchangeably; in most of the literature quoted, pasture trend has not been taken into account.

It is also useful to define a critical stocking rate as that stocking rate at which no higher gain/animal is possible (Sandland and Jones 1975).

The relationships between stocking rate and both individual performance and liveweight gain/ha have been reviewed and discussed by a number of authors. Mott (1960) suggested the following relationship:

$$y' = k - ab^{x'}$$

where y' = product per animal/product per animal at the optimum grazing pressure

x' = stocking rate/stocking rate at the optimum grazing pressure and k, a and b are constants.

(The use of ratios was to overcome the wide variation in magnitude and stocking rate found in the literature.) It was pointed out that stocking the pasture at more than optimum rates reduces production per animal, as selective grazing is no longer possible and forage available per head falls rapidly. Thus, it could be inferred by extrapolation that stocking the pasture with 50% more animals than the optimum reduced the feed supply to that required for maintenance only and thus allowed no gain. Unfortunately, few results approaching zero gain were included in Mott's test of his theoretical curves and the use of the quadratic equation implies a more rapid fall-off in production per head than perhaps is justified. However, it was emphasized that no statistical analysis of the results seemed appropriate and actual values for the constants were not suggested.

Jones and Sandland (1974) analysed data from both tropical legume and temperature pastures (including those examined by Mott 1960) and concluded that a linear relationship fitted the available data better than a quadratic one at stocking rates below the critical. One hundred and fourteen sets of data from thirty-three pastures were used to calculate values of the regression equation relating gain *per capita* with stocking rate, and the values of r^2 ranged from 0.72 to 0.99 for individual trials. However, results from all experiments were combined so as better to detect any departure from linearity. It was postulated that zero gain *per capita* would occur at twice the stocking rate at which maximum gain/ha occurred, i.e.

$$y = 2.00 - 1.00x$$

where y = gain per animal and x = stocking rate. Analysis of the collected data resulted in the overall linear regression,

$$y = 1.999 - 0.999x \ (r^2 = 0.98)$$

which is indistinguishable from the predicted equation. The authors suggest that one application of the linear equation would be to simplify the design of stocking rate trials in that even if only two rates were used these could give a good indication of the optimum.

In an attempt to verify that the linear equation is applicable to grazing trials in the tropics and subtropics, the information compiled in Table 11.1 was analysed according to the method suggested by Jones and Sandland (1974). Of a total of 176 sets of data from 53 pasture situations in 10 countries, 119 sets were used in the analysis (clearly, data from the optimum stocking rate could not be included when using the ratio method).

The equation derived was:

$$y = 1.76 - 0.82x \ (r^2 = 0.79)$$

About 80% of the total variation was attributable to the linear equation, indicating that this may be a useful approximation to estimate optimum stocking rates on tropical and subtropical pastures.

In the Jones and Sandland (1974) equation, when x = 1, y = 1. In the above equation, when x = 1, y = 0.94, sufficiently close to confirm the Jones and Sandland model. However, the value of b indicates that zero gain/head occurs at slightly more than twice (i.e. 2.15) the optimum stocking rate and that maximum gain/head is reached at 1.76 times the optimum stocking rate, indicating that cattle utilize forage more efficiently at lower rates of gain.

However, there was too much residual variation to conclude that the linear equation necessarily represents the true relationship between the two variables. The data were unevenly distributed (many fell in the range between 0.4 and 0.8 of the optimum stocking rate and few above this value) which reduces confidence in the validity of b. There was in fact an indication that a semi-logarithmic curve would fit the data as well as the linear form as the value of r^2 for the equation log y = a + bx was also about 80%.

Such a relationship could be justified by the improved efficiency of forage utilization by animals at reduced rates of gain and also supports the concave response curve suggested by Conniffe et al. (1970).

Connolly (1976) has considered the errors that will arise if linearity is assumed when the true relationship is quadratic. He concludes that the use of at least three stocking rates is justified. However, if this is impossible, then the use of two stocking rates which approximately span the true stocking rate would tend to obviate the dangers of assuming linearity. It appears that one of the main objections of Connolly (1976) to the work of Jones and Sandland (1974) is that this claims to be a true relationship whilst at the same time having simple properties for experimental purposes. However, it would appear to the present author that if the assumption of linearity fulfils the latter objective whilst approximating to the first, it provides a useful technique for handling data from tropical and subtropical pastures.

A more sophisticated relationship was suggested by Hart (1972), who proposed a model which took forage yield (and therefore grazing pressure) as well as stocking rate and liveweight gain into account. Values of R^2 (the multiple correlation coefficient) of between 0.65 and 0.95 were obtained when using the three parameters. The author also emphasized the necessity to consider the effects of sex, initial weight and previous management of cattle so as to incorporate these terms into a model. The variation in genetic ability to utilize forage was also mentioned. Unfortunately, few data on these variables are available from experiments carried out in tropical and subtropical areas and no attempt has been made to investigate how much variability they would contribute, but the general principles involved should be taken into account when planning future stocking rate trials.

Several workers have measured the yield of forage available to the animal (see, e.g. Jones 1974) but without incorporating these values into the stocking rate productivity model. Blunt (1978) estimated that availability accounted for 27% of the total variation of seasonal liveweight change, but this was only studied at the central stocking rate 'because of the strong effect of stocking rate on the variables'.

Yates et al. (1964) plotted monthly increments in liveweight against the amount of green *Sorghum almum* (excluding stem) available at different stocking rates. The data were grouped according to stocking rate and time of year, and during the late summer–autumn–winter period weight gains were logarithmically related to pasture availability with an asymptote at approximately 700 kg available forage/ha.

Mears and Humphreys (1974b) also related cattle growth to the availability of green Kikuyu grass (leaf plus stem) and found a value of 600 kg green Kikuyu DM/head during the autumn–winter period which was identified as the critical time determining overall stocking rate. However, again, all three sets of values were not incorporated into a model.

Shaw (1978b) calculated regressions of gain/head for whole years on $\log_{10}$ transformed yields of total forage, Townsville Stylo, Spear grass and 'other grasses'. However, in most cases, correlation coefficients were low and no attempt was made to incorporate forage availability and stocking rate into a unifying concept with liveweight gain.

Perhaps the only workers to have done this under tropical conditions were Edye et al. (1978), who used the results of Winter et al. (1977b) to calculate a multiple regression relating stocking rate, pasture yield in the natural logarithmic form and liveweight gain. The response surfaces were quadratic during the wet season and linear during the dry but, on an annual basis, effects attributable to stocking rate cancelled out. It was submitted that this was due to antagonistic effects which in part could be attributed to compensatory growth but also to different grazing behaviour of animals grazing tall-grass-dominant pastures with the formation of 'mosaics' of relatively ungrazed areas with a consequent higher 'effective' stocking rate than that imposed. This effect was also observed by Stobbs (1970), but it is extremely common in situations where grazing pressure is insufficient to remove rank growth of even palatable species. As I write, there is an appreciable area of ungrazed Guinea grass in the pasture at the bottom of my garden whilst the surrounding 'dambo' pasture has been heavily utilized (Bunda College, Central Region, Malawi). The extent of undergrazing because of the build-up of unpalatable material of low digestibility may lead to the anomalous situation where liveweight gain actually declines as stocking rate is reduced. This would account for the quadratic form of the curve during the wet season.

As it is generally agreed that liveweight gains decline as stocking rate increases, then it is logical that the take-off also decreases (as has already been noted when fertilizer application enables stocking rates to be increased). Yates et al. (1964) and Winks et al. (1970) have noted that killing-out percentages are also lower at high stocking rates and that this is associated with insufficient 'finish' on the carcass. Bisset and Marlow (1974) found that only animals grazed at low stocking rates were ready for slaughter at the end of their trial, and Jones (1974) reported that animals from highly-stocked pastures were not ready for slaughter until between 2.6 and 3.2 years. It is also a common stockman's observation that animals do not fatten as well on pastures which carry more cattle as they do on less intensive pastures (cv. Pangola and Guinea grasses on the Gulf Coast of Mexico).

The advantages of high stocking rates must therefore be weighed against the reduced turn-off rate, and consequent longer time on inventory, and higher interest charges for cattle. Alternatively, high stocking rates may be used for 'stocker' cattle where another method of finishing is possible.

It is also perhaps worth mentioning again at this point the importance in the management of tropical grasses of effectively utilizing the 'flush' of grass available at the onset of the rains whilst at the same time maintaining enough forage, either standing or conserved, for the dry or winter season.

It has already been observed that 'optimum' has commonly been used to indicate 'maximum' liveweight gain/ha (without necessarily implying optimum take-off rates in economic terms) without consideration of the long-term effects on the botanical composition of the pasture. However, some results are available which indicate the effects of higher stocking rates on individual species. Before discussing the subject further, it is appropriate to redefine some terms used in ecological succession. Commonly, when high stocking rates (and thus high grazing pressure) are practised, overgrazing (continued overuse) occurs, which in its turn gives rise to retrogression. This is 'the process of vegetational deterioration whereby the same area becomes successively occupied by different plant communities of lower ecological order'. When this occurs, certain plant species increase in relative amount, at least for a time. These are known as increasing plant species and are commonly termed 'increasers'. At the same time, some plant species of the original vegetation decrease with continuing overuse, and these are commonly known as 'decreasers'. It is often implied that the decreasers are more palatable then the increasers and of a higher nutritive value but this may not necessarily be the case in natural grasslands in the tropics and subtropics.

There appears to be several common patterns in the reactions of natural pastures to high stocking rates. Perennial species tend to be replaced by ephemeral pioneer annual species and by woody brush species (bush encroachment). At the same time, a reduction in the amount of forage produced and available tends to reduce the intensity of fires which normally serve to control bush (Kennan 1969). There also may be a decrease in total cover, further exposing land to erosion, particularly in areas where bunch grasses predominate (Edwards and Nel 1973a; Shaw and 'tMannetje 1970). Numerous authors have reported that decumbent and procumbent species increase, and this is particularly true for *Cynodon dactylon* strains (Rodel and Boultwood 1967; Smith 1966; Thornton 1970; Moir et al 1969), and *Brachiaria decumbens* (Smith 1966; Thornton 1970; Thornton and Harrington 1971; Harrington and Pratchett 1974a; Winter et al. 1977b). Other examples of increasers and decreasers are given in Table 11.9.

It is interesting to note that many of the increasers are regarded as good grazing species, e.g. *C. dactylon, Panicum maximum, B. decumbens* and *Setaria sphacelata*. In this connection, it has also been noted that animals on a high stocking rate treatment are able to select a better diet than those on a lower rate treatment (Harrington and

Table 11.9 Some increasing and decreasing grass species in tropical and subtropical pastures

Increasers	*Country*	*Reference*
Cynodon dactylon	Uganda	Fiennes (1940)
Cynodon dactylon	Zimbabwe	Rodel and Boultwood (1967)
Cynodon dactylon	Zambia	Smith (1966)
Cynodon dactylon	Uganda	Thornton (1970)
Cynodon dactylon	Australia	Moir et al. (1969)
Panicum maximum	Uganda	Thornton (1970)
Brachiaria decumbens	Zambia	Smith (1966)
Brachiaria decumbens	Uganda	Thornton (1970)
Brachiaria decumbens	Uganda	Thornton and Harrington (1971)
Brachiaria decumbens	Uganda	Harrington and Pratchett (1974a)
Brachiaria decumbens	Australia	Winter et al. (1977b)
Setaria sphacelata	Zambia	Smith (1966)
Setaria sphacelata	Uganda	Thornton (1970)
Sporobolus pyramidalis	Uganda	Fiennes (1940)
Sporobolus pyramidalis	Uganda	Hosking and Stephens (1941)
Sporobolus pyramidalis	Uganda	Stobbs (1969b, 1970)
Heteropogon contortus	Zambia	Smith (1966)
Eremochloa bimaculata	Australia	Shaw and 'tMannetje (1970)
Chrysopogon fallax	Australia	Shaw and 'tMannetje (1970)
Tripogon loliiformis	Australia	Shaw and 'tMannetje (1970)
Brachiaria miliiformis	Australia	Shaw and 'tMannetje (1970)
Bothriochloa decipiens	Australia	Shaw and Dale (1978)
Sida acuta	Australia	Shaw and Dale (1978)
Digitaria decumbens	Australia	Bryan and Evans (1973)
Axonopus affinis	Australia	Bryan and Evans (1973)
Pennisetum clandestinum	Australia	Mears and Humphreys (1974b)
Digitaria didactyla	Australia	Jones (1974)
Imperata cylindrica	Uganda	Fiennes (1940)
Cymbopogon excavatus	Uganda	Fiennes (1940)
Decreasers	*Country*	*Reference*
Hyparrhenia spp.	Zambia	Smith (1966)
Hyparrhenia spp.	Uganda	Stobbs (1969b)
Pennisetum purpureum	Uganda	Thornton (1970)
Imperata cylindrica	Uganda	Fiennes (1940)
Imperata cylindrica	Uganda	Thornton (1970)
Themeda triandra	South Africa	Edwards and Nel (1973a)
Andropogon amplectens	South Africa	Edwards and Nel (1973a)
Loudetia simplex	South Africa	Edwards and Nel (1973a)
Themeda australis	Australia	Shaw and Dale (1978)
Heteropogon contortus	Australia	Shaw and 'tMannetje (1970)
Paspalum scrobiculatum	Uganda	Hosking and Stephens (1941)
Paspalum scrobiculatum	Australia	Bryan and Evans (1973)
Chloris gayana	Australia	Bryan and Evans (1973)
Paspalum dilatatum	Australia	Bryan and Evans (1973)
Paspalum dilatatum	Australia	Moir et al. (1969)

Pratchett 1974a; Winter et al. 1977b). Part of this effect is no doubt due to the stimulation of young growth and stolon formation in heavily-grazed areas but part of the effect must be attributed to the superior nutritive value of some of the increasers, e.g. *Brachiaria decumbens, Digitaria decumbens.*

In this respect, the grazing behaviour of heavily-stocked animals is of interest; Smith (1966) has observed that they spend a longer time grazing than more lightly stocked cattle, whilst Smith (1966), Thornton and Harrington (1971) and Edye et al. (1978) have remarked on the uneven grazing pattern of animals at a light stocking rate leading to higher effective grazing pressure.

Summary

Experiments investigating the influence of varying stocking rate on tropical and subtropical pastures have been reviewed and it has been shown that this factor exerts an extremely important effect on beef production. After defining some of the terms used in discussing the various relationships involved, it was shown that the equation relating gain/animal and stocking rate was approximately linear and negative and suggested that this equation could be useful in estimating optimum stocking rates on tropical and subtropical pastures. However, it was also noted that there was too much residual variation to conclude that the linear equation necessarily represented the true relation between the two variables and that a semi-logarithmic curve would also fit the data. This could be explained by the increased efficiency of forage utilization by animals at reduced rates of gain. More complicated models, which included pasture yield, were also discussed. The existence of antagonistic effects caused by compensatory growth were noted, as was the effect of different grazing behaviour patterns which tended to increase grazing pressure over that originally imposed.

Adverse effects of increased stocking rates may include reduced turn-off rate and thus higher interest charges on cattle as well as less carcass finish.

Finally, some of the effects of high stocking rates on botanical composition of pastures have been discussed. The role of individual species as increasers or decreasers in heavily-stocked pastures has been noted.

12 Deferred and rotational grazing

It has been suggested that the resting of pastures in various forms may have a positive effect on animal production. Before discussing this method of management it may be as well to define some of the more important terms used in this connection.

The American Society of Range Management (ASRM 1964) has defined deferred grazing as 'the discontinuance of grazing by livestock on an area for a specified period of time during the growing season to promote plant reproduction, establishment of new plants or restoration of vigor (sic) by old plants'. From previous discussion, it is clear that this will also promote plant succession whereby the area becomes 'successively occupied by different plant communities of a higher ecological order' and will therefore be useful in the regeneration of range which has deteriorated by misuse, i.e. continued overgrazing. Photosynthetic leaf area increases, with a consequent increase in the reserves of available carbohydrate in the plant, as well as a more extensive root system. It would be expected that, at least in the long term, such deferment would increase the number of more palatable plants of a better nutritive value and would lead to higher stocking rates and consequent higher productivity per unit area. It would also be expected that deferment would lead to greater availability of herbage for burning and therefore to a greater potentiality for control of undesirable woody bush species. Deferment tends therefore to be of long-term benefit, largely manifested in range trend, and few data are available on its effect on productivity of beef cattle under tropical and subtropical conditions.

On the other hand, rotational (or rotation) grazing is defined as an 'orderly sequence of use when each sub-division is both grazed and deferred during the same grazing season or calendar year' (ASRM 1964). A definition which might better suit common practice in tropical regions would be an 'orderly . . . deferred *throughout* the same grazing season'.

The practice of rotational grazing clearly requires that a grazing area be divided into a number of sub-divisions or camps, and systems based on rotational grazing have come to be known as 'multi-camp' systems. The various supposed advantages and disadvantages of these systems as compared with conventional (closed, fixed, continuous or 'paucicamp', an appalling invention) have been widely debated.

The pros and cons of the 'multi-camp controversy' have been reviewed by Roberts (1969). He has suggested that advantages might include: the stimulation of tiller growth by greater light intensity with consequent encouragement of vegetative growth and spread of tufted spp.; increased mineral content of grass regrowth; improved soil surface and seed establishment; less woody and more nutritious shrub regrowth; and minimization of selective grazing. These advantages would be expected to lead to higher animal production per unit area. On the other hand, disadvantages might be decreased root growth, low rate of shoot regrowth and decreased seed production, accompanied by lower production from the plant and consequent reduction of animal production both *per capita* and per unit area. This would be accompanied by greater soil damage by splash erosion, soil compaction, evaporation losses and deterioration of soil surface conditions. However, all these advantages or disadvantages would be expected from any system which either rationally or overexploited grazing resources and cannot be directed at multi-camp systems alone.

Probably the most important constraint to the implementation of such systems in practice is the danger of overcapitalization (which, indeed, was mentioned by the *Farmers Handbook* as early as 1929). Acocks (1966) advocated a system based on 'non-selective grazing' for which, in its ideal form, twelve or more paddocks were required per herd, whilst Howell (1976) has suggested that as many as forty-two sub-divisions might be advantageous. Either of these systems would require large investments in terms of fencing materials, watering points, etc.

What advantages, then, may be expected to accrue in terms of animal production by the adoption of a rotational grazing system? Despite the amount of discussion on the merits and demerits of such systems, little critical information is available from our area, and real advantages in terms of livestock production are difficult to quantify.

Most of the information available has come from Africa, and only isolated trials have been carried out in other parts of the tropics and subtropics. It must be emphasized at this point that it is not the principal objective of this review to discuss the advantages or disadvantages of rotational or deferred grazing insofar as the effects on vegetation or soil are concerned, but to investigate beef production from rotational compared with more conventional systems. For this reason, much of the early work where no results of livestock production are given will not be included.

Finally, this chapter has been used a 'catch-all' to include several other forms of pasture management which have not been mentioned elsewhere, such as bush control, shade provision and the effects of night kraaling.

Africa

Zimbabwe

The first findings on the livestock production from rotational as opposed to continuous grazing seem to be from Grasslands, where West (1952) compared systems with one herd to two or three paddocks with continuous grazing. Stocking rate was the same for the various systems. The cattle on free range performed a little better than those on the other systems, mean performance over the four years of the experiment being 257 vs. 234 kg/head. These findings were confirmed in later work (Grasslands 1967/68, 1968/69) when a system utilizing three herds to four pastures (continuous) was compared with a rotational system employing one group of animals grazing six paddocks in quick succession (two weeks' grazing followed by ten weeks' rest). The production of steers over the two years for which results are available were 120 and 93 kg/head for the two systems respectively. Stocking rates were again the same on both treatments.

Early work was also carried out at Matopos by Kennan (1962), who showed that over a twelve-year period, cattle grazing at a stocking rate of 0.25 beast/ha without rotation gained an average of 199 kg/head whilst those grazing a one-herd-in-four-paddocks system gained only 150 kg/head. Production/unit area was 49 and 28 kg/ha respectively. Subsequent evaluation of later results of this trial, however, showed that the continuously grazed pastures were more subject to bush encroachment and that their early advantage was eroded thereby. It was to avoid such deterioration that West (1964) suggested 'the three herds to four paddocks system' as a compromise measure to benefit the veld whilst at the same time maintaining high levels of animal production (this system involves rest and burning every fourth year).

A comprehensive series of trials to assess the possible benefits which might be obtained by the use of 'multi-camp' non-selective grazing systems propounded by Acocks (1966) were established by Denny and his colleagues near Bulawayo.

The treatments used in the first trial at Matopos Research Station were as follows:

12-paddocks-to-one-herd 'quick graze-long rest' (stocking rate: 1 beast/3.4 ha)
12-paddocks-to-one-herd 'complete graze-long rest' (stocking rate: as above)
Four-paddocks-to-three herds (effective stocking rate: 1 steer/4 ha)

Unfortunately, the fact that different stocking rates were used for the intensive systems and the four-paddock system make the results for liveweight gain difficult to interpret. It is also unfortunate that the experiment had to be concluded after three years because of drought-induced shortage of forage. However, total gains varied from season to season but were consistently higher on the continuous grazing treatment, being 98, 68 and 127 kg/head respectively (this was equivalent to 29, 20 and 31 kg/ha in the same order as given above). Performance of the steers in the twelve-paddock systems was depressed except during the early growing season and this was attributed to the reduced opportunity for selective grazing. There were no differences in botanical composition or density of perennial grasses among treatments. It may be concluded that improved livestock production due to this multi-paddock system were unlikely and that neither of the intensive systems used were likely to bring about substantial veld improvement, at least in the short term (Denny et al. 1977).

A further trial was carried out at the Nyamandhlovu Experiment Station, 50 km north-west of Bulawayo, to investigate the effects of two rotational grazing systems on the production of breeding cows. The treatments were sixteen-paddocks-to-one-herd intensive rotational grazing system and a four-paddocks-to-one-herd rotational resting system. After rapid rotation in both systems during the early growing season, the grazing period in the former system varied from 7–14 days and in the latter from six weeks during the growing season to twelve weeks during the dormant season. With the exception of the first year of the experiment (when it was expected that the carrying capacity of the intensive system would be higher than the other), stocking rates for the two treatments were the same (between 4.9 and 5.8 ha/cow calf unit). The fertility of cows on both treatments was satisfactory with weaning percentages of 91 and 85 for intensive and less intensive respectively. Mean mass of weaned calf/cow/year was 186 and 175 kg, a difference of 6% in favour of the intensively-managed cows, but this was not significant. There were no effects of treatment on botanical composition, although there was some evidence of lower density of perennial species on lithosols in the sixteen-paddock system (Denny and Steyn, 1977).

Denny and Barnes (1977) carried out a further trial at Matopos to investigate the possibility that stocking rate may be increased by intensive rotational grazing over that acceptable with less intensive systems. The experiment with steers compared six grazing procedures comparing different periods grazing each paddock (5, 10 or 20 days) with number of paddocks/head (4 and 8) at two stocking rates, one rate being double the other (stocking rates ranged from 1.8 to 2.2 ha/head at the high rate and 3.6 to 5.9 ha/head at the low rate). The experiment continued for five years. In general, the use of eight paddocks resulted in lower productivity than that of four during the mid and late growing season. The twenty-day grazing period resulted in a marked reduction in gain steer. Gains steer were higher at the lower stocking rate (as would be expected; see Chapter 10). Grazing intensity (500 kg livestock unit days/ha) was shown to be

Table 12.1 Liveweight gain (kg/head and kg/ha) of steers grazing veld with different grazing procedures at two stocking rates (Matopos) (adapted from Denny and Barnes 1977)

Grazing duration (days)	*Stocking rate*			
	High		*Low*	
	No. of paddocks/herd		*No. of paddocks/herd*	
	8	4	8	4
5	53 (24)	75 (34)	72 (16)	73 (17)
10	56 (25)	70 (32)	63 (14)	75 (17)
20	34 (15)	48 (22)	54 (12)	68 (16)

kg/ha in brackets

highly correlated with mean body weight gain/steer ($R^2 = 0.86$). It was concluded, therefore, that valid comparisons of effects of grazing frequency could only be made at similar levels of grazing intensity and 'because of the overriding influence of grazing intensity on gain per steer . . . individual gains will be similar irrespective of the number of paddocks used per head'. Liveweight gains per steer and per unit area are shown in Table 12.1.

Grazing procedure had no apparent effect during the dormant season. Botanical measurements were carried out on all treatments and there was no indication of any change attributable to grazing procedure (although there was a significantly higher basal cover of some increaser species in the high stocking rate treatments). The main conclusions from this experiment appear to be that increased number of paddocks reduce liveweight gains both per head and per hectare (although this latter only marginally at low stocking rates) and that extending the duration of grazing to twenty days results in decreased gains, presumably because of reduced opportunity for selective grazing. Preliminary results of this trial were published by Denny et al. (1974).

The above experiment did not provide firm evidence that the grazing capacity of veld is higher with rotational than with continuous grazing, and a further experiment was set up at Matopos to examine this possibility. The twenty-day grazing period of the previous experiment was discarded and the treatments were a four-paddock system with a ten-day period of stay and an eight-paddock system with a five-day period of stay compared with continuous grazing. All treatments were repeated at two stocking rates, one yearling steer to 1.8 ha (high) or 2.5 ha (low). Summarized results are given in Table 12.2.

No differential effects due to grazing treatments could be demonstrated. Botanical analysis showed that the rotational grazing was

Table 12.2 Body weight changes kg/head of yearling and kg/ha) steers with rotational or continuous grazing on veld (Matopos) (adapted from Denny and Yuba 1979)

Stocking rate	*Grazing procedure*		
	Four paddocks (10 days)	*Eight paddocks (5 days)*	*Continuous*
Low	91 (36)	87 (34)	97 (38)
High	87 (49)	87 (49)	89 (50)

kg/ha in brackets

associated with a slight increase in desirable grass species at the high stocking rate and a slight decrease at the low stocking rate, whereas the continuous grazing treatment was accompanied by the reverse effect (i.e. decrease in desirable species at high stocking rates and increase at low stocking rates). No explanations are offered for this unexpected effect and, in any event, changes were probably not significant (Denny and Yuba 1979).

The situation with regard to rotational grazing may be expected to be different in the case of artificially established pastures from that of veld. Pastures are usually more homogenous, there is less opportunity for preferential grazing and stocking rates are usually higher. It is interesting, therefore, to consider the effects of rotational v. continuous grazing under these conditions.

Early trials with Ermelo Love grass (*Eragrostis curvula*) at Grasslands indicated that liveweight gains were lower in a rotational than in a continuous grazing system, although few details are given (Grasslands 1967/68, 1968/69).

A more comprehensive trial was set up at Grasslands Research Station to assess the effects of eight grazing treatments on the production of steers grazing irrigated Star grass (*Cynodon aethiopicus*) pastures. The pastures were fertilized at the rate of 650 kg/ha N and 450 kg/ha superphosphate, with dolomitic lime as required. The treatments applied and the results for carcass weight gain/ha and mean herbage yield/ha are given in Table 12.3 (Parkin et al. 1976; Parkin et al. 1977, 1978).

The above results (which were abstracted by the author from annual reports and have not been statistically analysed) indicate that treatments involving rotational grazing produce, at best, as much carcass gain/ha as the continuous treatment. There is, however, a strong indication that more forage was available on the rotational grazing systems. In order to determine whether this increased amount of available forage would support a higher carrying capacity, a further trial was initiated (Parkin and Price 1979). The number of paddocks used were one (continuously grazed) and six (2, 4 and 6 days' stay and

Table 12.3 Carcass weight gains (kg/ha) and herbage present in the paddocks of different grazing treatments (Grasslands) (adapted from Parkin et al. 1976, 1977, 1978)

Number of paddocks	*Period of stay (days)*	*Period of absence (days)*	*Carcass gains (kg/ha)*	*Herbage yield (kg DM/ha)*
1	Set stocked		997	2 661
2	6	6	976	2 808
3	6	12	997	2 798
6	4	20	968	3 298
6	2	10	890	2 849
6	6	30	887	3 419
18	1	17	967	3 345
18	2	34	902	3 653

10, 20 and 30 days absence, respectively) at three stocking rates (15, 18 and 21 yearling steers/ha). At the time of writing (March 1981), only preliminary results were available, but these indicated that the greatest carcass gains both per steer and per hectare were from the continuously grazed treatment stocked at 15 steers/ha.

More systematic work has been carried out in Zimbabwe than anywhere else in Africa or the tropics or subtropics. However, it appears that no consistent advantage has been demonstrated for rotational grazing under either veld or improved pasture conditions.

Tanzania

Walker and Scott (1968a and b) carried out a series of trials at Ukiriguru where rotational was compared with continuous grazing. In the first experiment, the stocking rate was 1 steer/ha and the grazing period for the three-paddock rotational grazing system was three weeks (i.e. six weeks' rest). There was an additional treatment, rotational plus reserve grazing, where one paddock was rested from April to September each year to provide standing hay for grazing during the late dry season. Pastures were predominantly *Bothriochloa insculpta*. Weight gains over the 111 weeks of the experiment were 195, 143 and 126 kg/head for animals on the continuous, rotational plus reserve and rotational grazing treatments respectively, a clear advantage for continuous grazing. Similar results were obtained in a second trial where the only difference was that the stocking rate was increased to 1.23 animals/ha. The results for the three treatments in the same order as before were 136, 133, and 110 kg for the seventy-four weeks of the experiment, again a clear indication that there was no advantage to be gained from rotational grazing. A botanical survey indicated that the rotational grazing treatment had more of the tall grasses, *Hyparrhenia* spp and *Panicum infestum*, and less *Bothriochloa insculpta* than the continuous grazing treatment. It was considered that the differences in liveweight gains between treatments were due to the effects of different grazing pressures and frequency of grazing on the quantity and quality of forage available as well as these differences in botanical composition.

In order to evaluate these factors, a further experiment was carried out in which an extra-period Latin-square change-over design was used. The experimental treatments were continuous grazing on a predominantly *B. insculpta* ('short grass') pasture, continuous grazing on a pasture containing larger amounts of *Hyparrhenia* spp and *P. infestum* ('tall grass') and rotational grazing on a tall grass pasture. The stocking rate was 1.6 steers/ha. Six trials were carried out, each lasting eight weeks. The summarized mean weight gains were 2.3 and 2.8 kg/week/head for continuous and rotational grazing respectively, equivalent to 3.8 and 4.7 kg/week/ha. It was suggested that this disparity of results was attributable to an inter-relationship of stocking rate and grazing management systems. Results were higher from continuous than with rotational grazing systems at low stocking rates but, at higher stocking rates, the reverse was the case. It was also suggested that this was associated with differences in both quantity and quality of herbage available to the animal. The theory was put forward that under continuous grazing at low stocking rates, the nutritional value is high, whereas on the rotational grazing system, herbage has more time to mature and therefore its nutritive value declines. At a higher stocking rate, however, the total amount of dry matter available becomes more important and this tends to be less on the continuously grazed treatment because of continued defoliation. In any event, this latter experiment by Walker (1968) is one of the few in which an apparent advantage for rotational grazing is demonstrated.

Uganda

A series of trials were carried out in Uganda to compare rotational and continuous grazing, first by Stobbs at Serere and later by Thornton and his colleagues at Mbarabara in Ankole, western Uganda. The Serere work was carried out on artificial pastures, whilst trials at Ankole were on natural veld.

Stobbs (1969e) compared rotational grazing at two grazing frequencies: low, i.e. 7 days' grazing and 28 days' resting; and high, i.e. 3.5 days' grazing and 14 days' resting for each pasture treatment. Pastures were *Hyparrhenia rufa* sown with either *Stylosanthes gracilis* or *Centrosema pubescens*, and stocking rate was common for the various combinations of pasture and grazing frequency but was varied according to seasonal availability of forage. High-frequency rotation produced 5.7% more liveweight gain than the lower-frequency treatment, pasture averages being 785 and 743 kg/ha for the two frequencies respectively. However, this slight difference was non-significant.

In another trial carried out at about the same time, Stobbs (1969f) carried out a comparison of continuous grazing with either rotational grazing using three paddocks, moving cattle every two weeks, or a six-paddock system where cattle were moved weekly. Pastures were composed of *Panicum maximum* sown with *Macroptilium atropurpureum* and the stocking rate used was 5 beasts/ha until the twenty-first grazing cycle when half of each treatment was increased to 5.9 beasts/ha. In the first twenty-one grazing periods (i.e. 882 days), most liveweight gain was produced from the continuous grazing treatment and the least from the six-paddock system. During the period, cycles 22–28, continuous grazing was superior to rotational except for the heavy stocking rate treatment in the three-paddock system. Results are summarized in Table 12.4.

Table 12.4 Liveweight gaihs (kg/ha) on continuous and rotational grazing systems (Serere) (Adapted from Stobbs 1969f)

Grazing period (six weeks each)	*Stocking rate (beasts/ha)*					
	Continuous		*3-paddock*		*6 paddock*	
	5.0	*5.9*	*5.0*	*5.9*	*5.0*	*5.9*
Cycles 1–21	1 178	—	1 127	—	1 001	—
Cycles 22–28	506	517	503	643	466	461
Cycle 29	−157	−246	−101	−152	−91	−169

The superior performance from this treatment might have been due to greater forage production from rotational grazing systems, as has been suggested previously (see, e.g. Parkin et al. 1978). This might also account for the greater weight loss on the continuous grazing treatment in cycle 29.

The grazing treatment utilized by Thornton and Harrington (1971) was deferred rather than rotational grazing. The available grazing area was divided into two and, for six months in every year, the cattle all grazed one paddock whilst the other was rested and burnt. Stocking rates of 0.8, 0.4 and 0.2 beast/ha were compared. Average liveweight gains for the three stocking rates on continuous and deferred grazing treatments were identical (three separate periods of approximately one year each).

In a further experiment, Harrington and Pratchett (1974a) compared continuous with rotational grazing on natural veld described as a fire-induced *Acacia/Cymbopogon/Themeda* wooded savannah grassland. The rotational treatment involved five paddocks, four of which were grazed rotationally, being moved when subjective assessment indicated that all the available forage had been used and the fifth being reserved for dry season grazing. The stocking rate on both continuous and rotational grazing treatments was 1.25

beasts/ha. *Cymbopogon afronardus*, an unpalatable species, whose presence demonstrably reduced the productivity of the pasture, was removed before the experiment was stocked.

Rotational grazing over four years was significantly less productive both per ha and per head than the continuous grazing treatment. Mean liveweight gains were 105 and 91 kg/head and 131 and 114 kg/ha for continuous and rotational treatments respectively. It was suggested that this difference might at least partially be accounted for by the fact that the *Cymbopogon afronardus* was better able to re-establish on the rotational than the continuously grazed paddocks because of rest periods.

It may be concluded, from the Ugandan experiments that whilst there was no consistent advantage to be gained from rotational grazing, a more rapid transfer from paddock to paddock conferred an advantage over a slower system.

Malagasy Republic

Granier and Gilibert (1974) have reported one of the few studies in our area which have demonstrated the advantages of deferred (as opposed to rotational) grazing over continuous grazing. The experiment was carried out on natural savannah vegetation principally comprising *Hyparrhenia* and *Heteropogon* spp. at Kianjasca (mean annual precipitation 1600 mm).

The deferment practised involved three pastures grazed in the early and late wet season and the dry season, with a burn in the early wet season every third year. The overall stocking rate was 0.30 beast/ha. Various classes of the Renitelo breed (comprising Limousine, Afrikander and Malagasy Zebu) were used to evaluate the grazing systems from 1967 to 1972. Various measurements were used to assess performance, including weaning weight, weight of males and females at two years and weight of females at three years. In every case the deferred grazing treatment was advantageous, the weight differences in the various parameters being 17.7 kg (males at weaning), 17.9 kg (females at weaning), 58.2 kg (males at two years), 31.1 kg (females at two years) and 65.6 kg (females at three years). The technique was said to increase basal cover with desirable grass species and also to increase the protein content of the forage.

This experiment provides a rare but convincing example of the long-term benefits of deferred grazing combined with periodic controlled burning on tropical savannah grassland.

Ethiopia

O'Donovan and his colleagues compared the performance of steers managed on rotational or continuous grazing systems at Bako, (mean annual precipitation 1250 mm) in Ethiopia. *Cenchrus ciliaris* (Rhodes) grass pastures (which contained Pokot, Masaba and Rongai cultivars) were compared under rotational and continuous grazing systems at a stocking rate of between 5.3 and 7.0 steers/ha (although these were not the same for both treatment for the entire experimental period from June to November). Average daily gains for the entire period showed a significant superiority of the continuously over the rotationally grazed treatment, average daily gains being 0.50 and 0.38 kg/head respectively (O'Donovan et al. 1978), further evidence in Africa that there are no advantages to rotational grazing systems.

West Africa

Pasture management systems in West Africa were reviewed by Crowder and Chheda (1977), who mention that de Leeuw (1971) could demonstrate no advantage of either deferred or rotational over continuous grazing in northern Nigeria, liveweight gains for the three treatments bing 15.8, 12.1 and 11.2 kg/ha in the order given. On the other hand, the same author (loc. cit.) mentioned a trial in the southern Guinea zone where rotational grazing was shown to be superior to continuous (104 v. 86 kg/ha at a stocking rate of 1 beast/ha), one of the few trials to indicate a slight advantage for rotational over continuous grazing. This advantage was attributed to a greater degree of species selectivity and increased time for recovery of preferred grasses.

America and the Caribbean

Only a few relevant trials have been carried out in this area and two of these are properly not in the subtropics but have been included because of the similarities they bear to other work.

Creek and Nestel (1964) reported the effects of different grazing cycle duration on the liveweight gains and chemical composition of Pangola grass in Jamaica. Although the trial was not a comparison of continuous and rotational grazing systems, it is relevant to the subject. The cycles used were 32 and 40 days over a total of 416 days with an average stocking rate of 4.3 beasts/ha. The average daily gains were 645 and 549 g/head for the 32- and 40-day cycles respectively, or an advantage of 17% for the shorter cycle. This was associated with an increased crude protein content (15.9 v. 14.7%), a reduced dry matter content (15.2 v. 15.8%) and an increased daily production of dry matter (36 v. 32 kg/ha).

A similar superiority of a short grazing cycle over a longer one was reported by Horrell (1972), although the former was slightly inferior to continuous grazing. Experiments were carried out at Gualaca (mean annual precipitation 4000 mm) in Panama using pasture of Faragua (*Hyparrhenia rufa*) stocked at a rate of 2.5 beasts/ha. Daily liveweight gains were low on unfertilized pastures, being 0.20, 0.19

and 0.07 kg/day for continuous grazing, rotational (21-day rest) and rotational (42-day rest) grazing respectively (few details are given in this mimeographed report).

Neither of the other two trials which will be described here fall within our geographical limits but have been included because they were both carried out with Coastal Bermuda grass (*Cynodon dactylon*) which has been successfully used in many parts of the tropics.

Oliver (1972) obtained better results from rotational than from continuous grazing at Homer, Louisiana. In the first year of the trial, yearling steers were grazed at a stocking rate of 3 steers/ha on both systems and the rotational system employed four pastures with approximately three days' grazing and ten days' recovery. Liveweight gains were 368 and 318 kg/ha for rotational and continuous grazing respectively. In the second year, the treatments were similar, but stock used were weaned calves at a stocking rate of approximately 7 beasts/ha and liveweight gains were 360 and 318 kg/ha. Yearling steers were used in the third year at a stocking rate of 3.5 beasts/ha and liveweight gains were 430 and 403 kg/ha, the only year in which the advantage of rotational grazing was non-significant.

This experiment is unusual in that it is one of the few recorded where a clear advantage of rotational over continuous grazing has been demonstrated. It was suggested that this was attributable to delayed plant maturity in the rotational grazing system.

This superiority could not be demonstrated in Tifton, Georgia, however, in an experiment where six systems of Coastal Bermuda grass exploitation were compared in terms of steer gains. The rotational system used was a four-paddock, one week grazed, three weeks' rest and the stocking rate used was 7.4 steers/ha. Over the three years of the trial, steer gains averaged 600 and 469 kg/ha for continuous and rotational respectively (a 'strip' grazing treatment was also included and comparable gain from this management system was 487 kg/ha, i.e. little difference from the rotational system and inferior to the continuous grazing treatment). It was concluded that the systems of rotational and strip grazing were clearly inferior to continuous grazing under the conditions of this experiment (Hart et al. 1976).

Advantages attributable to rotational grazing were also demonstrated in Colombia where pastures of *Paspalum notatum* were managed in an area, Jamudi, with an annual precipitation of more than 2000 mm. The rotation used was of six or seven days' grazing followed by forty-two or forty-nine days' rest. Experimental cattle used were Zebu heifers and the daily gain (over an experimental period of 218 days) was 118 kg for those continuously grazed v. 133 kg for those grazed rotationally. This more rapid rate of gain resulted in the latter animals reaching breeding weight seventy-seven days before the continuously grazed heifers and pregnancy rates of 81 and 34% in the two groups respectively. (Escobar et al. 1971b).

An early trial was carried out on Napier grass (*Pennisetum purpureum*) in Hawaii. Hereford steers or Holstein heifers varying greatly in weight and age were allocated to three grazing treatments: immature, short grazing period (paddocks were mowed to ground level after each grazing period); mature grazing short period (four animals for three weeks); and mature grazing (two animals for six weeks). The results are difficult to interpret, confounded as they are by different stocking rates, but it appeared that the highest productivity was reached on the mature, long grazing period (549 kg/ha) whilst the immature and mature, short grazing periods gave similar results (345 and 391 kg/ha). These results were a four-year average (1943–8) and showed that there was no advantage to be obtained from the provision of immature grass under the conditions of the experiment (Takahashi et al. 1966).

Australia

Compared with the large volume of work carried out on rotational grazing in Southern Africa, very little has been published from Australia. However, results from the South Johnstone Research Station indicated that rotational grazing was 16% more effective in terms of liveweight gain than continuous grazing (Grof and Harding 1970).

The two grazing systems were compared on a Guinea grass (*Panicum maximum*)/Centro (*Centrosema pubescens*) pasture at Utchee Creek where the rotation was two weeks' grazing and two weeks' rest at an average stocking rate of 3.5 beasts/ha. The cumulative liveweight gains for two years were 1073 and 932 kg/ha for the rotational and continuous grazing treatments respectively.

Alternative pasture management systems

Many other techniques have been employed in an effort to improve productivity of range and improved pasture but, unfortunately, little information is available as to their effect on the grazing animal. One of the effects of overstocking (as has been pointed out in a previous chapter) is bush encroachment and this clearly affects liveweight gain unit area, both by inhibiting grass growth and by physical interference with the grazing animal. Although many trials on control of bush have been carried out, and clear responses with regard to herbage availability have been obtained, few cattle data are available from tropical and subtropical grasslands.

In an experiment carried out at the Tuli Breeding Station in

Southern Matabeleland in Zimbabwe, it was shown that cleared areas (bush was largely *Acacia* spp., *Combretum* spp., *Terminalia sericiea* and *Colophospermum mopane*) supported a higher stocking rate and liveweight gain unit area than uncleared areas. Two grazing systems were used, short-duration grazing and split season, for both treatments. The average stocking rate for the cleared veld was 1 animal/4.8 ha whilst that for the uncleared areas was 1 animal/11.7 ha. Total gain was 264 kg for the former and 227 kg for the latter and the return ha was Z$10.59 v. 3.19 (Jarvis 1978).

A bush-clearing treatment was included in an experiment on grazing management carried out on an area typical of the semi-arid country west of the Great Dividing Range in northern Queensland (Gillard 1970). The woody vegetation comprised mainly *Eucalyptus crebra*, *E. papuana* and *E. dichromophloia* with a density of about 100 trees/ha and this was cleared using a bulldozer. Other treatments included fertilizer application and different stocking rates, but these have been dealt with elsewhere. The most overwhelming treatment effect was clearing and this was equivalent to an increase in liveweight gain of 75.5 kg/head (179 v. 103 kg/head for the cleared and uncleared treatments respectively).

The removal of dense bush has a great emotional appeal, but Harrington (1973) has sounded a note of caution. Trials carried out at the Muko Range Experimental Station at Ankole, Uganda on pastures heavily encroached by *Acacia hockii* indicated that liveweight gains on paddocks where 2400 bushes/ha had been removed were not maintained. The dominating grass species was *Brachiaria decumbens*, with *Themeda*, *Hyparrhenia* and *Cybopogon* spp., and stocking rates of 1.66 and 1.25 beasts/ha. During the first year of the trial there was a significant advantage to the stock grazing the cleared treatment at the higher stocking rate, amounting to 32 kg/head (115 v. 83 kg/head for the two treatments respectively), but in the second and third years this was −3 and 3 kg respectively, the difference in both cases being non-significant. At the lighter stocking rate, there were no significant differences between treatments. Harrington concluded that any increase in grazing area provided by the bush control was offset by lost advantages, which included enrichment of the soil by leaf fall, reduction in the rate of moisture loss and shade and browse for the grazing animals. It was also pointed out that there was a considerable loss of insects and birds which account for a large proportion of total cycling energy and nutrients. Many bush species are legumes which would contribute to the nitrogen available to the ecosystem.

Fire is an accepted method of bush control as well as being useful for the removal of mature, unpalatable herbage. Nevertheless, although the use of fire has been fully reviewed elsewhere (see, e.g. West 1969; Funes 1975), no work has been carried out in our area on its effect on cattle production. Thornton and Harrington (1971) attempted to study the effects of fire with deferred and continuous grazing. However, it was not possible to compare burnt deferred paddocks with unburnt continuous paddocks as had been planned because this 'was frustrated by incendiarists'.

However, Paladines (1974) has compared the weight gains of steers on the *Trachypogon* savannahs of Carimagua, Colombia, during 1971–2 when there was no burning and 1972–3 when the savannahs were burnt. Stocking rates of 0.2, 0.35 and 0.5 steer/ha were used on both occasions. Weight gains were 28.3, 38.2 and 1.5 kg/animal/year and 92.0, 94 and 74 kg/animal/year respectively for the two periods. Although no data are given for rainfall etc., which aid in the interpretation of the data, there is a strong indication that productivity is improved by the use of fire. This is to be expected because removal of the intensely unpalatable nature of the mature *Trachypogon* would allow regrowth and selection of more palatable and nutritious material.

Paladines (1974) also quoted results to show that sequential burning was superior to single burning at the same three stocking rates. Sequential burning involved dividing the area by fireguards (fire breaks?) into eight plots, each of which was burned at different times of year. The advantage of the sequential burning was 27, 16 and 4 kg/head for the three stocking rates referred to above, equivalent to 29, 17 and 5% respectively.

The effects of shade have not been widely studied in tropical and subtropical areas, most results having been obtained in more extreme climates such as Imperial Valley, California (see, e.g. Ittner et al. 1954, 1958). However, whilst McDaniel (1956) working in Louisiana reported that gains of cows and calves on pastures with abundant natural shade were higher than those grazing pastures without shade, this effect could not be confirmed by McCormick et al. (1963) working at Coastal Plains Experimental Station, Tifton, Georgia. Trials were carried out on both growing and fattening cattle over a period of four years. For the first two years of the trial, metal and straw shade were compared with no shade. There were no significant differences among treatments, indicating that animals provided with shade did not grow more than those which had no shade even during the hottest period of the year (June–August) when maximum and minimum temperatures were 33 and 21 °C respectively.

The practice of kraaling cattle at night is widely used in Africa and elsewhere to protect cattle from predators and thieves. However, it would seem that this would reduce grazing time and result in lowered rates of gain. Kyomo et al. (1972) carried out an experiment to investigate the magnitude of this effect at the Malya Livestock Breeding Station in Tanzania. At the end of 120 weeks, the yarded cattle weighed 216 kg whilst those that had had access to pasture for most of each twenty-four-hour period weighed 305 kg, an advantage for the latter of 89 kg. Studies of grazing behaviour indicated that part of this effect could be attributed to a longer period spent grazing during the night in the cool dry season.

Owen found that the enclosure of Boran steers during the night reduced weight gains 11.3 and 5.1% on improved and unimproved pastures respectively (Owen 1968). In a later trial, Boran and Angus Boran steers given access to night grazing gained 5.5 and 6.1% more than confined animals. However, it was pointed out that this small improvement had to be balanced against 'the loss through lion attack of two steers in the night grazing paddock'! (Wigg and Owen 1973).

Discussion

Few experiments have conclusively demonstrated that positive effects on animal production have been obtained by using rotational grazing. Unequivocal results were obtained by Oliver (1972), Escobar, et al. (1971b) and Grof and Harding (1970). Stobbs (1969f) could demonstrate beneficial effects at one stocking rate only and Walker (1968) demonstrated advantages to rotational grazing but with short-term experiments only. Granier and Gilibert (1974) have provided evidence of beneficial effects from deferred grazing.

The major factors which are involved in animal production from rotational and continuous grazing are both the quantity and quality of forage available to the animal, as well as the opportunity for choice of diet.

There is some evidence that more forage is produced from rotationally than from continuously grazed pastures. For example, in the cases of rotational grazing mentioned by Parkin et al. (1976, 1977, 1978), forage production was invariably greater than in continuous treatments and this difference could amount to almost 30%. In considering this increase in production, it may be useful to consider some aspects of the process of tillering in grasses. A new tiller originates as a growing point or bud which is composed of an apical meristem and which is found near the base of a parent shoot near the surface of the soil. The apical bud is at some stage (depending on the habit of the species concerned) elevated some distance above the soil surface by means of elongation of the growing point. This is associated with a morphological change in which a transition occurs from the vegetative to the reproductive stage when the formation of new leaves from this apical bud ceases. The consequence of the elevation of the apical bud is to render it more susceptible to defoliation by grazing than when it was at ground level. The time of change from the vegetative to the reproductive state, furthermore, marks the end of the useful period of the shoot insofar as leaf production is concerned. The differences among species with respect to the time of elevation of the shoot apex will influence their reaction to defoliation by grazing. Bridgens (1968) has given examples which make this clear. *Themeda triandra* is highly susceptible to defoliation as the shoot apex remains in a vulnerable position for some eight to ten months before inflorescence and whilst still in the vegetative state. On the other hand, *Eragrostis curvula* is much less susceptible as internodal elongation is only of the order of 10 to 15 mm. Bridgens has described the reactions of various common South African veld grasses to grazing systems (two, three and four camp) in terms of time of shoot apex elevation, change from vegetative to reproductive condition and the times and conditions of lateral tiller production.

Although there is some difference of opinion as to whether removal of the shoot apex by grazing when in the reproductive state is beneficial to grass productivity under all conditions (see, e.g. Humphreys 1966), it is clear that differences in productivity in mixed swards as a response to different grazing systems may be partially accounted for by habits of the component species.

The accumulation and maintenance of high levels of available or reserve carbohydrates in the crown and roots of the grass plant have been emphasized as one of the important factors to determine recovery after defoliation, as has the photosynthetic ability of the plant as expressed by the leaf area remaining or Leaf Area Index. The role of these factors has been reviewed by Humphreys (1966).

It has been clearly established that the nutritive value of tropical pastures decreases with increasing maturity (or time elapsed since the last defoliation). This is indicated by lower levels of crude protein, lower digestibility of crude protein and crude fibre, increased crude fibre and lignin levels (Butterworth 1965) and decreased intake (Elliot and Topps 1963b). It would be expected, therefore, that rotational grazing would lead to reduced nutritive value as a result of the grass maturing during the resting period. This is also consistent with observations by several authors that when rotational systems are compared, those with a shorter resting period are generally superior in terms of animal production to those with a longer period (Creek and Nestel 1964; Horrell 1972; Denny and Steyn 1977; Denny and Barnes 1977; Parkin et al. 1978; but not Takahashi et al. 1966). Creek and Nestel (1964) also reported increased crude protein content of the short rotation pasture.

Stocking rate is the most important management factor influencing the production of beef from tropical pastures. However, it appears that an interaction between stocking rate and grazing system might be possible. Walker (1968) has suggested that although the nutritive value of continuously grazed pasture is higher than that which is rotationally grazed, the increased productivity of the latter becomes more important at higher stocking rates and enables higher levels of beef production to be maintained.

Frequent defoliation of certain plants would enable nutritive value to be maintained by ensuring regrowth during continuous grazing. However, little critical work has been carried out to ascertain whether this is in fact the case. Gammon and Roberts (1978b) carried out a series of experiments at Matopos Research Station in Matabeleland, Zimbabwe, on patterns of defoliation during continuous and rotational grazing on sandveld. Defoliation on continuously grazed pastures was relatively infrequent at the stocking rates used and the reduction of multiple defoliations under rotational grazing amounted to only 20% of tillers grazed more than three times. It was also noted that distribution of defoliation frequencies could, in many cases, be accounted for by chance alone.

Other factors affecting the distribution of defoliation frequency included selection of previously grazed tillers because of their palatability, but conversely rejection of previously grazed tillers because of their shortness. It was concluded that the timing and severity of defoliation under two systems of management were similar and therefore that considerable differences in animal production could not be expected from such manipulation of grazing systems. It also appeared that in continuous grazing of a mixed sward at relatively low intensity, as in this study, cattle imposed their own rotational grazing system of different grass species according to their stage of growth during the grazing period. Thus during the early growing period, veld in which *Sporobolus ioclades* predominated was more heavily grazed, whilst during autumn–early winter period *Themeda triandra* had the highest percentage of multiple defoliations. This is in agreement with the aspects of shoot apex morphogenesis reviewed by Bridgens (1968) and already referred to.

Increased selection of the *S. ioclades* was associated with a higher content of crude protein, lower crude fibre and greater leaf : stem ratio, whilst the development of coarse culms in *Hyperthelia dissoluta* negatively affected its acceptability (this is one of the major problems of managing this type of veld in large areas of Africa). Preference was also related to moisture content of veld in that grasses with lowest dry matter content were more readily selected.

It should be pointed out that the grazing pressure during the experiments carried out by Gammon and Roberts (1978b) was not high and that this was also associated with 'good' rainfall years. It was also interesting that selection of taller tillers ensured protection for shorter tillers against excessive defoliation.

Important factors which influence livestock production from rotationally and continuously grazed pastures are the opportunity for selection of forage of a high nutritive value and the ease of harvesting this forage. Both these are likely to be adversely affected by the high grazing intensities which may occur in rotationally grazed systems. Some comments have already been made on selective grazing from mixed swards but the order of preference from non-specific swards is leaf first, followed by leaf-bearing stem and, finally, if no leaf is available, stem (it is well known that the nitrogen content is higher in the leaf than in the stem of tropical grasses; see, e.g. Juko and Bredon 1961). Stobbs (1975) has emphasized the importance of ease of harvesting to efficient animal production from pasture and has pointed out that leaf bulk densities are low in tropical pastures and that it is therefore more difficult for an animal to prehend adequate quantities of leaf in each bite and that grazing time, and thus number of bites, is limited by fatigue. He has suggested that grazing management practices should be directed towards maintaining swards in optimum condition for easy defoliation. It is also relevant that although pastures have higher bulk densities when mature than immature, however, much of this leaf may be in the lower strata of the canopy and thus be inaccessible because of stem.

Summary

Few experiments have demonstrated positive effects on livestock production by the use of rotational grazing systems. There is some evidence that production of forage from rotationally grazed may be greater than that from continuously grazed pastures, but there is also evidence that its nutritive value may be inferior. Wide differences in tiller morphogenesis and recovery after defoliation are found from one grass species to another but it is uncertain how these account for differences among grazing systems. Ability to select forage of high nutritive value is an important aspect of grazing management, as is the ability to harvest such material easily. In any event, there is no doubt that the aspect of any grazing system with most bearing on animal production is the regulation of stocking rate. This is also the most important factor in maintenance of the vigour of the sward, although systems of deferred grazing may also be useful in this respect.

Bush clearing may be a useful technique to increase pasture productivity in badly encroached areas, but little information is available as to the magnitude of effects on weight gains in stock.

There seems to be little advantage in providing cattle with shade, but confining cattle during the night reduces grazing time and consequently weight gains.

13 *The use of growth-promoting substances and other techniques*

This chapter presents the results of a number of trials involving the use of hormone-like and other compounds which might be expected to exert an effect on the growth rate of beef cattle grazing tropical and subtropical pastures.

The effects of such techniques as dehorning, castrating and vasectomy have also been included.

It has been known for some time that certain compounds exert an activity similar to that of hormones on the growth rate of cattle both in feedlot and on pasture. The most commonly-used compound is diethylstilboestrol (stilboestrol, stilbestrol, DES) which has many properties of the naturally occurring oestrogenic hormone, oestradiol. However, other hormone-like substances which have been found to be effective in this respect include hexoestrol and resorcylic acid lactone, whilst the naturally occurring hormones, progesterone and oestradiol, have also been used. It has been shown that these exert a positive effect on growth rate when given either orally or implanted subcutaneously. The latter is clearly convenient for grazing animals.

The majority of trials on hormone-like substances in the tropics have been carried out in the Americas. A series of experiments was initiated by Quinn and his colleagues at Ibec in Brazil, workers at Turrialba in Costa Rica as well as other parts of Central and South America.

Certain compounds have been shown to exert an effect on the ratio of volatile fatty acids produced in the rumen and a limited amount of work has been carried out to investigate the use of these under subtropical pasture conditions.

The content of carotene in tropical forages is drastically reduced whilst the plant is dormant and whilst the liver is able to store large quantities of vitamin A; the application of this vitamin might be beneficial under certain circumstances.

The use of hormone-like compounds

Brazil

Experiments on the use of stilboestrol implants for Zebu steers were made by Quinn et al. (1958) at Fazenda Jangada and Matão, Sao Paulo. In the first experiment, cattle grazed high-quality Coloniao (*Panicum maximum*) pastures (no stocking rate is given) for a period of 140 days, and the advantage conferred by stilboestrol implantation amounted to an average of 21 kg/head during this period, equivalent to 40%. The second experiment was carried out on *Hyparrhenia rufa* pastures over a period of 122 days, and the advantage attributable to the implants was 11 kg, equivalent to 26%. The effects of the compound appeared to be greater with two- and three-year-old animals than with yearlings.

In a later series of experiments (Quinn et al. 1960b), the influence of stilboestrol implantation on grazing Zebu steers and male suckling calves was investigated. In the first experiment, two- and three-year-old Zebu steers were grazed on *Panicum maximum* pastures fertilized at different rates and stocked according to the amount of forage available. The use of stilboestrol implants conferred considerable advantage amounting to 21 kg/head during the winter and 36 kg/head during the summer period. It was noted that less growth stimulation was obtained from stilboestrol at lower levels of nutrition.

A further experiment was designed to investigate the influence of single or double implants of stilboestrol on the growth rate of two- or three-year-old steers grazing inferior pastures comprising a mixture of common Bahia grass (*Paspalum notatum*) and Jaragua grass *Hyparrhenia rufa*. As would be expected, the advantage attributable to the hormone analogue in this experiment was not as great as in the first trial because of the inferior quality of the pastures used. There appeared to be no advantage to the use of the double implant, although the authors comment that had the period been longer, this might have been the case. There appeared to be a slightly greater response to implantation in the older animals which gained an average of 82 g/day more than control animals (the corresponding value for two-year-old steers was 54 g/day). The third experiment of the series was carried out to investigate the response of four- and five-year-old steers to single and double implants during the winter period. The animals grazed *Panicum maximum* pastures at a stocking rate of approximately 1 beast/ha. Scales were not available to weigh the animals during the experiment and only carcass weights were obtained at time of slaughter. However, despite the lack of individual information, this trial must count as one of the most extensive quoted in this review as a total of 4100 head of cattle were used. The increase in carcass weight attributable to a single stilboestrol implant was 17 kg and to a double implant was 11 kg. There was a greater response to implantation in older animals.

Finally, the effect of single and double implants on suckling male calves was investigated. Calves implanted with 12 mg stilboestrol at two to four weeks of age showed an average growth increase of 6 kg over controls at six months. A second implantation at this stage was highly beneficial and produced an additional 20 kg/calf over a period of about four and a half months.

In a further experiment to investigate the response of male Zebu calves to castration, creep feeding, stilboestrol and supplemental feeding on pasture (Quinn et al. 1966), it was shown that the response attributable to stilboestrol was equivalent to the detrimental effects of castration. Calves were grazed on uniform *Panicum maximum* pastures. The overall effect of stilboestrol was to increase the rate of gain by 53 g/head/day. Stilboestrol-treated steers had a slightly higher killing-out percentage than non-treated animals (55.1 v. 54.4%).

Quinn et al. (1967) carried out a further series of trials using a total of 1086 steers in ten separate experiments. Implanted steers gained an average of 111 g/day more than controls, equivalent to a 22% increase. The average increase in daily gain was 100 and 121 g during the dry and wet seasons respectively. The carrying capacity of the *P. maximum* pastures was increased by nitrogen fertilization from 1.2 to 3.0 steers/ha, and stilboestrol-treated steers produced 117 and 33 kg/ha at high and low stocking rates respectively.

Costa Rica

Carrera and Soikes (1958) investigated the effects of implanting both male and female animals with stilboestrol at Turrialba. Cattle were between one and a half and two and a half years old and 30 mg implants were administered to heifers and 60 mg to steers. Grazing consisted of Para grass (*Brachiaria mutica*) pastures and groups were either pastured without supplement or given supplements containing 12 or 15% crude protein. The advantage due to the hormone implantation was significant in both steers and heifers and amounted to 36% on pasture alone, 26% with the 12% protein supplement and 42% in the case of animals given the 15% protein supplement. Nevertheless, the results should be viewed with caution as some of the groups, originally consisting of five animals each, were reduced in number due to accidents. An interesting observation was made that the growth rate of cattle without supplement but with implant was equivalent to unimplanted cattle given approximately 2.2 kg supplement and it was pointed out that the implant cost considerably less than the concentrate.

In a later experiment at Turrialba, de Alba and Maltos (1963) investigated the use of progesterone and estradiol benzoate compared with stilboestrol as treatments. Steers were given implants of 200 mg progesterone plus 20 mg oestradiol (Synovex S Squibb) or 30 mg diethylstilboestrol. Untreated animals were used as controls. Two experiments were carried out: the first on pastures of Gordura (*Melinis minutiflora*) and the second on Para grass. Both hormone treatments resulted in significantly higher gains than those obtained from control animals (410 v. 502 g) but there were no apparent differences between treatments. There was an indication that hormone application could give detrimental results in conditions where the growth rate of control animals was less than 300 g/day.

Mexico

Working at Paso del Toro in the state of Vera Cruz, Carrera et al. (1962) investigated the efficacy of hexoestrol as a growth-promoting substance. Implants of 36 mg were made every seventy days to two-year-old steers grazing *Panicum maximum* pastures during both wet and dry seasons. Treated animals gained 29 and 19% more than controls during wet and dry seasons respectively.

United States

Chapman and his colleagues, working at the Everglades Station in Florida, carried out a series of trials to investigate the use of oral and implanted stilboestrol for beef cattle fattened on pasture and in drylot following promising results obtained by Baker (Chapman et al. 1964a).

Steers grazing Roselawn St. Augustine grass pastures with limited supplementation did not gain more than controls when stilboestrol was given orally. Results with implants were equivocal: in one experiment, steers on pasture receiving limited supplementation gained more (872 g) when implanted with stilboestrol than control animals (700 g), but this could not be repeated in a subsequent trial. When no supplementary feed was given, implanted animals grew marginally better than controls (168 v. 200 g/head/day). There was some indication that animals which received stilboestrol had a slighter greater cooler shrink than those that had not.

The performance of steers, steers implanted with zeranol and bull calves was investigated at the Coastal Plain Station, Tifton, Georgia, by Utley and McCormick (1976). Zeranol (zearalanol, 6-(6, 10 dihydroxy-undecyl)-β-resorcylic acid-μ-lactone) is a compound isolated from the fungus *Gibberella zea* which has oestrogen-like activity and has been widely used since undesirable side-effects attributable to stilboestrol were demonstrated in the early 1970s.

Calves were assigned to treatments at three months of age and grazed Coastal Bermuda grass pastures until weaning at seven to eight months. Calves left entire grew faster than castrated calves and there was no advantage to implantation. These tendencies continued during a subsequent feeding period and it was concluded that implantation with zeranol was not justified.

Honduras

The effect of implanting steers with diethylstilboestrol was investigated at the Pan American Agricultural School 'El Zamorano'. Rios and Osegueda (1967) implanted steers grazing *Hyparrhenia rufa* pastures with 24 mg of stilboestrol and compared their growth over a period of 150 days with that of untreated control animals. Daily gains of treated and untreated animals were 617 and 522 g/head respectively. The effects of the implant were apparently greater when forage of a high nutritive value was available.

Colombia

Results obtained on the use of hormone implants in beef cattle in Colombia have not been consistent, although they indicate better growth rates in treated cattle.

Escobar and his colleagues carried out an experiment to determine the effect of different hormones on the fattening rate of Zebu × Polled Sinu steers grazed rotationally. Treatments were 200 mg progesterone plus 200 mg oestradiol benzoate, 35 mg stilboestrol, 120 mg testosterone plus 24 mg stilboestrol or no implantation (control). Average daily gain for the groups in the order given above were 808, 911, 846 and 714 g/day but the differences were not significant. The carcasses of treated animals contained 2.9% more water and 2.0% less fat than the controls (Escobar et al. 1971a).

Gutierrez (1973) carried out an experiment at Palmira to compare the effects of both testosterone and zeranol on the growth rate of heifers.

Two-year-old Zebu cross-bred heifers were rotationally grazed on pastures of Bahia grass (*Paspalum notatum*) at a stocking rate of 3.6 beasts/ha. Animals were given 36 mg of zeranol or 200 mg testosterone plus 20 mg estradiol benzoate, whilst untreated heifers were used as a control. Although analysis of the data did not reveal any statistically significant differences between treatments, untreated animals gained 31 kg during the 125 days of the experiment whilst zeranol- and testosterone-treated animals gained 42 and 41 kg respectively. The authors attributed the lack of response to the poor quality and insufficiency of the pasture during the period of the experiment.

However, Rubio (1973), also working at Palmira, obtained a positive respond to implantation with zeranol from steers grazing Para grass (*Brachiaria mutica*) pastures. The trial was carried out over 224 days and treated animals were implanted at the start of the experiment and after 112 days. The stocking rate used was 4 beasts/ha. The daily weight gains were 427 and 589 g for untreated and treated animals respectively, equivalent to increased production for the latter of approximately 38% over the whole period. There was a significant response to both implants but that to the second was lower than the first (30 v. 43% respectively).

Apparently an earlier trial was carried out at Turipana by Cardona et al. (1966), which demonstrated that Romosinuano steers treated with stilboestrol gained 0.57 kg/day whilst untreated controls gained 0.46 kg/day, but I have not been able to obtain the original reference (F. Duran and A. F. Llano, unpublished seminar).

Venezuela

Chicco and his colleagues investigated the use of both zeranol and stilboestrol on cross-bred Criollo × Brahman steers during the wet season on improved savannah pastures mainly comprising Pangola grass (*Digitaria decumbens*). During the first trial, the effect of zeranol implants was investigated in both pasture-fed and supplemented steers. Daily gains for supplemented cattle were 781 and 716 g for implanted and control animals respectively and the difference did not reach significance. However, in the case of the animals which were fed pasture alone, the corresponding values were 751 and 580 g, indicating a significant superiority of the implanted animals.

In the second trial, steers were given 36 mg implants of either stilboestrol of zeranol or were left untreated as controls. Daily gains over the ninety-day period of the experiment were 1293, 1043 and 876 g for the three treatments respectively and all differences were significant.

Both stilboestrol and zeranol improved performance of pasture-fed steers, but the advantage was not significant when supplements were given in addition to pasture. Stilboestrol was significantly more effective than zeranol (Chicco et al. 1974).

Jamaica

The effects of hormone implantations on the growth rate of grazing steers in Jamaica was investigated by Nestel and Clay (1961). Experiments were carried out to determine the relative merits of hexoestrol and progesterone/oestradiol, the optimum dose of hexoestrol and the effects of hormone implantation on the dressing percentage of the carcass. The effects of breed type and size of animal were also observed. Trials were carried out on both Pangola and Guinea grass pastures.

The effects of the progesterone/oestradiol implant (200 mg progesterone plus 20 mg oestradiol benzoate) were measured in three experiments. Responses to the hormone were 35, 9 and 13%, but only the treated animals in the last experiment were significantly better than the controls (this was probably due to lack of replication; in the first experiment only five steers were used per treatment).

On the other hand, hexoestrol gave consistently improved performance amounting to some 20%, which was statistically significant in three out of the five experiments. There was a small (but non-significant) fall in dressing percentage in treated animals which accompanied undesirable side-effects such as elevation of the sacrum and coccygeal vertebrae and enlargement of the teats at the higher levels of hexoestrol (60 mg). The use of the level did not confer benefits insofar as growth was concerned, however, and it was concluded that a smaller dose (e.g. 30 mg) was adequate. There was no indication that the effects of hormone implantation were affected by the size or breed type of the animal (so far as proportion of either *Bos indicus* or *B. taurus* was concerned).

Australia

Experiments on the use of synthetic oestrogens for the promotion of weight gains in beef steers were initiated at Brian Pastures Beef Cattle Pasture Research Station near Gayndah. Pastures consist mainly of *Heteropogon contortus* (black Spear grass) and blue grasses (*Dichanthium and Bothriochloa spp.*). The treatments used were 30 or 60 mg implants of hexoestrol whilst unimplanted animals were used as controls. Implanted animals gained significantly more than unimplanted controls, daily gains being 817, 772 and 726 g for the three groups in the order as above. Implanted cattle had greater depths of eye-muscle than unimplanted animals (5.0 v. 4.8 cm) although this difference was not significant. However, fat cover was greater on control cattle (2.2 v. 1.8 cm) and this resulted in considerably lower grading of the implanted animals due to lack of finish. A second trial was carried out using a 30 mg implant and weight gains of the treated animals were again superior to the controls (862 v. 726 g/day) by some 19% (Burns and Sutherland 1960).

Sutherland and his colleagues later carried out a comprehensive series of trials on the use of hormones in yearling to mature bullocks. Fifteen trials were carried out (one further trial was carried out in a feedlot which does not concern us) involving 377 treated animals and 295 controls at experimental sites ranging from Warwick in south Queensland to Utchee Creek in the north. All implants were of either 5, 30 or 60 mg hexoestrol, except two which employed 200 mg progesterone plus 25 mg oestradiol benzoate.

Pastures used ranged from good-quality improved pastures to native range of poor nutritive value. In all trials, the gains of treated animals were greater than those of the controls, although these differences were not always significant. The responses ranged from 64 to 327 g/head/day and these were apparently higher at better nutritional planes. Several defects in the conformation of treated animals were observed. These included elevation of the tail head, depression in the back line ('sway-back'), lateral sloping of the rump and abnormal mammary development, but carcasses were not downgraded. However, in one trial, significant down-grading of treated animals occurred because of insufficient fat cover. Nevertheless, eye-muscle depth was significantly greater in treated animals than in controls. Conformation defects became more frequent at levels of hormone implantation greater than 30 mg and no corresponding benefits were obtained at these levels. The authors therefore suggest that this level of implantation not be exceeded (Mawson et al. 1962).

An experiment was carried out at Yeerongpilly to determine the effects of implantation with progesterone plus oestradiol benzoate, with or without the addition of cholesterol plus hydrocortisone acetate, on weight gains and carcass composition of steers. The composition of the implants was 200 mg progesterone, 20 mg oestradiol benzoate and, where appropriate, 25 mg cholesterol and 4.9 mg hydrocortisone acetate. A group of animals was slaughtered at the start of the experiment to serve as a control for carcass composition. The experimental animals grazed pastures comprising *Paspalum dilatatum* (66%) and *Cynodon dactylon* (15%) for a period of 105 days.

Mean daily gain (shrunk) for the two hormone-treated groups was 1.27 kg compared with 1.04 kg for the cattle which had not been implanted. There were no differences in killing-out percentages among groups, but there was a strong indication that the hormone-treated cattle contained a higher percentage of water (50.6 v. 48.9%) and protein (14.6 v. 13.9%) than the controls. The authors conclude that the effect of hormone treatment was to increase the quantity of the protein in the carcass and that this was associated with approximately three times its weight of water. The only difference which could be attributed to the addition of cholesterol to the implant was a slight increase in the quantity of fat (expressed as ether extract) in the separable portion of the carcass (19 v. 17%). The authors commented that this effect might be of importance in minimizing possible downgrading of hormone-treated carcasses (Beames and O'Bryan 1962).

Southern Africa

Responses to zeranol implantation in Southern Africa have generally been positive, with the exception of one treatment in an experiment carried out by Grant (1976b) at the Grasslands Research Station at Marandellas in Zimbabwe. Grazing steers were implanted with zeranol or untreated and given no supplement, monocalcium phosphate plus salt or 'Rumevite 12P' (a commercial phosphorus supplement). The response to implantation during the 215 days of the trial was 6.9 and 8.5 kg (9.5 and 7.9%) for the control (no supplement) and Rumevite 12P-supplemented cattle respectively, but there was no response by the cattle supplemented with monocalcium phosphate plus salt. No explanation could be offered for this lack of response and the author commented that a similar phenomenon had been reported when steers were offered Rumevite 12P in a previous experiment.

Shorrock and his colleagues carried out trials on pasture and lot-fed steers to determine the effect of zeranol on the performance of steers in Botswana. Trials were carried out over a period of three years at a number of ranches over a fattening period of 180 days during each year. Implanted steers showed a significant improvement of liveweight gain of 11.9, 24.5 and 12.3% over control steers for the three years during which the trial was carried out (Shorrock et al. 1978).

Trials on the use of zeranol in steers fattened on grass were also carried out in the Swaziland low veld. In the first experiment, gains of implanted cattle were 704 compared with 566 g/day for untreated controls. In a second trial, application of zeranol in either winter or summer or both indicated a strong and significant superiority of the implanted animals (470 v. 406 g/day), but statistical analysis confirmed that the time of implantation did not make any significant difference to results and that there was no advantage to a double implantation (FAO 1978b).

In most of the trials which have been carried out with hormone-like substances on cattle grazing tropical and subtropical pastures, positive results on growth rate have been obtained, although it has been noted by several authors that these are more marked at higher nutritional levels (Carrera et al. 1962; Rios and Osegueda 1967; Mawson et al. 1962). These results are in line with those obtained elsewhere as reviewed by Heitzman (1980) and Buttery et al. (1978).

There have been indications, however, of adverse effects on carcass quality. Chapman, Palmer, Kidder, Carpenter and Haines (1964a) noted a slightly higher cooler shrink in treated animals than in controls, and this is consistent with the findings of Escobar et al. (1971a) and Beames and O'Bryan (1962) that treated carcasses contained a higher percentage of water. The latter authors also noted that this increase in water content was associated with an increase in area of the rib-eye muscle (Burns and Sutherland 1960). Hormone treatment also reduces the fat content (Escobar et al. 1971a) and cover of the carcass which may result in lower grading (Burns and Sutherland 1960). Effects on killing-out or dressing percentage have been variable as Nestel and Clay (1961) obtained a lower dressing percentage with treated animals than with controls, as would be expected from the previous remarks on fat content, whilst Quinn et al. (1960b) obtained the opposite results; this could well have been because of different weighing times of the dressed carcass, i.e. before it had completely shrunk.

The use of higher doses of hormones has been observed to confer female characteristics on the carcass, such as elevation of the tail head and increased mammary development, but these apparently occur only at levels above those which are necessary to obtain maximum growth benefits (Nestel and Clay 1961; Mawson et al. 1962).

At this point it may be pertinent to comment that the use of some hormone-like susbstances has been banned in some countries (e.g. the use of DES in the US) because of the possibility of health hazards on the consumer. This would seem to be unlikely providing that hormones are properly used and that limitations on time of slaughter, place of implantation, etc. are observed. In view of the necessity for animal protein, it is unfortunate that limitations are placed on substances that promote improved growth and food conversion efficiency.

Supplementation with monensin

Monensin sodium is a biologically-active substance produced by fermentation of the micro-organism *Streptomyces cinnamonensis*. It is not a hormone but appears to act by increasing the proportion of propionic acid relative to the other volatile fatty acids in rumen fermentation (Richardson et al. 1974).

Apparently no work has been done on the use of this compound in the tropics or subtropics, but a trial carried out in north Louisiana is mentioned here as it was carried out on Coastal Bermuda grass pastures similar to those encountered in many parts of our area.

Oliver (1975) compared different levels of monensin fed with corn to steers and calves grazing Coastal Bermuda grass pastures at a stocking rate of 6.2 steers/ha. Levels of monensin fed were 0, 25, 50, 100 and 200 mg/day. A summary of the results obtained during the 140 days of the experiment is given in Table 13.1.

Table 13.1 The effects of different levels of monensin on steers grazing Coastal Bermuda grass pastures (Homer, north Louisiana) (Adapted from Oliver 1975)

Monensin level (mg)	*Gain/head (kg)*	*Gain/ha (kg)*
0	79	489
25	98	608
50	102	632
100	109	675
200	99	613

All levels of monensin increased weight gains on both a per head and per hectare basis, but the 100 mg/day level was the most effective, giving an additional 30 kg/head gain during the 140-day period, equivalent to some 38% over the control animals.

It is generally agreed that higher levels of propionic acid increase the efficiency of utilization of energy for fattening (see, e.g. Blaxter

1962) and that the fermentation of tropical grasses, which are high in fibre, generally leads to a high molar proportion of acetic acid (see, e.g. Butterworth and Butterworth 1965). It would seem, therefore, that the use of monensin offers a promising possibility for the improved utilization of such forages. Howver, cf, Ørskov and McDonald (1980).

Dehorning and tipping

There are several evident disadvantages to the presence of horns on cattle. These include increased incidence of bruising of slaughter cattle during transportation etc., greater difficulty in handling, necessity for additional trough space in feedlots and pastures. However, although both dehorning or tipping are recognized as causing a setback to weight gains of adult cattle, there is little information available to quantify the extent of any losses involved.

Winks et al. (1977a) carried out a trial at Swan's Lagoon on the effects of tipping (reduction of horn length to 10–15 cm) and complete dehorning on mature Brahman cross-bred steers. Guillotine-type dehorners were used and haemorrhage occurred in all cattle subjected to horn alteration. Severe setbacks were suffered by dehorned steers and, despite some degree of compensation, differences were still apparent at the end of the grazing season. Horned steers gained on average 11.9 kg more than dehorned steers during the five months after the operations. Tipped steers gained less than horned steers in one experiment only. The effect of sinus opening on weight gains was significant during the immediate post-operative period, but compensation occurred subsequently and overall gains were independent of size of opening.

Reduced growth rate could be compensated for by reduced damage due to bruising in the carcass. However, that was not the case in this experiment nor in other cases quoted by the authors (Anderson 1973; Meischke et al. 1974). It was concluded that heavy tipping is also of questionable value and that neither practice should be carried out on adult Brahman cattle.

The question also arises whether dehorned cattle are not at a disadvantage when compared with horned cattle insofar as shipping to slaughter is concerned. Dehorning may be a disadvantage when horned cattle are sold or transported in the same lot.

Castration

Castration has traditionally been practised to render animals more tractable, to obtain a better fat cover on the carcass and to prevent uncontrolled breeding. However, there has recently been a move towards bull meat in many countries both because of the faster rate of gain of the entire male and the leaner carcass that results.

An experiment to determine the effects of several treatments on the weight gain of male Zebu calves grazing *Panicum maximum* pastures was carried out at Fazenda Jangada in Araĉatuba in the state of Sao Paulo. Responses were measured to creep-feeding, castration, implantation with diethylstilboestrol and supplemental feeding. Average daily gain for castrated animals was 545 v. 601 g for entire males. Dressing percentages were 57.4 and 56.8% (U S standards), fat thickness (presumably over the twelfth rib) 3.3 and 1.5 mm and rib-eye area 31 and 34 sq. cm for castrated and non-castrated animals respectively. All differences were significant. Lean bull meat was softer, darker, coarser and less well marbled than corresponding steer meat, but the authors concluded that this would not be discriminated against provided bulls were sold at less than two years of age (Quinn et al. 1966).

Similar results for weight gains were obtained in a later experiment carried out in the state of Minas Gerais (Pereira et al. 1977). Animals were castrated at 3, 8, 12, 24 or 30 months by the use of a burdizzo or removal of the testes. Bulls were heavier than steers regardless of the time or method of castration but, in this experiment, there were no significant differences in dressing percentage.

An experiment was carried out at Linares in the state of Nuevo Leon in Mexico on the development of cross-bred Charolais bulls and steers on Buffel grass pastures was observed. After 153 days, bulls had gained 36 kg whilst steers only gained 29 kg. There were, however, no differences in the carcasses of the two groups (Temple et al. 1973).

In contrast to the above results where there was a definite superiority of the entire males in terms of liveweight gains, Macfarlane (1966) working at Tanga in Tanzania could demonstrate no difference in liveweight gains between entire males and castrates to one year of age. There was a trend for males to be superior to castrates up to three years but this was not significant and not maintained during the fourth year. Average gain/day to four years was 230 g.

Workers in Zambia and Zimbabwe have investigated methods of making the entire male unable to breed rather than carrying out complete castration. This has the effect of preventing uncontrolled breeding without reducing the production of androgens from the testes, thus maintaining the growth rate of the entire animal. These methods have included vasectomy and 'short scrotum' induced cryptorchidism.

The effects of castration and vasectomy on Zebu bulls grazing on veld were compared at Mazabuka and Mount Makulu in Zambia. In view of the fact that the improved growth rate of bulls may be limited by nutritional plane, two nutritional treatments were imposed, These were 'good veld', which consisted of upland veld during the wet season and 'dambo' grazing during the dry season with a winter supplement. 'Poor veld' animals grazed dry-land veld pasture which had been heavily grazed before the start of the experiment. The stocking rate for the former group was 0.2 head/ha and, for the latter, 0.5 head/ha. The overall growth rates for the duration of the trial (twenty-nine months including a three-month finishing period) were 0.29, 0.33 and 0.29 kg/day for the good veld and 0.22, 0.22 and 0.23 kg/day for the poor veld for bulls, vasectomized bulls and steers respectively. Castration or vasectomy of animals grazing poor veld did not influence depth of eye-muscle, but castration of animals on good grazing caused a reduction in eye-muscle depth. Castration led to an increase of back-fat thickness on both nutritional planes and this was greater in the case of low-plane animals (1.0 v. 2.4 mm and 1.1 v. 3.0 mm for non-castrated v. castrated animals on good and poor grazing respectively). The authors (Hale and Oliver 1972) concluded that even at the highest rates of gain (i.e. good grazing during the summer; 0.5–0.6 kg/day), intact animals grew no faster than castrates (cf. Macfarlane 1966).

The principle of the short-scrotum method is that when the testes are confined within the body cavity, body temperature inhibits spermatogenesis but the production of testosterone is not affected. The operation is carried out using elastrator rings which confine the testes whilst occluding as much of the scrotum as possible.

Hopley and Holness (1977) compared short-scrotum bulls, steers and heifers grazed on veld after weaning (with a winter protein supplement) and slaughtered at eighteen months of age after receiving a supplement for 104 days. The short-scrotum bulls gained more weight than both the castrates and the heifers, and the final carcass weights for the three groups were 185, 166 and 157 kg respectively; i.e. an advantage for the short-scrotum animals over the castrates of 20 kg and over the heifers of 28 kg. No significant differences in dressing percentages could be demonstrated.

A subsequent trial (Holness and Hopley 1979) was designed to show whether there was any difference between short-scrotum and entire males when slaughtered before two years of age, a comparison which was not included in the previous experiment. There were no differences in growth rate during the period on pasture between the two groups (although entire males gained slightly more than short-scrotum animals over the fattening period). Classification according to fleshing index and fat cover did not differ between treatments.

Hopley and Holness (1977) investigated the effects of the age of castration on the body weight gain of calves from birth to weaning. Calves were castrated using a knife within ten days of birth, at three months or at six months. Differences in body weight were not significant at weaning or at more than six months. However, the authors point out that castration is more difficult at six months, causes more stress and is more subject to infection (cf. Pereira et al. 1977).

It appears then that castration reduces liveweight gains from those to be expected in the entire animal, increases fat cover on the carcass and possibly dressing percentage but reduces the depth (or area) of the rib-eye muscle. However, these changes are not marked at lower rates of growth and may in fact not be apparent. Alteration of males by the short-scrotum method appeared to maintain the advantages of the entire animal, whilst eliminating the problems associated with uncontrolled breeding. The choice of the appropriate system of management for any particular set of circumstances must depend upon the priorities given to carcass quality and ease of management as well as increase in rate of gain.

Vitamin A administration

The content of β-carotene in tropical grasses fluctuates widely according to the stage of maturity of the plant and, as this is the main source of vitamin A in the grazing animal, it would be expected that the status of the animal with regard to this vitamin would also be subject to wide variation. Values for the content of β-carotene in tropical forages are relatively few, but the average of the eighty-eight determinations cited by McDowell et al. (1974) is 102 International Units (IU)/g. However, the standard deviation is high, amounting to 123 IU/g. Even more indicative of the wide variation is the range, i.e. from a minimum of 1 IU/g in a sample of *Digitaria decumbens* hay to the maximum of 516 IU/g in a sample of green *Panicum altissimum* (the highest value given for *D. decumbens* is 444 IU/g. Values are quoted as IU vitamin A activity calculated from carotene where the conversion of this precursor to the active vitamin is assumed to be 1 mg β-carotene equivalent to 400 IU vitamin A. Values for the few forage legumes quoted are generally higher, although the standard deviation is also high. The mean of eighteen values is 235 ± 184 IU/g.

Gartner and his colleagues working at 'Brian Pastures' Pasture Research Station near Gayndah in south-eastern Queensland have accumulated a considerable amount of data on the degree of liver storage of vitamin A in cattle under tropical conditions grazing unimproved pastures. These consisted mainly of bunch Spear grass (*Heteropogon contortus*) and blue grasses (*Dichanthium* and *Bothriochloa spp.*). Samples were taken by biopsy from cows at different stages of lactation in May, August, November and February during a two-year period. Calves were also sampled. Mean hepatic concentrations of vitamin A in cows were some five times the mean values given in the literature by Moore (1957) and equivalent to 256 μg vitamin A/g liver. Levels were highest in August and lowest in February and May, i.e. lower in the summer than in the winter. Variation did not appear to be attributable to differences in the carotene content of the herbage.

Level of vitamin A in the livers of calves was adequate in most cases but lower than that found in the case of the dams. This was explained by the fact that calves are born with low hepatic reserves of vitamin A. The authors concluded that year-long grazing was the reason for the high reserves and that supplementation was not indicated.

Experiments designed to investigate the effects of supplementation of breeding cows with either vitamin A alone or in combination with vitamins D and E (because this is a common method of merchandising these compounds) have yielded equivocal results. Large doses of vitamin A administered by intramuscular injection to Brahman and Criollo cattle grazing improved pastures in the Venezuelan llanos were without effect on corpus luteum development or ovulation, and birth weight and subsequent development of calves (Mueller-Haye et al. 1968).

However, Carrazoni (1974), working in subtropical Argentina, reported that 70% of cows treated with vitamins A, D and E showed signs of oestrus during the month after the application as contrasted with only 55% of untreated control animals. Working at Tupancireta in Brazil, Leal and his colleagues gave 2 million units 30 to 100 days before calving and 4 million units 24 hours after calving. The incidence of abortion was decreased in the treated animals (0 v. 8%) and the calving interval was markedly reduced (566 v. 672 days). The incidence of retained placenta in treated animals, however, was higher than that in controls (Leal and da Silva 1977).

The effect of administering vitamins A, D and E to suckling calves was investigated by da Silva and his colleagues in São José do Rio Prêto in the state of São Paulo. Calves grazed *Digitaria decumbens* and *Panicum maximum* pastures throughout the experiment and no beneficial effect of the vitamins could be demonstrated (da Silva et al. 1974).

It would appear that under the conditions of the experiments reviewed here, there may be slight advantages in terms of reproductive efficiency in the administration of vitamin A to mature cows but that, normally, hepatic reserves would be sufficient to carry stock over all but the most severe dry season. Although the values of carotene in mature tropical grasses are low, these are compensated by high values which occur in the plant during the active growing period as well as by the high levels in legumes and browse plants which may form an important part of the diet of the animal grazing mixed pastures or range land.

14 *Beef cattle production from tropical pastures – present and future*

Research findings on the nutrition of cattle on tropical pastures have been reviewed in the preceding chapters and it has been shown that improved productivity may be brought about by the employment of different techniques of nutrient supplementation and pasture management.

Present levels of productivity of tropical and subtropical pastures will now be considered and an attempt will be made to identify the constraints which prevent an increase in productivity from being attained. Finally, suggestions will be made as to how the *status quo* may be improved.

This book does not attempt to deal with either disease or breeding of beef cattle and, whilst the importance of these factors as constraints to beef cattle production is fully recognized, they are given only scant attention in the following pages.

Similarly, no attempt has been made to carry out a comprehensive review of the present situation of the beef cattle industry in the tropics, but examples are cited as appropriate. This has been done with the aid of data from both statistical and case studies.

Present levels of production

Overall productivity

The use of national statistics to estimate production is fraught with uncertainty. The values given by such compilations as that of FAO (1978a) are only as good as the field systems used to collect the data. In countries where all cattle are dipped regularly, estimates of population may be relatively accurate, but it is unlikely that such values are reliable where cattle are not subject to periodic official scrutiny. Apart from the intrinsic difficulties in the estimation of cattle populations, cattle owners may be reluctant to disclose herd numbers to officials because of the fear that such information may lead to higher taxes, etc.

The number of cattle slaughtered may be estimated accurately where centrally-located government-controlled slaughterhouses account for the majority of slaughter stock or where an efficient meat inspection service exists. However, the number of cattle slaughtered for home consumption or for clandestine sale is difficult to estimate. Similarly, the number of cattle which are slaughtered before they die of disease or emaciation is impossible to estimate, as is the number of cattle which are used for human consumption after such a death.

Estimates of carcass weight are also subject to considerable doubt. FAO (1978a) quote carcass weights for fifty-three African countries; in 42% of these examples, there was no change in carcass weight during the period 1969–78, a not inconsiderable feat, even had an optimum carcass weight been decided upon in advance. Nevertheless, three estimates of productivity have been tabulated for selected countries in Africa, Latin America and Asia (Tables 14.1, 14.2 and 14.3). These are:

Carcass weight per unit of area (kg/ha)
Slaughter cattle as a percentage of the total herd (% extraction rate)
Carcass weight per head of the total national herd (kg)

Carcass weight/unit of area ranges from 1.3 kg/ha in Africa to 6.1 kg/ha in Asia. Such values are subject to considerable variation because of the wide range of environmental conditions which prevail. These variables include climatic (principally precipitation and evaporation) and edaphic factors, as well as the type of land use employed. FAO (1978a) has classified this into:

Arable land
Land occupied by permanent crops
Permanent pasture
Forest and woodland
Other land

However, livestock utilize categories of land other than permanent pasture to a greater or lesser degree. Thus, arable land includes land under 'temporary meadows for moving or pasture' and pastures may be included in arable rotations (in Australia, the arable area includes some 27 000000 ha of cultivated grassland). Similarly, arable land is used by livestock which consume crop residues after harvesting.

The phenomenal stocking density to be found in India (i.e. 1.64 ha/head) could only be achieved by the widespread utilization of crop residues. Land under permanent crops includes areas where cattle may be grazed under tree crops, e.g. coconuts, and many naturally forested areas may also be used for grazing.

Payne (1976) has further classified the major systems of tropical agriculture in which cattle are important. These include:

Migratory shifting cultivation
Sedentary shifting cultivation

Table 14.1 Land area, cattle population and offtake of selected countries of Africa (Source FAO 1978a)

	Total land area ('000 ha)	*Cattle population ('000 head)*	*Cattle slaughtered ('000 head)*	*Extraction rate (%)*	*Slaughter weight (kg)*	*Offtake*	
						kg/head	*kg/ha*
Africa (total)	2 964 612	168 864	19 064	11.3	136	11.3	0.87
Tropical humid Africa							
Burundi	2 565	808	91	11.3	130	14.6	4.61
Cameroon	46 944	2 972	318	10.5	150	15.8	1.00
Congo	34 150	70	11	15.7	156	24.5	0.05
Central Africa Rep.	62 298	660	134	20.3	160	32.5	0.34
Benin	11 062	730	95	13.0	110	14.3	0.94
Gambia	1 000	275	33	12.0	115	13.8	3.80
Ghana	23 002	900	108	12.0	115	13.8	0.54
Guinea	24 586	1 600	110	6.9	100	6.9	0.44
Guinea Bissau	2 800	262	26	9.9	110	10.9	1.02
Nigeria	91 077	11 566	1 639	14.2	120	17.0	2.16
Rwanda	2 495	610	105	17.2	104	17.9	4.38
Sierra Leone	7 162	260	56	21.5	100	21.5	0.78
Togo	5 360	245	42	17.1	125	21.4	0.98
Tropical dry West Africa							
Mali	122 000	4 263	369	8.7	92	7.9	0.28
Mauritania	103 040	1 500	130	8.7	120	10.4	0.15
Niger	126 670	2 990	220	7.4	110	8.1	0.19
Senegal	19 200	2 671	260	9.7	140	13.6	1.90
Chad	125 920	4 012	230	5.7	120	6.9	0.22

Tropical dry East Africa							
Ethiopia	110 000	27 500	1 925	7.0	110	7.7	1.92
Kenya	56 925	9 100	1 190	13.1	108	14.1	2.26
Somalia	62 734	4 000	430	10.8	110	11.8	0.75
Sudan	237 600	16 567	1 032	6.2	165	10.3	0.72
Tanzania	88 604	15 272	1 414	9.3	92	8.5	1.47
Uganda	19 971	5 321	600	11.3	150	16.9	4.51
Tropical dry South Africa							
Angola	12 460	3 070	310	10.1	160	16.2	0.40
Botswana	58 537	3 000	200	6.7	207	13.8	0.71
Malawi	9 408	760	77	10.1	140	14.2	1.15
Mozambique	76 553	1 370	245	17.9	150	26.8	0.48
Zambia	74 072	1 800	178	9.9	155	15.3	0.37
Mean				11.52		14.7	1.33
S.D.				±4.18		±6.09	±1.36

O.P.C. ABERYSTWYTH
DYFERGELL

Table 14.2 Land area, cattle population and offtake of selected countries of Central–South America and the Caribbean (Source FAO 1978a)

	Total land area ('000 ha)	Cattle population ('000 head)	Cattle slaughtered ('000 head)	Extraction rate (%)	Slaughter weight (kg)	Offtake	
						kg/head	kg/ha
Central America							
Costa Rica	5 066	2 062	279	13.5	256	34.6	14.10
Guatemala	10 789	2 417	463	19.2	168	32.2	7.21
Honduras	11 189	1 700	350	20.6	145	29.8	4.54
Nicaragua	11 875	2 774	453	16.3	188	30.7	7.14
Panama	7 505	1 396	258	18.5	261	37.1	6.91
Caribbean							
Cuba	11 542	5 700	841	14.8	170	25.1	12.48
Dominican Rep.	4 838	2 050	225	11.0	167	18.3	7.77
Haiti	2 756	900	102	11.3	180	20.4	6.66
Jamaica	1 083	285	68	23.9	176	42.0	11.05
South America							
Bolivia	108 547	3 772	490	13.0	156	20.3	7.04
Brazil	845 651	89 000	11 500	12.9	196	25.3	2.67
Chile	74 880	3 492	715	20.5	257	52.6	2.45
Colombia	103 870	25 294	2 460	9.7	175	17.0	4.14
Ecuador	27 684	2 874	481	16.7	159	26.6	2.76
Paraguay	39 730	5 800	676	11.7	198	23.1	3.37
Peru	128 000	4 167	685	16.4	120	19.7	0.64
Venezuela	88 205	10 231	1 468	14.4	192	27.6	3.20
Mean				15.5		28.38	6.13
S.D.				3.96		±9.35	±3.75

Table 14.3 Land area, cattle population and offtake of selected Asian countries and Australia (source FAO 1978a)

	Total land area ('000 ha)	*Cattle population ('000 head)*	*Cattle slaughtered ('000 head)*	*Extraction rate (%)*	*Slaughter weight (kg)*	*Offtake*	
						kg/head	*kg/ha*
Asia (total)	2 676 993	362 259	26 629	7.35	140	10.3	1.39
Southern Asia							
Bangladesh	13 391	27 007	1 732	6.41	97	6.2	12.55
India	297 319	181 651	890	0.49	80	0.4	0.24
Pakistan	77 872	14 946	1 480	9.90	113	11.2	2.15
South-East Asia							
Burma	65 888	7 865	647	8.23	120	9.9	1.18
Indonesia	181 135	6 167	890	14.43	170	24.5	0.84
Kampuchea	17 652	1 300	143	11.00	120	13.2	0.97
Laos	23 080	534	43	8.05	120	9.7	0.22
Malaysia	32 655	430	63	14.65	117	17.1	0.23
Philippines	29 817	1 820	651	30.82	157	48.4	2.95
Sri Lanka	6 474	1 542	190	12.32	68	8.4	2.00
Thailand	51 177	4 650	690	14.84	200	29.7	2.70
Viet Nam	32 536	1 700	265	15.59	125	19.5	1.02
Mean				12.23	123.92	16.51	2.25
S.D.				±7.30	±36.93	±12.88	±3.38
Australia (total)	761 793	29 379	12 557	42.74	170	72.66	2.80

Sedentary subsistence rain-fed or irrigated cultivation
Regulated ley farming
Perennial crop cultivation
Nomadic herding
Transhumance, or semi-nomadic herding
Commercial ranching

Payne has estimated that the most important systems with respect to cattle numbers are sedentary subsistence cultivation and commercial ranching with 46 and 29% of the total population of bovines respectively. Clearly the system used would affect productivity as expressed on a per hectare basis. A common method of expressing the productivity of beef herds is extraction rate (i.e. slaughter cattle as a percentage of total herd) and this value has also been included in the tables. Average values vary between 11.5 and 15.6% for the three groups of selected developing countries. Values for North America, Western Europe and Oceania, on the other hand, are 38, 35 and 42% respectively, a level which is only approached by the Philippines.

Perhaps the most meaningful relationship to express productivity is the quantity of carcass produced per animal in the total herd. Values for this parameter are quite similar in Africa and Asia, being approximately 15 and 17 kg carcass weight/head respectively. The values for Latin America and the Caribbean, on the other hand, are somewhat higher, being equivalent to approximately 28 kg carcass weight/head. This latter value is to be expected in view of the higher proportion of commercially-orientated enterprises in this area and the amount of development that has taken place, e.g. establishment of improved pastures. Jasiorowski (1973) calculated similar values for Latin America, the Near and Far East and Africa based on figures for the years 1950 (average of 1947/48–51/51) and 1970. These were, for Latin America, 28.5 and 29.4 kg carcass weight/head and for Africa, 13.9 and 13.6 kg carcass weight/head. The Latin American value is similar to that obtained in the present analysis, whilst a comparable value for Africa from FAO (1978a) is 15.4 kg carcass/head national herd.

Jasiorowski (1973) also calculated comparable values for Europe, North America and Oceania for 1950–70. These were 38 and 60, 59 and 88, and 42 and 46 kg carcass/head respectively, a considerably higher productivity than that obtained from the herds of the less-developed countries. It is noteworthy that in the more developed countries, productivity increased significantly over the period 1950–70, whilst this was not the case in Latin America and Africa.

Calculation of current levels of productivity (FAO 1978a) for North America, Western Europe and Oceania are 96, 79 and 70 kg carcass/head respectively, a further increase in productivity. Jasiorowski's values are not directly comparable to those extracted in the present work as it is not clear which countries were chosen by him.

The comparison of developed and developing countries is also not entirely valid because of the high levels of grain inputs which are used for beef cattle production in both North America and Europe. Nevertheless, it is an important indication that the productivity of the national herd when expressed on a *per capita* basis is substantially lower in developing than in more developed countries.

It has been established therefore that productivity of the beef cattle herd in tropical countries is generally lower than that in more developed temperate regions although, exceptionally, high levels of production may be achieved in certain areas. It is now appropriate to examine in more detail the herd characteristics that lead to these low levels. Relatively few figures are available from un-improved herds, but some of these will be quoted here to establish the pattern of performance to be expected.

Calving rate

Except where accurate dip tank records are made, few data are available from long-term observation of unimproved herds. Most information has been obtained as a result of sample surveys carried out over a relatively short period or even a single observation. It is impossible with such data to assess the effects of different years, which clearly play an important role insofar as the availability of forage and therefore reproductive performance is concerned. Nevertheless, survey must play an important role in the assessment of unimproved cattle populations until the results of more critical studies become available.

Reed and his colleagues of the University of Edinburgh Veterinary Expedition to Botswana 1971–2 carried out a survey in six widely separate areas of the country to investigate bodily conditions and reproductive performance under different management systems. Cattle at two of the locations chosen were given improved management but at the other four locations, management was poor. This poor management reflects the 'ignorance, apathy or financially embarrassed state of the owners'. Between 40 and 100% of cattle in poor management areas were pastured on overgrazed land. The average calving rate of cows was 40% on poor management areas v. 52% on better management areas. The authors note that the low calving rates recorded reflect long calving intervals and high age at first calving. Low calving percentages were recorded in cows between two and three years of age at the time of the survey (18%) and the highest calving rate was observed in cows of eight years of age (48%) Reed *et al* (1974a).

A further study was carried out in Botswana by the Animal Production Research Unit (APRU 1977) who determined that the average calving percentage under the traditional 'cattle post' system of management in Eastern Botswana was 47.3 as compared with 74.8% to be found in ranches run under a 'reasonably acceptable' level of management. In a later study (APRU 1979), results of reproductive behaviour in the western areas of Botswana are reported. Cows were monitored on cattle posts at Nojane and Kalfontein and calving rates were 46.2 and 22.3% respectively; the latter low value reflects 'very scarce grazing and harsh environmental conditions'. Cows in ranches developed as part of the first Livestock Development Project at Nojane were also observed and calving percentage was found to be 40.8%. The authors comment that the survey method employed did not consistently detect stillbirths and early mortality and thus values were somewhat lower than might have been expected. This conclusion was supported by low values for calf mortality after birth.

In Swaziland, Butterworth and Presswood (1978) reported the number of calves surviving to one year of age in figures extracted from censuses carried out at dip tanks which are probably reasonably reliable. Births were classified according to whether they were from the small farmer (sedentary subsistence cultivation) or from commercial ranches. The values for the two sectors were 33 and 48% respectively. These values are composites involving both calving rate and mortality to one year, but they indicate the low level of reproductive efficiency to be found in both the subsistence and commercial sectors.

Elsewhere in Southern Africa, Brown (1969, cited by Hamburger et al. 1980) stated that the calving percentage in the Ciskei was between 20 and 80% with an average of about 35%, whilst Jacobs (1972, cited by Hamburger et al. 1980) showed that the average calving percentage in the Transkei was 44.3%.

Oyenuga (1966) estimated that the annual calf crop in Nigeria varies between 40 and 60% and observed that reproduction seldom starts before the age of four to six years. Warnick (1976) cited Plasse, who estimated that the birth rate in the Latin American tropics was between 35 and 60%. In the eastern Plains of Colombia, Stonaker et al. (1976) estimated annual calving rate in the Meta plain, a poorer area of the llanos and better pasture in the Piedmont foothills. The method used was that of Reed et al. (1974a) and the values for the two areas were 52 and 55% respectively.

Groenewold and Crossing (1975) estimated that calving percentage in India was 60% but that high calf mortality reduced weaning rate to only 48%.

Most of the above results have dealt with non-commercial herds. However, surveys carried out on commercial herds in Australia and Florida indicate that considerable improvements could be made.

In a study carried out in north-eastern Australia, Lamond (1969) observed pregnancy rates averaging 63% over a total of more than 14 000 cows. Pregnancy rates in the poorest third, however, averaged only 51%. Losses between pregnancy diagnosis and branding ranged from 6 to 40% in a small sample of the herds.

In Florida, Warnick and his colleagues carried out a survey on thirteen commercial ranches with more than 7000 cows. The overall pregnancy rate was 60% whilst that of the poorest five ranches was only 53% (Warnick et al. 1969).

It is clear from the foregoing examples that calving rates in the developing countries of the tropics and subtropics are low and that this is true not only for areas where subsistence farming is practised, but also in many commercial areas.

The venereal diseases (brucellosis, vibriosis, leptospirosis, etc.) can undoubtedly limit reproductive performance, but poor management is a major limitation to high calving rates. Poor management includes failure to cull non-productive cows; for example, 19% of adult cows in the study by Reed et al. (1974) had never calved. Lack of a weaning policy also reduces calving rate. Reed et al. (1974) state that in their four poor management areas, virtually none of the cattle owners interviewed weaned calves at less than one year of age. Stonaker et al. (1976) noted that only 13 and 9% of nursing cows in the Piedmont and Plains zones respectively were pregnant, whilst Lamond (1969) observed that only 37.1 and 55.5% of lactating heifers and adult cows respectively were pregnant.

There is no doubt, however, that the principal factor involved in poor fertility is inadequate nutrition. Aspects of the nutrition of the breeding cow have been discussed in earlier chapters. Poor fertility may be associated with specific nutrients, e.g. protein or minerals, but a deficiency of feed *per se* is probably the most frequent limiting factor. This also leads to slow rates of gain in heifers and a consequent delay in their becoming sexually mature. This was noted by Reed et al. (1974a) and by Butterworth and Presswood (1978).

Herd structure

The number of animals of different classes in a herd has an important effect upon its productive capacity. At the same time, the principal function of the herd will also affect its composition. Several examples have been drawn from both ranching (or cattle post) and sedentary cultivation situations.

The work of Reed et al. (1974a) covered a wide range of herds, from those close to towns where cattle were kept for trading and work purposes but were barely involved in the commercial sector (area I) to the commercial producer (area VI). Some of the results obtained are summarized in Table 14.4.

In areas I and II, herds were small, characterized by relatively few breeding cows, of which a large number were of an advanced age (i.e. eight years old or more) as well as castrate males (i.e. work oxen) and mature and immature bulls. In areas IV and V, owners were more commercially orientated and this was reflected by a higher number of mature cows and breeding heifers; the number of bulls, however, was low. Owners in area III bought in steers for fattening and area VI was wholly commercial.

Table 14.4 Composition of selected herds in different parts of Botswana (Adapted from Reed et al 1979a)

	Area I	*Area II*	*Area III*	*Area IV*	*Area V*	*Area VI*
Average herd size	21.3	34.9	342.3	61.4	53.1	514.8
Calves less than 1 year (%)	6.6	12.0	8.6	15.6	11.7	18.3
Heifers, 1–3 years (%)	18.9	21.5	18.1	17.1	20.1	22.0
Breeding cows (%)	33.0	35.5	29.5	48.7	47.8	32.3
Castrates, 1–3 years (%)	10.8	8.0	20.2	8.9	8.6	14.4
Castrates, older than 3 years (%)	19.8	13.5	19.2	6.0	7.0	10.0
Mature bulls (%)	4.7	2.6	0.9	1.0	0.6	0.5
Immature bulls, 1–3 years (%)	6.1	6.9	3.3	2.6	4.1	2.4

Later estimates of herd structure made in the western areas also reflect increased commercialism; cows represented 40% of the herd and calves 19%, bulls between 2 and 3% and young stock (1–3 years) between 26 and 33% (APRU 1979).

Surveys in other parts of southern and central Africa have been more concerned with the sedentary subsistence sector. Butterworth and Presswood (1978) compared Swazi nation and commercial cattle populations in the kingdom of Swaziland using data collected by the Department of Veterinary Services. The proportion of cows was similar in both populations, being 34.9 and 35.4% respectively. The number of oxen in both cases was high in Swazi nation cattle (20%), reflecting use as draught oxen, and 18% in commercial herds representing slaughter cattle. Followers constituted a large proportion of the herd, even in the commercial sector, because of the extended time required to reach slaughter weight. Numbers of bulls were relatively high in both groups, equivalent to approximately 10% of cows in both cases. The reason for this in the small farmer sector (overall average herd size, approximately twenty head) is the need for each herd to include a bull.

A similar structure of the sedentary subsistence farmer's herd is given by le Roux and Stubbs (1980). The percentages of bulls, cows and oxen are 5.5, 30.8 and 21.4% respectively, whilst the rest of the herd comprises younger followers. The authors comment that in view of the significance of oxen in the agricultural context, it is surprising that they constitute such a small part of the herd. However, in view of the low calving rates and high mortality, it is doubtful whether more oxen could be sustained.

A lower percentage of oxen (12.8%) was reported by van Winden and Beale (1979) from a herd study carried out in the north of

Malawi. The number of cows (35.7%) was similar to that in other comparable studies.

Cruz (1974) carried out a detailed study of herd composition on traditional farms in south-western Angola. Areas were classified as to whether rains were consistent enough for annual crop production, or whether the predominant system was pastoral or mixed. The average percentage of cows in the farming and mixed areas was about 41% whilst draught oxen accounted the 14.9 and 7.5% respectively. Cows made up 49% in pastoral areas whilst oxen accounted for less than 1%.

Meyn (1970) has indicated that cows may make up as much as 40–60% of the total herd in East Africa. This is so as to provide milk for large families as well as to ensure that the family herd may be rapidly replaced after decimation by disease or drought.

It may be inferred from the study of Stonaker et al. (1976) that the number of cows and heifers in the herds in the Piedmont and Plains areas of Colombia was about 50 and 40% respectively (although the figures given referred only to cattle examined and some animals may have been omitted). Williams (1976) has remarked on the high proportion of non-productive females in a herd in India: of a total of 453 females, only 123 were in milk.

Although there are minor differences in herd composition because of different emphases placed on the function of the herd, e.g. production of beef or oxen, and from region to region, it is clear that many herds are characterized by a large number of non-productive animals. This is an important factor in explaining the low productivity of herds noted previously.

Mortality

Low calving percentage combined with high mortality can reduce productivity considerably. Mortality varies widely depending on incidence of disease and availability of forage. In their study in Botswana, Reed et al. (1974a) encountered levels of mortality of calves less than one year old of 19.2 and 11.8% and of 22.4 and 13.5% for calves aged six to seventeen months for areas I and II respectively. Although values for other areas were lower, the authors commented that this was one aspect of the survey where results were inaccurate as values had been minimized. Values of 10.2 and 8.5% calf mortality were quoted by APRU (1977) for cattle post and ranch conditions in eastern Botswana. Overall mortality in western Botswana was 4.9% (in this case, the exceedingly low value for calf mortality of 0.9 was recorded and the authors commented that this was almost certainly depressed by the omission of neonatal deaths).

Butterworth and Presswood (1978) quoted an overall mortality of 5.8% for the national herd in Swaziland. This value is low because of efficient disease-control measures. On the other hand, van Winden and Beale (1979) state that about 30% of calves in the northern region of Malawi died in their first year whilst about 12% of the rest of the herd died annually.

High calf losses have also been reported by Wilson (1963) from Ankola in western Uganda where losses of heifer calves were 21.7% and bull calves 46.7%. It is suggested that heifer calves get better attention than bull calves because of their value as potential members of the breeding herd.

High mortality occurs when forage resources are depleted by drought. Temple and Thomas (1973) reported that death losses for the Sahelian region as a whole amounted to 25% during the dry season of 1973 whilst some districts reported losses as high as 80%. Plasse (1973, cited by Warnick 1976) estimated the death loss to breeding age as between 10 and 25% in the Latin American tropics. Groenewold and Crossing (1975) estimated that calf and adult mortality in India were 15 and 10% respectively.

Whilst mortality is variable in tropical and subtropical countries, it is apparent from the examples given above that it can frequently equal or exceed offtake.

Slaughter weight

The low slaughter weights attained in many tropical and subtropical countries are evident from Tables 12.1, 12.2 and 12.3. Slow growth rates are partly attributable to the characteristic fluctuating feed supplies causing weight gains during the rainy season followed by checks or losses during the dry season. Typical growth patterns were quantified by Norman and Arndt (1959) and Norman (1965) for steers grazing native pastures at Catherine in the Northern Territory of Australia. It was shown that cattle did not reach 450 kg (or approximately 225 kg carcass weight) before six years of age. These cattle were under reasonable management conditions, i.e. adequate stocking rates, spraying for tick and fly control and supplemented with bone meal and salt during the dry season. Plasse (1973, cited by Warnick 1976) has estimated that cattle in the Latin American tropics are between 3.5 and 5 years at slaughter. The slow rate at which maturity is attained is often exacerbated by the fact that cattle are frequently sold to satisfy the cash requirements of the farmer rather than when they are in suitable condition. The slaughter of emaciated cows also contributes to the low slaughter weight.

Constraints on higher productivity

Low calving rates, large numbers of non-productive animals, high mortality and low slaughter weights are therefore important contributing factors to the low offtakes typical of tropical and subtropical areas. Some of the underlying causes of the unsatisfactory situation will now be discussed, together with their implications. Sociological aspects will be considered first because of their interrelationships with other constraints.

Alternative reasons for cattle keeping

In examining the constraints to beef production from the cattle herds of tropical and subtropical countries, it is important to consider the reasons that motivate farmers to keep cattle and the expectations that they have from them.

In the economically-developed countries, we are accustomed to think of cattle entirely as producers of milk and meat. In many parts of the developing world, however, these, particularly the latter, are of secondary importance. Traction, manure, milk, hides, etc. are primary products whilst social, ceremonial and financial considerations are major reasons for keeping cattle.

The use of cattle for power

The extent to which animals are used for the power requirements of farm cultivation have been indicated by McDowell (1977). Animals (other than man) account for 82, 99, 88 and 75% of power used in agriculture in Africa, Far East, Near East and Latin America respectively. Boshoff (1968) has stated that the use of animals for pulling machinery has been particularly successful where the following conditions have been fulfilled:

1. Cattle are part of the normal farming activities of the area and animal disease and pests do not present a great problem.
2. The land is reasonably flat and the soils are light.
3. The vegetation is relatively short and light.
4. The extension service is adequate

The use of animal power is justified when it is considered that there is a correlation between available power per hectare and crop yield. Typically, under African conditions, one adult can work about 0.5 ha with a power input of about 0.1 kW/ha. If the desirable power input is assumed to be 0.4 kW/a, then a pair of work oxen could provide the necessary additional power to cultivate about 3 to 4 ha of land (Inns 1980). It is further assumed that the use of oxen can improve the timeliness of the various arable operations involved in cropping and thus improve performance. Inns (1980) has estimated that 3.2 'feed units' (sic.) are required per day whilst supplementary feed must be provided for the work performed. These feed requirements must be provided either from crop residues produced from the arable land itself or from nearby land which is unsuitable for cultivation (range).

A point which is frequently ignored when discussing the use of work oxen is the number of cattle which are required to produce or replace them. Thus, Kidner (1970) has estimated that twenty-two head (including the oxen themselves) are necessary to produce and maintain a 'span' or team of six work oxen (six oxen are commonly used to plough dry land in Southern Africa). The estimate was based on a five-year working life for oxen, 40% calving rate and 15% mortality.

The author has made an estimate of the number of cattle required to support two oxen, probably a more common number for Central African conditions, and this amounts to some twelve animals or about ten livestock units (the difference is largely attributable to the proportionally increased requirement for bulls and bull replacements). The latter calculation is consistent with the frequently-made observation that oxen constitute 15 to 20% of the herd.

In any event, it is clear that much greater forage resources are required for the whole supporting herd than for the oxen alone and that the provision of this forage must be taken into account when advocating the use of work oxen.

With common levels of reproductive and growth performance, it is unlikely that there will be store or slaughter animals available for sale from the supporting herd other than the oxen themselves at the end of their working lives. The farmer when given the alternatives of selling his young male stock as feeder animals or keeping them for work oxen generally prefers the latter unless in urgent need of cash (see later section on marketing).

An additional advantage favouring the keeping of oxen is as a source of off-farm income. In areas where only the minority of farmers are cattle keepers, cash may be obtained by hiring oxen for purposes of transport or cultivation.

More male animals might be available for sale if cows were to be used for draught purposes. The author has seen cows in-spanned in Swaziland, and preliminary work at Debre Zeit in Ethiopia indicated that work for 2–3 hours/day caused only small reductions in milk production (ILCA 1980).

Manure

Without the application of manure, crop production from sedentary subsistence agriculture would be considerably lower than it is at present in many parts of the tropics and subtropics. In many farms, it is the only fertilizer which is applied with any regularity. The amount of manure that one livestock unit will produce in a year may be estimated. If we assume that one livestock unit weights 350 kg and eats 2% of its body weight daily of forage dry matter which is 50% digestible, daily manure production will be equal to 3.5 kg or 1277 kg/year. If forage dry matter contains 5% crude protein which is also 50% digestible, the production of nitrogen in manure will be approximately 10 kg/year. Whilst this may seem negligible from the point of view of additional nutrients, the application of organic matter to the land improves the physical condition of the soil, slightly increases soil pH and level of exhangeable bases, and the relatively slow decomposition of the organic matter results in an even supply of nutrients.

The advantages and disadvantages of the use of kraal manure have been reviewed by Webster and Wilson (1966), who have suggested that dressings of the order of about 2.5 tonnes/ha/year are generally necessary to maintain soil fertility (a large part of East Africa was specifically mentioned) whilst in areas of lower fertility, as much as 12.50 tonne/ha/year might be required under systems of continuous cropping. It should be emphasized from the point of view of grassland management, however, that when cattle are not fed on crop residues alone, they are depleting nutrients from range to supply arable land and that this can only increase the rate of range deterioration, already excessive in many areas.

Dried manure constitutes an important energy source for cooking, particularly for nomadic and pastoral peoples and in the Indian subcontinent in areas where other fuels are not available. Manure is also widely used as a building material either as a component of bricks, as a plaster or as a flooring material.

The adverse effects of kraaling on beef production have been mentioned earlier. It should perhaps be pointed out that the primary reason for the practice was to avoid losses caused by wild animal predation. Reed et al (1974a) state that this did not appear to be justified even in the north of Botswana where wild predators are frequent. A more valid reason for kraaling in sedentary subsistence areas, however, would seem to be that animals must be attended by herdboys during the day to prevent damage to neighbours' crops and kraaling at night stops animals straying as well as reducing theft and predation dangers.

Milk

The majority of cows in the systems mentioned by Payne (1976) are milked at some time or another. This is particularly true of the sedentary subsistence sector where about 50% of tropical cattle are to be found. Milking is normally seasonal to provide milk for the family. Milk may be either drunk fresh or fermented (e.g. maas or emasi in Southern Africa). This seasonal production has also justified the establishment of routes to collect milk for processing in various countries.

Cows conceive after the rainy season has already started, when their nutritional status has improved (Butterworth 1983). Calving is in the early part of the following rainy season and milk is collected during the short period when forage is readily available. The sale or consumption of milk has probably little effect on the growth of the calf which is able to extract sufficient for its requirements by restricted suckling (Preston 1976). Yields are low; van Winden and Beale (1979) estimated that cows produced an average of 147 litres in a lactation period averaging three months (i.e. about 1.6 litres/diem), although

Hamburger et al. (1980) have suggested that the average daily yield of Nguni cows in Southern Africa may be as high as 9 litres, and this concurs with personal observations by the author of cross-bred and Criollo cattle in the Zulia area of Venezuela. Nevertheless, this milk provides a useful nutritional supplement to the family or, in the case of sale, a regular cash income.

Other products

Many other articles are derived from cattle but these are usually by-products after the animal has been slaughtered or died.

Hides are widely used for clothing, bedding, tents and the manufacture of leather artefacts. In some parts of Africa, shields are made out of hides and the skin of a particular animal may be much prized because its colour pattern is suitable for making the shield of a particular regiment (e.g. Zululand and Swaziland).

Bones are used for buttons, ornaments and utensils, etc., whilst horn is used for similar purposes as well as for containers, musical instruments, etc. These examples indicate uses to which non-edible parts of the carcass are frequently put by the small farmer and do not include such industrial applications as manufacture of fertilizer, animal feed, glue, gelatin, pharmaceuticals, etc.

It is unlikely that these products influence offtake but they are mentioned to emphasize the great variety of uses to which parts of the carcass are put by small farmers in different parts of the world.

Cultural aspects of cattle keeping

Cattle play an important role in the lives of many peoples whether they are nomads or pastoralists, or sedentary subsistence farmers who have traditionally kept stock. It is not surprising, therefore, that they have become associated with the cultural practices of the peoples concerned. These practices may be ceremonial and religious, as well as sometimes witchcraft and magic. Cattle are used for paying the bride price and are frequently, although less quantifiably, a symbol of social prestige. They are also associated with sports and recreations.

It is not the intention to provide a complete account of cultural aspects of cattle keeping in this work. Instead, some examples have been given which may be typical of practices followed, to a greater or lesser degree, elsewhere. The African examples have mainly been taken from Swaziland where the author has drawn freely from the ethnographic account of the Swazi nation prepared by Marwick (1940) as well as his own observations.

It has already been pointed out that the King's cattle kraal is the focus of Swazi national life; similarly, the family kraal (*sibaya*) is the focus of the village and of the life of the extended family. It forms the centre of the village and the other huts are grouped round it in the form of a crescent. The cattle in the *sibaya* are normally from different sources and belong to various owners. In addition to those that may have been directly acquired by the farmer, they may include cattle which have been inherited (and which therefore are held in trust for his ancestors), those acquired as *lobolo* or bride price for a daughter and those which belong to his wife. The latter may include those given by the groom's family as a compensation after a daughter has married (*msulamnyembeti* to wipe away the tears) and the beast (*lipakhelo*) which is given to her by her in-laws to indicate that she may drink sour milk (*emasi*) in her husband's house (this animal and its get are normally inherited by the woman's eldest son).

In addition, the *sibaya* may also house cattle which are kept for a relation or friend under an arrangement known as *sisa*. This enables an owner who lives elsewhere to ensure that his cattle are looked after. The keeper has the usufruct of the cattle, including labour, milk, manure and perhaps some, or all, of thc progeny. Such systems are widespread in Africa (in Malawi, *kupazga*, for example). Cruz (1974) has commented that the practice ensures that livestock products are more equitably distributed throughout the community. Herds are kept at a manageable size and true ownership is disguised, affording protection from an administration anxious to impose group ranch schemes, increase cattle sales or levy higher taxes. *Sisa* is also a means whereby migrant labourers may invest earned capital and accumulate their own herd. This is important in Southern Africa where a large work force is employed in the mines of the reef and Zambia. It is clear from the foregoing that many different interests may be represented in a herd of cattle which appears to be that of a single owner.

The main reason for cattle leaving the *sibaya* is as *lobolo*. *Lobolo* is a payment made by the groom to the family of the bride and is traditionally paid in cattle, although a cash equivalent is now becoming more common. Its function is to legalize the children of the marriage and to act as a guarantee for the good conduct of the man and his wife. The traditional amount of *lobolo* is ten head but higher payments are common for brides of royal or other esteemed lineage as well as for educated women. In addition to the cattle actually involved in *lobolo*, the groom is also responsible for the provision of an animal (*lugege*), which is slaughtered when the arrangements have been completed and the *lobolo* has been paid in full, and for the payment of the *msulamnyembeti* beast mentioned above (*mkhuzi* in Malawi). In addition, various beasts are usually slaughtered as part of the marriage festivities.

Meyn (1970) has summarized the number of cattle which normally constitute the bride price in some pastoral tribes in East Africa and these range from two head in the case of the Nandi to as much as fifty head in the Turkana and Jie. The author comments that the family needs more cows because it has been made greater by a new wife. The newly-married wife gets an allocation of cattle from which she has to maintain herself and her children.

Van Winden and Beale (1979) found that the payment of *lobolo* was the most important reason for keeping cattle in the northern region of Malawi, and the custom emphasizes the importance of cattle as a source of wealth and security. Security is conferred by a large family which looks after the father when he is no longer able to work. The possession of cattle to pay for additional brides ensures that more children will be born to the family. Barrenness may be a reason for divorce and the return of the *lobolo* cattle.

Cattle may also be sold (after consultation with the ancestors) to pay taxes or school fees or to satisfy other urgent cash requirements. They may also be sacrificed to the ancestors, used in death ceremonies (the eating of meat of the slaughtered beast is equivalent to ritual eating of the deceased) and for the payment of witch doctors. They may also be used for the payment of fines (five head for adultery).

Cattle are also used for ritual purposes on a national level. On the occasion of *Incwala*, the annual first fruits ceremony, a black bull is overpowered and rendered insensible by young men after which it is killed without bloodshed. This is associated with the rain-making powers of the king.

The above is a brief account of some of the cultural aspects of cattle keeping among the Swazi. Whilst important differences occur from one culture to another, it indicates the range of values commonly found in African cattle-keeping societies.

Religious attitudes to cattle in India are somewhat different to those in Africa. One of the beliefs of Hinduism is the transmigration of souls. A corollary of this is that all living beings are equal. However, Williams (1976) points out that 'only the wealthy can afford to oppose cow (sic.) slaughter as they are financially able to purchase their food needs at any time' (the slaughter of cows as opposed to cattle is forbidden by the constitution). He further comments that one of the factors leading to cow respect was the reduction of cattle populations to an unacceptably low level and there was therefore a need for urgent conservation measures to be introduced. The basic reasons are, however, very similar to those already discussed, i.e. the prime importance of cattle is not as a source of meat but of milk, fuel, fertilizer and draught power. In fact, old and surplus cattle are sold for slaughter and abattoirs operate in many cities.

Finally, the use of cattle for sport and recreational purposes will be mentioned. Bullfights are popular in parts of Latin America (and in Mozambique under Portuguese rule) and *corridas* are held frequently in many cities and towns in Mexico and less often in Venezuela, Peru and Colombia. For *corridas* and most *novilladas*, bulls specially bred for fighting (*toros de casta*) are used whilst cross-bred (*de media casta*) or local (*criollo*) cattle may be used for local festivals. The oldest ranch dedicated to the breeding of fighting bulls in

Mexico is Atenco near Toluca which was founded by Juan Gutierrez Altamirano, a cousin of Cortés, in the sixteenth century with cattle imported from Navarra. Later importations of stock came from the *ganaderias* of Saltillo and Parladé. There are about thirty major ranches dedicated to raising fighting bulls in Mexico and many smaller ranches, some of which may only supply one group (encierro) of bulls per year. The interested reader may find a detailed account of these ranches in the definitive work by de Cossio (1961) which also includes many details of the management of *toros de lidia*. In comparison with the large industry which has been developed in Mexico, there are few ranches which are exclusively devoted to raising fighting bulls in the other Latin American countries, although the recent establishment of the Ranch 'Los Aranguez' at Carora in Venezuela is worthy of mention.

Several sports in Latin America are based on traditional methods of working cattle. The most elaborate of these is the *charreria*, the national sport of Mexico, which combines several skills both on horseback and on foot. Those involving cattle include throwing a bull by roping horns and hind feet from horseback (*lazar en plaza*), riding a bareback bull (*jinetear*) and throwing a bull by its tail from horseback (*coleadero de toros*). This last event is also very popular in Venezuela during fairs and festivals. Full details of the *charreria* are given by Rincón Gallardo (1939), who also gives much useful information on horsemanship.

The above examples indicate that economically-rational production may be of less important than cultural practices when motives for cattle keeping are considered. This must be taken into account when schemes for increasing productivity and offtake are formulated.

It is also clear that the traditional cattle population is not a beef herd but rather one that subserves numerous functions of which beef production usually has a relatively low priority.

Financial implications of cattle keeping

Cattle function as wealth, as insurance and as a source of income. Their role in the acquisition of wives has already been mentioned ('purchase' is not an appropriate word in this context as it implies that it would be possible to sell the wife, which is not the case). Additional wives bear daughters who may be exchanged for a further bride price and thus additional wealth is accreted. Similarly, extra income may be used to buy cattle which provide an investment. This is done by migrant labourers with cash surplus to their requirements (as mentioned previously), and farmers participating in rural development programmes which encourage the production of cash crops or disposable surpluses (often these rural development programmes are also endeavouring to reduce stock numbers!).

Although it is easy to criticize the practice of stock purchase as traditionalist, it is often for lack of any other attractive investment opportunity. Cattle keep pace with inflation (although cattle prices have not increased uniformly in recent years, even a low rate of natural increase assures an adequate return on investment) and their running costs are minimal when kept on communal grazing areas. Savings banks have been introduced in many countries, some of which provide mobile services to country areas. However, despite relatively attractive rates of interest as well as publicity and educational campaigns to explain their value, savings schemes have commonly been slow to take off. The author issued a newsletter to extension workers in Swaziland in which it was shown that the proceeds of the sale of an ox from a fattening ranch (see later section) could be invested in the savings and loan bank so that the interest obtained would pay for tractor hire to plough the average size holding and still leave appreciable change to spare; the response in terms of cattle sales was minimal.

The possession of adequate numbers of cattle is also seen as a form of insurance. In case of crop failure, for example, cattle may be sold in order to purchase staple crops. In some societies particularly at risk because of unpredictable rains and high drought frequency, or disease, a large herd is insurance that some animals will survive as a nucleus breeding herd although the majority may have perished.

Cattle may be used as a source of income by sale of milk or of draught power. Milk over and above the requirements of the family may be sold to neighbours or taken to the nearest milk collection centre if such is available. Similarly, oxen may be rented to neighbours for ploughing or for transport. Such income may form a substantial part of the total family cash requirement. Cattle may also be sold for such seasonal expenditure as payment of taxes or school fees. However, it is unusual for cattle to be sold at other times in many communities, and it has been reported by van Winden and Beale (1979) that a number of farmers did not like to have money 'lying around the house after cattle sales and thus encouraging unnecessary consumption expenditure'. It is suggested, however, that greater opportunities for investment could give rise to increased cattle sales and greater interest in improved husbandry to improve offtake.

It is clear from the above that whilst cattle may be sold to defray a particular expense, it is not a deliberate policy to produce cattle specifically for sale as a continual source of income. The sale of milk and draught power, however, provides a useful cash supplement.

Land availability and tenure

Human population is increasing at a rate of between 2 and 3.5%/year in most of the developing world. This leads to an increased requirement for staple crops and increased pressure on arable land. Land use for arable farming and permanent crops in Africa was 188.1 million ha in 1961–5 and 208.7 million ha in 1977, i.e. an increase of approximately 11%. The corresponding value for South America is 23.6%. Suitable arable land is still available in some countries, but elsewhere lands are cultivated which would be more appropriately used for grazing. The establishment of large irrigation schemes for non-food crop production also utilizes large tracts of land previously only suitable for grazing. There is thus a strong tendency for the amount of land available for grazing to decrease.

At the same time there is a tendency for the number of cattle to rise. The increase of cattle in Africa in the period mentioned above was approximately 7% whilst the corresponding figure for the developing countries of Latin America was 20% and for the Near East, 21%. The increase in Africa would possibly have been higher had it not been for the drought conditions in the Sahelian zone.

Various factors contribute to the increase in cattle numbers. The importance of cattle in many societies has been emphasized in the previous section. It is clear that as the population increases, so does the number of aspirant cattle owners. This provides a market for cattle which otherwise might have gone for slaughter. Butterworth (1981) calculated that even with the modest calving rates and high mortality found in most tropical countries, a cattle population may be expected to increase by 62% over a ten-year period. Disease control also plays an important role in population increase. If mortality is decreased and calving rate increased at a rate which would be expected by the improvement of management, herd size would increase by 114% over the ten-year period.

Disease control and improved management programmes will result in higher cattle numbers if increased offtake is not achieved at the same time.

Le Roux and Stubbs (1980) have illustrated the combined effects of reduced grazing land and increased cattle numbers on tribal trust lands in Zimbabwe. Over the period 1961/2 to 1976/7, the gross area cultivated and the number of cultivators increased by between 80 and 100% whilst the number of cattle increased by about 70%. Grazing area (ha/head) and grazing arable ratio were reduced by about 40% during the same period.

If these trends continue, it is inevitable that many areas will become seriously overstocked and overgrazed with consequent reduced carrying capacity.

Problems of land availability may be aggravated by the system of land tenure customary in the area. In many African communities an individual acquiries rights over arable land by direct grant by a chief, another individual or by inheritance. These rights entitle the farmer to produce crops to feed himself and his family as well as surplus cash crops for sale. Although the farmer has no formal title to the land, he normally has effective security of tenure. The tenure system in Swaziland serves as an example and a detailed account has been given by Hughes (1972).

Grazing lands, on the other hand, are often considered to be communal. The traditional Swazi belief is that all grazing is available to everyone and therefore anyone has a right to graze his cattle anywhere in the kingdom (although this is limited by the restrictions on building the cattle byre, *sibaya*, which can only be done with the permission of the local authorities; nevertheless, also see the practice of *sisa* already referred to).

The communal grazing system leads to considerable misuse of range land. As has been mentioned earlier, costs of cattle keeping on communal land are negligible (dipping fees may be charged, but grazing fees are not acceptable because of the 'rights' referred to above). There is therefore no constraint to holding large numbers of cattle. On the other hand, large herds are desirable for sociological and other reasons referred to above. Enlightened individuals may realize that range condition is deteriorating, but there is no incentive to reduce stock numbers. Other farmers will not necessarily follow suit and may indeed benefit by the sacrifice of the individual, to increase their own herds. Similar situations have been provoked in other parts of the tropics and subtropics where land reform has been carried out without adequate control. Many of the *ejidos* or communal farms established in Mexico have suffered severe overgrazing because no system to control stock numbers has been implemented. It is a common occurrence for farmers to observe that conditions have deteriorated; bush encroachment is frequently remarked upon, as is higher carrying capacity in earlier years. The connection between this deterioration and increased cattle numbers is less frequently made. Farmers may comment that 'this pasture carried so many head in my father's day', the inference being that this situation should continue indefinitely, rather than that the original stocking rate was too high.

Agricultural systems which attempt to adjust seasonal availability of grazing to stocking are nomadism and transhumance (or semi-nomadic herding). These practices are widespread in West and East Africa. Nomads migrate both in order to obtain better grazing and to escape flooding and pests such as the tsetse fly. Transhumants combine migratory herding with subsistence agriculture at a 'base' village. Whilst these systems would appear at first sight to make efficient use of available grazing lands because some resting of pastures is involved, this is not necessarily the case. Nomadism is a high-risk system because of the drought hazards involved, and greater security is conferred by large numbers of animals, thus ensuring that a nucleus herd survives the drought to replenish the cattle population. At the same time, the area of land available is diminishing for reasons indicated above and considerable degradation of range resources is apparent in many areas.

The aetiology of reduced carrying capacity caused by overgrazing has been covered in some detail elsewhere. Nevertheless, several points should be re-emphasized here in view of the importance and widespread nature of the phenomenon.

Butterworth and Brandl (1981) have calculated that the forage requirements of a (productive) beef cow weighing 450 kg and rearing a calf amounts to some 2.5 tonnes of dry matter/year. The optimum stocking rate of any site may therefore be calculated, bearing in mind this production of forage as well as an allowance to enable the individual forage plant to lay down sufficient root reserves to maintain vigour. In the event of overgrazing, the productivity of the site is reduced and the *effective* stocking rate is increased, thus further reducing the carrying capacity. If the number of animals on the site is not reduced, overgrazing will continue at an accelerating rate.

The extent of the problem is difficult to estimate because of the lack of accurate (or any) monitoring of range conditions in most developing countries. However, Presswood (1978) has quoted a land-planning survey in Swaziland which estimated that approximately one third of the grazing land required immediate measures to prevent further soil loss whilst a similar area had a life expectancy of less than fifty years if soil loss was not checked. Le Houerou (1977) has estimated that desert expansion in the Western Sudan is taking place at a rate of 820 000 ha/year.

Although the subject of overgrazing has recently caused concern, it is by no means a new issue. The Kenya Land Commission reported that by 1920 there were signs of overstocking, whilst in 1932 'large areas where there used to be grass had been replaced by thorn scrub or reduced to bare earth and the soil washed away by gulley and sheet erosion'. Similarly, the use of rotational or deferred grazing to combat soil erosion was advocated as early as 1931 (Staples 1941). It is unfortunate that despite repeated warnings, so little has been done to combat what must surely be one of the most important problems confronting livestock production in developing countries.

Marketing constraints

Some of the financial and sociological implications of cattle keeping have been mentioned in earlier sections and these should be borne in mind when reading the present discussion of marketing.

An absence of effective marketing channels has commonly been supposed to constitute a major constraint to increasing cattle sales and offtakes. Where there are no markets, cattle are often sold to middlemen who profit from farmers who need finance. The establishment of more efficient marketing systems has been an important part of livestock development programmes. Such an infrastructure has usually been based on the establishment of local cattle saleyards at strategic points where cattle sales are held at intervals which depend on the local response. There is usually a residual buyer who sets the floor price for stock, frequently on a weight and grade basis, and therefore a market is guaranteed even for old or emaciated stock. Markets can be of different degrees of sophistication. Sales may be stratified into sales of mature stock for processing or local markets and sales of immature stock for subsequent finishing in feedlot or as stall feeders. Price structures may be based on classification of the live animals according to weight, age and grade or the farmer may be paid according to the cold-dressed weight of the animal after slaughter. Prices may be determined by free market factors or, more frequently, may be subject to some degree of government intervention. Some examples of cattle marketing systems in developing countries have been mentioned by Simpson (1976).

There is some discussion, however, as to whether the seller in pastoral or sedentary subsistence agricultural societies responds to supply/price relationships in the same way as would be expected in a more market-orientated community.

It has been argued that attitudes to cattle holding may be significantly influenced by the fact that cattle are regarded more as a source of wealth rather than a means of generating income. Doran et al. (1979) have used data from Swaziland to calculate a multiple regression between percentage offtake from the traditional Swazi herd, and the real cattle price and summer rainfall as independent variables. A significant negative relationship was observed between offtake and short-term price movements which accounted for 65% of the total variation. It is inferred that if it is necessary to sell cattle to meet specific cash needs, the minimum number will be sold. Factors that increase the market value of cattle will therefore enable the owner to meet these cash needs by selling less animals. Further, a farmer will seek highest-priced markets so that he can sell less cattle and 'thereby maximize his relative wealth'. Sales will increase in times of drought because there is a higher risk of crop failure and the risk attached to cattle increases because of lack of feed.

Other authors have argued that this perverse supply (or inverse supply/price relationship) caused by the predomination of non-monetary motivations is not necessarily typical of the types of community mentioned above. In an analysis of cattle supply and price changes among the Baggara tribes in the Sudan, Khalifa and Simpson (1972) showed that price change accounted for 95% of the variation of supplies in the inter-war period and 79% during the post-war period (until 1965). The elasticity of coefficients was positive and greater than unity in both cases, indicating that an increase in price was associated with a comparable increase in supply. Although Low (1978) has questioned the validity of the methodology used by Khalifa and Simpson, largely on the grounds of inadequate account being taken of the clearly discernible time trend associated with supply and price in the inter-war period, he has failed to provide any alternative analysis and his objections must be conjectural.

Evidence indicating that normal supply/price relationships prevail in sedentary subsistence agriculture has been provided by Hamburger (1978). He suggests that stock farmers in the developing territories in Southern Africa behave like 'normal economic men' in that the total number of cattle sold at sales held by the Department of Plural Relations and Development bore a strong relationship to the prevailing prices at the time of the sale. Hamburger concedes, however, that 'increased prices do not necessarily lead to the marketing of more stock. In some communities and cases, cattle have such an important and intrinsic value that stockowners would be reluctant to part with them under almost any circumstances'. He then goes on to mention the role that can be played by a 'highly competent and motivated extension officer', a point that will be returned to later.

In the tribal trust lands (TTL) of Rhodesia, Reid (1978) reported that marketing for slaughter was variable, depending on surplus cattle in the TTL, market (i.e. external) prices and TTL prices. In a subsequent clarification, Reid (personal communication) made the important point that in many communities, cattle are required for draught and for reproduction and only after these two requirements have been met can a decision be taken on which cattle, and how many, are surplus to prime needs. It is these cattle which are marketed. The perverse case occurs when there is no surplus and owners are under pressure to obtain cash, in which case they will sell fewer cattle the higher the price. This does not represent economic irrationality but an emergency sale. The extension to this explanation is that when productivity can be increased (by improving the various characteristics mentioned earlier in this chapter), then more animals will be available for sale and offtake will be increased. This African example is reflected by the Asian case mentioned by Williams (1976) where the farmer will readily sell surplus and non-productive cattle for slaughter.

The above examples indicate the complexity of cattle marketing in rural communities. This is confounded in many cases by the fact that cattle apparently belonging to a homestead may in fact be the property of a relative or friend and thus are not available for sale without due consultation. Similarly, the householder may be employed away from home and the woman who is left in charge of the homestead is not traditionally empowered to take decisions of such a nature.

It seems that whilst traditional motivations are important, they do not over-ride commercial considerations provided that surplus cattle are available for sale when requirements for draft power, etc. have been met. What constitutes an acceptable minimum herd can only be judged by detailed examination of local conditions.

The previous discussion has been based on the reactions of the small farmer under African conditions. However, Valdes (1974) noted that negative supply elasticity had been reported in short-term sales in Latin America. It had been deduced from this that such a response would also hold for long-term sales and therefore that cattle raisers did not respond to economic incentives. However, further analysis indicated that when beef prices increase, producers tend not to sell so as to allow cattle weights to increase and that sales of females are also decreased. After a stabilizing period of four to five years, the response becomes positive and sales of male animals are again increased. It is clear, then, that too much reliance cannot be placed on correlations based on short-term fluctuations. The purchase of cattle as a hedge against inflation is also noted, but this also leads to an increase in sales in the long run.

Perhaps this discussion may best be summed up by Ruthenberg (1978) who states that 'nothing will work unless price relations are sufficiently attractive from the farmer's point of view'.

Availability of trained staff

An important constraint to the development of the livestock industry is the shortage of staff trained in animal husbandry. The original emphasis of veterinarians in many colonial territories was disease eradication and control whilst husbandry aspects were generally disregarded. Comprehensive dipping and vaccination systems were set up as well as facilities to isolate areas from such diseases as foot and mouth. Training efforts were dedicated to the preparation of veterinary assistants whose main *raison d'être* was the implementation of such disease-control measures. Government professional establishments were staffed by veterinarians rather than animal husbandry officers. Whilst such methods led to effective control of many of the major diseases of cattle, they led to the situation where 'cattle were saved from the ravages of disease only to die a lingering death of starvation'.

Not only was the veterinary profession to blame in this situation, but agricultural colleages and faculties were usually more orientated towards the teaching of crop than animal husbandry. This resulted in a field officer who was uncertain of his abilities in livestock production. Specialist animal husbandry officers were few and far between and general agriculturalists found it easier to concentrate on straightforward crop-extension techniques rather than face the complex sociological aspects of livestock extension. When technical aspects of livestock production were incorporated into general agricultural degrees, the essential aspects of management of livestock were often given little emphasis or even ignored altogether, as was training in animal handling and simple practical techniques. It is a common experience to find students, particularly for the diploma and degree in agriculture, who are unwilling to carry out the menial chores associated with cleaning, feeding, milking, etc. Unfortunately, in many countries, the career structure is such that specialization at the undergraduate level is not justified as manpower planning is not sufficiently advanced to provide an accurate estimate of requirements. Such requirements, when available, are frequently based on theoretical grounds (e.g. one extension worker per given number of farmers) rather than on what the future recurrent budget will support.

This lack of communications insofar as manpower planning is concerned is reflected in the lack of interaction between the university and the 'consumer' with regard to course content. This may be aggravated by the fact that in many countries the universities, although theoretically autonomous, are more closely associated with the Ministry of Education than with that of Agriculture. In some cases, certificate and diploma courses are the responsibility of technical colleges run by the Ministry of Agriculture whilst only degree courses are given by the university (e.g. Zimbabwe), whereas in others (e.g. Malawi) both diploma and degree courses are university-taught. The 'two-tier' system whereby degree courses are taught to a student only after he has received the diploma (as in the University of Malawi, for example) is of interest in this connection.

This is not the place for general exposition of the deficiencies of universities in developing countries; however, it is appropriate to mention some of those which are most widespread. External examiners are not employed to ensure that reasonable academic standards are maintained. Staff are recruited on the basis of nepotism, or political belief, rather than academic competence. Staff are selected because they are from the same school or university, which leads to 'in-breeding' and lack of new ideas. Universities have often been established on the basis of the imagined needs of a region or state; this proliferation has led to inefficient use of the limited resources available.

Despite the foregoing somewhat negative comments, it is the author's experience that well-planned and implemented projects involving support to educational institutions are usually the most effective and long-lasting form of development. The positive influence of a good teacher lasts the whole of the student's professional life whilst relevant counterpart training, both on site and abroad, can form the nucleus of a qualified national teaching cadre.

Suitable research programmes

Much of the technology required to improve the productivity of livestock enterprises in tropical areas has already been developed elsewhere and it has been argued that basic research must occupy a low priority in the allocation of resources in developing countries. However, the minimum requirements in any region must be to collect information as to the performance of indigenous breeds, availability of locally available pastures, by-products and other feeds, etc. and the formulation and testing of systems suitable for local adoption.

Some countries have established efficient research organizations which obtain information about the forms of livestock production appropriate to the conditions of the country concerned. Notable examples include Australia, Zimbabwe, South Africa and, more recently, Botswana. However, these are the exceptions. Although it is easy to plead that this is because of inadequate funding, there are other contributory factors.

The lack of trained staff has already been mentioned in connection with extension programmes. This lack is more critical for research, as a higher level of training is required. Students are often trained in developed countries in techniques which are not appropriate to the conditions to which they will return. Frustration is caused by lack of sophisticated equipment (which, even if available, is not functional because of lack of service), skilled technical assistance, etc.

Postgraduate training of research workers should be carried out in the country where they will carry out their research. When this is not the case, it is an important factor limiting the quality of university research. The only way that overloaded teaching staff can sustain a valid research programme is with the aid of postgraduate students. Undergraduate dissertations are no substitute because of the limited time available for their completion within the already overloaded curriculum. University research is thus often fragmented and not relevant to the needs of the country.

Even when staff are trained, there is no motivation to carry out research. This may be because there is no financial incentive and little professional competition. It may be due to lack of confidence in younger research workers or lack of positive leadership by older ones. Whatever the cause, lack of motivation is tangible in many developing countries in the tropics and subtropics.

Development strategies to improve productivity

Many solutions to low productivity are implicit in the earlier discussion of nutritional deficiencies and management. It is not the purpose of this chapter to go into these in any further detail, but rather to investigate organizational and sociological changes which might be expected to lead to increased output from sectors which are at present insufficiently exploited in terms of contribution to the beef industry. The main factor limiting the productivity of cattle on tropical and subtropical pastures is inadequate nutrition. This may be caused by seasonal variations in forage production as well as nutritional deficiencies which have been discussed in previous chapters. Possibly the most important reason for malnutrition in already exploited areas, however, is a shortage of forage caused by inappropriate stocking rates which lead to overstocking with consequent overgrazing.

Development may therefore take several forms, depending on the existing population of both people and cattle on the area concerned as well as its productive capacity. In all cases, one of the principal factors to be considered is how to control cattle so as to maintain a stocking rate which will allow profitable livestock production without accompanying pasture deterioration.

Land tenure is also an important aspect of development. Sociological problems are usually more acute where communal land ownership is combined with individual ownership of cattle. This is equivalent to the situation where a group of people have rights to the usufruct of the land which may be held in trust for them. Communal land use has commonly led to overgrazing and may be extremely difficult to alter. The reasons which led people to aspire to cattle ownership have already been referred to. When this aspiration is combined with a traditional right to use grazing land without any obligation to maintain it, it constitutes a powerful political force to combat change. This is particularly so when senior civil servants, politicians and a newly-emerging middle class have the purchasing power to acquire cattle and stand to benefit directly from the system.

On the other hand, this tendency to adhere to the traditional way of life amidst the uncertainties of a developing country is understandable. Sibisi (1980) has examined the traditional securities of the Swazi as well as the response to 'modern' economic opportunities. She states that the traditional system 'still ensures for all Swazis the essentials of survival' and that 'it would be foolhardy for a Swazi to enter the 'modern system' without ensuring that his rural base is secure, and in present circumstances there is indeed no obvious reason why he should seek to abandon or discard it'.

The developmental approach for any area therefore depends on several factors, including the distribution of human and cattle population, system of land tenure and the productivity of the land itself (i.e. rainfall, soil, adaphic and topographic factors, etc.). As might be expected, many different approaches have been suggested in different parts of the tropics and subtropics. Meyn (1970) has given details of organizational structures employed to promote ranch development to suit the local conditions in East Africa and these include: large-scale development enterprises, corporate leasehold ranches, directed companies, commercial co-operative societies, communal co-operative societies, group ranches, individual ranches, council ranches, grazing associations and rehabilitation areas (Kenya), large-scale para-statal ranches, ranching associations and satellite ranches (Tanzania) and large-scale ranches and tenant-occupied smaller (1200 ha) ranches (Uganda).

Clearly an examination of each of these systems is beyond the scope of this chapter and the interested reader is referred to the exhaustive treatment given by Meyn. However, examples of some of the different systems which have been proposed for already-populated as well as sparsely-populated areas will be considered, together with some of their advantages and disadvantages.

Developmental projects for populated areas may or may not change land tenure. It has been pointed out that communal ownership of grazing land has caused considerable range degradation in the past. Projects which involve a change to private property should therefore be in the best interests of the maintenance of renewable resources.

Davis (1971) has given an early account of the formation of group ranches on trust lands in Kenya. All members of a tribe possessing traditional claims to land were given the right to receive land by a committee which was advised by an adjudication team. The rationale for this programme was both to increase the productivity of Kenya's range land, which was low under traditional management practices, as well as to conserve range resources. It was assumed that the granting of tenure to a group made the consequences of range misuse or the benefits of good management specific to that group.

A group may behave as a corporate body eligible to receive a loan on electing group representatives, adopting a constitution and electing a committee which has responsibility for the day-to-day management of the ranch. The representatives of the group function as trustees. It is stated that the formation of group ranches enables the economies of scale to be achieved whilst 'allocating land in economic units without creating a landless class'. Nevertheless, livestock are still individually owned and the problems of persuading cattle owners to limit the size of their individual herds remains the responsibility of the group representatives and the managerial committee. If they are able to do this, then there is no reason why the group ranch system should not be successful. If not, then because they are no longer able to change their grazing areas so as to share the lot of more fortunate sections of the tribe, the advantages which had previously accrued to the Masai would no longer be available and overgrazing would inevitably result. Jahnke and Ruthenberg (1976) have noted that the provision of water and veterinary services leads to periods of overgrazing being dubbed 'resource degrading development activities'. However, the advantages of systems in which the group has been given title to the land after adjudication are that boundaries are legally fixed and therefore it is easier to keep non-entitled cattle out of the ranch and the land title can be used as a 'power lever' to enforce stock control and debt repayment. The authors also predict that group ranching is unlikely to persist as such for any length of time because 'the groups are too heterogenous to reach agreement on a system of stock control'.It is likely. therefore, that grazing rights would be exchanged, leading to quota ownership and absentee membership. It is a problem in any ranching system on occupied land that it is not labour intensive and cannot be expected to provide employment for all the original inhabitants of the area.

Unless, therefore, the group ranch has been formed in a sedentary subsistence agricultural area where the stockowner is also a farmer, alternative job opportunities must be found either within or away from the area. This has been taken a stage further in certain countries, e.g. Zimbabwe, where it is foreseen that agricultural development will be in the hands of those who farm to produce surpluses whilst the present subsistence farmers will be found jobs off the land. This would require high capital investment in order to create industrial jobs. Jahnke and Ruthenberg (1976) go so far as to suggest that income gained by rural communities which is in excess of reasonable limits of individual necessity (and which could therefore be used to buy more cattle; see earlier section) should be collected as taxes in order to finance new job opportunities in other areas.

Group ranches have been started elsewhere without attempting to change the land tenure system. Clearly, for such an enterprise to be successful, a considerable extension input is required to convince participants that such measures as stock control etc. are necessary in order to maintain grazing land. In Swaziland, improved bulls were used as an incentive to such change. However, with the improved pasture generated by destocking and the improved rate of growth of the cross-bred progeny, it was often difficult to convince farmers to sell surplus cattle and the overgrazing cycle started once more.

Land reform has been carried out in Mexico by 'affecting' large landowners and constituting *Ejidos ganaderos* in their place. The constitution of such settlements made them eligible to receive loans which are administered by banks having a large complement of technical staff. Nevertheless, such attempts to increase productivity have frequently failed because of a lack of appreciation of the essentials of ranch management by the participants.

Many similar examples may be given of the systems involving the change of land tenure, but one of the most interesting is that which is in the process of implementation in Botswana.

The objectives of the programme are to stop degradation of the veld by overgrazing, to promote greater equality of incomes in rural areas and 'to allow growth and commercialization of the livestock industry on a sustained basis'. This can be achieved by the implementation of simple methods of management which include rotational grazing, controlled breeding, early weaning (sic., presumably appropriate weaning), daily watering and bone meal feeding. It is emphasized that farmers who are able to invest must have an incentive as well as complete control over the areas where their animals graze. To meet these aims, tribal grazing areas are divided into three zones: commercial farming areas, communal grazing areas and reserved areas. In commercial farming areas, groups and individuals are given exclusive rights to defined areas of grazing land and are given leases on which rents are paid. Existing communal grazing areas near villages are maintained and no rent is paid. Reserved areas are set aside for the future and constitute safeguards for poorer members of the community with few cattle (Botswana 1975).

Great efforts were made to ensure that this policy was understood by the public and feedback from all concerned was encouraged. These were exemplified by the speech given by the late President, Sir Seretse Khama, to Parliament on the launching of grazing land policy in July 1975 in which he emphasized that government proposals were ecologically and economically sound as well as socially just. This was supported by a widespread public consultation involving both face-to-face interviews as well as a broadcast campaign supported by a system of group leaders and radio learning groups (Colclough and Crowley 1974).

Despite this preparation, the project has not been problem-free. It has been suggested that it has caused range deterioration, concentrated on large herds at the expense of smaller ones and failed in communication with farmers and appreciation of the real requirements of increased production (Anon 1981). Supporting government departments lacked the resources and coordination to provide the necessary support facilities. It appears that the transition from traditional to commercial production was more complicated than had been foreseen.

An alternative strategy for the development of large areas of sparsely-populated land is the development of ranches by the government or by para-statal organizations. These may be either for breeding or fattening or a combination of the two aspects of production. An example of the para-statal approach is the National Ranching Company Ltd. in Tanzania. Fattening ranches to finish immature cattle purchased in from local farmers are given priority, but an increase in numbers of breeding cattle is also an important aspect of the programme (Meyn 1970). It has beeen noted, however, that such projects may encounter considerable problems during development. Weaknesses occur in the overall financial structure of the implementing organization which in turn affect project management. Excessive reliance may be placed on other development projects to provide service in crucial areas of the project (as previously noted for the Botswana example) and delayed procurement of materials needed for infrastructural development may cause increased expenditure due to inflation as well as delays in achieving planned stocking rates. (Akilimali et al. 1978).

Governments cannot normally provide the management required to achieve maximum benefits from a commercial organization. The bureaucracy inherent in government causes delays in implementation and management decisions. It is also difficult to incorporate incentive schemes into civil service salary scales. These are essential if a manager is to make difficult decisions, otherwise the tendency will be to take the decision with the least risk.

The first objective of any change in land use or tenure is to attain

Table 14.5 Comparison of productivity of two management systems (eastern Botswana) (Adapted from APRU 1977)

	Unimproved (Cattle post)	*Improved (Ranch)*
Calving percentage	47.3	74.8
Calf mortality (%)	10.2	8.5
Weaning percentage	42.5	68.4
Weaning weight (kg)	123.5	180.4
Post weaning gain (kg) 7–18 months	89.7	105.9
Weight of weaner (kg) calf/cow/year	52.5	123.4
Weight of 18-month (kg) calf/cow/year	90.6	195.8

'reasonably acceptable levels of management'. These include some fencing to control stock numbers and grazing pressure, control of diseases and mineral deficiencies, and adequate water supplies (Trail and Fisher 1971).

The Botswana workers (APRU 1977) have indicated the magnitude of improvement in animal performance to be expected when traditional cattle keeping is replaced by these measures (Table 14.5). Other workers have reported similar improvements (CIAT 1978).

Campion (personal communication) has pointed out that when reasonably acceptable levels of management have been adopted, more advanced technology can be introduced. This may be classified according to capital requirements.

Low-cost management changes include introduction of breeding and weaning seasons (fencing to control bulls and weaned calves is necessary); identification of individual animals, castration and weighing allow selection programmes to be carried out when data are recorded.

Medium-cost developments include fencing programmes for deferment, controlled burning, bush control, etc. These promote range regeneration.

High-cost programmes include more sophisticated methods of bush control, further pasture sub-division to facilitate cattle management, pasture improvement (fertilization, use of improved grass species, legume introduction), better handling facilities and water supplies, etc.

Stratification and subsequent intensification may also increase overall productivity.

Stratification generally involves the separation of the beef cattle operation into breeding and fattening sectors which may be carried out by the same organization or, more commonly, by different interests in different ecological zones. Examples of stratification include cattle breeding in the Venezuelan llanos and fattening in areas with better grassland, e.g. Valencia; the export of weaner cattle from the north of Mexico for fattening in feedlots in the United States, purchase of cattle from the Northern Frontier District of Kenya for subsequent fattening in areas more accessible to markets, etc.

Several systems for the intensification of beef production in the tropics have been suggested by Preston (1976). He proposed a dual-purpose cow herd which would produce weaned calves of 180 kg at 200 days of age plus 4 litres of milk/day throughout a 180-day lactation.

Weaned animals would be finished as soon as possible on locally-available feeds. The complementary value of starch when fed with sugar was pointed out and a cropping policy was suggested based on appropriate areas of sugar cane and cassava or bananas for the dry season and pasture for the wet season.

It is customary to think of intensification as involving large numbers of animals, but this is not necessarily the case. Thomas and Addy (1977) have described a stall-feeder system developed in Malawi where small-holder farmers fatten steers on crop residues.

Squire (1976) has pointed out that beef production systems based on range and pasture in developing countries have failed to produce adequate and reliable supplies of beef for local consumption and export, to realize the potential productivity of the national herd and to prevent extensive range destruction. Thus, 'intensive systems are complementary to both existing and improved pasture and range systems, not competitive alternatives'.

The subject of credit provision and infrastructural development have not been considered in the foregoing discussion. Whilst acknowledging their importance, it is felt that these may more properly be dealt with in a more general work on rural development *per se* rather than the present rather more specialized monograph.

Summary

The available information as to cattle populations, extraction rate and carcass weights in developing countries in the tropics and subtropics are subject to large sources of error and do not inspire confidence. Nevertheless, there is little doubt that productivity falls far short of its full potential. Cattle herds are often characterized by low calving rates, high mortality and a large proportion of unproductive animals. Levels of management are minimal and nutritional levels are less than adequate during long periods of the year. Previous chapters have shown that technical measures are available which could substantially increase the amount of beef produced from the majority of tropical pasture areas.

However, it must be borne in mind that the cattle population in many major tropical agricultural systems cannot be considered primarily as a beef herd but must fulfil a number of functions, of which the production of meat is often the lowest of a long list of priorities. These functions include the provision of milk, manure and draught power, which in some cases provide a cash income, but the role of cattle as a means of investing capital and as a form of insurance is often more important. Cattle also play a major part in many social responsibilities and customs which have been evolved as part of a way of life often associated with insecurity. Alleged reluctance to market cattle can often be explained by the fact that only cattle surplus to these multiple requirements are available for sale and, unless there is a substantial increase in productivity, such surpluses are limited.

There is a general tendency for the land available for grazing to decrease. This may be attributed to several factors, including an increased demand for land for arable farming, due in its turn to pressure from an increasing population, as well as pasture degradation. Similarly, an increased human population is accompanied by a consequent increase in aspirant cattle keepers and a significant increase in cattle numbers. This leads to increased stocking rates, reduced carrying capacity and range deterioration, leading to bush encroachment, sheet and gulley erosion and, finally, desertification.

Increased stock numbers may also be associated with improved animal health or management programmes where no provision has been made for the limitation of stock numbers by improved marketing. Increased numbers and reduced forage resources lead inevitably to lower levels of nutrition, the primary cause of reduced productivity.

Emphasis in development in such areas must therefore be on the adjustment of stock numbers to meet the grazing resources available, and it is vital when such schemes are proposed that the inhabitants are well aware of the reasons behind the schemes and that they are prepared to co-operate.

Several systems that have been tried to improve productivity in different parts of the developing world have been described as well as the resulting improvements which may be expected. Many of these involve the formation of ranch-type enterprises which are probably more suitable for areas where there is a relatively low human population; ranching is not labour intensive and alternative employment must be found for farmers who cannot, or will not, fit into the system. Payne (1976) has pointed out that some 46% of the total number of bovines in tropical agriculture are reared under conditions of sedentary subsistence agriculture, and it is in this sector that the greatest increases of productivity are to be expected. These may involve stratification to obtain optimum use of grasslands, combined with the efficient use of available crop residues and locally-produced agro-industrial by-products.

Gazetteer

Abingdon Downs, Queensland, Australia, 17.34 S 143.10 E

Adelaide River, N. Terr., Australia, 13.14 S 131.08 E

Agua Limpa, near Coronel Pacheco, Brazil, 21.35 S 43.16 W

Aldama, Mexico; 22.54 N 98.05 W

Ambahikely, Malagasy Republic, 21.37 S 43.40 E

Aracatuba, Brazil, 21.12 S 50.24 W

Archers Post, Kenya 0.39 N 37.40 E

Armidale, N.S.W., Australia, 30.32 S 151.40 E

Armoedsvlakte, near Vryburg S. Africa 26.57 S 24.44 E

Atherton, Queensland, Australia 17.05 S 145.40 E

Ayr, Queensland, Australia, 19.32 S 147.25 E

Bako, Ethiopia, 5.51 N 36.32 E

Banana, Queensland, Australia, 24.30 S 150.08 E

Barretos, Brazil, 20.37 S 48.38 W

Beerwah, near Brisbane, Queensland, Australia, 27.30 S 153.00 E

Belle Glade, Florida, U.S.A. 26.41 N 80.41 W

Belmont, near Rockhampton, Queensland 23.22 S 150.32E

Berrimah, near Darwin N.T., Australia, 12.23 S 130.44 E

Bloemfontein, S. Africa, 29.07 S 26.14 E

Bodles, near Old Harbour, Jamaica, 17.56 N 77.07 W

Bouaké, Ivory Coast, 7.42 N 5.00 W

Brian Pastures, near Gayndah, Queensland, Australia, 25.36 S 151.32 E

Brisbane, Queensland, Australia, 27.30 S 153.00 E

Bulawayo, Zimbabwe, 20.10 S 28.43 E

Burnett, Queensland, Australia, 24.43 S 152.22 E

Cairns, Queensland, Australia, 16.51 S. 145.43 E

Calabozo, Venezuela, 8.58 N 67.28 W

Cape Town, S. Africa, 33.56 S 18.28 E

Casamance, Senegal, 12.40 N 16.45 W

Charters Towers, Queensland, 20.02 S 146.20 E

Chiang Mai, Thailand, 18.48 N 98.59 E

Chitedze, near Lilongwe, Malawi, 13.58 S 33.49 E

Coolum, near Brisbane, Queensland, Australia, 27.30 S 153.00 E

Dohne, S. Africa, 32.32 S 91.51 W

East London, S. Africa, 33.00 S 27.54 E

Eidsvold, Queensland, Australia, 25.23 S 150.08 E

El Real, Panama, 8.06 N 77.42 W

Enkeldoorn, Zimbabwe, 19.01 S 30.53 E

Ermelo, S. Africa, 26.32 S 29.59 E

Everglades, Florida, U.S.A. 26.52 N 81.22 W

Fort Rixon, Zimbabwe, 20.01 S 29.18 E

Frankenwald, S. Africa, 25.45 S 28.12 E

Gainsville Florida U.S.A. 29.40 N 82.20 W

Gandajika, Zaire, 6.46 S 23.58 E

Glen Innes, N.S.W. Australia, 29.42 S 151.45 E

Glen, S. Africa, 28.58 S 26.21 E

Goondiwindi, Queensland, Australia, 28.30 S 150.17 E

Grafton, N.S.W., Australia, 29.40 S 152.56 E

Gualaca, Panama, 8.33 N 82.19 W

Guanabara, Brazil, 8.53 S 71.07 W

Guararapes, Brazil, 21.17 S 50.38 W

Gympie, Queensland, Australia 26.10 S 152.35 E

Habana, Cuba, 23.07 N 82.25 W

Henderson, near Mazoe, Zimbabwe, 17.30 S 31.03 E

Highbury, Queensland, Australia, 16.25.143.09 E

Homer, Louisiana, U.S.A. 32.48 N 93.02 W

Hueytamalco, near Puebla, Mexico, 1903 S 98.10 W

Ibadan, Nigeria, 7.23N 3.56 E

Impala, near Manzini, Swaziland, 26.30 S 31.22 E

Ipswich, Queensland, Australia, 27.38 S 152.40 E

Iquatama, Brazil, 20.10 S 45.39 W

Jamundi, Colombia, 3.16 N 76.31 W

Johannesburg, S. Africa, 26.10 S 28.02 E

Kabanyolo, near Kampala, Uganda, 0.19 N 32.35 E

Kairi, near Atherton, Queensland, Australia, 17.05 S 145.40 E

Kampala, Uganda, 0.19 N, 32.35 E

Kangaroo Hills, near Townsville, Queensland, Australia, 19.13 S 146.48 E

Katherine, N. Terr., Australia, 14.29 S 132.20 E

Kimberley, Near Kununurra, W. Australia, 15.425 128.50 E

Kingsville, Texas, U.S.A., 27.32 N 97.53 W

Kluang, Johore, Malaysia, 2.01 N 103.18 E

Kongwa, Tanzania, 6.13 S 36.28 E

Koopmansfontein, S. Africa, 28.15 S 24.03 E

Lansdown, near Townsville, Queensland, Australia, 19.13 S 146.48 E

Lavumisa, Swaziland, 27.18 S 31.55 E

Lawes, near Brisbane, Queensland, Australia, 27.30 S 153.00 E

Legume, N.S.W. near Killarney, Queensland, Australia, 28.18 S 152.15 E

Lilongwe, Malawi, 13.58 S 33.49 E

Linares, Mexico, 24.54 N 99.38 W

Los Bãnos near Manila, Luzon, Philippines 14.37 N 120.58 E

Machakos, Kenya, 1.32 S 27.16 E

Mackay, Queensland, Australia, 21.10 S 149.10 E

Makaholi, near Fort Victoria, Zimbabwe, 21.10 S 30.49 E

Malkerns, Swaziland, 26.35 S 31.10 E

Malya, Tanzania, 2.59 S 33.32 E

Mara, near Pietersburg, S. Africa, 23.54 S 29.23 E

Maracay, Venezuela, 10.20 N 67.28 W

Marandellas, Zimbabwe 18.10 S 31.36 E

Marlborough, Queensland, Australia, 22.51 S 149.50 E

Masbate, Philippines, 12.21 N 123.36 E

Matanzas, Cuba, 23.04 N 81.35 W

Mato Grosso, Brazil, 15.05 S 59.57 W

Mayagüez, Puerto Rico, U.S.A. 18.13 N 67.09 W

Mazabuka, Zambia, 15.50 S 27.47 E

Mbarara, Uganda, 0.36 S 30.40 E

Mbulu, Tanzania, 3.51 S 35.30 E

Melsetter, Zimbabwe, 19.48 S 32.50 E

Mercedes, Argentina, 29.15 S 58.05 W

Merluna, Queensland, Australia, 13.00 S 142.30 E

Mlezu, near Que Que, Zimbabwe, 18.55 S 29.51 E

Mt Garnet, Queensland, Australia, 17.39 S 145.08 E

Muguga, Kenya 1.14 S 36.40 E

Murray Upper, near Cardwell, Queensland, Australia, 18.11 S 146.00 E

Mwanza, Tanzania, 2.30 S 32.54 E

Nakuru, Kenya, 0.16 S 36.04 E

Nakyesasa, near Kampala, Uganda, 0.19 N 32.35 E

Nanyuki, Kenya, 0.01 N 37.05 E

Narayen, near Mundubbera, Queensland, Australia, 25.35 S 151.18 E

Narok, Kenya, 1.04 S 35.53 E

Naro Moru, Kenya, 0.10 S 37.01 E

New Iberia, Louisiana, U.S.A. 30.00 N 91.51 W

Njoro, Kenya, 0.20 S 35.57 E

Ntabamhlope, S. Africa, 29.06 S 29.36 E

Nyamandhlovu, Zimbabwe, 19.50 S 28.16 E

Oakwood, near Gympie, Queensland, Australia, 26.10 S 152.35 E

Ol Joro, Kenya, 0.04 S 36.22 E

Ol Kalou, Kenya, 0.16 S 36.22 E

Onderstepoort, near Pretoria, S. Africa 24.45 S 28.12 E

Ord Valley, W. Australia, 16.10 S 128.38 E

Palm Beach, Florida, U.S.A., 26.41 N 80.02 W

Palmira, Colombia, S.SS N 76.17 W

Parada, near Atherton, Queensland, Australia, 17.05 S 145.40 E

Paso del Toro, near Cotaxtla, Mexico, 19.11 N 96.10 W

Pietermaritzburg, S. Africa, 29.36 S 30.24 E

Playa Vicente, near Veracruz, Mexico, 19.11 N 96.10 W

Potchefstroom; S. Africa, 26.42 S 27.06 E

Pucallpa, Peru 8.21 S 74.33 W

Que-Que, Zimbabwe, 18.55 S 29.51 E

Quincy, Florida, U.S.A. 30.34 N 84.35 W

Ribeirão Prêto, Brazil, 21.09 S 47.48 W

Rio de Janeiro, Brazil, 22.53 S 43.17 W

Rio Piedras, Puerto Rico, U.S.A. 18.25 N 66.03 W

Rocklea, near Brisbane, Queensland, Australia, 27.30 S 153.00 E

Rodds Bay, near Gladstone, Queensland, Australia, 23.52 S 151.16 E

Rodriguez Clara, near Veracruz, Mexico, 19.11 N 96.10 W

Rongai, Kenya, 0.11 S 35.52 E

Samford, near Brisbane, Queensland, Australia, 27.30 S 153.00 E

San Joaquin, California, U.S.A., 36.35 N 120.13 W

Saõ José do Rio Prêto, Brazil, 20.50 S 49.20 W

Serere, Uganda, 1.31 N 33.27 E

Sertãozinho, Brazil, 21.08 S 47.59 W

Shangani, Rhodesia, 19.41 S 29.20 E

Shika, near Zaria, Nigeria, 11.01 N 7.44 E

Sigatoka, near Nadi, Fiji, 17.47 S 177.29 E

Solai, Kenya, 0.03 N 36.09 E

South Johnstone, near Innisfail, Queensland, Australia, 17.30 S 146.00 E

St. Ignatius, Guyana, 3.15 N 59.50 W

Swan's Lagoon, near Millaroo, Queensland Australia, 19.45 S 147.15 E

Tampa, Florida, U.S.A., 27.58 N 82.38 W

Tanga, Tanzania, 5.07 S 39.05 E

Thomsons Falls, Kenya, 0.04 N 36.22 E

Tifton, Georgia, U.S.A. 31.27 N 83.31 W

Timau, Kenya, 0.05 N 37.15 E

Tizimín, Mexico, 21.10 N 88.09 W

Tuli, Zimbabwe, 21.59 S 29.15 E

Tully, Queensland, Australia, 17.55 S 145.59 E

Tupanicireta, Brazil, 29.05 S 53.50 W

Turrialba, Costa Rica 9.56 N 83.40 W

Ukiriguru, near Mwanza, Tanzania, 2.30 S 32.54 E

Umbogintwini, S. Africa, 30.01 S 30.55 E

Utchee Creek, near Innisfail, Queensland, Australia, 17.30 S 146.00 E

Vaalhartz, S. Africa, 28.07 S 24.57 E

Vryburg, S. Africa, 26.57 S 24.44 E

Walkamin, near Mareeba, Queensland, 17.00 S 145.28 E

Warmbaths, S. Africa, 24.55 S 18.41 E

Warwick, Queensland, Australia 28.12 S 152.00 E

West Mzimba, Malawi, 11.55 S 33.39 E

Westwood, Queensland, Australia, 23.37 S 150.12 E

Woolongbar, near Lismore, N.S.W., Australia 28.48 S 153.17 E

Wrotham Park, near Weipa, Queensland, 12.35 S 141.56 E

Yeerongpilly, near Brisbane, Queensland, Australia, 27.30 S 153.00 E

Zaria, Nigeria, 11.01 N 7.44 E

The above list includes experiment stations where sufficient information was available to enable the location to be found in the Times Atlas of the World, Comprehensive edition (1980) London: Times Books

References

Acocks, J. P. H. (1966) Non-selective grazing as a means of veld reclamation, *Proc. Grasl. Soc. South Africa* **1**, 33–9

Acocks, J. P. H. (1975) Veld types of South Africa, *Memoirs of the Botanical Survey of South Africa*, No. 40, 128 pp.

Addison, K. B. (1970) Management systems on Spear grass country, *Proc. XI Intnl Grassl. Cong.*, 789–93

Addy, B. L. and Thomas, B. (1976) *Rhodes grass pastures: some aspects of management and utilisation*. Ministry of Agriculture and Natural Resources, Malawi: Research Bull. 3/76. 20 pp.

Addy, B. L. and Thomas, D. (1977) Intensive fattening of cattle on Rhodes grass pastures on the Lilongwe plain, Malawi, *Trop. Anim. Hlth, Prod.* **9**, 99–106

Addy, B. L. and Thomas, D. (1978) Stocking rate and the productivity of Rhodes grass pastures on the Lilongwe plain, Malawi, *Trop. Anim. Hlth, Prod.* **10**, 11–17

Adegbola, A. A., Onayinka, B. O. and Eweje, J. K. (1968) The management and improvement of natural grasslands in Nigeria, *Nigerian Agric. J.* **5**, 4–6

Aguiar de Mattos, J. C., Tundisi, A. G. A., Velloso, L., Dupas, W. and Soliva, R. (1967) Contribuição para o estudo da alimentação de bovinos durante o periodo de sêca, *Bol. Ind. Anim.* **24**, 17–26

Aii, T. and Stobbs, T. H. (1980) Solubility of the protein of tropical pasture species and the rate of its digestion in the rumen, *Anim. Feed Sci. and Tech.* **5**, 183–92

Akilimali, B. M., Fayd'herbe, J. C. and Calo, L. L. (1978) *Implementation of livestock development programmes: experience on parastatal ranches in Tanzania*, IV World Conf. Anim. Prod. Summaries, p. 30

de Alba, J., (1958) *Alimentación del ganado en la América Latina*, México La Prensa Medica Mexicana. 336 pp.

de Alba, J. and Davis, G. K. (1957) Minerales en la nutrición animal en la América Latina, *Turrialba* **7**, 16–33

de Alba, J. and Maltos, J. (1963) Valor de hormonas estrogénicas y progesterona en ceba de novillos, *Turrialba* **13**, 28–30

Alexander, G. I. (1972) Non protein nitrogen supplements for grazing animals in Australia, *World Anim. Rev.* (FAO) **4**, 11–14

Alexander, G. I., Harvey, J. M., Lee, J. H. and Stubbs, W. C. (1967)

Studies on copper and cobalt therapy of cattle in central coastal Queensland, *Aust. J. Agric. Res.* **18**, 169–81

Alexander, G. I., Sullivan, J. J. and Stokoe, J. (1970) Cottonseed meal and grain as supplements for grazing beef cattle, *Proc. Aust. Anim. Prod.* **8**, 45–9

Allan, P. J., Litschner, J. C., and Swain, A. J. (1972) Factors affecting the milk yield of beef cows in northern Queensland, *Proc. Aust. Soc. Anim. Prod.* **9**, 271–5

Allen, G. H. and Cowdry, W. A. R. (1961) Beef gains from irrigated pastures in the Burdekin delta, *Queensland Agri. J.* **87**, 175–9

Allman, R. T. and Hamilton, T. S. (1949) Nutritional deficiencies in live-stock, *FAO Agricultural Studies* No. 5, Washington

Alpizar, J. and Vohnout, K. (1973) *Crecimiento de bovinos supplementados con banano 1. Efecto de presión de pastoreo.* Compendios de Trabajos presentados en la IV Reunión Latinoamericana de Producción Animal (ALPA) Guadalajara, México, p. R35

Altona, R. E., Rose, C. J. and Tilley, T. J. (1960) Urea as a supplementary protein for bulk feeds, *South African J. Agric. Sci.* **3**, 69–81

Ammerman, C. B., Loaiza, J. M., Blue, W. G., Gamble, J. F. and Martin, G. F. (1974) Mineral composition of tissues from beef cattle under grazing conditions in Panama *J. Anim; Sci.* **38**, 158–162

Ammerman, C. B. and Miller, S. M. (1975) Selenium in ruminant nutrition; a review, *J. Dairy Sci.* **58**, 1561–77

Ammerman, C. B., Loaiza, J. M., Blue, W. G., Gamble, J. F. and Martin, F. G. (1974) Mineral composition of tissues from beef cattle under grazing conditions in Panamá, *J. Anim. Sci.* **38**, 158–62

Ammerman, C. B., Miller, S. M. and McDowell, L. R. (1979) Selenium in ruminant nutrition in *Proc. Latin American symposium on mineral nutrition research with grazing ruminants* (J. H. Conrad and L. R. McDowell, eds). University of Florida: Gainesville, Florida. 199 pp.

Anderson, A. W. (1933) Problems of animal nutrition and animal husbandry in northern Nigeria, *Imp. Bur. Anim. Nutr. Tech. Comm.* No. 4

Anderson, B. (1973) Study on cattle bruising, *Queensland Agric. J.* **99**, 234–40

Anon (1968) Supplementary feeding trial, *Annual Report of the Research Branch*, Republic of Zambia. Ministry of Rural Development, Department of Agriculture, pp. 324–6

Anon (1981) A sad end to Botswana's dreams of cattle ranching, *New Scientist* **92**, 151

APRU (1972) *Review of beef cattle production research in Botswana 1970–1972.* Animal Production Research Unit: Gaborones, Botswana. 58. pp. (litho)

APRU (1975) *A five-year programme of integrated beef cattle and range research in Botswana 1970–1975.* Part 1, Technical Results. Animal Production Research Unit: Gaborones, Botswana. 182 pp. (litho)

APRU (1976) *An integrated programme of beef cattle and range research in Botswana 1970–1976.* Animal Production Research Unit: Gaborones, Botswana. 160 pp. (litho)

APRU (1977) *An integrated programme of beef cattle and range research in Botswana 1970–1977.* Animal Production Research Unit: Gaborones, Botswana. 141 pp. (litho)

APRU (1979) Livestock and range research in Botswana, *Annual Report 1979*. Animal Production Research Unit: Gaborones, Botswana. 186 pp. (litho)

ARC (1965) *The nutrient requirements of farm livestock* No. 2, Ruminants Technical Reviews. Agricultural Research Council: London. 264 pp.

Archibald, K. A. E. and Cowlishaw, S. J. (1972) Calf grazing trials in Trinidad: I – Growth and development of calves, *Trop. Agric.* (*Trinidad*) **49**, 297–303

Arnold, R. M. (1969) The interaction of calcium, magnesium, phosphorus, other minerals and vitamin D in the aetiology of Manchester wasting disease, *Trop. Anim. Hlth Prod.* **1**, 75–84.

Arnold, R. M. and Bras, G. (1956) Observations on the morbid anatomy and histology of Manchester wasting disease of cattle in Jamaica, and related conditions in the countries of the Americas, *Amer. J. Vet. Res.* **17**, 630–9

Arnold, R. M. and Fincham, I. H. (1950) Cardiovascular and pulmonary calcification apparently associated with dietary imbalance in Jamaica, *J. Comp. Path. and Therap.* **60**, 51–64

Aronovich, S., Serpa, A. and Ribeiro, H. (1970) Effect of nitrogen fertilizer and legume upon beef production of Pangola grass pasture, *Proc. XI Intnl Grassland Congress*, 796–800

ASRM (1964) *A glossary of terms used in range management* (D. L. Huss, ed.). American Society of Range Management: Portland, Oregon. 32 pp.

Atkinson, F. J. B. (1963) Improved pastures the key to meat production. Rhodesia. *Min. of Agric. Bulletin* 2223. 4 pp.

Australian Meat Research Commission (1967) *First Annual Report* p. 20

Auty, J. H. (1964) M.V.Sc. Thesis, University of Queensland, cited by Alexander et al. (1970)

Baker, F. S., Jr (1966) *Steer wintering rations in north Florida*. Florida Agric. Exp. Sta. Circular S-174

Baker, P. R. (1976) The social importance of cattle in Africa and the influence of social attitudes on beef production in *beef production in developing countries* (A. J. Smith, ed). University of Edinburgh, Centre for Tropical Veterinary Medicine: Edinburgh. 487 pp.

Barnes, D. L. (1966) Studies on the use of *Sabi Panicum* for forage, *Rhod. Agric. J.* **63**, 144–54

Barr, N. C. E. (1971) *Reproduction in Queensland beef cattle*. Research report of Beef Cattle Husbandry Branch, Queensland Dept. of Primary Industries. 37 pp.

Barr, N. C. E. and Burns, M. A. (1972) Supplementation to increase fertility of beef cows in a drought, *Proc. Aust. Soc. Animal Prod.* **9**, 159–64

Barrie, N. and Clark, R. (1955) The effects of a) soluble phosphate and b) cobalt on the utilisation of urea nitrogen by cattle, *J. South African Vet. Med. Asso.* **30**, 457–8

Bauer, M. (1965) Five years study of ranch breeding stock, *Rhodesia Agric. J.* **62**, 28–33

Beames, R. M. (1959) Molasses and urea as a supplement to low quality pasture hay for cattle, *Queensland J. Agri. Sci.* **16**, 223–32

Beames, R. M. (1960a) The supplementation of low quality hay and pasture with molasses and molasses urea mixtures, *Proc. Aust. Soc. Animal Prod.* **3**, 86–92

Beames, R. M. (1960b) A note on the preservation of a solution of molasses and urea in water, *Queensland J. Agric. Sci.* **17**, 205–6
Beames, R. M. (1963) Provision of urea to cattle in a salt/urea/molasses block, *Queensland J. Agric. Sci.* **20**, 213–30
Beames, R. M. and O'Bryan, M. S. (1962) Effect of hormone implantation on growth rate and carcass composition of cattle grazing *Paspalum*-dominant pasture, *Queensland J. Agric. Sci.* **19**, 485–91
Becker, J. G. (1932) Studies on mineral metabolism XXIV on the administration of phosphorus to animals through their water supply, *18th Rep. Dir. Vet. Serv. Anim. Ind. Union of South Africa*, 751–97
Becker, R. B., Dix Arnold, P. T., Kirk, W. G., Davis, G. K. and Kidder, R. W. (1957) *Minerals for dairy and beef cattle.* Florida Agric. Exp. Sta. Bull. 513
Becker, R. B., Henderson, J. R. and Leighty, R. B. (1965) *Mineral malnutrition in cattle.* Florida Agric. Exp. Sta. Bull. 699 (technical). 54 pp.
Becker, R. B., Neal, W. M. and Shealy, A. L. (1933) *Stiffs or Sweeney (phosphorus deficiency in cattle).* Florida Agric. Exp. Sta. Bull. 264
Bedrak, E., Warnick, A. C., Hentges, J. F. and Cunha, T. J. (1964) *Effect of protein intake on gains, reproduction and blood constituents of beef heifers.* Florida Agric. Exp. Sta. Tech. Bull. 678. 30 pp.
Bembridge, T. J. (1963) Protein supplementary feeding of breeding stock proves profitable under watershed ranching conditions, *Rhodesia Agric. J.* **60**, 98
Bembridge, T. J. (1964) High calving rates with limited feed, *Rhodesia Agric. J.* **61**, 112–13
Bembridge, T. J. (1966) Aphosphorosis trial with beef cattle, *Rhodesia Agric. J.* **63**, 109–12
Bembridge, T. J. (1970) Influence of two levels of supplementary feeding and three stocking rates on the lifetime performance of beef cows, *Rhodesia Agric. J.* **67**, 2–6
Benintendi, R. P., Nogueira de Freitas, E. A. and Biondi, P. (1974) Efeito da administração de suplemento de milho desintegrado durante a seca a novilhas Gir e Sindi, *Bol. Ind. Anim.* **31** (2), 221–7
Bevan, L. E. W. (1906–07) *Rhodesian Agric. J.* **4**, 53
Bewg, W. P., Kyneur, G. W. and Hart, I. B. (1970) Beef production from tropical legume-grass pastures in the coastal ranges of south eastern Queensland, *Proc. Aust. Soc. Anim. Prod.* **8**, 433–8
Bingley, J. B. and Carrillo, B. J. (1966) Hypocuprosis of cattle in the Argentine, *Nature* **209**, 834–5
Bishop, E. J. B. (1957) Combat losses with urea and molasses, *Farming in South Africa* June 1957, 30–2
Bishop, E. J. B. (1959) The value of urea and molasses proved at Dohne, *Farming in South Africa* **35**, 27–9
Bisschoff, W. V. A., Quinn, L. R., Mott, G. O. and da Rocha, G. L. (1967) *Supplemental feeding of steers on pasture with protein-energy supplements.* IRI Research Institute, Bulletin 35. 47 pp. (also *Pesq. Agrotec. Bras.* **2**, 421–36)
Bisschop, J. H. R. (1932) Dosing bonemeal to cattle, *Farming in South Africa* **7** 1141–3
Bisschop, J. H. R. (1964) *Feeding phosphates to cattle.* Department of Agricultural Technical Services: Pretoria. 267 pp.
Bisschop, J. H. R. and Groenewald, J. W. (1963) Limitations to efficiency

imposed by inadequate nutrition with particular reference to feed resources, *Proc. First World Cong. Animal Prod.* (*Rome*) **1**, 47–72

Bisschop, J. H. R., Malan, A. I., Steyn, H. P. and Laurence, G. (1939) On the feeding of a phosphatic supplement to mother reared calves under open range conditions in British Bechuanaland, *Onderstepoort J. Vet. Sci.* **13**, 321–46

Bisset, W. J. and Marlow, G. W. C. (1974) Productivity and dynamics of two Siratro based pastures in the Burnett coastal foothills of south eastern Queensland, *Trop. Grassl.* **8**, 17–24

Black, W. H., Tash, L. H., Jones, J. M. and Kleberg Jr., R. J. (1943) *Effect of phosphorus supplements on cattle grazing on range deficient in this mineral* USDA Tech. Bull. 856. 23 pp.

Blaser, R. E., Glasscock, R. S., Killinger, G. B. and Stokes, W. E. (1952) *Carpet grass and legume pastures in Florida.* Florida Agric. Exp. Sta. Bull. 453. 36 pp.

Blaser, R. E., Stokes, W. E., Warner, J. D., Ritchey, G. E. and Killinger, G. B. (1945) *Pastures for Florida.* Florida Agric. Exp. Sta. Bull. 409. 73 pp.

Blaxter, K. L. (1962) *The energy metabolism of ruminants.* Hutchinson: London. 329 pp.

Blaxter, K. L., Wainman, F. W. and Wilson, R. S. (1961) The regulation of food intake by sheep, *Anim. Prod.* **3**, 51–61

Blaxter, K. L. and Wilson, R. S. (1962) The voluntary intake of roughages by steers, *Anim. Prod.* **4**, 351–8

Blunt, C. O. (1976) Preliminary cattle grazing trials on irrigated *Leucaena leucocephala* and Pangola grass in the Ord Valley, N. W. Australia, *Proc. Aust. Soc. Anim. Prod.* **11**, 10

Blunt, C. O. (1978) Production from steers grazing nitrogen fertilised irrigated Pangola grass in the Ord Valley, *Trop. Grassl.* **12**, 90–6

Blydenstein, J. (1965) The natural grasslands of the Columbian llanos, *Proc. IX Intnl Grassl. Congr.*, 1205–6

Bogdan, A. V. (1977) *Tropical pasture and fodder plants.* Longman: London. 475 pp.

Bohmann, V. R. (1955) Compensatory growth of beef cattle. The effect of hay maturity, *J. Animal Sci.* **14**, 249–55

Boshoff, W. H (1968) *Using field machinery.* Oxford University Press: London. 53 pp.

Botswana (1975) *National policy on tribal grazing land* Government Paper No. 2 of 1975. Republic of Botswana: Government Printer: Gaborone. 19 pp.

Bowen, E. J. and Rickert, K. G. (1979) Beef production from native pastures sown to fine-stem stylo in the Burnett region of south-eastern Queensland, *Aust. J. Exp. Agric. and Anim. Husb.* **19**, 140–9

Bowhill, T. (1906) Stock diseases of the Eastern coastal districts, *Agric. J. Cape of Good Hope* **28**, 550–67

Boyazoglu, P. A. (1970) The correction of mineral imbalance *Proc. sem. trace elements in animal and plant nutrition.* Swaziland Agricultural College and University Centre, 31–7.

Boyazoglu, P. A. (1976) A review of mineral imbalances of grazing animals in Southern Africa, *J. South African Vet. Ass.* **47**, 129–32

Bredon, R. M., Harker, K. W. and Marshall, B. (1963) Nutritive value of grasses grown in Uganda when fed to Zebu cattle 1. Relation between the percentage of crude protein and other nutrients, *J. Agric. Sci. (Cambridge)* **61**, 101–8

Bridgens, A. B. (1968) *Aspects of shoot apex morphogenesis, development and behaviour in grasses with reference to the utilisation and management of natural grassland.* Department of Agricultural Technical Services: Republic of South Africa. Technical Communication No. 67, 44 pp.

Briggs, M. H. (1967) *Urea as a protein supplement.* Pergamon Press: Oxford. 466 pp.

British Guiana (1962) Director of Agriculture, *Annual Report (1962)*, 87–92

Brody, S. (1945) *Bioenergetics and growth.* Reinhold: New York. 1023 pp.

Bryan, W. W. (1968a) Grazing trials on the Wallum of south-eastern Queensland 1. A comparison of four pastures, *Aust. J. Exp. Agric. and Anim. Husb.* **8**, 512–20

Bryan, W. W. (1968b) Grazing trials on the Wallum of south-eastern Queensland 2. Complex mixtures under common grazing, *Aust. J. Exp. Agric. and Anim. Husb.* **8**, 683–90

Bryan, W. W. and Evans, T. R. (1971) A comparison of beef production from a nitrogen fertilised Pangola grass and from a Pangola grass legume pasture, *Trop. Grassl.* **5**, 89–97

Bryan, W. W. and Evans, T. R. (1973) Effects of soil, fertilizers and stocking rates on pastures and beef production on the Wallum of south-eastern Queensland 1 Botanical composition and chemical effects on plants and soils, *Aust. J. Exp. Agric. and Anim. Husb.* **13**, 516–29

Bryan, W. W., Thorne, P. M. and Andrew, C. S. (1960) Cobalt status of the Wallum, *J. Aust. Inst. Agric. Res.* **26**, 273–75

Buck, N. G., Light, D., Rutherford, A., Miller, M., Rennie, T. W., Pratchett, D., Capper, B. S. and Trail, J. C. M. (1976) Environmental factors affecting beef cow reproductive performance in Botswana, *Anim. Prod.* **23**, 357–63

Bulcock, F. W. (1938) Cattle fattening on coastal pastures, *Queensland Agric. J.* **49**, 460–8

Buller, R. E. (1960) Evaluation of the forage resources of the tropical grasslands of Mexico, *Proc. VIII Intnl. Grassl. Cong.*, 374–7

Buller, R. E., Aronovich, S., Quinn, L. R. and Bisschoff, W. V. A. (1970) Performance of tropical legumes in the upland savannah of central Brazil, *Proc. XI Intnl Grassl. Cong.*, 143–6

Bumpus, E. F. (1957) Legume nodulation in Kenya 1. Exploratory field experiments, *E. Afr. Agric. J.* **23**, 91–9

Bunge, G. (1873) *Z. Biol.* **9**, 104

Burns, M. A. (1971) Care of beef cattle herds in drought Part 1, *Queensland Agric. J.* **97**, 202–9

Burns, M. A. and Sutherland, D. N. (1960) Use of synthetic oestrogens for promoting weight gains in beef steers grazing on native pastures, *Queensland J. Agric. Sci.* **17**, 39–42

Burt, R. L. and Williams, W. T. (1975) Plant introduction and the Stylosanthes story, *Australian Meat Research Committee Review*, No. 25, 1–26

Burton, G. W. (1947) Breeding Bermuda grass for the south-eastern United States, *J. Am. Soc. Agron.* **39**, 551–69

Burton, G. W., Hart, R. H. and Lowry, R. S. (1967) Improving forage quality in Bermuda grass by breeding, *Crop. Sci.* **7**, 329–32

Butterworth, M. H. (1961) Studies on Pangola grass at I.C.T.A., Trinidad II The digestibility of Pangola grass at various stages of growth, *Trop. Agric. (Trinidad)* **38**, 189–93

Butterworth, M. H. (1963) Digestibility trials on forages in Trinidad and their use in the prediction of nutritive value, *J. Agric. Sci.* (*Cambridge*) **60**, 341–6

Butterworth, M. H. (1964) The digestible energy content of some tropical forages, *J. Agric. Sci* (*Cambridge*) **64**, 319–21

Butterworth, M. H. (1965) Some aspects of the utilisation of tropical forages 1. Green Elephant grass at various stages of growth, *J. Agric. Sci.* (*Cambridge*) **65**, 233–9

Butterworth, M. H. (1966) The utilization of phosphorus in some Venezuelan forages, *Turrialba* **16**, 253–56

Butterworth, M. H. (1967a) The digestibility of tropical grasses, *Nutr. Abs. Rev.* **37**, 349–68

Butterworth, M. H. (1967b) *Recent developments relating to animal nutrition and their application in animal production; deficiencies in protein and energy supplies.* Sixth FAO Inter-American Conference on Animal Production and Health: Gainesville, Florida. Working paper No. 2

Butterworth, M. H. (1981) Calculation of cattle herd projections using a programmable hand calculator, *World Anim. Rev.* No. 39, 41

Butterworth, M. H. (1983) Seasonality of calving amongst unimproved cattle in Swaziland, *Trop. Agric.* (Trinidad) **60**, 218–20

Butterworth, M. H., Aguirre, E. Ll., Aragón, A. C. and Huss D. L. (1973) Use of molasses containing urea as a supplement to Pangola grass pastures in northeast Mexico, *J. Range Management* **26**, 269–71

Butterworth, M. H. and Brandl, F. I. (1981) Forage consumption of wild and domestic herbivores, *World Anim. Rev.* No. 37, 48

Butterworth, M. H. and Butterworth, J. P. (1965) Some aspects of the utilisation of tropical forages, 2. Pangola and Coastal Bermuda hays, *J. Agric. Sci* (*Cambridge*) **65**, 389–95

Butterworth, M. H. and Carrera, G. (1969) El uso de la melaza como suplemento para novillos en potrero, *México Ganadero* June 1969, 68–9

Butterworth, M. H. and Diaz, L. J. A. (1970) Use of equations to predict the nutritive value of tropical grasses, *J. Range Management* **23**, 55–8

Butterworth, M. H., Groom, C. C. and Wilson, P. N. (1961) The intake of Pangola grass (*Digitaria decumbens, Stent.*) under wet and dry season conditions in Trinidad, *J. Agric. Sci.* (*Cambridge*) **56**, 407–10

Butterworth, M. H. and Presswood, J. (1978) *Beef production handbook.* Ministry of Agriculture: Swaziland. 204 pp. (litho)

Buttery, P. J. (1976) Protein synthesis in the rumen: its implication in the feeding of non-protein nitrogen to ruminants in *Principles of cattle production* (H. Swan and W. H. Broster, eds). Butterworth: London. 438 pp

Buttery, P. J., Vernon, B. G. and Pearson, J. T. (1978) Anabolic agents – some thoughts on their mode of action, *Proc. Nutr. Soc.* **37**, 311–15

Cabaret, J. (1975) Premiers essais d'embouche de zébus en Haute-Volta, *Rev. élev Méd. Vét. Pays Trop.* **38** (new series), 409–17

Calvet, H., Friot, D. and Gueye, I. S. (1976) Supplémentations minérales, alimentaires et pertes de poids des zébus sahéliens en saison sèche, *Rev. Élev. Méd. Vét. Pays Trop.* **29**, 59–66

Camargo, W. V. de A., Fernandez, N. dos S. and Santiago, A. M. H. (1979) Mineral research projects for ruminants in the Amazonian region of Brazil in *Proc. Latin American Symposium on mineral nutrition research with grazing ruminants* (J. H. Conrad and L. R. McDowell, eds). University of Florida: Gainesville, Florida. 199 pp.

Campion, E. J. (1972) *The use of high levels of molasses with three different proteins for growing bulls.* Technical Report of FAO/UNDP Project DOM/65/503 (mimeo). 25 pp.

Campion, E. J. (1978) *The role of molasses in the intensification of livestock production in the Dominican Republic.* Technical Report of FAO/UNDP Project DOM/65/503 (mimeo). 21 pp.

Carlo, I., Mendoza, R. and Vélez, J. (1972) Evaluation of different feeding systems in beef cattle production, *J. Agric. Univ. Puerto Rico* **56**, 327–36

Carnevali, A. A., Chicco, C. F., Shultz, T. A. and Rodriguez C. S. (1970) Efecto de la suplementación con melaza y urea para bovinos a pastoreo, *Agronomía Tropical (Maracay)* **20**, 433–43

Carnevali, A. A., Chicco, C. F., Shultz, T. A., Rodriguez, S. and Shultz, E. (1968) Efecto de suplementación con melaza y urea para bovinos en pastoreo, *Asociación Latinoamericana de Producción Animal Memorias* **3**, 169–70

Caro Costas, R., Abruña, F. F. and Vicente Chandler, J. (1972a) Comparison of heavily fertilised Pangola and Star grass pastures in terms of beef production and carrying capacity in the humid mountain region of Puerto Rico, *J. Agric. Univ. Puerto Rico* **56**, 104–9

Caro Costas, R., Abruña, F. and Vicente Chandler, J. (1973) Comparison of heavily fertilised Pangola grass and Star grass pastures under humid tropical conditions, *Agronomy J.* **65**, 132–3

Caro Costas, R. and Vicente Chandler, J. (1961) Effect of fertilization on carrying capacity and beef produced by Napier grass pastures, *Agronomy J.* **53**, 204–5

Caro Costas, R. and Vicente Chandler, J. (1972) Effect of heavy rates of fertilization on beef production and carrying capacity of Napier grass pastures over 5 consecutive years of grazing under humid tropical conditions, *J. Agric. Univ. Puerto Rico* **56**, 223–7

Caro Costas, R., Vicente Chandler, J., and Abruña, F. (1972b) Effect of four levels of fertilization on beef production and carrying capacity of Pangola grass pastures in the humid mountain region of Puerto Rico, *J. Agric. Univ. Puerto Rico* **56**, 219–22

Caro Costas, R., Vicente Chandler, J. and Burleigh, C. (1961) Beef production and carrying capacity of heavily fertilized, irrigated Guinea, Napier and Pangola grass pastures on the semi-arid south coast of Puerto Rico, *J. Agric. Univ. Puerto Rico* **45**, 32–6

Caro Costas, R., Vicente Chandler, J. and Figarella, J. (1965) Productivity of intensively managed pastures of five grasses on steep slopes in the Humid Mountains of Puerto Rico, *J. Agric. Univ. Puerto Rico* **49**, 99–111

Carrazzoni, J. A. (1974) *Relación entre las Vitaminas A,* D_3 *y E y el celo.* Ganaderia sub-tropical Argentina: Experiencias con vacunos: Buenos Aires, Argentina, 131–4

Carrera, C., Muñoz, C. and Solares, L. (1962) Efecto del hexoestrol sobre sovinos cebados con Zacate Guinea, *Agric. Téc. Méx* **2**, 38–40

Carrera, C., Muñoz, H. and Solares, L. (1963) Melaza de caña como suplemento en el engorde de bovines en zacate guinea (*Panicum maximum*), *Téc. Pec. Méx.* **1**, 34–7

Carrera, M. C., Muñoz, C. H. and Solares, T. L. (1969) Uso de melaza urea en el engorde de novillos en pastoreo y corral, *Rev. Méx. Prod. Anim.* **2**, 19–26

Carrera, C. and Soikes, R. (1958) Efecto de implantaciones de estilbestrol sobre ceba de bovines en pasto Para (*Panicum purpurascens*, Raddi), *Turrialba* **8**, 96–103

Carrillo, B. J. and Worker, N. A. (1967) Enteque seco: arteriosclerosis y calcificación metastásica de origen tóxico en animales a pastoreo, *Rev. Invest. Agropecuar.* Ser. 4, **4**, 9–30

Chacón, E. and Stobbs, T. H. (1976) Influence of progressive defoliation of a grass sward on the eating behaviour of cattle, *Aust. J. Agric. Res.* **27**, 709–27

Chadhokar, P. A. (1977) Establishment of Stylo (*Stylosanthes guianensis*) in Kunai (*Imperate cylindrica*) pastures and its effect on dry matter yield and animal production in the Markham Valley Papua New Guinea, *Trop. Grassl.* **11**, 263–72

Chapman, H. L. Jr and Haines, C. E. (1960) *Fattening steers on pasture and in dry-lot.* Everglades St. Mim. Rep. 60–17

Chapman, H. L. Jr., Haines, C. E. and Kidder, R. W. (1961) *Feeding value of limited-fed mixed feed, citrus pulp, ground snapped corn, and blackstrap molasses for fattening steers on pasture.* Everglades St. Mim. Rep. 16–19

Chapman, H. L. Jr. and Kidder, R. W. (1964) *Copper and cobalt for cattle.* Florida Agric. Exp. Sta. Bull. 674

Chapman, H. L. Jr., Palmer, A. Z., Kidder, R. W., Carpenter, J. W. and Haines, C. E. (1964a) *Oral and implanted stilbestrol for beef cattle fattened on pasture and in drylot.* Florida Agric. Exp. Sta. Bull. 666. 16 pp.

Chapman, H. L. Jr., Peacock, F. M., Kirk, W. G., Shirley, R. L. and Cunha, T. J. (1964b) *Supplemental feeding of beef cattle on pasture in south Florida.* Florida Agric. Exp. Sta. Bull. 665. 28 pp.

Chesters, J. K. (1978) Biochemical functions of zinc in animals in *World Review of Nutrition and Dietetics* (G. H. Bourne, ed.) **32**, 135–56

von la Chevallerie, M. K. S. L. (1959) Urea as nutrient for ruminants: injudicious use dangerous, *Farming in South Africa* **35**, 44–5

von la Chevallerie, M. K. S. L. (1965) Waarnemings met die voer van ureum-melasse mengsels aan beeste, *Proc. S. Afri. Soc. Anim. Prod.* **4**, 120–3

Chicco, C. F. (1972) *Estudio de la nutrición mineral del ganado de la región occidental de Panamá.* Programa de Pastos y Forrajes Proyecto UNDP/SF No. 232: David, Panamá

Chicco, C. F. and French, M. H. (1959) Observaciones sobre deficiencias del calcio y fósforo en los animales de las regiones ganaderas del centro y este de Venezuela, *Agronomia Tropical* (*Maracay*) **9**, 41–62

Chicco, C. F., Rios, J., Bodisco, V. and López, A. (1973) Frecuencia de suplementación con concentrado de bovines en sabanas tropicales, *Second World Congress Animal Feeding* **5**, 247–52

Chicco, C. F., Rios, J., Schultz, E. and Shultz, T. A. (1975) Biuret y la urea como suplementos proteicos para novillos a pastoreo, *Agronomia Tropical* (*Maracay*) **25**, 561–7

Chicco, C. F. and Schultz, T. A. (1975) *Sistemas extensivos de producción de carne y leche usando pastos tropicales con y sin supplementación*. Organisación Panamericana de la Salud 7° Reunión Interamericana sobre al control de Fiebre Aftosa y otras Zoonoses No. 295, 43–76

Chicco, C. F., Shultz, T. A., Carnevali, A. and González, R. (1974) Implantación hormonal en novillos a pastoréo en sabanas tropicales, *Agronomia Tropical Maracay* **24**, 471–6

Chicco, C. F., Shultz, T. A., Rios, J., Plasse, D. and Burguera, M. (1971) Self feeding salt supplement to grazing steers under tropical conditions, *J. Anim. Sci.* **33**, 142–6

Christian, G. S. and Shaw, N. H. (1952) A study of two strains of Rhodes grass (*Chloris gayana* Kunth.) and of lucerne (*Medicago sativa* L.) as components of a mixed pasture at Lawes in south-east Queensland, *Aust. J. Agric. Res.* **3**, 277–99

CIAT (1974) Utilisacion eficiente de los forrajes en los tropicos, *Informe Annual 1974*. Centro Internacional de Agricultura Tropical, Cali, Columbia

CIAT (1977) Pasture utilisation, *Annual Report 1977*, A66–A70. Centro Internacional de Agricultura Tropical, Cali, Columbia.

CIAT (1978) *Beef program 1978 report*. Centro Internacional de Agricultura Tropical, Cali, Columbia. 174 pp.

Clark, R. (1952) Use of urea and molasses in winter feeding, *Farming in South Africa* **27**, 453–4

Clark, R. and Barrie, N. (1954) Winter feeding experiments, *Farming in South Africa* **29**, 135–8

Clatworthy, J. N. (1975a) Effects of legume sown in reverted land on animal production, Div. of Livestock and Pastures: Rhodesia. *Ann. Rep. 1974–5*, 152–3

Clatworthy, J. N. (1975b) An assessment of animal production from a dryland pasture of *Desmodium intortum*. Div. of Livestock and Pastures: Rhodesia. *Ann. Rep. 1974–5*, 153–4

Clatworthy, J. N. (1975c) An exploratory trial to assess animal production from an irrigated grass-clover pasture containing tropical species. Div. of Livestock and Pastures: Rhodesia. *Ann. Rep. 1974–5*, 154–6

Clatworthy, J. N. (1976) Effects of legumes sown in reverted land on animal production. Div. of Livestock and Pastures: Rhodesia. *Ann. Rep. 1975–6*, 184–5

Clatworthy, J. N. and Boultwood, J. N. (1978) Effect of veld reinforcement with Silverleaf Desmodium on dry season performance of weaner steers. Div. of Livestock and Pastures: Rhodesia. *Ann. Rep. 1977–8*, 62–5

Clatworthy, J. N. and Holland, D.G.E. (1978) Effects of legumes sown in reverted land on animal production. Div. of Livestock and Pastures: Rhodesia. *Ann. Rep. 1976–*7, 179–81

Clatworthy, J. N. and Holland, D.. G.. E. (1978) Effects of legumes sown in reverted land on animal production. Div. of Livestock and Pastures: Rhodesia. *Ann. Rep. 1977–8*, 57–9

Clatworthy, J. N. and Holland, D. G. E. (1979) Effects of legume reinforcement of veld on the performance of beef steers, *Proc. Grassl. Soc. Sth. Africa.* **14**, 111–14

Coaldrake, J. E. and Smith, C. A. (1967) Estimates of animal production from pastures on Brigalow land in the Fitzroy Basin, Queensland, *J. Aust. Inst. Agric. Sci.* **33**, 52–4

Coaldrake, J. E., Smith, C. A., Yates, J. J. and Edye, L. A. (1969) Animal production on sown and native pastures on Brigalow land in southern Queensland during drought, *Aust. J. Exper. Agric. and Anim. Husb.* **9**, 47–56

Cohen, R. D. H. (1972) Phosphorus nutrition of beef cattle: 1. Effect of supplementation on liveweight of steers and digestibility of diet, *Aust. J. Exper. Agric. and Anim. Husb.* **12**, 455–9

Cohen, R. D. H. (1973) Phosphorus nutrition of beef cattle: 2. Relations of pasture phosphorus to phosphorus content of blood, hair and bone of grazing steers, *Aust. J. Exper. Agric. and Anim. Husb.* **13**, 5–8

Cohen, R. D. H. (1974) Effect of molasses-urea supplements on digestibility of mature carpet grass (*Axonopus affinis*) and liveweight change of beef steers, *Aust. J. Exper. Agric. and Anim. Husb.* **14**, 589–92

Cohen, R. D. H. and O'Brian, A. D. (1974) Beef production from three pastures at Grafton, N. S. W., *Trop. Grassl.* **8**, 71–9

Colclough, M. and Crowley, D. (1974) *Setshaba le togamaano: The people and the plan*: Volume I *A report on the Botswana Government's educational project on the five year national development plan 1973–78*. University of Botswana, Lesotho and Swaziland, Division of Extra Mural Services: Gaborones, Botswana. 187 pp.

Colditz, P. J. and Kellaway, R. C. (1972) The effect of diet and heat stress on feed intake, growth and nitrogen metabolism in Friesian, FI Brahman × Friesian and Brahman heifers, *Aust. J. Agric. Res.* **23**, 717–25

Cole, J. W., Ramsey, C. B., Hobbs, C. S. and Temple, R. S. (1964) Effects of type and breed of British, Zebu and dairy cattle on production, palatability and composition III Percent wholesale cuts and yield of edible portion as determined by physical and chemical analysis, *J. Anim. Sci.* **23**, 71–7

Condon, M. J., Smith, P. C. and Winks, L. (1970) *Molasses-urea supplements for beef weaners*. Dept. of Primary Industries, Div. of Animal Industry, Australia. Advisory Leaflet No. 674.

Conniffe, D., Browne, D. and Walshe, M. J. (1970) Experimental design for grazing trials, *J. Agric. Sci.* (*Cambridge*) **74**, 339–42

Connolly, J. (1976) Some comments on the shape of the stocking rate curve, *J. Agric. Sci.* (*Cambridge*) **86**, 103–9

Conrad, H. R., Pratt, A. D., and Hibbs, J. W. (1964) Regulation of feed intake in dairy cows 1. Change in importance of physical and physiological factors with increasing digestibility, *J. Dairy. Sci.* **47**, 54–62

Conrad, J. H., Sousa, J. C., Mendes, M. O., Blue, W. G. and McDowell, L. R. (1978) *Fe, Mn, Na and Sn Interrelationships in a tropical soil plant and animals system*. Fourth World Conference on Animal Production: Summaries 80

Coombe, J. B. and Tribe, D. E. (1962) The feeding of urea supplements to sheep and cattle; the results of pen feeding and grazing experiments, *J. Agric. Sci.* (*Cambridge*) **59**, 125–41

Copeman, D. B., Sumbung, F. P., Poothongtong, P., Haq, S. and Armstrong, J. R. (1977) Effect of molasses supplementation and copper, cobalt and anthelmintic therapy on weight gain of beef yearling steers grazing improved pastures in a wet tropical environment, *Aust. J. Exper. Agric. and Anim. Husb.* **17**, 538–44 de Cossio, J. M. (1960) *Los toros: tratado tecnico e historico*. Espasa Calpe, S. A: Madrid. 1017 pp.

Cowlishaw, S. J. and Archibald, K. A. E. (1972) Calf grazing trials in Trinidad; II–Intake and utilization of nutrients from pasture and meal supplement, *Trop. Agric.* (*Trinidad*) **49**, 305–10

Creek, M. J. and Nestel, B. L. (1964) The effect of grazing cycle duration on liveweight output and chemical composition of Pangola grass *(Digitaria decumbens* Stent.) in Jamaica, *Proc. IX Intnl Grassl. Cong.*, 1613–18

Crowder, L. V. and Chheda, H. R. (1977) *Forage and fodder crops in food crops of the lowland tropics* (C. L. A. Leakey and J. B. Wills, eds). Oxford University Press: Oxford. 345 pp.

Cruz de Carvalho, E. (1974) 'Traditional' and 'modern' patterns of cattle raising in southwestern Angola: a critical evaluation of change from pastoralism to ranching, *J. Developing Areas* **8**, 199–226

Cubillos, G., Vohnout, K and Jiménez, C. (1975) Intensive cattle feeding under grazing systems in *Proceedings of the seminar on Potential to increase beef production in Tropical America*. Centro Internacional de Agricultura Tropical: Cali, Colombia. 315 pp.

Cunha, T. J., Shirley, R. L., Chapman, H. L. Jr., Ammerman, C. B., Davis, G. K., Kirk, W. G. and Hentges, J. F. (1964) *Minerals for beef cattle in Florida*. Florida Agric. Exper. Sta. Bull. 683. 60 pp.

Davis, R. K. (1971) Some issues in the evolution, organization and operation of group ranches in Kenya, *East Africa J. Rural development* **4**, 22–33

Deans, H. D., Chopping, G. D., Sibbick, R., Thurbon, P. N., Copeman, D. B. and Stokoe, J. (1976) Effect of stocking rate, breed, grain supplementation, nitrogen fertiliser level and anthelmintic treatment on growth rate of dairy weaners grazing irrigated Pangola grass, *Proc. Aust. Soc. Anim. Prod.* **11**, 449–52

Delhaye, R. E. (1967) Quelques enseignements d'un essai de charge mené a la Station de Recherches Agronomiques du Lac Alactra, *L'Agronomie Tropicale*, No. 5, 487–95

Denny, R. P. and Barnes, D. L. (1977) Trials of multi-paddock grazing systems on veld 3. A comparison of six grazing procedures at two stocking rates, *Rhodesia J. Agric. Res.* **15**, 129–41

Denny, R. P., Barnes, D. L. and Franklin, M. F. (1974) Some effects of varying the duration, frequency and intensity of grazing on the growth of steers on veld, *Proc. Grassl. Soc. South Africa* **9**, 133–7

Denny, R. P., Barnes, D. L. and Kennan, T. C. D. (1977) Trials of multi-paddock grazing systems on veld 1. An exploratory trial of systems involving twelve paddocks and one herd, *Rhodesia. J. Agric. Res.* **15**, 11–23

Denny, R. P. and Steyn, J. S. H. (1977) Trials of multi-paddock grazing systems on veld 2. A comparison of a 16 paddocks-to-one-herd system with a 4 paddocks-to-one-herd system using breeding cows, *Rhodesia. J. Agric. Res.* **15**, 119–27

Denny, R. P. and Yuba, T. (1979) A comparison of continuous and rotational grazing on open granite sandveld. Div. Livestock and Pastures: Rhodesia. *Ann. Rep. 1978–9*, 188–91

Devendra, C. (1977) The mineral content of Caribbean feeding stuffs, *Trop. Agric.* (*Trinidad*) **54**, 29–38

Devendra, C., Mohd Nordin Hassan, Hodge, R., Lee Kok Choo, T. and Pathamasingam, M. (1973) Kedah-Kelantan cattle of Malaysia, *Malays. Agric. J.* **49**, 25

Dick, A. T. and Bull, L. B. (1945) Some preliminary observations on the effect of molybdenum on copper metabolism in herbivorous animals, *Aust. Vet. J.* **21**, 70–2

Dick, A. T., Dewey, D. W. and Gawthorne, J. M. (1975) Thiomolybdates and the coppermolybdenum-sulphur interaction in ruminant nutrition, *J. Agric. Sci. (Cambridge)* **85**, 567–8

Dirksen, G., Plank, P., Spiess, A., Hänichen, T. and Dämmrich, K. (1970) Über ein enzootische 'Kalzinose' beim rind 1. Klinische Beobachtungen und Untersuchungen, *Deutsch. Tierarzt. Wschr.* **77**, 321–37

Dirven, J. G. P. (1963) The nutritive value of the indigenous grasses of Surinam, *Netherlands J. Agric. Sci.* **11**, 295–307

Döberreiner, J., Tokarnia, C. H., de Costa, J. B. D., Campos, J. L. E. and Dayrell, M. D. (1971) 'Espichamento' poisoning of cattle by *Solanum malacoxylon* in the marshy area of Mato Grosso, *Pesq. agropec. Bras. Ser. Vet.* **6**, 91–117

Dodge, J. R. (1874) *Diseases of farm animals.* Report of the Commissioner of Agriculture for the year 1872 (Florida), p. 36

Donaldson, L. E. (1962) Some observations on the fertility of beef cattle in north Queensland, *Aust. Vet. J.* **38**, 447–54

Donaldson, L. E., Harvey, J. M., Beattie, A. W., Alexander, G. I. and Burns, M. A. (1964) Effects of copper and cobalt supplementation on growth rate and fertility of Shorthorn heifers in northern coastal Queensland, *Queensland J. Agric. Sci.* **21**, 167–79

Doran, M. H., Low, A. R. C. and Kemp. E. L. (1979) Cattle as a store of wealth in Swaziland: Implications for livestock development and overgrazing in Eastern and Southern Africa, *American J. Agric. Economics.* **61**, 41–7

Dougall, H. W. and Bogdan, A. V. (1958a) Browse plants of Kenya with special reference to those occurring in south Baringo, *E. Afr. Agric. For. J.* **23**, 236–45

Dougall, H. W. and Bogdan, A. V. (1958b) The chemical composition of the grasses of Kenya 1, *E. Afr. Agric. J.* **24**, 17–23

Duckworth, J. (1946) A statistical comparison of the influence of crude fibre on the digestibility of roughage by *Bos indicus* (Zebu) and *Bos taurus* cattle, *Trop. Agric. (Trinidad)* **23**, 4–8

Duvall, V. L. (1969) Comparison of supplementation methods for cow herds grazing pine-bluestem range, *J. Range Management* **22**, 182–7

Duvall, V. L. and Hansard, S. L. (1967) Responses of southern range cattle to protein supplementation, *J. Range Management* **20**, 153–7

Duvall, V. L. and Whitaker, L. B. (1963) *Supplemental feeding increases beef production on bluestem-longleaf pine ranges.* Louisiana Agr. Exper. Sta. Bull. 564. 18 pp.

Dyer, I. A., Olive, F. R. and Nunez, E. (1954) A comparison of Elephant, Pará, Pangola and Guinea grass for native and native × Brahman yearling bulls *J. Anim. Sci.* **13**, 1007–8 (abs)

Dzowela, B. J. (1980) *Pasture agronomy and improvement research.* Ministry of Agriculture and Natural Resources: Malawi (mimeo). 27 pp.

Ebersohn, J. P. and Lee, G. R. (1972) The impact of sown pastures on cattle numbers in Queensland, *Aust. Vet. J.* **48**, 217–23

Edwards, P. J. (1974) Estimates of carrying capacity and production from herbage yields and grazing days, *Proc. Grassl. Soc. South Africa* **9**, 139–43

Edwards, P. J. and Nel, S. P. (1973a) Short term effects of fertiliser and stocking rates of the Bankenveld. 1. Vegetational changes, *Proc. Grassl. Soc. South Africa* **8**, 83–8

Edwards, P. J. and Nel, S. P. (1973b) Short term effects of fertilization and stocking rates on the Bankenveld. 2. Animal production, *Proc. Grassl. Soc. South Africa* **8**, 89–96

Edye, L. A., Ritson, J. B. and Haydock, K. P. (1972) Calf production of Droughtmaster cows grazing a Townsville Stylo-Spear grass pasture, *Aust. J. Exp. Agric. and Anim. Husb.* **12**, 7–12

Edye, L. A., Ritson, J. B., Haydock, K. P. and Griffiths Davies, J. (1971) Fertility and seasonal changes in liveweight of Droughtmaster cows grazing a Townsville Stylo-Spear grass pasture, *Aust. J. Agric. Res.* **22**, 963–77

Edye, L. A., Williams, W. T. and Winter, W. H. (1978) Seasonal relations between animal gain, pasture production and stocking rate on two tropical grass legume pastures, *Aust. J. Agric. Res.* **29**, 103–13

Elliott, R. C. (1964) Some nutritional factors influencing the productivity of beef cattle in Southern Rhodesia. University of London, PhD Thesis

Elliott, R. C. and Fokkema, K. (1960) Protein digestibility relationships in ruminants, *Rhodesia Agric. J.* **57**, 301–4

Elliott, R. C., Mills, W. R. and Reed, W. D. C. (1966) Survival feeding of Africander cows, *Rhodesia Zambia Malawi J. Agric. Res.* **4**, 69–75

Elliott, R. C., Reed, W. D. C., Cmelik, S. H. W. and Topps, J. H. (1965) The effect of season and supplementary feeding on rumen contents of African cattle grazing sub-tropical herbage, *J. Agric. Sci. (Cambridge)* **64**, 387–96

Elliott, R. C. and Topps, J. H. (1963a) Studies of protein requirements of ruminants 1. Nitrogen balance trials on two breeds of African cattle given diets adequate in energy and low in protein, *Brit. J. Nutr.* **17**, 539–47

Elliott, R. C. and Topps, J. H. (1963b) Studies of protein requirements of ruminants 2. Protein requirement for maintenance of three breeds of cattle, *Brit. J. Nutr.* **17**, 549–56

Eng, P. K., Kerridge, P. C. and Mannetje, L. 't. (1978a) Effects of phosphorus and stocking rate on pasture and animal production from a Guinea grass-legume pasture in Johore, Malaysia. 1. Dry matter yields, botanical and chemical composition, *Trop. Grassl.* **12** 188–97

Eng, P. K., Mannetje, L. 't. and Chen, C. P. (1978b) Effects of phosphorus and stocking rate on pasture and animal production from a Guinea grass-legume pasture in Johore, Malaysia. 2. Animal liveweight change, *Trop. Grassl.* **12**, 198–207

Ernst, A. J. (1973) Some observations on behaviour of beef cattle at urea-molasses lick feeders. *J. Aust. Inst. Agric. Sci.* **39**, 266–8

Ernst, A. J., Limpus, J. F. and O'Rourke, P. K. (1975) Effect of supplements of molasses and urea on intake and digestibility of native pasture hay by steers, *Aust. J. Exper. Agric. and Anim. Husb.* **15**, 451–5

Escobar, L. G., Ayala, C. H., Gómez, S. J., Castro, H. A. and Patiño, H. O. (1971a) Effect of hormones on the fattening of Zebu-Polled Sinú steers on rotational pasture, *Proc. III Reun. Ass. Lat. Am. Prod. Anim.*

Escobar, L. G., Ramirez, R. A., Michielin de P. A. and Gómez, S. J. (1971b) *Comportamiento de novillas cebú en pastoreo continuo y rotacional en pasto trenza.* Boletin Técnico No. 15 CNIA Palmira. 81 pp.

Evans, T. R. (1969) Beef production from nitrogen fertilized Pangola grass (*Digitaria decumbens*) on the coastal lowlands of southern Queensland, *Aust. J. Exper. Agric. and Anim. Husb.* **9**, 282–6

Evans, T. R. (1970) Some factors affecting beef production from sub-tropical pastures in the coastal lowlands of southeast Queensland, *Proc. XI. Intnl Grassl. Cong.*, 803–8

Evans, T. R. (1973) Animal production from nitrogen fertilized pastures, *Trop. Grassl.* **7**, 345

Evans, T. R. and Biggs, J. (1979) Breeding performance of Hereford cows grazing tropical pastures at Beerwah, southeast Queensland, *Trop. Grassl.* **13**, 129–33

Evans, T. R. and Bryan, W. W. (1973) Effects of soils, fertilizers and stocking rates on pastures and beef production on the Wallum of south-eastern Queensland: 2. Liveweight change and beef production, *Aust. J. Exper. Agric. and Anim. Husb.* **13**, 530–6

Evans, T. R., McLean, R. W. and Winter, W. H. (1979) *Grazing systems on Townsville Stylo pastures 34–35.* CSIRO Tropical Crops and Pastures Divisional Report 1977–8: Brisbane. 150 pp.

Falvey, L. (1976) Sabi grass (*Urochloa mozambicensis*) as a component of Townsville Stylo (*Stylosanthes humilis*) pasture, *Proc. Aust. Soc. Anim. Prod.* **11**, 337–40

Falvey, L. (1977) Response of steers to dry season protein supplementation, *Aust. J. Exper. Agric. and Anim. Husb.* **17**, 724–7

Falvey, L. (1979) Crude protein percentage of the diet and liveweight change of steers on improved pastures, *J. Aust. Inst. Agric. Sci.* **30**, 267–8

Falvey, L. and Mikled, C. (1978) Dry season supplementation of cattle in northern Thailand, *Proc. Aust. Soc. Anim. Prod.* **12**, 175

FAO (1978a) *FAO production yearbook* No. 32. 361 pp.

FAO (1978b) *Swaziland livestock production and extension: Report on project results, conclusions and recommendations.* 54 pp. (litho)

Ffoulkes, D. H. M. (1977) *Studies in beef production in the lowveld of Swaziland.* M. Agric. Sci. Tropical Agricultural Development: Reading University. 93 pp.

Fiennes, R. N. T. W. (1940) Grasses as weeds of pasture land in northern Uganda, *E. Africa Agric. J.* **5**, 255–8

Flores, J. F., Stobbs, T. H. and Minson, D. J. (1979) The influence of the legume *Leucaena leucocephala* and formol casein on the production and composition of milk from grazing cows, *J. Agric. Sci. (Cambridge)* **92**, 351–7

Follett-Smith, R. R. (1930) Minerals in the pasture grasses of British Guiana, *Agric. J. British Guiana* **3**, 39

French, M. H. (1933) *Ann. Rep. Dept. Vet. Sci. Anim. Husb. Tanganyika.* 88 pp

French, M. H. (1945) Geophagia in animals, *E. Africa. Med. J.* **22**, 103–10 and 153–61

French, M. H. and Chicco, C. F. (1961) Observaciones sobre deficiencias de calcio y fósforo en los animales de los Andes y sus immediaciones, *Agronomia Tropical* **11**, 157–73

Friot, D. and Calvet, H. (1971) Étude complémentaire sur les carences minérales rencontrées dans les troupeaux du Nord Sénégal, *Rév. Élev. Méd. Vét. Pays. Trop.* **24**, 393–407

Frisch. J. E. (1973) Comparative drought resistance of *Bos indicus* and *Bos taurus* cross-bred herds in Central Queensland 2. Relative mortality rates, calf birth rates and weight changes of breeding cows, *Aust. J. Exper. Agric. and Anim. Husb.* **13**, 117–26

Frisch, J. E. and Vercoe, J. E. (1969) Liveweight gain, food intake and eating rate in Brahman, Africander and Shorthorn × Hereford cattle, *Aust. J. Agric. Res.* **30**, 1189–95

Frisch, J. E. and Vercoe, J. E. (1977) Food intake, eating rate, weight gains, metabolic rate and efficiency of feed utilization in *Bos taurus* and *Bos indicus* crossbred cattle, *Animal Production* **25**, 343–58

Frisch, J. E. and Vercoe, J. E. (1979) Changes in fasting metabolism of cattle as a consequence of selection for growth rate in energy metabolism, *Proc. Eighth Symposium on Energy metabolism* (L. E. Mount ed.). Butterworths: London and Boston. 431–4

Funes, F. (1975) Effect of fire and grazing in the maintenance of tropical grasslands, *Cuban J. Agric. Sci.* **9**, 379–95

Fuquay, J. W. (1981) Heat stress as it affects animal production, *J. Anim. Sci.* **52**, 164–73

Gammon, D. M. and Roberts, B. R. (1978a) Characteristics of the herbage on offer during continuous and rotational grazing of the Matopos sandveld of Rhodesia, *Rhodesia J. Agric. Res.* **16**, 3–22

Gammon, D. M. and Roberts, B. R. (1978b) Patterns of defoliation during continuous and rotational grazing of the Matopos sandveld of Rhodesia. 1. Selectivity of grazing, *Rhodesia J. Agric. Res.* **16**, 117–31

Gardener, C. J. (1980) Diet selection and liveweight performance of steers on *Stylosanthes hamata* – Native grass pastures, *Aust. J. Agric. Res.* **31**, 379–92

Gartner, R. J. W., Alexander, G. I. and Bewg, W. P. (1968a) Seasonal fluctuations of hepatic vitamin A reserves in beef cattle grazing unimproved pastures, *Queensland J. Agric. Anim. Sci.* **25**, 225–33

Gartner, R. J. W., McLean, R. W., Little, D. A. and Winks, L. (1980) Mineral deficiences limiting production of ruminants grazing tropical pastures in Australia. *Trop. Grassl.* **14**, 266–272

Gartner, R. J. W., Young, J. G. and Pepper, P. M. (1968b) Hepatic copper concentration of steers grazing pastures on developed wet heath land in south-eastern Queensland, *Aust. J. Exper. Agric. and Anim. Husb.* **8**, 679–82

Garza, T. R. (1973) Potencial anual del zacate Guinea fertilisado y bajo pastoreo rotacional en clima Am, *Téc. Pec Méx.* **21**, 26

Garza, T. R. and Arroyo, R. D. (1971) Efectos de la fertilisación con nitrógeno en pasto Guinea bajo pastoreo en el trópico humedo Aw, *Téc. Pec. Méx.* **19**, 60–1

Garza, T. R., Arroyo, R. D. and Pérez, S. A. (1970) Producción de carne con los zacates Pangola and Jaragua fertilisados en el trópico Aw, *Téc. Pec. Méx.* **14**, 20–4

Garza, T. R., Portugal, A. and Ballesteros, H. (1973a) Evaluación en pastoreo de associaciones de zacates y leguminosas utilisando vaquillas de razas europeas en clima tropical, *Téc. Pec. Méx.* **23**, 7–11

Garza, T. R., Portugal, A. and Ballesteros, W. H. (1975a) Efecto de la supplementación a novillonas en pastoreo de zacate Pangola fertilizado y con riego II. Época de secas, *Téc. Pec. Méx.* **27**, 5–10

Garza, T. R., Portugal, G. A., Tijerina, R. V. H. and Ballesteros, W. H. (1975b) Efectos de la suplementación a novillonas en pastoreo asi como de la fertilisación al pasto Pangola 1. Época de lluvias, *Téc. Pec. Méx.* **26**, 41–6

Garza, T. R., Treviño, M and Chapa, O. (1973b) Producción de carne en ganado bovino bajo pastoreo rotacional en seis zacates tropicales, con y sin la adición de nitrogeno en el trópico húmedo 1. Época de lluvias, *Téc. Pec. Méx.* **25**, 40–9

Giesecke, D. and van Gylswyk, N. O. (1975) A study of feeding types and certain rumen functions in six species of South African ruminants, *J. Agric. Sci.* (*Cambridge*) **85**, 75–83

Gillard, P. (1970) Pasture development in the dry tropics of north Queensland, *Proc. XI. Intnl Grassl. Cong.* 807–10

Gillard, P., Edye, L. A. and Hall, R. L. (1980) Comparison of *Stylosanthes humilis*, with *S. Hamata* and *S. subsericea* in the Queensland dry tropics; Effects on pasture composition and cattle liveweight gain, *Aust. J. Agric. Res.* **31**, 205–20

Gitter, M., Boarer, C., Howard, D. A. and Hebert, C. N. (1975) Blood mineral levels of cattle in Kenya, *Trop. Anim. Hlth Prod.* **7**, 95–104

Glover, J., Duthie, D. W. and French, M. H. (1957) The apparent digestibility of crude protein by the ruminant I. A synthesis of the results of digestibility trials with herbage and mixed feeds, *J. Agric. Sci.* (*Cambridge*) **48**, 373–8

Glover, J. and French, M. H. (1957) The apparent digestibility of crude protein by the ruminant IV. The effect of crude fibre, *J. Agric. Sci.* (*Cambridge*) **49**, 78–80

Goering, H. K. and van Soest, P. J. (1970) Forage fiber analyses (apparatus, reagents, procedures and some application), *Agricultural handbook 379*, USDA, Washington

Göhl, B. (1975) *Tropical feeds.* Food and Agriculture Organization of the United Nations: Rome: 661 pp.

Goldschmidt, W. B. (1938) Preliminary report on the nitrogen problem in pasture soils of the Natal sourveld, *South Africa J. Sci.* **35**, 244–5

Gombe, S. and Hansel, W. (1973) Plasma luteinizing hormone (LH) and progesterone levels in heifers on restricted energy intake, *J. Anim. Sci.* **37**, 725–33

González, J. E., Parra, R. R. and Combellas, J. (1972) Composición y valor nutritivo de los forrajes producidos en el trópico 3. Consumo y digestibilidad de materia seca, *Agronomia Tropical* (*Maracay*) **22**, 613–21

González, M. H. (1966) Reducción de nutrientes en los pastizales de Chihuahua durante los meses de sequia II. Fósforo, *Téc. Pec. Méx.* **7**, 19–24

González, R. J. (1976) 'Ácido fosfórico y fosfato disódico como Suplemento a vaquillas en pastoreo de zacate Buffel'. Instituto Tecnológico y de Estudios Superiores de Monterrey, unpublished thesis

Granier, P. and Gilibert, J. (1974) Contribution á l'étude de l'exploitation par rotation des pâturages de savane soudanienne techniques et résultats, *Rév. Elév. Méd. Vét. Pays. Trop.* **27**, 223–33

Grant, J. L. (1975) Winter supplementation of young stock with urea containing foods. Div. of Livestock and Pastures: Rhodesia: *Ann. Rep. 1974–5*, 56–9

Grant, J. L. (1976a) Winter supplementation of young stock with urea containing foods. Div. of Livestock and Pastures: Rhodesia: *Ann. Rep. 1975–6*, 56–9

Grant, J. L. (1976b) Response of steers to phosphorus supplementation and zeranol. Div. of Livestock and Pastures: Rhodesia: *Ann. Rep. 1975–6*, 68–70

Grant, J. L. (1979) Urea in supplements for growing young cattle on veld grazing in the dry season, *South Africa J. Anim. Sci.* **9**, 33–9

Grant, R. J., Van Soest, P. J., McDowell, R. E. and Perez, C. B. (1974) Intake, digestibility and metabolic loss of Napier grass by cattle and buffaloes when fed wilted, chopped and whole, *J. Anim. Sci.* **39**, 423–34

Grasslands (1967/68) Ministry of Agriculture: Rhodesia. *Grassland Res. Sta. Ann. Rep. 1967/68.* 61 pp.

Grasslands (1968/69) Ministry of Agriculture: Rhodesia. *Grasslands Res. Sta. Ann. Rep. 1968/69.* 42 pp.

Groenewald, J. W. and van der Merwe, P. K. (1941) The value of molasses in stock farming, *Farming in South Africa* **16**, 233–4

Groenewald, H. H. and Crossing, P. R. (1975) The place of livestock in small farm development, *World Anim. Rev.* No. 15, 2–6

Grof, B. and Harding, W. A. T. (1970) Dry matter yields and animal production of guinea grass (*Panicum maximum*) on the humid tropical coast of north Queensland, *Trop. Grassl.* **4**, 85–95

Grossman, J., Aronovich, S. and Campello, E. do C. B. (1966) Grasslands of Brazil, *Proc. IX. Intnl Grassl. Cong.*, 39–47

Grunow, J. O., Pienaar, A. J. and Breytenbach, C. (1970) Long term nitrogen application to veld in South Africa, *Proc. Grassl. Soc. South Africa* **5**, 75–90

Grünwald, O., Ortiz, D. S., Romero, S. S. and Pollak, E. S. (1947) *Una contribución al estudio de las plantas forrajeras de Venezuela.* Ministerio de Agricultura y Cria: Caracas: Bol 5

Gutierrez, C. F. (1973) Efectos de zeranol y testosterona como promotores del crecimiento y engorde en novillas en pastoreo rotacional, *Acta agronómica* (*Colombia*) **23**, 27–39

Hagemann, O. (1891) Landwirtschaft, *Jahrb* **20**, 261–91

Haggar, R. J. (1971) The production and management of *Stylosanthes gracilis* at Shika, Nigeria I. In sown pastures, *J. Agric. Sci.* (*Cambridge*) **77**, 427–36

Haggar, R. J., de Leeuw, P. N. and Agishi, E. (1971) The production and management of *Stylosanthes gracilis* at Shika, Nigeria II. In savannah grassland, *J. Agric. Sci.* (*Cambridge*) **77**, 437–44

Haines, C. E., Chapman, H. L. Jr., Kidder, R. W. and Green, R. E. L. (1965) *Effects of feeding limited amounts of concentrate to stocker steers on pasture.* Florida Agric. Exper. Sta. Bull. 693

Hale, D. H. and Holness, D. H. (1977) Low plane nutrition and sexual activity in Africander and Mashona cows. Div. of Livestock and Pastures: Rhodesia. *Ann. Rep. 1976–7*, 28–34

Hale, D. H. and Oliver, J. (1972) The effect of castration and vasectomy on male Zebus which grazed veld under two systems of management, *South Africa J. Anim. Sci.* **2**, 27–31

Hall, T. D. (1932) Intensive grazing on veld, *South Africa J. Sci.* **28**, 389–413

Hall, T. D., Meredith, D. and Altona (1950) Production from grassland in South Africa (fertiliser treatments and livestock gains on veld), *Empire J. Exper. Agric.* **18**, 8–18

Hall, T. D., Meredith, D. and Altona, R. E. (1955) The role of fertilizers in pasture management in *The grasses and pastures of South Africa* (D. Meredith, ed.). Central News Agency.

Hamburger, H. (1978) The importance of the human factor – some social aspects of cattle farming in developing territories in Southern Africa, *Summaries: IV World Conference on Animal Production*, 26

Hamburger, H., Armstong, C. W. B. and Swanepoel, J. (1980) *Adaptability and reproductive efficiency; the value of indigenous Sanga cattle in the national states of South Africa and Namibia.* S.A.R.C.C.U.S., Standing committee for animal production: Swaziland (mimeo) 17 pp.

Hancock, J. and Payne, W. J. A. (1955) The direct effect of tropical climate on the performance of European-type cattle 1. Growth, *Emp. J. Exper. Agric.* **23**, 55–74

Harker, K. W. (1963) A stocking rate trial on rough grazing in Buganda, *E. Africa. Agric. For. J.* **29**, 63–6

Harrington, G. N. (1973) Bush control. A note of caution, *E. Africa. Agric. For. J.* **39**, 95–6

Harrington, G. N. and Pratchett, D. (1974a) Stocking rate trials in Ankole, Uganda. Weight gain of Ankole steers at intermediate and heavy stocking rates under different managements, *J. Agric. Sci. (Cambridge)* **82**, 497–506

Harrington, G. N. and Pratchett, D. (1974b) Stocking rate trials in Ankole, Uganda II. Botanical analysis and oesophageal fistula sampling of pastures grazed at different stocking rates, *J. Agric. Sci. (Cambridge)* **82**, 507–16

Harrington, G. N. and Thornton, D. D. (1969) A comparison of controlled grazing and manual hoeing as a means of reducing the incidence of *Cymbopogon afronardus* Stapf. in Ankole pastures, Uganda, *E. Africa. Agric. For. J.* **35**, 154–9

Hart, E. B., Bohstédt, G., Deobald, H. J. and Wegner, M. I. (1939) The utilization of simple nitrogenous compounds such as urea and ammonium bicarbonate by growing calves, *J. Dairy Sci.* **22**, 785–98

Hart, R. H. (1972) Forage yield, stocking rate and beef gains on pasture, *Herbage abs.* **42**, 345–53

Hart, R. H., Marchant, W. H., Butler, J. L., Hellwig, R. E., McCormick, W. C., Southwell, B. L. and Burton, G. W. (1976) Steer gains under six systems of coastal Bermuda grass utilisation, *J. Range Management* **29**, 372–5

Harvey, J. M. (1952) Copper deficiencies in ruminants in Queensland, *Aust. Vet. J.* **28**, 209–16

Harvey, J. M., Beames, R. M., Hegarty, A. and O'Bryan, M. S. (1963) Influence of grazing management and copper supplementation on growth rate of Hereford cattle in south-east Queensland, *Queensland J. Agric. Sci.* **20**, 137–59

Harvey, J. M., Ryley, J. W., Beames, R. M. and O'Bryan, M. S. (1961) Studies on the cause of a low copper status in cattle in south-east Queensland, *Queensland J. Agric. Sci.* **18**, 85–104

Harwin, G. O. and Venter, H. A. W. (1970) An investigation of the creep feeding of beef calves in the mixed bushveld, *Proc. South Africa. Soc. Anim. Prod.* **9**, 161–2

Harwin, G. O., Lamb, R. D. and Bisschop, J. H. R. (1967) Some factors affecting reproductive performance in beef females, *Proc. South Africa Soc. Anim. Prod.* **6**, 171

Heitzman, R. J. (1980) Manipulation of protein metabolism with special reference to anabolic agents in *Protein deposition in animals* (P. J. Buttery and D. R. Lindsay, eds). Butterworths: London. 305 pp.

Hely, F. C. and Hallsworth, E. G. (1947) The nutritive value of some north Australian grasses, *J. Aust. Inst. Agric. Sci.* **13**, 71

Henke, L. A. (1936) Beef cattle feeding trials, *Hawaii Agric. Exper. Sta. Ann. Rep.*, 72

Henke, L. A. (1958) Value of *Leucaena glauca* as a feed for cattle, *Proc. 8th Pacific Science Congress*, 591–600

Henke, L. A. and Work, S. H. (1940) Pasture fattening, *Hawaii Agric. Exper. Sta. Ann. Rep.*, 18

Henke, L. A. Work, S. H. and Burt, A. W. (1939) Cane molasses for pasture fattening of beef cattle, *Hawaii Agric. Exper. Sta. Ann. Rep.* 26

Hennessy, D. W., Nolan, J. V., Norton, B. W., Ball, F. M. and Leng, R. A. (1978) Response of beef cattle to infused supplements of urea and of urea-molasses when offered a low quality grass hay, *Aust. J. Exper. Agric. and Anim. Husb.* **18**, 477–82

Hennessy, D. W. and Robinson, G. G. (1978) Oat forage and grain supplements for the production of yearling beef from pastures on the Northern Tablelands of New South Wales, *Aust. J. Exper. Agric. and Anim. Husb.* **18**, 183–9

Hennessy, D. W. and Sundstrom, B. (1975) Na : K ratio in saliva of beef cattle grazing Kikuyu dominant pastures, *J. Aust. Inst. Agric. Sci.* **41**, 59–60

Henrici, M. (1928) The phosphorus content of the grasses of Bechuanaland in the course of their development, *13th and 14th Rep. Dir. Vet. Serv. Anim. Ind. Union of South Africa*, 1075–1208

Henrici, M. (1930) The phosphorus content of grasses in the eastern Transvaal, *16th Rep. Dir. Vet. Serv. Anim. Ind. Union of South Africa*

Henzell, E. F. (1977) Nitrogen nutrition of tropical pastures in *Tropical forage legumes* (P. J. Skerman, ed.). FAO: Rome. 609 pp.

Heydenrych, H. J. (1962) Urea in winter rations of young beef cattle, *Farming in South Africa* **37**, 33–7

Heyns H. (1960) The growth of the Afrikaner calf in relation to the production and composition of the milk of its dam 2. The milk production of the dam and growth of the calf, *South Africa J. Agric. Sci.* **3**, 517–30

Hill, R., Thambyah, R., Wan, S. P. and Shanta, C. S. (1962) The copper status of cattle and buffalo in Malaya, *J. Agric. Sci. (Cambridge)* **59**, 409–14

Hirschfeld, E. (1940) The phosphorus problem in Western Queensland, *Queensland Agric. J.* **54**, 196–207

Hodges, E. M., Jones, D. W. and Kirk, W. G. (1953) *Winter clovers in central Florida*. Univ. Florida Agric. Exper. Sta. Bull 517. 23 pp.

Hodges, E. M., Jones, D. W. and Kirk, W. G. (1958) *Grass pastures in Central Florida*. Florida Agric. Exper. Sta. Bull 484 A. 30 pp.
Hodges, E. M., Kirk, W. G., Davis, G. K., Shirley, R. L., Peacock, F. M., Easley, J. F., Breland, H. L. and Martin, F. G. (1968) *Effect of phosphate fertilizers on Pangola grass pastures and beef production*. Florida Agric. Exper. Sta. Circular S-190
Hodges, E. M., Kirk, W. G., Jones, D. W. and Fulford, H. J. (1949) Effects of fertilization and seeding on grazing value of flatwoods pastures, *Florida Agric. Exper. Sta. Ann. Rep. 1949*, 236–7
Hodges, E. M., Kirk, W. G., Peacock, F. M., Jones, D. W., Davis, G. K. and Neller, J. R. (1964) *Forage and animal response to different phosphatic fertilizers on Pangola grass pastures*. Florida Agric. Exper. Sta. Bull 686. 28 pp.
Hodges, E. M., Peacock, F. M. Chapman, H. L. and Crane, D. (1977a) Forage systems for cow-calf herds in south central Florida, *Proc. Soil Crop Soc. Florida*, 66–8
Hodges, E. M., Peacock, F. M., Chapman, H. L. and Martin, F. G. (1977b) Grazing studies with Pensacola bahia grass and warm season annual legumes, *Proc. Soil Crop Soc. Florida* **36**, 173–5
Holmes, J. H. G. (1976) Growth of Brahman cross heifers grazing Leucaena, *Proc. Aust. Soc. Anim. Prod.* **11**, 453–6
Holmes, J. H. G., Takken, A. and Seifert, G. W. (1968) Milk production and calf growth rate in Hereford, Africander-Shorthorn and Brahman-Shorthorn cattle in Queensland, *Proc. Aust. Soc. Anim. Prod.* **7**, 163–71
Holness, D. H., Hale, D. H. and Hopley, J. D. H. (1976) Effects of plane of nutrition, body mass, temporary weaning and breed on the occurrence of oestrus in beef cows during the post partum period. Div. of Livestock and Pastures: Rhodesia *Ann. Rep. 1975–6*, 23–31
Holness, D. H. and Hopley, J. D. H. (1979) Growth, food conversion efficiency and carcass classification of entire and 'short-scrotum' bulls. Div. of Livestock and Pastures: Rhodesia. *Ann. Rep. 1978–9*, 27–31
Holness, D. H., Hopley, J. D. H. and Hale, D. H. (1978) The effects of plane of nutrition, liveweight, temporary weaning, and breed on the occurrence of oestrus during the *post partum* period, *Animal Production* **26**, 47–54
Holroyd, R. G., Allan, P. J. and O'Rourke, P. K. (1977) Effect of pasture type and supplementary feeding on the reproductive performance of cattle in the dry tropics of north Queensland, *Aust. J. Exper. Agric. and Anim. Husb.* **17**, 197–206
Hopley, J. D. H. and Holness, D. H. (1977) Age of castration and growth rates of male beef calves, Ministry of Agriculture, Rhodesia. *Ann. Rep. Div. of Livestock and Pastures, 41–2*
Hoppe, P. P., Qvortrup, S. A. and Woodford, M. H. (1977) Rumen fermentation and food selection in East African sheep, goats, Thomson's gazelle, Grant's gazelle and Impala, *J. Agric. Sci.* (*Cambridge*) **89**, 129–35
Horrell, C. R. (1972) *Tropical pasture production and its development in Central America*. FAO Special Conference on Pasture and Forage Crops in Tropical America (Cali, Colombia), Document No. 1 (mimeo). 15 pp.
Hosking, H. R. and Stephens, A. L. (1941) Pastures and grasses on the Serere Exper. Sta. Uganda, *E. Africa Agric. J.* **6**, 213–19
Le Houerou, H. N. (1977) The scapegoat, *Ceres* **10**, 14–18

Howard, D. A. (1963) Notes on animal diseases. XXIII Mineral deficiency diseases, *E. African Agric. For. J.* **28**, 191–4

Howard, D. A. (1970) The effects of copper and cobalt treatment on the weight grains and blood constituents of cattle in Kenya, *Vet. Records* **87**, 771–4

Howard, D. A., Burdin, M. L. and Lampkin, G. H. (1962) Variation in mineral and crude protein contents of pastures at Muguga in the Kenya Highlands, *J. Agric. Sci.* (*Cambridge*) **59**, 251–6

Howell, L. N. (1976) The development of multi-camp systems on a farm in the southern Orange Free State, *Proc. Grassl. Soc. South Africa* **11**, 53–7

Hudson, J. R. (1944) Notes on animal diseases XXIII Deficiency diseases, *East African Agric. J.* **10**, 51–5

Hughes, A. J. B. (1972) *Land tenure, land rights and land communities on Swazi Nation land in Swaziland: a discussion of some inter-relationships between the traditional tenurial system and problems of agrarian development.* University of Natal, Institute for Social Research: Durban. 351 pp.

Humphreys, L. R. (1966) Pasture defoliation practice: a review, *J. Aust. Inst. Agric. Sci.* **32**, 93–105

Hungate, R. E. (1966) *The rumen and its microbes.* Academic Press: New York and London. 533 pp.

Hunter, R. A., Siebert, B. D. and Webb, C. D. (1979) The positive response of cattle to sulphur and sodium supplementation while grazing *Stylosanthes guianensis* pastures in north Queensland, *Aust. J. Exper. Agric. and Anim. Husb.* **19**, 517–21

Hurrell, A. D. and Dugdale, R. S. (1958) Aphosphorosis in ranch cattle, *Rhodesia Agric. J.* **55**, 474–9

Husband, A. D. (1932) Report for the branch of chemistry for the year ending 1932, *Rhodesia Agric. J.* **30**, 453–68

Husband, J. H. (1934) Rep. Dir. Agric: Southern Rhodesia

Husband, A. D. (1935) Report for the branch of chemistry for the year ending 1935, *Rhodesia Agric. J.* **33**, 262–72

Husband, J. H. and McChlery, R. (1936) *Rep. Dir. Agric. Southern Rhodesia*

Hutcheon, D. (1892) Lamsiekte, Adynamia or paralysis in cattle, *Agric. J. Cape Colony* **5**, 173–7

Hutcheon, D. (1903) *Reports of the Colonial Vet. Surgeon Cape of Good Hope*

Ibach, D. B. and Adams, W. E. (1965) An economic analysis of fertility experiments with Coastal and Common Bermuda grasses (*Cynodon dactylon* (L) Peps.), *Agronomy J.* **57**, 15–20

Ibrahim, K. (1975) *Glossary of terms used in pasture and range survey, research, ecology, management.* FAO: Rome. 153 pp.

ILCA (1980) *ILCA The First Years.* International Livestock Centre for Africa: Addis Ababa, Ethiopia. 127 pp.

Inns, F. M. (1980) Animal power in agricultural production systems, *World Anim. Rev.* No. 34, 2–10

I'ons, J. H. (1968) The development of tropical pastures for the Swaziland Middleveld, *Proc. Grassl. Soc. South Africa* **3**, 67–73

Ittner, N. R., Bond, T. E. and Kelly, C. F. (1954) Increasing summer gains of livestock, *J. Anim. Sci.* **13**, 866–77

Ittner, N. R., Bond, T. E. and Kelly, C. F. (1958) Methods of increasing beef production in hot climates, *Calif. Agric. Exper. Sta. Bull.* 761, 18–34

Ivory, D. A., Stobbs, T. H., McLeod, M. N. and Whiteman, P. C. (1974) Effect of day and night temperatures on estimated dry matter digestibility of *Cenchrus ciliaris* and *Pennisetum clandestinum*, *J. Aust. Inst. Agric. Sci.* **40**, 156–8

Jahnke, H. E. and Ruthenberg, H. (1976) Organizational aspects of livestock development in the dry areas of Africa in *Policy and practice in rural development* (G. Hunter, A. H. Bunting and A. Bottrall, eds). Groom Helm in association with the Overseas Development Institute: London. 520 pp.

Jamaica: Department of Agriculture (1950) *Investigations Bulletin* No. 47 (New Series). 127 pp.

Jamaica: Department of Agriculture (1953) *Investigations Bulletin* No. 53 (New Series). 163 pp.

Jamaica (1957) Ministry of Agriculture and Lands *Investigations, Bulletin* No. 57 (New Series), 115–18

Jardim, W. R., Moraes, C. L. and Peixoto, A. M. (1952) Contribuição para o estudo de composição e valor nutritivo de plantas forrageiras. Escola Superior de Agricultura, 'Luiz de Quieros' Piracicaba, Brazil *Anais* **9**, 31–8

Jarvis, P. J. (1978) *Report on Tuli grazing trials, July 1976–June 1978.* Rhodesia Ministry of Agriculture: Department of Conservation and Extension (mimeo)

Jasiorowski, H. A. (1973) Twenty years with no progress? *World Anim. Review* No. 5, 1–5

Johnson, J. I., Hatton, G. I. and Goy, R. W. (1969) The physiological analysis of animal behaviour in *The behaviour of domestic animals* (E. S. E. Hafez, ed.). Bailliere, Tindall and Cassell: London. 647 pp.

Jones, D. I. H. (1963) The mineral content of six grasses from a *Hyparrhenia* dominant grassland in Northern Rhodesia, *Rhodesia J. Agric. Res.* **1**, 35–9

Jones, D. I. H. (1964) Mineral content of some cultivated grasses grown in Northern Rhodesia, *Rhodesia J. Agric. Res.* **2**, 57–60

Jones, D. W., Hodges, E. M. and Kirk, W. G. (1960) *Year-round grazing on a combination of native and improved pasture.* Florida Agric. Exper. Sta. Bull. 554. 13 pp.

Jones, R. J. (1966) The effect of urea-salt-molasses supplements on the winter performance of beef cattle on improved pastures at Samford, south-eastern Queensland, *Aust. J. Exper. Agric and Anim. Husb.* **6**, 145–9

Jones, R. J. (1974) The relation of animal and pasture production to stocking rate on legume based and nitrogen fertilised subtropical pastures, *Proc. Aust. Soc. Anim. Prod.* **10**, 340–3

Jones, R. J. and Jones, R. M. (1978) The ecology of Siratro-based pastures in *Plant relations in pastures* (J. R. Wilson, ed.), pp. 353–67. CSIRO: Melbourne, Australia

Jones, R. J. and Sandland, R. L. (1974) The relation between animal gain and stocking rate. Derivation of the relation from the results of grazing trials, *J. Agric. Sci.* (*Cambridge*) **83**, 335–41

Jones, R. M. (1979) Effect of stocking rate and grazing frequency on a Siratro (*Macroptilium atropurpureum*) *Setaria anceps* cv Nandi pasture, *Aust. J. Exper. Agric. and Anim. Husb.* **19**, 318–24

Joubert, D. M. (1954a) The influence of winter nutritional depressions on the growth, reproduction and production of cattle, *J. Agric. Sci.* (*Cambridge*) **44**, 5–66

Joubert, D. M. (1954b) The influence of high and low nutritional planes on the oestrus cycle and conception rates of heifers, *J. Agric. Sci.* (*Cambridge*) **45**, 164–72

Joubert, D. M. (1963) Puberty in farm animals, *Anim. Breeding Abstracts* **31**, 295–306

Joubert, D. M. (1972) Plane of nutrition and reproductive performance in the cow, *J. South Africa Biol. Soc.* **13**, 39–43

Juko, C. D. and Bredon, R. M. (1961) The chemical composition of leaves and whole plants as an indicator of the range of available nutrients for selective grazing by cattle, *Trop. Agric.* (*Trinidad*) **38**, 179–87

Kaiser, A. G. (1975) Response by calves grazing Kikuyu grass pastures to grain and mineral supplements, *Trop. Grassl.* **9**, 191–8

Kayongo-Male, H., Thomas, J. W., Ullrey, D. E., Deans, R. J. and Arroyo-Aguilu, J. A. (1976) Chemical composition and digestibility of tropical grasses, *J. Agric. Univ. Puerto Rico* **60**, 186–200

Kemp, A., Deijs, W. B., Hemkes, O. J. and van Es, A. J. H. (1961) Hypomagnesaemia in milking cows; intake and utilisation of magnesium from herbage by lactating cows, *Netherlands J. Agric. Sci.* **9**, 134–49

Kennan, T. C. D. (1962) *Veld management in the semi-intensive, semi-extensive and extensive farming areas of Southern Rhodesia*. Proc. 1st Intdep. Dep. Conf. of Pasture workers, Bulawayo, Jan–Feb 1962 (cyclostyled)

Kennan, T. C. D. (1969) *A review of research into the cattle-grass relationship in Rhodesia*. Proc. Veld Management Conf. Bulawayo 5–26. 131 pp.

Khalifa, A. J. and Simpson, M. C. (1972) Perverse supply in nomadic societies, *Oxford Agr. Stud.* **1**, 47–56

Kidder, R. W. and Beardsley, D. W. (1952) *Protein and carbohydrate supplements for fattening steers on Everglades pasture*. Florida Agric. Exper. Sta. Bull. 193

Kidner, E. M. (1970) *A compendium of animal husbandry notes*. Swaziland Agricultural College and University Centre: Swaziland (mimeo). 232 pp.

Kirk, W. G., Hodges, E. M., Peacock, F. M., Yarlett, L. L. and Martin, F. G. (1974) Production of cow-calf herds: effect of burning native range and supplemental feeding, *J. Range Management* **20**, 153–7

Klock, M. A.. Schank, S. C. and Moore, J. E. (1975) Laboratory evaluation of quality in sub-tropical grasses. III Genetic variation among *Digitaria* species in *in vitro* digestion and its relationship to plant morphology, *Agron. J.* **67**, 672–5

Koger, M., Blue, W. G., Killinger, G. B., Green, R. E. L., Harris,

H. C., Myers, J. M., Warnick, A. C. and Gammon, N (1961) *Beef production, soil and forage analyses and economic returns from eight pasture programs in north central Florida.* Florida Agric. Exper. Sta. Bull. (Technical) 631. 76 pp.

Koger, M., Blue, W. G., Killinger, G. B., Green, R. E. L., Myers, J. M., Gammon. N. Jr., Warnick, A. C. and Crockett, J. R. (1970) *Production response and economic returns from five pastures programs in north central Florida.* Florida Agric. Exper. Sta. Bull. 740. 45 pp.

Koger, M., Green R. E. L., Killinger, G. B., Blue, W. G. and Myers, J. M. (1977) *Pasture programs and beef cattle breeding systems for beef production in north central Florida.* Florida Agric. Exper. Sta. Bull. 789. 47 pp.

Kreft, H. W. (1963) Feeding urea to cattle, *Proc. South Africa Soc. Anim. Prod.* **2**, 43–4

Kreft, H. W. (1966) Urea and biuret as nitrogen supplements for cattle, *Proc. South Africa Soc. Anim. Prod.* **5**, 66–70

Kyomo, M. L., Hutchison, H. G. and Salehe, I. (1972) Effect of yarding cattle at night without supplementary feeding on the growth of Zebu heifers, *E. Africa Agric. For. J.* **37**, 279–85

Lamond, D. R. (1969) Sources of variation in reproductive performance in selected herds of beef cattle in north eastern Australia, *Aust. Vet. J.* **45**, 50–8

Lampkin, G. H. (1961) Studies on the production of beef from Zebu cattle in East Africa III. The value of feeding a phosphatic supplement, *J. Agric. Sci.* **57**, 39–47

Langlands, J. P. and Bowles, J. E. (1976) Nitrogen supplementation of ruminants grazing native pastures in New England, New South Wales, *Aust. J. Exper. Agric. and Anim. Husb.* **16**, 630–5

Laredo, M. A. and Minson, D. J. (1973) The voluntary intake, digestibility and retention time by sheep of leaf and stem fractions of five grasses, *Aust. J. Agric. Res.* **24**, 875–88

Leach, R. M. and Lilburn, M. S. (1978) Manganese metabolism and its function in *World review of nutrition and dietetics* (G. H. Bourne, ed.) **32**, 123–32

Leal, T. C. and da Silva, B. O. (1977) Effects of vitamin A, D_3 and E administered parentarally on the reproductive performance of Charolais cattle at Tupanciretá, *An. Tech. Inst. Pesq. Zoot* **4**, 431–43

Leche, T. F. (1977a) Effects of sodium supplement on lactating cows and their calves on tropical native pastures, *Papua New Guinea Agric. J.* **28**, 11–15

Leche, T. F. (1977b) Effects of sodium and phosphorus supplements for steers on native pastures in the tropical highlands, *Papua New Guinea Agric. J.* **28**, 19–25

Ledger, H. P., Rogerson, A. and Freeman, G. H. (1970) Further studies on the voluntary food intake of *Bos indicus, Bos taurus* and cross bred cattle, *Anim. Prod.* **12**, 425–31

Ledger, H. P. and Sayers, A. R. (1977) The utilization of dietary energy by steers during periods of restricted food intake and subsequent realimentation 1. The effect of time on the maintenance requirements of steers held at constant liveweights, *J. Agric. Sci.* (*Cambridge*) **88**, 11–26

Lee, H. J. and Moule, G. R. (1947) Copper deficiency affecting sheep in Queensland, *Aust. Vet. J.* **23**, 303–9

Leeuw, P. N. de (1971) Inst. Agric. Res. Ahmadu, Bello University, Zaria, Nigeria, *Shika Agric. Res. Sta. Biennial Rep. 1969–1971*

Legg, S. P. and Sears, L. (1960) Zinc sulphate treatment of parakeratosis in cattle, *Nature U.K.* **186**, 1061–2

Leng, R. A., Kempton, T. J. and Nolan, J. V. (1977) Non-protein nitrogen and by-pass proteins in ruminant diets, *Aust. Meat Res. Comm. Rev.* No. 33, 1–22

Lesch, S. F., von la Chevallerie, M. K. S. L. and van Schalkwyk, A. P. (1969) *Proc. South Africa Soc. Anim. Prod.* **8**, 129–30

Lesch, S. F., Pieterse, P. J. S. and Oosthuizen, F. J. (1963) Utilisation of the energy in mature veld hay by steers: effect of urea supplementation, *Proc. South Africa. Soc. Anim. Prod.* **2**, 45–7

Lesch, S. F., Preller, J. H. and van Schalkwyk, A. P. (1968) Winter feeding and beef production, *Farming in South Africa* **44**, 60–1

Lesch, S. F. and van Schalkwyk, A. P. (1967) The effect of level of urea on the utilization of energy in mature veld hay by steers, *Proc. South Africa Soc. Anim. Prod.* **6**, 96–8

Levine, H., Remington, R. E. and Kolnitz, H. (1933) Studies on the relation of diet to goitre 2. The iodine requirement of the rat, *J. Nutrition* **6**, 347–54

Lignieres, J. (1912) *Rev. Zoot.* **4**

Lima, F. P., Martinelli, D., Sartini, H. J., Fonseca, M. and Biondi, P. (1969a) Pastejo competitivo entre 4 gramineas tropicais am latossolo roxo na engorda de bovinos da raça Nelore, *Bol. Ind. Anim.* **26**, 189–97

Lima, F. P., Martinelli, D. and Werner, J. C. (1968) Produção de carne de bovinos em pastagens de gramineas em região de terras roxas (Latosol Roxo), *Bol. Ind. Anim.* **25**, 129–37

Lima, F. P., Sartini, H. J., Martinelli, D., Biondi, P. and Fonseca, M. (1969b) Utilização de 4 gramineas tropicais na produção de carne em um solo tipico latosolo roxo, *Bol. Ind. Anim.* **26**, 199–214

Little, D. A. (1967) Effect of supplements to dry season pasture on the performance of mature lactating droughtmaster cows, *CSIRO Anim. Physiology Ann. Rep. 1966–67*, 73

Little, D. A. (1970) Factors of importance in the phosphorus nutrition of beef cattle in Northern Australia, *Aust. Vet. J.* **46**, 241–8

Little, D. A. (1972) Bone biopsy in cattle and sheep for studies of phosphorus status, *Aust. Vet. J.* **48**, 668–70

Little, D. A. (1975) Effects of dry season supplements of protein and phosphorus to pregnant cows on the incidence of first post-partum oestrus, *Aust. J. Exper. Agric. and Anim. Husb.* **15**, 25–31

Little, D. A. and Minson, D. J. (1977) Variation in the phosphorus content of bone samples obtained from the last three ribs of cattle, *Res. Vet. Sci.* **23**, 393

Little, D. A. and Shaw, N. H. (1979) Superphosphate and stocking rate effects on a native pasture oversown with *Stylosanthes humilis* in central coastal Queensland. 5. Bone phosphorus levels in grazing cattle, *Aust. J. Exper. Agric. and Anim. Husb.* **19**, 645–51

Llewelyn, D., Kempton, T. J. and Nolan, J. W. (1978) Liveweight response in heifers fed a meatmeal-molasses supplement, *Proc. Aust. Soc. Anim. Prod.* **12**, 174

Lombard, P. E. and Greeff, H. M. (1964) Supplement veld pasturage in winter, *Farming in South Africa* **40**, 54–5

Long, M. I. E., Marshall, B., Ndyanabo, W. K. and Thornton, D. D. (1972) Mineral status of dairy farms in eastern Uganda, *Trop. Agric. (Trinidad)* **49**, 227–34

Long, M. I. E., Ndyanabo, W. K., Marshall, B. and Thornton, D. D. (1969) Nutritive value of grasses in Ankole and the Queen Elizabeth National Park, Uganda. 4. Mineral content, *Trop. Agric. (Trinidad)* **46**, 201–9

Long, M. I. E., Thornton, D. D., Ndyanabo, W. K., Marshall, B. and Ssekaalo, H. (1970) The mineral status of dairy farms in the parts of Buganda and Busoga bordering Lake Victoria II. Nitrogen and mineral content, *Trop. Agric. (Trinidad)* **47**, 37–50

Louwrens, J. H. (1964) *Success achieved with cheap winter supplementary feeding*. Rhodesia: Ministry of Agriculture Bull. 2234

Low, A. R. C. (1978) Cattle supply response in Sudan and Swaziland. Motivational inferences and overgrazing implications, *Oxford Agr. Stud.* **7**, 62–74

Lowe, K. F. (1974) Sod-seeding. Siratro into native pastures, *Trop. Grassl.* **8**, 125–8

Lowe, K. F., Filet, G. F., Burns, M. A. and Bowdler, T. M. (1977) Effect of sod-seeded Siratro on beef production and botanical composition of native pasture in south-eastern Queensland, *Trop. Grassl.* **11**, 223–9

Lüdemann, C. J. F. (1971) *The improvement of body mass at weaning by means of feeding and management*. Transvaal (Republic of South Africa) Lecture Series No. 13, p. 20

Lüdemann, C. J. F. (1973) *Optimal land use: Manipulation of the nutritional level for increased beef production in the bushveld, and some management aspects*. Farmers Day Mara Agricultural Research Station Lecture Series No. 15, Transvaal Region 16–20

Macfarlane, J. S. (1966) Castration in farm animals, *Vet. Rec.* **78**, 436

Mackenzie, H. I. and Altona, R. E. (1964a) Biuret as a supplement to roughage rations for sheep and cattle, *J. South Africa Vet. Med. Ass.* **35**, 301–6

Mackenzie, H. I. and Altona, R. E. (1964b) Observations on a biuret-containing lick as a nitrogen supplement to low quality roughage, *J. South Africa Vet. Med. Ass.* **35**, 309–13

Magadan, P. B., Javier, E. Q. and Madamba, J. C. (1974) Beef production on native (*Imperata cylindrica* (1) Beauv.) and Para grass (*Brachiaria mutica* (Forsk.) Stapf) pastures in the Philippines, *Proc. XII Intnl. Grassl. Cong.*, 293–8

Mahadevan, P. (1966) *Breeding for milk production in tropical cattle*. Commonwealth Agricultural Bureaux: Farnham Royal, Bucks, England. 154 pp.

Malan, A. I., Green, H. H. and du Toit, P. J. (1928) Studies in mineral metabolism v. composition of bovine blood on phosphorus deficient pastures, *J. Agric. Sci.* (*Cambridge*) **18**, 376–83

Malan, A. I. and du Toit, P. J. (1932) Studies in mineral metabolism XXI. The comparison of phosphatic supplements for the prevention of aphosphorosis, *18th Rep. Dir. Vet. Serv. Anim. Ind. Union of South Africa*, 677–702

Mannetje, L. 't (1972) The effects of some management practices on

pasture production, *Trop. Grassl.* **6**, 260–3
Mannetje, L. 't (1974) Relations between pasture attributes and liveweight gain on a sub-tropical pasture, *Proc. XII Intnl Grassl. Cong.* 386–90
Mannetje, L. 't, Ajit Singh Sidhu and Murugaiah, M. (1976) Cobalt deficiency in cattle in Johore. Liveweight changes and response to treatments, *MARDI Res. Bull.* **4**, 90–8
Mannetje, L. 't. and Coates, D. B. (1976). Effect of pasture improvement on reproduction and pre-weaning growth of Hereford cattle in sub-coastal Queensland. *Proc. Aust. Soc. Anim. Prod.* **11**, 257–260
Mannetje, L. 't and Ebersohn, J. P. (1980) Relations between sward characteristics and animal production, *Trop. Grassl.* **14**, 273–9
Mannetje, L. 't and Nicholls, D. F. (1974) Beef production from pastures on granitic soils, *Ann. Rep. CSIRO Trop. crops and pastures Division 1973–1974*
Mannetje, L. 't and Nicholls, D. F. (1975) Beef production from pastures on granitic soils, *Ann. Rep. CSIRO Trop. crops and pastures Division 1974–1975*
Maree, C. and Harwin, G. O. (1971) Nutritional effects on ovarian activity, puberty and the early mating of beef heifers under extensive grazing conditions, *Agroanimalia* **3**, 103–6
Martin, P. C. and Alonso, F. (1978) Effect of restricted molasses-urea supplementation on bulls grazing Pangola grass (*Digitaria decumbens* Stent), *Cuban J. Agric. Sci.* **12**, 51–8
Martin, L. C., Ammerman, C. B., Burns, W. C. and Koger, M. (1976) Biuret as a source of supplemental nitrogen for wintering beef cows, *J. Anim. Sci.* **42**, 21–26
Martinez, J. R. (1974) Comparacion de harina de hueso y fosfato dicalcico como suplemento de fosforo a vaquillas en pastoreo, Instituto Tecnologico y de Estudios Superiores de Monterrey, unpublished thesis
Marwick, B. A. (1940) The Swazi: An ethnographic account of the natives of the Swaziland Protectorate. Frank Cass & Co. Ltd: London. 320 pp.
Mawson, W. F., Beattie, A. W. and Sutherland, D. N. (1962) Effect of synthetic hormone implants on grazing steers in Queensland, *Queensland J. Agric. Sci.* **19**, 467–84
Mba, A. U., Omole, J. O. and Oyenuga, V. A. (1976) Studies on the ruminal concentration of total volatile fatty acids (VFA's) and their molar proportions in African cattle, sheep and goats, *Livestock Prod. Sci.* **3**, 43–56
McCleod, M. N. and Minson, D. J. (1974) Predicting organic matter digestibility from *in vivo* and *in vitro* determinations of dry matter digestibility, *J. Bri. Grassl. Soc.* **29**, 17–21
McClymont, G. L. (1956) Response of stock to supplementary feeding on pasture, *Proc. Aust. Soc. Anim. Prod.* **1**, 63
McCormick, W. C., Givens, R. L. and Southwell, B. L. (1963) Effects of shade on rate of growth and fattening of beef steers, *Georgia Agric. Exper. Sta. Tech. Bull.* (New Series) **27**. 13 pp.
McDaniel, A. H. (1956) Beef cattle research, *Louisiana State Univ. Bull.* **502**, 10–16
McDonald, I. E. (1968) The nutrition of grazing ruminants, *Nutr. Abs. Revs.* **38**, 281–400
McDowell, L. R. (1976) Mineral deficiencies and toxicities and their effect on beef production in developing countries in *Beef cattle production in*

developing countries (A. J. Smith, ed.). University of Edinburgh, Centre for Tropical Veterinary Medicine: Edinburgh. 487 pp.

McDowell, L. R., Bauer, B., Galdo, E., Koger, M., Loosli, J. K. and Conrad, J. H. (1982) Mineral supplementation of beef cattle in the Bolivian tropics, *J. Anim. Sci.* **55**, 964–70

McDowell, L. R., Conrad, J. H., Thomas, J. E. and Harris, L. E. (1974) *Latin American table of feed composition.* University of Florida: Gainesville, Florida. 509 pp.

McDowell, L. R. Houser, R. H. and Fick, K. R. (1979) Iron, manganese and zinc in ruminant nutrition, *Latin American symposium on mineral nutrition research with grazing ruminants* (J. H. Conrad and L. R. McDowell eds). University of Florida: Gainesville, Florida. 199 pp.

McDowell, L. R., Kiatoko, M., Lang, C. E., Fonseca, H. A., Vargas, E., Loosli, J. K. and Conrad, J. H. (1978) Investigación mineral en América Latina – Costa Rica, *Fourth World Conference of Anim. Prod.: Summaries*, 79

McDowell, R. E. (1977) *Ruminant products: more than meat and milk.* Winrock report: Winrock Intnl Livestock Research and Training Centre. 28 pp.

McLeod, M. N. (1973) The digestibility and nitrogen, phosphorus and ash contents of the leaves of some Australian trees and shrubs, *Aust. J. Exper. Agric. and Anim. Husb.* **13**, 245–50

Mears, P. T. and Humphreys, L. R. (1974a) Nitrogen response and stocking rate of *Pennisetum clandestinum* pastures: I. Pasture nitrogen requirement and concentration, distribution of dry matter and botanical composition, *J. Agric. Sci. (Cambridge)* **83**,451–67

Mears, P. T. and Humphreys, L. R. (1974b) Nitrogen response and stocking rate of *Pennisetum clandestinum* pastures: II. Cattle growth, *J. Agric. Sci. (Cambridge)* **83**, 469–78

Medina, E. and Sarmiento, G. (1979) Tropical grazing land ecosystems of Venezuela 1. Ecophysiological studies in the *Trachypogon* savannah (Central Llanos) in *Tropical grazing land ecosystems* UNESCO/UNEP/ FAO: Paris: 655 pp.

Meischke, H. R., Ramsey, W. R. and Shaw, F. D. (1974) The effect of horns on bruising in cattle, *Aust. Vet. J.* **50**, 432

Mellor, W., Hibberd, M. J. and Grof, B. (1973) Beef cattle liveweight gains from mixed pastures of some Guinea grasses and legumes on the wet tropical coast of Queensland, *Queensland J. Agric. Anim. Sci.* **30**, 259–66

Meyn, K. (1970) *Beef production in East Africa.* Institut Für Wirtschafts forschung: Munich. 227 pp.

Mikled, C. (1977) Dry season supplementation of cattle grazing native grasslands in the north Thailand highlands, *Thailand J. Agric. Sci.* **10**, 135–42

Milford, R. and Minson, D. J. (1965) Intake of tropical pasture species, *Proc. IX Intnl Grassl. Cong.*, 815–22

Miller, C. P. and van der List, J. T. (1977) Yield, nitrogen uptake and liveweight gains from irrigated grass-legume pasture on a Queensland tropical highland, *Aust. J. Exper. Agric. and Anim. Husb.* **17**, 949–60

Mills, C. F. (1969) Mineral nutrition and metabolism in *International Encyclopaedia of Food and Nutrition, Vol. 17 Nutrition of animals of agricultural importance. Part I The science of nutrition of farm livestock* (Sir David Cuthbertson, ed.). Pergamon Press: Oxford. 592 pp.

Minson, D. J. (1971) The digestibility and voluntary intake of six varieties of Panicum, *Aust. J. Exper. Agric. and Anim. Husb.* **11**, 18–25

Minson, D. J. (1977) Chemical composition and nutritive value of tropical legumes in *Tropical forage legumes* (D. J. Skerman, ed.). FAO: Rome. 609 pp.

Minson, D. J. and McCleod, M. N. (1970) The digestibility of temperate and tropical grasses, *Proc. XI. Intnl Grassl. Cong.*, 719

Moir, K. W. (1966) Diagnosis of phosphorus deficiency in grazing beef cattle, *Queensland J. Agric. Anim. Sci.* **23**, 97–100

Moir, K. W., Ryley, J W., Pepper; P. M. and Middleton, C. H. (1969) Effect of nitrogen fertiliser on growth rate of Hereford heifers grazing *Paspalum dilatatum*-dominant pasture, *Queensland J. Agric. Anim. Sci.* **26**, 341–52

Moir, K. W., Wilson, J. R. and Blight, G. W. (1977) The *in vitro* digested cell wall and fermentation characteristics of grasses, as affected by temperature and humidity during their growth, *J. Agric. Sci.* (*Cambridge*) **88**, 217–22

Molina, S. I., Garza, T. R. and Torres, H. M. (1977) Producción de carne en los zacates Guinea y Buffel con dos niveles de fertilisación durante un año de pastoreo en Tizimín, Yuc. Clima AW, *Téc. Pec. Méx.* **31**, 17–20

Moore, A. W. (1962) The influence of a legume on soil fertility under a grazed tropical pasture, *Emp. J. Exper. Agric.* **30**, 239–48

Moore, T. (1957) *Vitamin A*. Elsevier: New York.

Moran, J. B. (1973a) Effect of meat-and-bone meal supplements on growth of weaner cattle on dry season improved pasture at Adelaide River, N.T., *Aust. J. Exper. Agric. and Anim. Husb.* **13**, 134–41

Moran, J. B. (1973b) Liveweight changes of Braham (*sic.*) Cross, Shorthorn, Banteng and Buffalo steers grazing improved pastures at Darwin, N.T., *J. Aust. Inst. Agric. Sci.* **39**, 68–70

Morciego, S., Muñoz, F. and Preston, T. R. (1970) Commercial fattening of bulls with molasses/urea and restricted grazing, *Rev. Cubana Ciencia Agricola* (Eng. edn) **4**, 97–100

Morgado, F. (1961) Influencia do po de ossos no desenvolvimento de novilhas mantidas na pastagem natural dos terrenos de 'managa' de Chobela, *Anais dos Serv. de Vet. de Mocambique* No. 7, 347–55

Morris, E. J. (1980) The cell wals of *Eragrostis tef*: variations in chemical composition and digestibility, *J. Agric. Sci.* (*Cambridge*) **95**, 305–11

Morris, E. J. and van Gylswyk, N. O. (1980) Comparison of the action of rumen bacteria on cell walls from *Eragrostis tef*, *J. Agric. Sci. (Cambridge)* **95**, 313–23

Morris, J. G. (1958) Drought feeding studies with cattle and sheep, *Queensland J. Agric. Sci.* **15**, 161–80

Morris, J. G. and Gulbranson, B. (1970) Effect of nitrogen and energy supplements on the growth of cattle grazing oats or Rhodes grass, *Aust. J. Exper. Agric. and Anim. Husb.* **10**, 379–83

Mortiss, P. D. (1974) Endeavour stylo . . . strengths and weaknesses, *Queensland Agric. J.* **100**, 285–8

Moses, D. (1934) Experiments with improved pastures in the Coastal Belt Part 2. Production under intensive management, *South Africa J. Sci.* **31**, 299–314

Mott, G. O. (1960) Grazing pressure and the measurement of pasture production, *Proc. IIX Intnl. Grassl. Cong.*, 606–11

Mott, G. O., Quinn, L. R. and Bisschoff, W. V. A. (1970) The retention of nitrogen in a soil-plant-animal system in Guinea grass (*Panicum maximum*) pastures in Brazil, *Proc. XI Intnl Grassl. Cong.*, 414–16

Mott, G. O., Quinn, L. R., Bisschoff, W. V. A. and da Rocha, G. L. (1967) *Molasses as an energy supplement for Zebu steers grazing nitrogen-fertilised Colonial Guinea grass pasture* IBEC Res. Inst. Bull. 54 pp. (Also: *Pesq. Agropec. Bras.* **2**, 441–59)

Mueller-Haye, B., Plasse, D., Gil, R. A., Koger, M., Butterworth, M. H. and Linares, T. (1968) Influencias genéticas sobre el peso al nacer su relación con ganancia diaria en becerros Criollos Brahman y sus cruces reciprocos, *Memória Associación Latino-Americano de producción animal* **3**, 89–102

Mukhtar, A. M. S. (1967) Effect of castration and feed supplementation on liveweight gains of W. Sudan range cattle, *Ind. Vet. J.* **44**, 496–500

Mulinos, M. G. and Pomerantz, L. (1940) Pseudohypophysectomy, a condition resembling hypophysectomy produced by malnutrition, *J. Nutr.* **19**, 493–504

Murphy, G. M. and Connell, J. A. (1970) A simple method of collecting saliva to determine the sodium status of cattle and sheep, *Aust. Vet. J.* **46**, 595–8

Murphy, G. M. and Plasto, A. W. (1972) Sodium deficiency in a beef cattle herd, *Aust. Vet. J.* **48**, 129

Murphy, G. M. and Plasto, A. W. (1973) Liveweight response following sodium chloride supplementation of beef cows and their calves grazing native pasture, *Aust. J. Exper. Agric. and Anim. Husb.* **13**, 369–74

Murray, C. A. (1938) Research work at Government Research Station, Rhodes Matopos, *Rhodesia Agric. J.* **35**, 513–23

Murray, C. A. (1940) Annual Report of the Senior Animal Husbandry Officer, *Rhodesia Agric. J.* **37**, 296–309

Murray, C. A. and Romyn, A. E. (1932) Matopos Agric. Exper. Sta. Rep.

Murray, C. A. and Romyn, A. E. (1937a) The feeding of different winter supplements to young growing steers and the effect of these supplements on the subsequent development and costs of production of the steers, *Rhodesia Agric. J.* **34**, 357–67

Murray, C. A. and Romyn, A. E. (1937b) The feeding of phosphorus supplements to growing cattle, *Rhodesia Agric. J.* **34**, 384–90

Murray, C. A. and Romyn, A. E. (1937c) Preliminary report on the feeding of winter supplements to young growing steers and the effect of the supplementary feeding on the subsequent development of these animals, *Rhodesia Agric. J.* **34**, 526–36

Murray, C. A. and Romyn, A. E. (1939) Urea as a possible substitute for peanut cake for wintering young stock, *Rhodesia Agric. J.* **36**, 554–9

Murray, C. A., Romyn, A. E., Haylett, D. G. and Ericksen, F. (1936) The supplementary feeding of mineral and protein to growing cattle in Southern Rhodesia and its relation to the production of beef steers, *Rhodesia Agric. J.* **33**, 422–40

Mutch, C. B. (1966) Handy molasses lick feeder, *Queensland Agric. J.* **92**, 630–2

Naga, M. A. and el-Shazley, K. (1969) Activities of rumen micro-organisms in water buffalo (*Bos bubalis* L.) and in Zebu cattle, *J. Dairy Res.* **36**, 1–9

Naik, A. H. (1965) The mineral status of grasses and soils of part of the Mbulu district of Tanzania in relation to animal health, *E. Africa Agric. J.* **31**, 175–82

National Research Council (1976) *Nutrient requirements of beef cattle* (5th rev. edn). National Academy of Science: Washington, D.C. 56 pp.

Neidhardt, R., Plasse, D., Weniger, J. H., Verde, O., Beltran, J. and Benavides, A. (1979) Milk yield of Brahman cows in a tropical beef production system, *J. Anim. Sci.* **48**, 1–6

Nel, L. O. and Bosman, S. W. (1979) Beef production potential of *Desmodium intortum* cv Greenleaf in the coastal belt to the north of East London, *Proc. Grassl. Soc. South Africa* **14**,109–10

Nestel, B. L. (1964) Animal production studies in Jamaica: IV. The costs of developing and maintaining Pangola grass pastures, *J. Agric. Sci.* **62**, 179–86

Nestel, B. L. and Clay, D. (1961) The effects of hormone implantations on the growth rates of European, Zebu, and hybrid beef steers grazing in Jamaica, *Brit. Vet. J.* **117**, 455–65

Nestel, B. L. and Creek, M. J. (1962) Pangola grass, *Herbage Abstracts* **32**, 1–7

van Niekerk, B. D. H. (1973) Supplementary feeding of range animals in grassveld areas of Southern Africa III. World Conference on Animal Production, *Pre-Conference Volume No. 2* 2(d)-17-2(d)-25

van Niekerk, B. D. H. (1974) *Supplementation of grazing cattle. Proc. Seminar potential to increase beef production in Latin America.* Centro Internacional de Agricultura Tropical: Cali, Colombia. 315 pp.

Noland, P. R., Broce, E. and Vergara, I. (1959) Uso de fertilizantes en potreros de zacate jaragua (*Hyparrhenia rufa* (Nees) Stapf) en Panamá, *Turrialba* **9**, 29–35

Norman, M. J. T. (1960) *Grazing and feeding trials with beef cattle at Katherine, N. T.* Div. Land Res. Reg. Survey Technical Paper No. 12. 15 pp.

Norman, M. J. T. (1963) The pattern of dry matter and nutrient content changes in native pastures at Katherine, N. T., *Aust. J. Exper. Agric. and Anim. Husb.* **3**, 119–24

Norman, M. J. T. (1964) The value of standing dry forage for winter cattle grazing at Katherine, Northern Territory, Australia, *Ann. Arid Zones* **3**, 38–43

Norman, M J. T. (1965) Seasonal performance of beef cattle on native pasture at Katherine, N. T., *Aust. J. Exper. Agric. and Anim. Husb.* **5**, 227–31

Norman, M. J. T. (1968) The performance of beef cattle on different sequences of Townsville lucerne and native pasture at Katherine, N. T., *Aust. J. Exper. Agric. and Anim. Husb.* **8**, 21–5

Norman, M. J. T. (1970) Relationships between liveweight gain of grazing beef steers and availability of Townsville lucerne, *Proc. XI Intnl Grassl. Cong.*, 829–32

Norman, M. J. T. (1974a) Beef production from tropical pastures: Part 1, *Australian Meat Research Committee Review* No. **16**, 1–23

Norman, M. J. T. (1974b) Beef production from tropical pastures: Part 2, *Australian Meat Research Committee Review* No. **17**, 1–18

Norman, M. J. T. and Arndt, W. (1959) *Performance of beef cattle on native and sown pastures at Katherine, N. T.* CSIRO Aust. Div. Land Res. Reg. Survey Technical Paper No. 4

Norman, M. J. T. and Phillips, L. J. (1970). Wet season grazing of Townsville stylo pasture at Katherine, N. T. *Aust. J. exp. Agric. Anim. Husb.* 710–15

Norman, M. J. T. and Stewart, G. A. (1964) Investigations on the feeding of beef cattle in the Katherine region, N. T., *J. Aust. Inst. Agric. Sci.* **30**, 39–46

Norman, M. J. T. and Stewart, G. A. (1967) Complementary grazing of native pasture and standing Townsville lucerne in the dry season at Katherine, N. T., *Aust. J. Exper. Agric. and Anim. Husb.* **7**, 225–31

Nuthall, P. L. and Whiteman, P. C. (1972) A review and economic evaluation of beef production from legume based and nitrogen fertilized tropical pasture, *J. Aust. Inst. Agric. Sci.* **38**, 100–8

Oakes, A. J. (1960) Pangola grass (*Digitaria decumbens* Stent.) in the Caribbean, *Proc. IIX Intnl Grasl Cong.* 386–9

O'Bryan, M. S. (1960) Observations on the utilisation by cattle of *Axonopus affinis* following foliar application of urea, molasses and monosodium phosphate, *Queensland J. Agric. Sci.* **17**, 135–46

O'Donovan, P. B., Woldegebriel, A., Taylor, M. S. and Gebrewolde, A. (1978) Liveweight performance of crossbred and Zebu cattle grazing on native or Rhodes grass pastures, *Trop. Anim. Hlth. Prod.* **10**, 243–9

O'Kelly, J. C. and Robinson, D. W. (1968) The effect of drought feeding with whole cottonseed and vitamin therapy on the serum lipids and liveweight changes of beef cattle in North Western Australia, *Aust. J. Agric. Res.* **19**, 657–64

Okorie, I. I., Hill, D. H. and McIlroy, R. J. (1965) The productivity and nutritive value of tropical grass legume pastures rotationally grazed by Ndama cattle at Ibadan, Nigeria, *J. Agric. Sci. (Cambridge)* **64**, 235–45

Oliver, W. M. (1972) *Grazing stocker cattle during the warm season.* Louisiana State University and Agric. and Mech. College, North Louisiana Hill Farm Exper. Sta. Hill Farm Facts, Beef cattle No. **14**

Oliver, W. M. (1975) Effect of monensin on gains of steers grazed on coastal Bermuda grass, *J. Anim. Sci.* **41**, 999–1001

Olubajo, F. O. and Oyenuga, V. A. (1971) The measurement of yield, voluntary intake and animal production of tropical pasture mixtures, *J. Agric. Sci. (Cambridge)* **77**, 1–4

Orr, J. B. and Holm, A. (1931) *Economic Advisory Council Sixth Report.* 14 pp

Ørskov, E. R. and McDonald, I. (1980) Utilisation of volatile fatty acids for maintenance and for energy retention. *Energy metabolism proceedings of the eighth symposium on energy metabolism* (L. E. Mount, ed.). Butterworth: London, Boston. 484 pp.

Otim, J. J. (1975) Comparative beef production from ley and improved natural pasture, *E. Africa Agric. For. J.* **41**, 45–51

Owen, M. A. (1968) Studies with beef steers on the Kongwa Plain, Central Tanzania, *Trop. Agric. (Trinidad)* **45**, 159–71

Oyenuga, V. A. (1966) The level of Nigeria livestock industry, *World Rev. Anim. Prod.* **1**, 91–104

Oyenuga, V. A. (1971) Biological productivity in West Africa, *J. West Africa Sci. Ass.* **16**, 93–115

Oyenuga, V. A. and Olubajo, F. O. (1966) Productivity and nutritive value of tropical pastures at Ibadan, *Proc. X Intnl Grassl. Cong.*, 962–9

Padmanabha, Aiyer, A. R. and Kayasth, R. N. (1931) Mineral composition of the fodders of central provinces and Berar and its bearing on animal nutrition, *Agric. Livestock in India* **1**, 526–33

Paisana, F. C. (1969) A accão de suplementos minerals em bovinos vivendo nos terrenos lavados de alta Zambézia, *An. Serv. Vet. Mocambique* No. **15**, 285–96

Paladines, O. (1974) Management and utilisation of native tropical pastures in America in *Proceedings seminar to increase beef production in tropical America*. Centro Internacional de Agricultura Tropical: Cali. 315 pp.

Paladines, O., Alarcon, E., Hilton, J., Spain, J. M., Grof, B. and Perez, R. (1974) Development of a pasture programme in the tropical savannah of Colombia, *Proc. XII Intnl Grassl. Cong.*, 388–401

Pallares, O. R. and Mufarrege, D. J. (1970) *Producción animal de pasturas subtropicales fertilisadas*. INTA Serie Técnica No. **6**. 23 pp.

Pallares, O. R. and Mufarrege, D. J. (1974) *Producción animal en pasto Pangola fertilisado INTA Estación Experimental agropecuaria Mercedes (Ctes)*. Serie Tecnica No. 9, 16 pp.

Pamp, D. E., Goodrich, R. D. and Meiske, J. C. (1976) A review of the practice of feeding minerals free choice, *World Rev. Anim. Prod.* **12**, 13–18

Parkin, D. D. and Price, B. C. R. (1979) Interactions between grazing procedures and stocking rates on carcass mass gains of steers grazing irrigated Star grass pastures. Div. of Livestock and Pastures: Rhodesia: *Ann. Rep. 1978–9*, 138–142

Parkin, D. D., Rodel, M. G. W. and Boultwood, J. N. (1979) The effect of stocking rate and applied nitrogen on the carcass mass gains of steers grazing dryland Star grass pastures during the growing season. Div. of Livestock and Pastures: Rhodesia. *Ann. Rep. 1978–9*, 131–7

Parkin, D. D., Rodel, M. G. W., Clatworthy, J. N. and Holland, G. (1976) Effects of grazing systems on carcass mass gains of steers grazing irrigated Star grass pastures. Div. of Livestock and Pastures: Rhodesia: *Ann. Rep. 1975–6*, 144–9

Parkin, D. D., Rodel, M. G. W., Clatworthy, J. N. and Price, B. C. R. (1977) Effects of grazing systems on carcass mass gains of steers grazing irrigated Star grass pastures. Div. of Livestock and Pastures: Rhodesia. *Ann. Rep. 1976–7*, 136–140

Parkin, D. D., Rodel, M. G. W., Clatworthy, J. N. and Price, B. C. R. (1978) Effects of grazing systems on carcass mass gains of steers grazing irrigated Star grass pastures. Div. of Livestock and Pastures: Rhodesia. *Ann. Rep. 1977–8*, 11–14

Parra, R., Combellas, J. and González, E. (1972) Composicion y valor nutritivo de forrajes producidos en el trópico. 2. Fracciones quimicas que afectan la disponibilidad de los componentes fibrosos, *Agronomia Tropical* (Maracay) **22**, 219–30

Partridge, I. J. (1979) Improvement of Nadi Blue grass (*Dichanthium caricosum*) pastures on hill land in Fiji with superphosphate and Siratro: effects of stocking rate on beef production and botanical composition, *Trop. Grassl.* **13**, 157–64

Partridge, I. J. and Ranacou, E. (1974) The effects of supplemental *Leucaena leucocephala* browse on steers grazing *Dichanthium caricosum* in Fiji, *Trop. Grassl.* **8**, 107–12

Pate, F. M., Haines, C. E., Green, R. E. L., Palmer, A. Z., Carpenter, J. W. and Beardsley, D. W. (1972) *Supplementing steers on pastures from weaning to feedlot and subsequent effect on finishing performance.* Florida Agric. Exper. Sta. Bull. 752. 21 pp.

Pate, F. M., Hayes, B. W., Beardsley, D. W., Green, R. E. L., Palmer, A. Z., Carpenter, J. W. (1975) *Supplemental feeding of steers on Everglades Pasture.* Florida Agric. Exper. Sta. Bull. 775. 17 pp.

Patterson, E. L., Milstrey, R. and Stokstad, E. L. R. (1957) Effect of selenium in preventing exudative diathesis in chicks, *Proc. Soc. Exper. Biol. Med.* **95**, 617–20

Payne, W. J. A. (1966) Nutrition of ruminants in the tropics, *Nutr. Abs. Revs.* **36**, 653–70

Payne, W. J. A. (1976) Systems of beef production in developing countries in *Beef cattle production in developing countries* (A. J. Smith, ed.). University of Edinburgh, Centre for Tropical Veterinary Medicine: Edinburgh. 487 pp.

Payne, W. J. A and Hancock, J. (1957) The direct effect of tropical climate on the performance of European-type cattle 2. Production, *Emp. J. Exper. Agric.* **25**, 321–38

Payne, W. J. A., Laing, W. I. and Raivoka, E. N. (1951) Grazing behaviour of dairy cattle in the tropics, *Nature (London)* **167**, 610–11

Peacock, F. M., Green, R. E. L., Kirk, W. G., Koger, M. and Crockett, J. B. (1976a) *Forage systems, beef production and economic evaluations, South Central Florida.* Florida Agric. Exper. Sta. Technical Bull. 783. 14 pp.

Peacock, F. M., Kirk, W. G., Hodges, E. M., Palmer, A. Z. and Carpenter, J. W. (1964) *The effect of winter gains of beef calves on subsequent feedlot performance.* Florida Agric. Exper. Sta. Technical Bull. 667. 12 pp.

Peacock, F. M., Koger, M., Kirk, W. G., Hodges, E. M. and Crockett, J. R. (1976b) *Beef production of Brahman, Shorthorn and their crosses on different pasture programs.* Florida Agric. Exper. Sta. Technical Bull. 780. 19 pp.

Pearson, H. A. (1974) Liquid supplements for cattle grazing native range, *Proc. Louisiana State Univ. Fourteenth Livestock Prod. Day*, 33–40

Penzhorn, E. J. and Meintjies, J. P. (1968) Preliminary results with regard to early reproduction in Afrikander heifers, *Proc. South Africa Soc. Anim. Prod.* **7**, 155

Penzhorn, E. J., Meintjies, J. P. and Naude, L. H. (1970) Die invloed van luzern hooi of n ureumlick as winter byvoeding op die kalfpersentasie van Afrikanerkoeie, *Proc. South Africa Soc. Anim. Prod.* **9**, 159

Pereira, J. C. C., Ribeiro, R. M. P., Val, L. J. L., do Andrade, V. J. and Sampaio, I. B. M. (1977) Efeito da idade e do metodo de castração sobre o desenvolvimento proderal e rendimento de carcaça de bovinos azebuados em regimen de pasto, *Arq. Esc. Vet. Un. Fed. Minas Gerais* **29**, 77–84

Perez Infante, F. (1969) Bull fattening on Pangola grass pastures supplemented with molasses/urea 3% and concentrates, *Rev. Cubana Cienc. Agricola* (Eng. edn) **3**, 36 (abs)

Pezo, D., Vohnout, K., Velarde, C. L. and Camargo, A. (1977)

Predicción del consumo de pastos en base a fracciones quimicas y parametros de digestion, *Turrialba* **27**, 157–62

Pieterse, P. J. S. (1961) Feeding urea to cattle on winter veld; new system successful, *Farming in South Africa* **37**, 15–17

Pieterse, P. J. S. (1964) Supplement the deficiencies in low-grade roughage, *Farming in South Africa* **40**, 37–40

Pieterse, P. J. S. (1967) Urea in winter rations for cattle in *Urea as a protein supplement* (M. H. Briggs, ed.). Pergamon Press: Oxford. 466 pp.

Pieterse, P. J. S. and de Kock, J. P. (1962) Urea proved effective in mineral lick, *Farming in South Africa* **39**, 84–7

Pieterse, P. J. S. and Lesch, S. F. (1963) The effect of urea on the nitrogen metabolism of steers fed on poor quality veld hay, *Proc. South Africa Soc. Anim. Prod.* **2**, 49–52

Pieterse, P. J. S., Lesch, S. F. and van Schalkwyk, A. P. (1963) Rate of passage of poor quality veld hay through the digestive tract of beef steers, *South Africa J. Agric. Sc.* **6**, 737–40

Pieterse, P. J. S. and Preller, J. H. (1965) Preliminary results of supplementing beef cattle on summer veld, *Proc. South Africa Soc. Anim. Prod.* **4**, 123–6

Piot, J. (1970) Pâturage aerien au Cameroun; utilisation des ligneux par les bovins, *Rév. Elév. Méd. Vét. Pays. Trop.* **23**, 503–17

Playne, M. J. (1970) The sodium concentration in some tropical pasture species with reference to animal requirements, *Aust. J. Exper. Agric. and Anim. Husb.* **10**, 32–5

Playne, M. J. and Kennedy, P. M. (1976) Ruminal volatile fatty acids and ammonia in cattle grazing dry tropical pastures, *J. Agric. Sci. (Cambridge)* **86**, 367–72

Playne, M. J., Siebert, B. D. and Edye, L. A. (1973) The use of non-protein nutrients with cattle grazing improved pastures in the Australian dry tropics, *Proceedings of the III World Conference on Animal Production, Melbourne 1973* (R. L. Reid, ed.), 575–9

Plucknett, D. L. (1970) Productivity of tropical pastures in Hawaii, *Proc. XI Intnl Grassl. Cong., A38–A49*

Pollard, D. (1970) *Mixing molasses licks . . . an easy way*. Australia Dept. of Primary Industries, Div. of Animal Industry, Advisory Leaflet No. 678

Porres, G. (1971) Molasses and molasses/urea as supplements for F_1 Holstein × Brahman bulls grazing highly fertilised Pangola (*Digitaria decumbens*) pasture, *Rev. Cubana Cienc. Agric.* (Eng. edn) **5**, 13–20

Presswood, J. (1978) *Overgrazing and livestock development: report of a seminar.* Ministry of Agriculture: Swaziland. 32 pp.

Preston, T. R. (1976) Prospects for the intensification of cattle production in developing countries in *Beef cattle production in developing countries* (A. J. Smith ed.). University of Edinburgh, Centre for Tropical Veterinary Medicine: Edinburgh. 487 pp.

Quinn, L. R., Mott, G. O. and Bisschoff, W. V. A. (1960a) *The influence of stilbestrol upon pasture-fed Zebu steers and male suckling calves.* IBEC Res. Inst. Bull. 23. 34 pp.

Quinn, L. R., Mott, G. O. and Bisschoff, W. V. A. (1960b) *Fertilisación de pastizales de hierba guinea y producción de carne con novillos cebú*. IBEC Res. Inst. Bull. 24. 35 pp.

Quinn, L. R., Mott, G. O. and Bisschoff, W. V. A. (1967) *The effect of diethylstilbestrol upon the performance of pasture-fed Zebu steers*. IRI Res. Inst. Bull. 34. 47 pp.

Quinn, L. R., Mott, G. O., Bisschoff, W. V. A. and de Freitas, L. M. M. (1970) Production of beef from winter vs. summer nitrogen-fertilised colonial Guinea grass (*Panicum maximum*) pastures in Brazil, *Proc. XI Intnl Grassl. Cong.*, 832–5

Quinn, L. R. C., Mott, G. O., Bisschoff, W. V. A. and Jones, M. B. (1964) Beef production of six tropical grasses in Central Brazil, *Proc. IX Intnl Grassl. Cong.*, 1015–20

Quinn, L. R., Mott, G. O., Bisschoff, W. V. A. and McClung, A. C. (1958) *Stilbestrol and its effect on pasture fed Zebu steers*. IBEC Res. Inst. Bull. 15. 19 pp.

Quinn, L. R., Mott, G. O., Bisschoff, W. V. A. and da Rocha, G. L. (1962) *Beef production of six tropical grasses*. IBEC Res. Inst. Bull. 28. 36 pp.

Quinn, L. R., Mott, G. O., Bisschoff, W. V. A. and da Rocha, G. L. (1966) *The response of male Zebu calves to creep feeding, castration, diethylstilbestrol and supplemental feeding on pasture*. IRI Res. Inst. Bull. 30. 37 pp.

Rakka, A. M. and Igboeli, G. (1972) Effects of nutrition season and age on the oestrous cycle of indigenous Central African cattle, *J. Anim. Sci.* **32**, 943–5

Ramirez, J. R. (1954) El pasto Rojas, una graminea promisora en el Paraguay, *Rev. Argentina Agron.* **21**, 84–101

Rattray, J. M. (1960) *The grass cover of Africa*. FAO of the United Nations. 168 pp.

Reed, J. B. H., Doxey, D. L., Forbes, A. B., Finlay, R. S., Geering, I. W., Smith, S. D. and Wright, J. D. (1974a) Productive performance of cattle in Botswana, *Trop. Anim. Hlth Prod.* **6**, 1–21

Reed, J. B. H., Smith, S. D., Doxey, D. L., Forbes, A. B., Finlay, R. S., Geering, I. W. and Wright, J. D. (1974b) Serum inorganic phosphate, calcium and magnesium levels of cattle in Botswana, *Trop. Anim. Hlth Prod.* **6**, 23–9

Reed, J. B. H., Smith, S. D., Forbes, A. B. and Doxey, D. L. (1974c) Inorganic phosphate, calcium and magnesium levels in the sera of Botswanan cattle receiving feed additives, *Trop. Anim. Hlth Prod.* **6**, 31–6

Reed, J. B. H., Smith, S. D., Forbes, A. B., Finlay, R. S., Geering, I. W., Wright J. D. and Doxey, D. L. (1974d) A note on seasonal changes in serum inorganic phosphate levels of cattle in Botswana, *Trop. Anim. Hlth Prod.* **6**, 37–8

Reid, M. G. (1978) Practice-change adoption in tribal agriculture, *Rhodesia Agric. J.* **75**, 145–9

Reinach, N. and Louw, J. G. (1952) The supplementation of phosphate to animals through their drinking water, *Farming in South Africa* **27**, 417–19

van Rensburg, S. W. J. (1961) Copper deficiency influences animal fertility, *Farming in South Africa* **36**, 50–3

Reveron, A. E. and Topps, J. H. (1972) The assessment of mineral disorders in young lambs affected with the gut parasite *Trichostrongylus colubriformis*, *Summaries IX Intnl. Nutr. Cong.* 212 pp.

Reyneke, J. (1976) Supplementation of energy and/or protein to steers grazing summer veld, *South Africa J. Anim. Sci.* **6**, 59–66

Reynolds, E. B., Jones, J. M., Jones, J. H., Fudge, J. F. and Kleberg Jr., R. J. (1953) *Methods of supplying phosphorus to range cattle in South Texas.* Texas Agric. Exper. Sta. Bull. 773. 16 pp.

Rhodes, F. B. (1956) The effect of feeding mineral, protein and carbohydrate supplements to growing beef cattle in Mashonaland, *Rhodesia Agric. J.* **53**, 969–82

Rhodesia Ministry of Agriculture (1967) *Ann. Rep. Matopos Res. Sta.* 53 pp.

Richards, J. A. (1965) Productivity of Guinea grass (*Panicum maximum*), *Proc. IX Intnl Grassl. Cong.*, 1032–4

Richardson, F. D., Hannah, P. E., Sithole, M. E. and Khaka. R. W. (1976) Stocking rate and the provision of different amounts of protein to growing cattle. Div. of Livestock and Pastures: Rhodesia. *Ann. Rep. 1975–6*, 45–9

Richardson, F. D., Kegel, L. and Khaka, R. V. (1977a) Protein source and the frequency of feeding young steers. Div. of Livestock and Pastures: Rhodesia. *Ann. Rep. 1976–7,* 59–61

Richardson, F. D., Kegel, L. and Khaka, R. V. (1978) Protein source and the frequency of feeding young steers. Div. of Livestock and Pastures: Rhodesia. *Ann. Rep. 1977–8*, 143–52

Richardson, F. D., Kegel, L. and Sithole, M. E. (1977b) Concentration of metabolites in the plasma of heifers in relation to the protein content of the diet and the consumption of energy. Div. of Livestock and Pastures: Rhodesia. *Ann. Rep. 1976–*7, 57–9

Richardson, F. D., Kegel, L., Sithole, M. E. and Khaka, R. V. (1977c) Stocking rate and the provision of different amounts of protein to grazing cattle. Div. of Livestock and Pastures: Rhodesia. *Ann. Rep. 1976–7*, 52–6

Richardson, F. D., Oliver, J. and Clarke, G. P. Y. (1975a) Analysis of some factors which affect the productivity of beef cows and their calves in a marginal rainfall area of Rhodesia 1. Factors affecting calving rate, *Anim. Prod.* **21**, 41–9

Richardson, F. D., Oliver, J. and Clarke, G. P. Y. (1977d) Analysis of some factors which affect the productivity of beef cows and their calves in a marginal rainfall area of Rhodesia 2. The yield and composition of milk of suckling cows, *Anim. Prod.* **25**, 359–72

Richardson, F. D., Ward, H. K. and Harvey, E. R. (1975b) Rumevite supplementation of growing veld-grazed steers. Div. of Livestock and Pastures: Rhodesia. *Ann. Rep. 1974–5*, 48–50

Richardson, L. F., Raun, A. P., Potter, E. L., Cooley, C. O. and Rathmacher, R. P. (1974) Effect of monensin on ruminal fermentation *in vitro* and *in vivo*, *J. Anim. Sci.* **39**, 250

Rincón Gallardo, C. (1939) *El libro del Charro Mexicano.* Editorial Porrúa S. A.: México. 335 pp.

Rios, C. and Osegueda, F. L. (1967) Uso de estilbestrol en engorde de novillos sobre pasto Jaragua, *Turrialba* **17**, 172–5

Risopoulus, S. A. (1966) *Management and use of grasslands.* FAO Pastures and Fodder Crop Studies No. 1: Democratic Republic of Congo.

Ritson, J. B., Edye, L. A. and Robinson, P. J. (1971) Botanical and chemical composition of a Townsville Stylo-Spear grass pasture in relation to conception rate of cows, *Aust. J. Agric. Res.* **22**, 993–1007

Rivera Brenes, L. (1947) The utilisation of grasses, legumes and other forage crops for cattle feeding in Puerto Rico I. Comparisons of Guinea grass, Para grass 'Malojillo' and a mixture of Para grass and tropical Kudzu as pasture crops, *J. Agric. Univ. Puerto Rico* **31**, 180–9

Rivera Brenes, L., Colon, Torres, E. N., Gelpi, F. and Torres Más, J. (1958) Influence of nitrogenous fertilizers on Guinea grass yield and carrying capacity in Lajas Valley, *J. Agric. Univ. Puerto Rico* **42**, 239–47

Rivera Brenes, L., Marchán, F. J. and Cabrera, J. I. (1949) The utilization of grasses, legumes and other forage crops for cattle feeding in Puerto Rico II. Comparison of fertilized Guinea grass, Para grass and tropical Kudzu and tropical Kudzu alone as forage crops, *J. Agric. Univ. Puerto Rico* **33**, 85–97

Rivera Brenes, L., Marchan, F. J. and Cabrera, J. I. (1950) The utilization of grasses, legumes and other forage crops for cattle feeding in Puerto Rico III. Comparison of fertilised Guinea grass, Para grass and tropical Kudzu and Guinea grass and tropical Kudzu, *J. Agric. Univ. Puerto Rico* **34**, 309–15

Roberts, B. R. (1969) The multi-camp controversy – a search for evidence, *Proc. Veld. Man: Conf. Bulawayo*, 41–55

Robinson, D. W. and Sageman, R. (1967) The nutritive value of some pasture species in north west Australia during the late dry season, *Aust. J. Exper. Agric. and Anim. Husb.* **7**, 533–8

Robinson, D. W. and Steward, G. A. (1969) The influence of protein supplements on growth of local and imported cattle in the Kimberley area of north western Australia, *Aust. J. Exper. Agric. and Anim. Husb.* **9**, 140–6

Robinson, I. B. and Sing, N. C. (1975) Estimated profitability of investment in legume pasture development in far north Queensland, *Trop. Grassl.* **9**, 199–207

Rodel, M. G. W. (1974); Effects of stocking rate on carcass mass gains of steers grazing star grass pastures. Div. Livestock and Pastures: Rhodesia: *Ann. Rep. 1973–74, 84–86*

Rodel, M. G. W. and Boultwood, J. N. (1967) *Recent pasture research at Henderson Research Station.* Rhodesia Ministry of Agriculture Bull. 2441. 6 pp.

Rodel, M. G. W., Groenewald, I. B., Scheerhorn, G., Boultwood, J. N. and Hopley, J. D. H. (1975) Effects of combinations of stocking rate and levels of fertilising with nitrogen on carcass mass gains of steers grazing Star grass pasture and subsequent performance in the feed lot. Div. of Livestock and Pastures: Rhodesia. *Ann. Rep. 1974–5*, 122–5

Roeder, P. L. (1980) Effects of copper and cobalt of cattle in the Ethiopian Rift valley, *Trop. Anim. Hlth. Prod.* **12**, 63–68

Rogerson, A., Ledger, H. P. and Freeman, G. H. (1968) Food intake and liveweight gain comparisons of *Bos indicus* and *Bos taurus* steers on a high plane of nutrition, *Anim, Prod.* **10**, 373–80

Romero, A. and Siebert, B. D. (1980) Seasonal variations of nitrogen and digestible energy intake of cattle on tropical pasture, *Aust. J. Agric. Res.* **31**, 393–400

Romero, A., Siebert, B. D. and Murray, R. M. (1976) A study on the effect of frequency of urea ingestion on the utilisation of low quality roughage by steers, *Aust. J. Exper. Agric. and Anim. Husb.* **16**, 308–14

Rose, C. J., Christie, G. J. and Conradie, A. P. (1963) The effect of early weaning on the reproductive efficiency of ranch cattle in Southern Rhodesia, *Proc. First World Conference of Anim. Prod.* **3**, 125–35

Rosenfeld, I. and Beath, O. A. (1964) *Selenium: Geobotany, biochemistry, toxicity and nutrition.* Academic Press: New York and London. 411 pp.

de Rouen, T. M., Reynolds, W. L., Meyerhoeffer, D. C., Harpel, R. A. and Poche, N. T. (without date) *Biuret for wintering beef hiefers in the Gulf Coast area.* Louisiana Agric. Exper. Sta. and USDA (mim. rep.) 3 pp.

Roux, H. and Parada, J. (1969) Efecto de cuatro forrajes en la utilisacion de una mezcla de melaza y urea, *Turrialba* **10**, 465–71

Roux, H. and Rodriguez, H. (1971) Utilizacion de melaza y urea en el mantenimiento de ganado bovino durante la estación seca en Panamá, *Turrialba* **21**, 137–44

le Roux, P. and Stubbs, A. T. (1980) *The contribution of peasant agriculture to the beef industry.* Paper delivered to the 9th Ann. Cong. of the Cattle Prod. Assoc.: Zimbabwe (mimeo). 15 pp.

Rubio, C. E. (1973) *Efecto de una lactona del ácido resorcilico (RAL) en el aumento de peso de novillos en pastoreo.* Trabajo presentado el el Séptimo Congreso Panamericano de Medicina Veterinaria y Zootecnia: Bogotá (mimeo). 10 pp.

Rudert, C. P. and O'Donovan, W. M. (1976) The productivity of beef cows and their progeny when offered different mineral supplements over three breeding seasons, *Rhodesia J. Agric. Res.* **14**, 3–10

Ruiz, M. E., Vohnout, K., Isidor, M. and Jimenez, C. (1973) *Crecimiento de bovino supplementados con banano II. Efecto de nivel de banano.* Compendios de trabajos presentados en la IV Reunión Latinoamericano de Producción Animal (ALPA): Guadalajara, México. pR36

Ruthenberg, H. (1978) Socio-economic variables within animal production systems, *Summaries: IV World Conf. on Anim. Prod.*, 8–9

Ryley, J. W., Harvey, J. M., Watson, J. W. and Levitt, M. S. (1961) A comparison of the copper status of sheep and cattle grazing predominantly *Paspalum dilatatum* pastures in southeast Queensland, *Queensland J. Agric. Sci.* **18**, 353–66

Sanchez, P. A. (1976) *Properties and management of soils in the tropics.* J. Wiley: New York. 618 pp.

Sandland, R. L. and Jones, R. J. (1975) The relation between animal gain and stocking rate in grazing trials: an examination of published theoretical models, *J. Agric. Sci. (Cambridge)* **85**, 123–8

Santhirasegaram, K. (1974) Legume-based improved tropical pastures in *Proceedings of the seminar on potential to increase beef production in tropical America.* Centro Internacional de Agriculture Tropical: Cali, Colombia. 315 pp.

Sartini, H. J., Santamaria, M., Lourenço, A. J., Caielli, E. L. and da Rocha, G. L. (1975) Ensaio de pastejo usando lotação fixa em uma consorciação entre capim-Gordura (*Melinis minutiflora*) e *Centrosema pubescens, Zootecnia sp* **13**, 219–27

van Schaffhausen, R. (1965) Weight increase of Zebu cattle grazing on the legumes *Dolichos lablab and Cajanus indicus, Proc. IX Intnl Grassl. Cong.*, 965–8

van Schalkwyk, A. and Lombard, P. E. (1969) The influence of phosphorus supplementation on body function and growth of young steers, *Agroanimalia* **1**, 45–52

Schank, S. C., Klock, M. A. and Moore, J. E. (1973) Laboratory evaluation of quality in sub-tropical grasses: II. Genetic variation among *Hemarthrias* in *in vitro* digestion and stem morphology, *Agron. J.* **65**, 256–8

Schmidt, H. (1926) *Feeding bonemeal to range cattle on the coastal plains of Texas.* Texas Agric. Exper. Sta. Bull. 344

Schofield, J. L. (1945) A comparison of soil nitrate nitrogen values under bare fallow and after ploughing in various perennial tropical legumes and cowpeas, *Queensland J. Agric. Sci.* **2**, 170–4

Schofield, J. L. (1946) Mineral content and yield of grasses in the wet tropics as influenced by seasonal productivity, frequency of cutting and species, *Queensland J. Agric. Sci.* **3**, 44–53

Schwartz, H. M. (1967) The rumen metabolism of non-protein nitrogen in *Urea as a protein supplement* (M. H. Briggs, ed.) Pergamon Press: Oxford. 466 pp.

Sen, K. C. and Ray, S. N. (1964) *Nutritive values of Indian cattle feeds and the feeding of animals.* Indian Council of Agric. Res. Bull. 25. 133 pp.

Shaw, N. H. (1961) Increased beef production from Townsville lucerne (*Stylosanthes sundaica* Taub.) in the Spear grass pastures of central coastal Queensland, *Aust. J. Exper. Agric. and Anim. Husb.* **1**, 73–80

Shaw N. H. (1978a) Superphosphate and stocking rate effects on a native pasture oversown with *Stylosanthes humilis* in central coastal Queensland 1. Pasture production, *Aust. J. Exper. Agric. and Anim. Husb.* **18**, 788–99

Shaw, N. H. (1978b) Superphosphate and stocking rate effects on a native pasture oversown with *Stylosanthes humilis* in central coastal Queensland 2. Animal production, *Aust. J. Exper. Agric. and Anim. Husb.* **18**, 800–7

Shaw, N. H. and Dale, M. B. (1978) Superphosphate and stocking rate effects on a native pasture oversown with *Stylosanthes humilis* in central coastal Queensland 3. The balance of native species in the pasture, *Aust. J. Exper. Agric. and Anim. Husb.* **18**, 808–16

Shaw, N. H. and Mannetje, L.'t (1970) Studies on a Spear grass pasture in central coastal Queensland – the effect of fertilizer, stocking rate and oversowing with *Stylosanthes humilis* on beef production and botanical composition, *Trop. Grassl.* **4**, 43–55

Shorrock, C., Capper, B. S., Light, D. and Mlambo, M. M. J. (1978) A note on the performance of fattening steers implanted with zeranol under grazing and the feedlot conditions in Botswana, *Anim. Prod.* **26**, 221–4

Sibisi, H. (1980) *Traditional securities and the response to 'modern' economic opportunities.* Ministry of Agriculture: Swaziland. (mimeo). 14 pp.

Siebert, B. D. and Field, J. B. F. (1975) Reproductive activity in beef heifers following post-weaning feeding on spear grass hay alone or with supplements, *Aust. J. Exper. Agric. and Anim. Husb.* **15**, 12–16

Siebert, B. D. and Kennedy, P. M. (1972) The utilization of spear grass (*Heteropogon contortus*) 1. Factors limiting intake and utilization by cattle and sheep, *Aust. J. Agric. Res.* **23**, 35–44

Siebert, B. D., Newman, D. M. R., Hart, B. and Michell, G. L. (1975) Effects of feeding varying levels of protein and phosphorus in relation to bone disorders in cattle, *Aust. J. Exper. Agric. and Anim. Husb.* **15**, 321–4

Siebert, B. D., Playne, M. J. and Edye, L. A. (1976) The effects of climate and nutrient supplementation on the fertility of heifers in north Queensland, *Proc. Aust. Soc. Animal Prod.* **11**, 249–53

Da Silva, D. J., Da. Cunha, P. G., Siqueira, A. C. M. F. and Roverso, E. A. (1974) Efeito da administração de vitamina A, D e E nos bovinos em fase de aleitamento, *Bol. Ind. Anim. SP* **31**, 205–12

Da Silva Lopes, H. O., Neto, J. M. F. and Sampaio, I. B. M. (1973a) Resposta à aplicação de farinha de osso nos teores séricos de fósforo, cálcio, magnésio e atividade da fosfatase alcalina em bovinos criados no cerrado, *Arq. Esc. Vet.* **25**, 299–303

Da Silva Lopes. H. O., Neto, J. M. F., and Sampaio, I. B. M. (1973b) Alguns fatores que influenciam os niveis de fósforo inorgânico, cálcio, magnésio e fosfatase alcaline no soro sangüineo de bovinos no cerrado, *Arq. Esc. Vet.* **25**, 309–27

Silveira, P. J. V., Barbosa, de M. H., Melotti, L. and Machado, de C. H. (1975) Estimativas da capacidade de supporte de capins consorciados com leguminoses, *Bol. Ind. Anim. SP* **32**, 281–92

Silvey, M. W., Coaldrake, J. E., Haydock, F. P., Radcliff, D. and Smith, C. A. (1978) Beef cow performance from tropical pastures on semi-arid brigalow lands under intermittant drought, *Aust J. Exper. Agric. and Anim. Husb.* **18**, 618–28

Simpson, M. C. (1976) Problems of marketing cattle in developing countries in *Beef production in developing countries* (A. J. Smith, ed.). University of Edinburgh, Centre for Trop. Vet. Med.: Edinburgh 487 pp.

Siota, C. M., Castillo, A. P., Moog, F. A. and Javier, E. Q. (1977) Productivity of native and native/stylo pastures, *Philippines J. Anim. Ind.* **31**, 101–8

Siota, C. M., Castillo, A. P., Moog, F. A. and Javier, E. Q. (1978) Beef production on native, native/Stylo and native/Centro pastures, *Philippines J. Anim. Ind.* **32**, 25–34

Skerman, P. J. (1977) *Tropical forage legumes.* Food and Agriculture Organization of the United Nations: Rome. 609 pp.

Skinner, J. D. (1964) Urea and molasses as a supplement in the sweet bushveld, *Proc. South Africa Soc. Anim. Prod.* **3**, 125–8

Smith, A. J. (ed.) (1974) *Beef cattle production in developing countries.* University of Edinburgh Centre for Trop. Vet. Med.: Edinburgh. 487. pp.

Smith, C. A. (1959) Studies on the Northern Rhodesia *Hyparrhenia* veld Part I. The grazing behaviour of indigenous cattle grazed at light and heavy stocking rates, *J. Agric. Sci.* **52**, 369–75

Smith, C. A. (1961) The utilization of *Hyparrhenia* veld for the nutrition of cattle in the dry season II. Veld hay compared with *in situ* grazing of the mature forage and the effects of feeding supplementary nitrogen, *J. Agric. Sci.* (*Cambridge*) **57**, 311–17

Smith, C. A. (1962a) The utilization of *Hyparrhenia* veld for the nutrition of cattle in the dry season III. Studies on the digestibility of the produce of mature veld and veld hay and the effect of feeding supplementary protein and urea, *J. Agric. Sci.* (*Cambridge*) **58**, 173–7

Smith, C. A. (1962b) Tropical grass/legume pastures in Northern Rhodesia, *J. Agric. Sci.* (*Cambridge*) **59**, 111–18

Smith, C. A. (1966) Studies on the *Hyparrhenia* veld of Zambia. VII. The effects of cattle grazing veld and dambo at different stockingrates, *J. Agric. Sci.* (*Cambridge*) **66**, 49–56

Smith, C. A. and Hodnett, G. E. (1962) Compensatory growth of cattle on the natural grasslands of Northern Rhodesia, *Nature (UK)* **195**, 919–20

van Soest, P. J. and Jones, L. H. P. (1968) Effect of silica in forages on digestibility, *J. Dairy Sci.* **51**, 1644–8

van Soest, P. J. and Robertson, J. B. (1979) *Systems of analysis for evaluating fibrous feeds.* Standardisation of Analytical Techniques, IDRC: Ottawa, Canada. 128. pp.

Sparke, E. J. and Lamond, D. R. (1968) The influence of supplementary feeding on growth and fertility of beef heifers grazing natural pastures, *Aust. J. Exper. Agric. and Anim. Husb.* **8**, 425–33

Squire, H. A. (1976) Intensive feeding systems for beef production in developing countries, *World Anim. Rev. (FAO)* No. 17, 42–3

Stangel, H. J. (1967) History of the use of urea in ruminant feeds in *Urea as a protein supplement* (M. H. Briggs, ed.). Pergamon Press: Oxford. 466 pp.

Staples, R. R. (1941) Anti-tsetse clearing and land use, *East Africa Agric. J.* **7**, 48–50

Steenkamp, J. D. G. and van der Horst, C. (1975) Genetic and environmental influences on performance of supplementary fed and not-fed grade Africander cattle, *Rhodesia Agric. J.* **72**, 9–13

Steenkamp, J. D. G., van der Horst, C. and Andrew, M. J. A. (1975) Reconception in grade and pedigree Africander cows of different sizes – postpartum factors influencing reconception, *South Africa J. Anim. Sci.* **5**, 103–10

Stobbs, T. H. (1965a) Animal production from pasture leys, *African Soils* **10**, 293–9

Stobbs, T. H. (1965b) Beef production from Uganda pastures containing *Stylosanthes gracilis and Centrosema pubescens, Proc. IX Intnl Grasl. Cong.*, 939–42

Stobbs, T. H. (1969a) The use of liveweight-gain trials for pasture evaluation in the tropics. 3. The measurement of large pasture differences, *J. Brit. Grassl. Sec.* **24**, 177–83

Stobbs, T. H. (1969b) The use of liveweight-gain trials for pasture evaluation in the tropics 4. Animal replication, *J. Brit. Grassl. Soc.* **24**, 258–63

Stobbs, T. H. (1969c) The use of liveweight-gain trials for pasture evaluation in the tropics 5. Type of stock, *J. Brit. Grassl. Soc.* **24**, 345–8

Stobbs, T. H. (1969d) The effect of grazing management upon pasture productivity in Uganda I – Stocking rate, *Trop. Agric. (Trinidad)* **46**, 187–94

Stobbs, T. H. (1969e) The effect of grazing management upon pasture productivity in Uganda II – Grazing frequency, *Trop. Agric. (Trinidad)* **46**, 195–200

Stobbs, T. H. (1969f) The effect of grazing management upon pasture productivity in Uganda III – Rotational and continuous grazing, *Trop. Agric. (Trinidad)* **46**, 293–301

Stobbs, T. H. (1969g) The effect of grazing management upon pasture productivity in Uganda IV – Selective grazing, *Trop. Agric. (Trinidad)* **46**, 303–9

Stobbs, T. H. (1969h) Animal production from *Hyparrhenia* grassland oversown with *Stylosanthes gracilis*, *East Africa Agric. For. J.* **35**, 128–34

Stobbs, T. H. (1970) The use of liveweight-gain trials for pasture evaluation in the tropics 6. A fixed stocking rate design, *J. Brit. Grassl. Soc.* **25**, 73–7

Stobbs, T. H. (1973a) The effect of plant structure on the intake of tropical pastures 1. Variation in the bite size of grazing cattle, *Aust. J. Agric. Res.* **24**, 809–19

Stobbs, T. H. (1973b) The effect of plant structure on the intake of tropical pastures 2. Differences in sward structure, nutritive value and bite size of animals grazing *Setaria anceps* and *Chloris gayana* at various stages of growth, *Aust. J. Agric. Res.* **24**, 821–9

Stobbs, T. H. (1975) Factors limiting the nutritional value of grazed tropical pastures for beef and milk production, *Trop. Grassl.* **9**, 141–50

Stobbs, T. H. and Joblin, A. D. H. (1966a) The use of liveweight-gain trials for pasture evaluation in the tropics 1. An 'animal' latin square design. *J. Brit. Grassl. Soc.* **21**, 49–55

Stobbs, T. H. and Joblin, A. D. H. (1966b) The use of liveweight-gain trials for pasture evaluation in the tropics 2. Variable stocking rate designs, *J. Brit. Grassl. Soc.* **21**,181–5

Stobbs, T. H. and Whitting, J. B. (1970) A note on the performance of cross-bred steers at different stocking rates in Uganda, *East Africa Agric. For. J.* **35**, 234–6

Stonaker, H. H., Villar, J., Osorio, G. and Salazar, J. (1976) Differences among cattle and farms as related to beef cow reproduction in the Eastern Plains of Colombia, *Trop. Anim. Hlth. Prod.* **8**, 147–54

Sutherland, A. K. (1956) Copper supplementation trials with cattle in Queensland, *Proc. Aust. Soc. Anim. Prod.* **1**, 18–19

Sutherland, D. N. (1952) Copper deficiencies in cattle in Queensland, *Aust. Vet. J.* **28**, 204–8

Sutherland, D. N. (1959) Factors affecting the performance of beef cattle on unimproved pastures, *Aust. Vet. J.* **35**, 129–

Sutmoller, P., Vahia de Abreu, A., van der Grift, J. and Sombroek, E. G. (1966) *Mineral imbalances in cattle in the Amazon Valley*. Communication No. 53 of the Department of Agricultural Research, Kononklijk Instituut voor de tropen. 133 pp.

Symington, R. B. and Hale, D. H. (1967) The stimulation of sexual activity in lactating Zebu cows, *Rhodesia Zambia Malawi J. Agric. Res.* **5**, 13

Tagari, H., Ben Ghadalia, D. and Shtern, Y. (1977). The effect of two feeding levels of diets containing fields-cured or frozen Rhodes grass on disgestibility and rumen metabolites in sheep, *J. Agric. Sci.* (*Cambridge*). **89**, 177–82

Takahashi, M., Moomaw, J. C. and Ripperton, J. C. (1966) *Studies of Napier grass III. Grazing management.* Hawaii Agric. Exper. Sta. Bull. 128. 47 pp.

Teleni, E., Siebert, B. D., Murray, R. M. and Nancarrow, C. D. (1977) Effects of phosphorus or phosphorus and protein on the ovarian activity of cows fed native pasture hay, *Aust. J. Exper. Agric. and Anim. Husb.* **17**, 207–13

Temple, R. S., Carrera, M. C. and Cenoz, F. B. (1973) Comparación del ritmo de crecimiento y calidad de las canales entre toretes y novillos

post destete, *XIII Inf. Inv. Inst. Tec. Est. Sup. Monterrey 1971–2*, 98–9
Temple, R. S. and Thomas, M. E. R. (1973) The Sahelian drought – a disaster for livestock populations, *World Anim. Rev.* No. **8**, 1–7
Terr. of Papua and New Guinea (1965) Dept. of Agriculture, Stock and Fisheries. *Ann. Rep. 1963–4*. 20. pp.
Theiler, A. (1920) The cause and prevention of lamsiekte, *J. Dept. Agric. Union of South Africa* **7**, 221–44
Theiler, A. (1927) Lamsiekte (Parabotulism) in cattle in South Africa, *11th and 12th Rep. Dir. Vet. Serv. Anim. Ind. 1926–7*, 821–1361
Theiler, A. and Green, H. H. (1932) Aphosphorosis in ruminants, *Nutr. Abs. Revs.* **1**, 359–85
Theiler, A., Green, H. H. and du Toit, P. J. (1924) Phosphorus in the livestock industry, *J. Dept. Agric. South Africa* **8**, 460–504
Theiler, A., du Toit, P. J. and Malan, A. I. (1937) Studies in mineral metabolism XXXVII. The influence of variations in the dietary phosphorus and the Ca:P ratio on the production of rickets in cattle, *Onderstepoort J. Vet. Sci. Anim. Ind.* **8**, 375–414
Theron, E. P. and Harwin, G. O. (1976) The future role of improved pastures for beef cattle, *South Africa J. Anim. Sci.* **6**, 139–46
Theron, E. P., Lesch, S. F. and Mappledoram, B. D. (1974) The potential in Natal for the radical improvement of the veld and the fortification of established pastures, *Proc. Grassl. Soc. South Africa* **9**, 175–8
Thomas, D. and Addy, B. L. (1977) Tropical pasture legumes and animal production in Malawi, *World Rev. Anim. Prod.* **13**, 47–52
Thornton, D. D. (1970) A stocking rate trial on rough grazing in Buganda Part III. Liveweight gains in the last two years, *East Africa Agric. For. J.* **35**, 331–5
Thornton, D. D. and Harrington, G. N. (1971) The effect of different stocking rates on the weight gain of Ankole steers on natural grasslands in Western Buganda, *J. Agric. Sci. (Cambridge)* **76**, 97–106
Thornton, R. F. and Minson, D. J. (1973) Effects of soils, fertilisers and stocking rates on pastures and beef production on the Wallum of south eastern Queensland 3. Relation of liveweight changes to chemical composition of blood and pasture, *Aust. J. Exper. Agric. and Anim. Husb.* **13**, 537–43
Tierney, T. J. (1973) Breeder performance on wallum pasture, *Trop. Grassl.* **7**, 348–50
Tiffin, J. de W. (1976) Plane of nutrition and fertility of Mashona heifers mated at 15 months of age. Div. of Livestock and Pastures: Rhodesia. *Ann. Rep. 1975–6*, 39–44
Tiharuhondi, E. R., Olsen, F. J. and Musangi, R. S. (1973) Application of nitrogen and irrigation to pasture to enhance cattle production during the dry seasons in Uganda, *East Africa. Agric. For. J.* **38**, 383–93
Tilman, A. D. (1970) *Deficiency diseases*. Paper presented to Animal Production Society of Kenya (mimeo). 26 pp.
du Toit, P. J. and Bisschop, J. H. R. (1929) The breeding of cattle on phosphorus-deficient veld, *15th Ann. Rep. Dir. Vet. Serve. Anim. Ind. Union of South Africa*, 1059–1165
du Toit, S. P., Lesch, S. F. and Pieterse, P. J. S. (1963) The effect of urea on the production of volatile fatty acids in the rumen of steers fed on poor quality veld hay, *Proc. South Africa Soc. Anim. Prod.* **2**, 52–4
du Toit, P. J., Louw, J. G. and Malan, A. I. (1940a) A study of the mineral

content and feeding value of natural pastures in the Union of South Africa, *Onderstepoort J. Vet. Sci. Anim. Ind.* **14**, 123–327

du Toit, P. J., Malan, A. I., Louw, J. G., Holzappel, C. R. and Roests, G. C. S. (1932) A study of the mineral content and feeding value of natural pastures in the Union of South Africa, *18th Rep. Dir. Vet. Serv. Anim. Ind.*, 525–77

du Toit, P. J., Malan, A. I., Louw, J. G., Holzappel, C. R. and Roets, G. C. S. (1934) A study of the mineral content and feeding value of natural pastures in the Union of South Africa, *Onderstepoort J. Vet. Sci. Anim. Ind.* **2**, 607–43

du Toit, P. J., Malan, A. I., Louw, J. G., Holzappel, C. R. and Roets, G. C. S. (1935a) A study of the mineral content and feeding value of natural pastures in the Union of South Africa, *Onderstepoort J. Vet. Sci. Anim. Ind.* **5**, 201–70

du Toit, P. J., Malan, A. I. and van der Merwe, P. K. (1939) The feeding of phosphate to animals in their drinking water, *Farming in South Africa* **14**, 429–33

du Toit, P. J., Malan, A. I., van der Merwe, P. K. and Louw, J. G. (1940b) Mineral supplements for stock, *Farming in South Africa* **15**, 233–48

du Toit, P. J., Malan, A. I. and Roets, J. (1935b) A study of the mineral content and feeding value of natural pastures in the Union of South Africa, *Onderstepoort J. Vet. Sci. Anim. Ind.* **5**,201–14

Toon, B. C. (1971) Cattle growth on Townsville stylo pasture, *Queensland Agric. J.* **97**, 442

Topps, J. H. (1972) Urea or biuret supplements to low protein in Africa, *World Anim. Rev.* (*FAO*) **3**, 14–18

Topps, J. H. and Elliott, R. C. (1964) Studies on natural herbage in the sub-tropics III. Concentration of volatile fatty acids and ammonia in the rumen of sheep fed fresh herbage, *J. Agric. Sci.* (*Cambridge*) **63**, 245–8

Topps, J. H., Reed, W. D. C. and Elliott, R. C. (1965) The effect of season and of supplementary feeding on the rumen contents of African cattle grazing sub-tropical herbage II. pH values and concentration and proportions of volatile fatty acids, *J. Agric. Sci.* (*Cambridge*) **64**, 397–402

Tothill, J. C. (1974) Experiences in sodseeding Siratro into native Speargrass pastures on granite soils near Mundubbera, *Trop. Grassl.* **8**, 128–32

Trail, J. C. M. and Fisher, I. L. (1971) *The operation of beef cattle breeding research under semi-arid extensive conditions in Botswana*. Anim. Prod. Res. Unit: Gaborones, Botswana. 124 pp.

Trail, J. C. M., Sacker, G. D. and Fisher, I. L. (1971) Crossing beef cattle in western Uganda 1. Performance of Ankole, Boran and Zebu cows, *Anim. Prod.* **13**, 127–41

Treviño, S. M., Garza, T. R. and Robles, B. C. (1976) Producción anual de carne/ha en gramas nativas (*Axonopus* y *Paspalum*) en pastoreo rotacional con fertilización al pastizal y suplementación al ganado, *Téc. Pec. Méx.* **30**, 7–11

Treviño, S. M., Garza, T. R., Torres, . M. and Robles, B. C. (1975) Producción anual de carne/ha en pastoreo rotacional en los zacates Ferrer, Estrella de Africa y Señal, con y sin fertilizantes en Hueytamalco, Pue, *Téc. Pex. Méx.* **29**, 7–11

Tronson, W. (1926) 'Soft Bone' in cattle, *Queensland Agric. J.* **26**, 466–7

Trowbridge, P. F., Moulton, C. R. and Haigh, L. D. (1915) *The maintenance requirement of cattle as influenced by condition, plane of nutrition, age, season, time on maintenance, type and size of animal.* Univ. Missouri Res. Bull. No. 18

Truter, G. J. and Louw, J. G. (1952) Defluorination of aqueous solutions of superphosphate for use in animal nutrition, *Onderstepoort J. Vet. Res.* **25**, 93–7

Tundisi, A. G. A., Lima, F. P., Kalil, E. B., Villares, J. B., Correa, A. and Vidal, M. E. P. (1966) Novas interpretações sobre a eficiência das provas de ganho de peso e a viabilidale da producão econômica de novilhos zebus próximo dos 24 meses de idade, *Bol. Ind. Anim.* **23**, 67–81

Turner, A. W., Kelley, R. B. and Dann, A. T. (1935) 'Peg leg' of cattle in north Queensland, *J. Counc. Sci. Ind. Res. Aust.* **8**, 120–32

Ugarte, J. and Preston, T. R. (1974) Hay or restricted grazing as roughage source for Holstein steers fed molasses/urea, *Cuban J. Agric. Sci.* **8**,151–5

Ugarte, J. and Preston, T. R. (1975) Effect of molasses/urea supplementation on Holstein calves grazing Pangola grass, *Cuban J. Agric. Sci.* **9**, 109–18

Underwood, E. J. (1962) *Trace elements in human and animal nutrition* (2nd edn). Academic Press: New York. 429 pp.

Underwood, E. J. (1966) *The mineral nutrition of livestock.* Food and Agriculture Organization of the United Nations and Commonwealth Agricultural Bureaux. 237 pp.

Underwood, E. J. (1971) *Trace elements in human and animal nutrition* (3rd edn). Academic Press: New York. 543 pp.

Underwood, E. J. and Filmer, J. F. (1935) The determination of the biologically potent element (cobalt) in limonite, *Aust. Vet. J.* **11**, 84–92

UNESCO/UNEP/FAO (1979) *Tropical grazing land ecosystems.* UNESCO: Paris. 655 pp.

Utley, P. R. and McCormick, W. C. (1976) *Performance and carcass characteristics of steers, steers implanted with zeranol and bull calves.* Dept. Anim. Sci. Coastal Plain Georgia Res. Rep. 221. 9 pp.

Valdés, A. (1974) Some economic aspects of the cattle industry in Latin America in *Proceedings of a seminar to increase beef production in tropical America.* Centro Internacional de Agricultura Tropical: Cali. 315 pp.

Vallee, B. L. (1962) in *Mineral Metabolism* (C. L. Comar and F. Bronner, eds). New York: Academic Press. 443 pp.

Veitia, J. L., Preston, T. R. and Delgado, S. (1972) Pasture for beef production 1. Different concentrations of urea in molasses as a supplement to grazing bulls during the wet season, *Rev. Cubana Cienc. Agric.* (*Habana*) **6**, 323–9

Veitia, J. L., Preston, T. R. and Delgado, S. (1974a) The use of pasture for beef production II. Effect of stocking rate and supplemental molasses/urea feeding on performance of grazing bulls during the wet season, *Cuban J. Agric. Sci.* **8**, 123–7

Veitia. J. L., Preston, T. R. and Delgado, S. (1974b) Effect of stocking rate and supplemental molasses/urea feeding on performance of grazing bulls during the wet season, *Proc. XII Intnl Grassl. Cong.*, 560–6

Vélez, C. A., Escobar, G. and Ayala, C. H. (1972) Efecto de la composición quimica y la disponibilidad de nutrientes del pasto Pará (Panicum purpurascens raddi.) sobre la producción de carne por animal, *Acta Agronómica (Colombia)* **22**, 11–25

Verbeek, W. A. and von la Chevallerie, M. K. S. L. (1958) *Die corwintering van vleisbeeste in die hoeveld.* Landboukollege, Potchefstroom. 12 pp.

Vicente Chandler, J., Abruña, F., Caro Costas, R., Figarella, J., Silva, S. and Pearson, R. W. (1974) *Intensive grassland management in the humid tropics of Puerto Rico.* Puerto Rico Agric. Exper. Sta. Bull. 233. 164 pp.

Vicente Chandler, J., Caro Costas, R., Pearson, R. W., Abruña, F., Figarella, J. and Silva, S. (1964) *The intensive management of tropical forages in Puerto Rico.* Puerto Rico Agric. Exper. Sta. Bull. 187. 152 pp.

Vilela, H., Bamberg, de C. O. and Siñoes, R. (1977a) Desempenho de novilhos zebuinos em pastagens naturais e cultivadas sob rotação, *Arq. Esc. Vet. U.F.M.G.* **29**, 295–300

Vilela, H., de Oliveira, S., Avila, P. J. A. and Possato, R. (1978) Desempenho de novilhos mesticos (Holandés-Zebu) em pastagens natural e melhorada establecidas em latossol vermelho oscuro (fase cerrado), *Arq. Esc. Vet. U.F.M.G.* **30**, 317–24

Vilela, H., de Oliveira, S. and Nascimento, C. H. F. (1976) Efeito de pastagens de graminea e de graminea e leguminosas sobre o ganho em peso de novilhos 1. Época daseca, *Rev. Soc. Bras. Zoot.* **5**, 236–7

Vilela, H., de Oliveira, S., Nascimento, C. H. F. and Gontijo, R. M. (1977b) Efeito de pastagens de graminea e de graminea e leguminosas sobre o ganho em peso de novilhos 1. época da 'seca', *Arq. Esc. Vet. U.F.M.G.* **29**, 11–17

Villaça, H. A., Moreira, H. A., Viana, J. de A., Carneiro, G. G., Souza, R. M. and Assis, A. G. (1976) Melaço e melaço mas 5% de uréia na recria de novilhos mestiços Holandés × Zebu em pastagem natural de capim Gordura (*Melinis minutiflora*), *Arq. Esc. Vet. U.F.M.G.* **28**, 153–67

Vohnout, K., Ettinger, A., Isidor, M. and Perla, F. (1973a) *Crecimiento de bovinos suplementados con melaza II. Efecto de la presión de pastoreo.* Compendios de trabajos presentados en la IV Reunión Latinoamericano de Producción Animal (ALPA): Guadalajara, México. pR33

Vohnout, K., Muñoz, H., Rios, S. and Valdes, F. (1973b) *Crecimiento de bovinos suplementados con melaza I. Efecto de nivel de melaza.* Compendios de trabajos presentados en la IV Reunión Latinoamericano de Producción Animal (ALPA): Guadalajara, México. pR32

Wagnon, K. A. and Goss, H. (1961) The use of molasses to increase the utilization of rank, dry forage and molasses-urea as a supplement for weaner calves, *J. Range Management* **14**, 5–9

Walker, B. (1968) Grazing experiments at Ukiriguru, Tanzania II. Comparisons of rotational and continuous grazing on natural pastures of hardpan soils using an 'extra-period, latin-square, change-over design', *E. Africa Agric. For. J.* **34**, 235–44

Walker, B. (1969a) Effects of feeding hay on the early wet season weight loss of cattle in western Tanzania, *Expl. Agric.* **5**, 53–7

Walker, B. (1969b) Effects of nitrogen fertiliser on natural pastures in western Tanzania, *Expl. Agric.* **5**, 215–22

Walker, B. (1977) Productivity of *Macroptilium atropurpureum* cv Siratro pastures, *Trop. Grassl.* **11**, 79–85

Walker, B. and Scott, G. D. (1968a) Grazing experiments at Ukiriguru, Tanzania I. Comparison of rotational and continuous grazing systems on natural pastures of hardpan soils, *E. Africa Agric. For. J.* **34**, 224–34

Walker, B. and Scott, G. D. (1968b) Grazing experiments at Ukiriguru, Tanzania III. A comparison of three stocking rates on the productivity and botanical composition of natural pastures of hardpan soils, *E. Africa Agric. For. J.* **34**, 245–55

Walker, C. A. (1948) *Rep. Dir. Agric. Southern Rhodesia*

Walker, C. A. (1957) Studies of the cattle of Northern Rhodesia 1. The growth of steers under normal veld grazing and supplemented with salt and protein, *J. Agric. Sci. (Cambridge)* **49**, 394–400

Walker, C. A. and Rhodes, F. B. (1951–4) *Grasslands Exper. Sta. Rep., Southern Rhodesia*

Walters, J. A. T. (1919) Improvement of the veld by artificial means, *Rhodesia Agric. J.* **16**, 32–6

Ward, H. K. (1964) 'The potential of indigenous cattle' Talk presented to the Northern Rhodesia Ministry of Agriculture Colloquy at Mount Makulu Res. Sta., 3 February 1964 (mimeo)

Ward, H. K. (1968) Supplementation of beef cows grazing on veld, *Rhodesia J. Agric. Res.* **6**, 93–101

Ward, H. K. and Tiffin, J. de W. (1975) Progress in beef cow productivity, *Rhodesia Agric. J.* **72**, 1–8

Warnick, A. C. (1976) Management factors affecting reproductive efficiency of beef cattle in developing countries in *Beef cattle in developing countries* (A. J. Smith, ed.). University of Edinburgh, Centre for Tropical Vet. Med.: Edinburgh. 487 pp.

Warnick, A. C., Koger, M., Martinez, A. and Cunha, T. J. (1965) *Productivity of beef cow as influenced by pasture and winter supplement during growth.* Florida Agric. Exper. Sta. Bull. 695. 15 pp.

Warnick, A. C., Meade, J. H. and Koger, M. (1969) *Factors influencing pregnancy rate in Florida beef cattle.* Florida Agric. Exper. Sta. Bull. 623. 10 pp.

Waters, H. J. (1908) *The capacity of animals to grow under adverse conditions.* 29th Proc. Soc. Prom. Agric. Sci.: New York. 71 pp.

Webster, C. C. (1954) The ley and soil fertility in Britain and Kenya, *East Africa Agric. J.* **20**, 71–4

Webster, C. C. and Wilson, P. N. (1966) *Agriculture in the tropics.* Longmans: London. 488 pp.

Weir, K. L., MacRae, I. C. and Allan, J. (1979) Nitrogen fixation associated with the root system of tropical grasses, *CSIRO Tropical crops and pastures Div. Rep. 1977–8*, 75 pp.

Weller, M. C. (1969) *Value of urea-molasses supplements.* Australia Dept. of Primary Ind. Div. of Anim. Ind., Adv. Leaflet No. 629

Wesley-Smith, R. N. (1972) Liveweight gains of shorthorn Steers on native and improved pastures at Adelaide River, Northern Territory, *Aust. J. Exper. Agric. and Anim. Husb.* **12**, 566–72

West, O. (1952) Grasslands Research Station, Pasture Section. Ann. Rep. for year ended 1951, *Rhodesia Agric. J.* **49**, 293–302

West O. (1956) Pasture improvement in the higher rainfall regions of Southern Rhodesia, *Rhodesia Agric. J.* **53**, 439–51

West, O. (1964) Pasture management and improvement in Rhodesia, *Rhodesia Agric. J.* **61**, 119–25

West, O. (1969) Fire: its effect on the ecology in Rhodesia and its application in grazing management, *Proc. Veld Management Conf. Bulawayo*, 59–64

Wheeler, J. L. and Hutchinson, K. J. (1973) Production and utilization of food for ruminants in *The pastoral industries of Australia* (G. Alexander and O. B. Williams, eds). Sydney University Press. 567 pp.

Whyte, R. O. (1968) *Grasslands of the monsoon*. Faber and Faber: London. 325 pp.

Whyte, R. O. (1974) *Tropical grazing lands*. Junk: The Hague. 222 pp.

Wigg, P. M. and Owen, M. A. (1973) Studies on water consumption, night grazing and growth of Boran and crossbred steers at Kongwa, Tanzania, *East Africa Agric. For. J.* **38**, 361–6

Williams, J. B. (1976) The social importance of cattle in Asia and the influence of social attitudes on beef production in *Beef cattle production in developing countries* (A. J. Smith, ed.). University of Edinburgh, Centre for Tropical Vet. Med.: Edinburgh. 487 pp.

Wilson, F. (1963) Calvings and calf mortality in Ankole Long Horned cattle, *East Africa Agric. For. J.* **29**, 178–81

Wilson, J. R. and Ford, C. W. (1971) Temperature influences on the growth, digestibility and carbohydrate composition of two tropical grasses, *Panicum maximum*, var. *trichoglume* and *Setaria sphacelata* and two cultivars of the temperate grass *Lolium perenne*, *Aust. J. Agric. Res.* **22**, 563–71

Wilson, P. N. (1961) Observations on the grazing behaviour of cross-bred Zebu Holstein cattle managed on Pangola grass pastures in Trinidad, *Turrialba* **11**, 55–71

Wilson, P. N. and Osbourn, D. F. (1960) Compensatory growth after malnutrition in mammals and birds, *Biol. Revs.* **35**, 324–63

van Winden, A. and Beale, C. I. A. (1979) Goals, motivations and practices among Northern Region smallholder cattle farmers, FAO/UNDP Project MLW/75/020, *Assistance to livestock development*: Working Paper No. 26: Lilongwe, Malawi (mimeo). 38 pp.

Winks, L. (1973) Townsville stylo research at Swan's Lagoon, *Trop. Grassl.* **7**, 201–8

Winks, L., Holmes, A. E. and O'Rourke, P. K. (1977a) Effect of dehorning and tipping on liveweight gain of mature Brahman crossbred steers, *Aust. J. Exper. Agric. and Anim. Husb.* **17**, 16–19

Winks, L. and Laing, A. R. (1972) Urea, phosphorus and molasses supplements for grazing beef weaners, *Proc. Aust. Soc. Anim. Prod.* **9**, 253–7

Winks, L., Laing, A. R., O'Rourke, P. K. and Wright, G. S. (1979) Factors affecting response to urea molasses supplements by yearling cattle in tropical Queensland, *Aust. J. Exper. Agric. and Anim. Husb.* **19**, 522–9

Winks, L., Laing, A. R., and Stokoe, J. (1972) Level of urea for grazing yearling cattle during the dry season in tropical Queensland, *Proc. Aust. Soc. Anim. Prod.* **9**, 258–61

Winks, L., Laing, A. R., Wright, G. S. and Stokoe, J. (1976) Effect of nitrogen, phosphorus and molasses supplements on the performance of weaner cattle during the dry season in North Queensland, *J. Aust. Inst. Agric. Sci.* **42**, 246–51

Winks, L. and Lamberth, F. (1968) Performance of steers in north

Queensland grazing native pasture and Townsville lucerne with and without fertiliser, *Proc. Aust. Soc. Anim. Prod.* **7**, 158–62

Winks, L., Lamberth, F. C. and O'Rourke, P. K. (1977b) The effect of a phosphorus supplement on the performance of steers grazing Townsville stylo-based pasture in north Queensland, *Aust. J. Exper. Agric. and Anim. Husb.* **17**, 357–66

Winks, L., O'Grady, P. and Chambers, G. (1975) *Beef from green panic-glycine on Atherton Tablelands.* Australia Queensland Dept. of Primary Ind., Div. of Anim. Ind. Adv. Leaflet No. 804

Winks, L., O'Grady, P., Edgley, W. and Stokoe, J. (1970) Performance of steers in north Queensland grazing a tropical legume-grass pasture at two stocking rates on two soil types, *Proc. Aust. Soc. Anim. Prod.* **8**, 450–4

Winter, W. H., Edye, L. A., Megarrity, R. G. and Williams, W. T. (1977a) Effects of fertilizer and stocking rate on pasture and beef production from sown pastures in northern Cape York Peninsula 1. Botanical and chemical composition of the pastures, *Aust. J. Exper. Agric. and Anim. Husb.* **17**, 66–74

Winter, W. H., Edye, L. A. and Williams, W. T. (1977b) Effects of fertilizer and stocking rate on pasture and beef production from sown pastures in northern Cape York Peninsula 2. Beef production and its relation to blood, faecal and pasture measurements, *Aust. J. Exper. Agric. and Anim. Husb.* **17**, 187–96

Winter, W. H., Mott, J. J. and McLean, R. W. (1979) *Pastures systems study – Manbulloo Station 33–34.* CSIRO Tropical crops and pastures Div. Rep. 1977–8: Brisbane. 150 pp.

Winter, W. H., Siebert, B. D. and Kuchel, R. E. (1977c) Cobalt deficiency of cattle grazing improved pastures in northern Cape York Peninsula, *Aust. J. Exper. Agric. and Anim. Husb.* **17**, 10–15

Woods, L. E. (1970) Beef production from pastures and forage crops in a tropical monsoon climate, *Proc. XI Intnl Grassl. Cong.*, 845–9

Worcester, V. D. M. (1889) *Agric. J. Cape Colony* **1**, 272 (letter); also see correspondence loc. cit. **7**, 62

Worker, N. A. and Carrillo, B. J. (1967) 'Enteque Seco' calcification and wasting in grazing animals in the Argentine, *Nature* **215**, 72–4

Yates, J. J., Edye, L. A., Davies, J. G. and Haydock, K. P. (1964) Animal production from a Sorghum almum pasture in southeast Queensland, *Aust. J. Exper. Agric. and Anim. Husb.* **4**, 326–35

Yeates, N. T. M. (1965) *Modern aspects of animal production.* Butterworth: London. 371 pp.

Young, J. G. and Chippendale, F. (1970) Beef cattle performance on pastures on heath plains in southeast Queensland, *Proc. XI Intnl Grassl. Cong.*, 849–52

Young, N. D. and Daly, J. J. (1967) A study of three important pasture mixtures in the Queensland sub-tropics: Progress Report 1958–1963, *Queensland J. Agric. Anim. Sci.* **24**, 149–58

Young, N. D., Fox, N. F. and Burns, M. A. (1959) A study of three important pasture mixtures in the Queensland sub-tropics: Progress Report 1955–1958, *Queensland Agric. J.* **16**, 199–215

Younge, O. R. and Plucknett, D. L. (1966) Beef production with heavy phosphorus fertilization in infertile wetlands of Hawaii, *Proc. IX Intnl Grassl. Cong.*, 959–63

Younge, O. R., Plucknett, D. L. and Rotar, P. P. (1964) *Culture and yield performance of* Desmodium intortum *and* D. canum *in Hawaii*. University Hawaii College of Tropical Agriculture, Hawaii Agric. Exper. Sta. Tech. Bull. No. 59. 22 pp.

Zavala, M. A., Galván, J. M. and Butterworth, M. H. (1972) Effects of two concentrations of phosphorus supplied to grazing heifers in the Mexican tropics, *Summaries: IX Intnl Cong. Nutr.*, 213

C.P.O. ABERYSTWYTH LLYFRGELL

Index

U.C.W. ABERYSTWYTH LIBRARY